Ecosystem Services – Concept, Methods and Case Studies

Karsten Grunewald
Olaf Bastian
Editors

Ecosystem Services – Concept, Methods and Case Studies

Springer

Editors
Karsten Grunewald
Leibniz Institute of Ecological Urban
and Regional Development
Dresden

Olaf Bastian
Leibniz Institute of Ecological Urban
and Regional Development
Dresden

ISBN 978-3-662-44142-8
DOI 10.1007/978-3-662-44143-5

ISBN 978-3-662-44143-5 (eBook)

Library of Congress Control Number: 2015940761

Printed on acid-free paper

Springer is part of Springer Science+Business Media (www.springer.com)

Preface

Humankind is a part of nature. It depends on nature for its existence, its well-being and its economic activity, and is connected with it by numerous ties. Nature provides food and water for our daily existence, the raw materials for handicrafts and industry and medicinal plants for healthcare. Forests not only supply us with wood, berries, mushrooms and wild game, but also protect us against soil erosion and flooding, create the oxygen we breathe and bind greenhouse gases that endanger our climate. Natural ecosystems act as water filters, and habitats for a large variety of plant and animal species, including the wild bees which are important for the pollination of our crops. People find spiritual inspiration and fulfillment in nature together with a esthetic pleasure, rest and recreation.

In recent years, the term 'ecosystem services' has become popular as the designation for all these benefits which are useful to people. Nature provides many effective, low-cost and sustainable solutions for human needs. Often however, people are not even aware of the role of natural resources or ecosystem services, or they see nature simply as an endlessly bubbling, never slacking fountain of human prosperity. Careful dealing with nature and investment in an intact natural environment is often considered a luxury, and conservation is generally a secondary issue. No wonder biodiversity is declining at a rapid pace worldwide–and also in Germany–and that the capacity of ecosystems to provide services is also being reduced to such a degree as to cause major concern.

Generally, growing economic use of nature involves a reduction of the regulatory and sociocultural services rendered. One goal of the concept of ecosystem services is to better demonstrate these contexts and move them into the public consciousness. It is therefore important to recognise and improve the standing of the non-marketable services of nature by improving the understanding for the systemic context and the dynamics between ecosystem properties, functions and services, natural capital and their beneficial effects in various spatial and temporal scales, and in connection with their multiple drivers. Valuating the services provided by ecosystems and landscapes–i. e. assigning economic/monetary value to them–is in accordance with the widespread tendency of our times. Often, the argument is raised that 'concrete' arguments need to be developed to persuade political leaders, and to gain broad acceptance by business and society at large. After all, monetary value and supposedly 'hard' figures are the language that is most easily understood, especially outside of the conservationist community. However, can we and should we really reduce nature, in all its complexity and its immeasurable significance for us human beings, to monetary values?

The reason and goal of the first comprehensive German-language discussion of this issue in 2013 was to present the multiple relationships between economics, ecology and ethics in a theoretically well-grounded manner, and to provide practical recommendations for the analysis, evaluation, control and communication of ecosystem services. We seek to address all those interested in building bridges and crossing borders between disciplines: both scientists and practitioners in the administrative, volunteer and professional spheres, especially those who deal with the environment, conservation and regional and land-use planning; experts from the business community, activists in politics, students, and all those interested in fundamental ecological, economic, ethical and environmentalist issues and issues which affect ecosystems and landscapes.

After a very positive reception of the German book, the English translation has now been completed. Springer-Spektrum as editor has initiated this project and made it possible; the organisation and the cooperative effort were carried out in a notably pleasant atmosphere. We would like to thank the numerous authors, from Dresden to Bonn and from Freiburg to Greifswald, for their contributions, and also apologise to those of our colleagues working in this and similar areas whom we were unable to accommodate for reasons of space. We hope that the present treatment will spark a constructive discussion with them. The length of the book was strictly limited, so that, in our view, while a number of very essential aspects of this highly complex topic have been addressed, others unfortunately have not.

Most of the authors provided their own translations; Phil Hill of Berlin translated the rest, and the publisher provided the final redaction. Our sincere thanks to all.

Phil Hill passed away suddenly on 22nd of December 2014. With the book we want say thank you to you, Phil, for the wonderful years of collaboration.

Karsten Grunewald and Olaf Bastian
Dresden, January 2015

Contents

Contributors

Dr. Kenneth Anders
Büro für Landschaftskommunikation,
Neutornow 54,
16259 Bad Freienwalde, Germany
k.anders@oderbruchpavillon.de

Dr. habil. Olaf Bastian
Leibniz Institute of Ecological Urban and
Regional Development, Weberplatz 1,
01217 Dresden, Germany
olaf.bastian@web.de

Dr. Claudia Bieling
University of Freiburg, Chair for Landscape
Management, Tennenbacher Straße 4,
79106 Freiburg, Germany
claudia.bieling@landespflege.uni-freiburg.de

Dr. habil. Benjamin Burkhard
Kiel University, Institute for Natural Resource
Conservation, Department Ecosystem
Management, Olshausenstr. 75, 24118 Kiel,
Germany
bburkhard@ecology.uni-kiel.de

Dr. Peter Elsasser
Thuenen Institute / WF,
Leuschnerstrasse 91,
21031 Hamburg, Germany
peter.elsasser@vti.bund.de

Hermann Englert
Thuenen Institute / WF,
Leuschnerstrasse 91,
21031 Hamburg, Germany
Hermann.englert@vti.bund.de

Malte Großmann
TU Berlin, FG Landschaftsökonomie,
EB 4-2, Straße des 17. Juni 145, 10623 Berlin,
Germany
grossmann@imup.tu-berlin.de

M.Sc. Anja Grünwald
Burgkstraße 30, 01159 Dresden,
Anja_gruenwald@gmx.de

Dr. habil. Karsten Grunewald
Leibniz Institute of Ecological Urban and
Regional Development, Weberplatz 1,
01217 Dresden, Germany
k.grunewald@ioer.de

Prof. Dr. Dagmar Haase
Humboldt-Universität Berlin, Department of
Geography, Rudower Chaussee 16, 12489 Berlin,
Germany and
Helmholtz Centre for Environmental Research -
UFZ, Permoserstraße 15, 04318 Leipzig, Germany
dagmar.haase@geo.hu-berlin.de

Prof. Dr. Volkmar Hartje
TU Berlin, FG Landschaftsökonomie,
EB 4-2, Straße des 17. Juni 145,
10623 Berlin, Germany
volkmar.hartje@tu-berlin.de

Dipl.-Ing. Michael Holfeld
Mary-Wigman-Straße 2,
01069 Dresden, Germany
michael-holfeld@gmx.de

Dr. Markus Leibenath
Leibniz Institute of Ecological Urban and
Regional Development, Weberplatz 1,
01217 Dresden, Germany
m.leibenath@ioer.de

Dr. Gerd Lupp
TU München, Chair for Strategic Landscape
Planning and Management,
Emil-Ramann-Str. 6, 85354 Freising, Germany
luppg@gmx.de

Prof. Dr. Karl Mannsfeld
Ahornweg 1, 01328 Dresden-Pappritz, Germany
karl.mannsfeld@googlemail.com

Prof. Dr. Bettina Matzdorf
Institute of Socio-Economics, Leibniz Centre
for Agricultural Landscape Research (ZALF),
Eberswalder Straße 84, 15374 Müncheberg,
Germany
matzdorf@zalf.de

Dr. Melanie Mewes
Helmholtz Centre for Environmental Research – UFZ, Dept. of Economics, Permoserstraße 15, 04318 Leipzig, Germany
melanie.mewes@ufz.de

Prof. Dr. Felix Müller
Kiel University, Institute for Natural Resource Conservation, Department Ecosystem Management, Olshausenstr. 75, 24118 Kiel, Germany
fmueller@ecology.uni-kiel.de

Bettina Ohnesorge
Warthestraße 17, 12051 Berlin, Germany
b.ohnesorge@posteo.de

Dr. Tobias Plieninger
Department of Geosciences and Natural Resource Management, University of Copenhagen, Rolighedsvej 23, 1958 Frederiksberg C, Denmark, Germany
tobias.plieninger@ign.ku.dk

Dipl.-Ing. (FH) Michaela Reutter
Institute of Socio-Economics, Leibniz Centre for Agricultural Landscape Research (ZALF), Eberswalder Straße 84, 15374 Müncheberg, Germany
reutter@zalf.de

PD Dr. habil. Irene Ring
Helmholtz Centre for Environmental Research – UFZ, , Dept. of Economics, Permoserstraße 15, 04318 Leipzig, Germany
irene.ring@ufz.de

Matthias Rosenberg
Leibniz Institute of Ecological Urban and Regional Development, Weberplatz 1, 01217 Dresden, Germany
m.rosenberg@ioer.de

Achim Schäfer
Ernst-Moritz-Arndt Universität Greifswald, Lehrstuhl für Allgemeine Volkswirtschaftslehre und Landschaftsökonomie, Soldmannstraße 15, 17487 Greifswald, Germany
schaefea@uni-greifswald.de

Dr. Christoph Schröter-Schlaack
Centre for Environmental Research – UFZ, Dept. of Economics, Permoserstraße 15, 04318 Leipzig, Germany
christoph.schroeter-schlaack@ufz.de

Dr. Christian Schleyer
Helmholtz Centre for Environmental Research – UFZ, Dept. of Environmental Politics, Permoserstraße 15, 04318 Leipzig, Germany
christian.schleyer@ufz.de

Dr. Burkhard Schweppe-Kraft
Federal Agency for Nature Conservation, unit I 2.1 "Legal Affairs, Economics and Ecologically Sound Regional Development", Konstantinstr. 110, 53179 Bonn, Germany
SchweppB@BfN.de

Dr. Ralf-Uwe Syrbe
Leibniz Institute of Ecological Urban and Regional Development, Weberplatz 1, 01217 Dresden, Germany
r.syrbe@ioer.de

Juliane Vowinckel
Lößnitzstraße 21, 01097 Dresden, Germany
juliane.vowinckel@gmx.de

PD Dr. habil. Ulrich Walz
Leibniz Institute of Ecological Urban and Regional Development, Weberplatz 1, 01217 Dresden, Germany
u.walz@ioer.de

Prof. Dr. Wolfgang Wende
Leibniz Institute of Ecological Urban and Regional Development, Head of Research Area Landscape Change and Management, Weberplatz 1, 01217 Dresden, Germany
and
Technische Universität Dresden, Chair Urban Development, Zellescher Weg 17, 01062 Dresden, Germany

Ecosystem Services (ES): More Than Just a Vogue Term?

K. Grunewald and O. Bastian

K. Grunewald, O. Bastian (eds.), *Ecosystem Services – Concept, Methods and Case Studies*,
DOI 10.1007/978-3-662-44143-5_1, © Springer-Verlag Berlin Heidelberg 2015

"Western civilization [is] the union of exquisitely sophisticated crowning achievements and a nervous, senselessly extravagant consumption."
Peter Høeg: *Miss Smilla's Feeling for Snow.*

During the 1990s, with the increasing demands of humankind upon the limited resources of the earth, and in view of the growing burdens upon the balance of nature, manifested, too, in biodiversity loss and in the problem complex of energy and the climate, the concept of ecosystem services (ES) entered into the international environmental discussion (e.g. de Groot 1992; Daily 1997; Costanza et al. 1997). Important milestones included the Millennium Ecosystem Assessment (MEA 2005), The Economics of Ecosystems and Biodiversity studies ('TEEB studies'; TEEB 2009), the RUBICODE project (Rationalising Biodiversity Conservation in Dynamic Ecosystems; e.g. Luck et al. 2009), the EASAC policy report (Ecosystem Services and Biodiversity in Europe; EASAC 2009) and the Strategic Plan for 2011–2020 adopted at the 10th Conference of the Parties of the Convention on Biodiversity (CBD 2010) in Nagoya (October 18–29, 2010), which used the term *ecosystem services* some 200 times.

The significance of the concept of ES is to enable greater account to be taken for the ecological services–the services provided to us by nature, free of charge–in decision-making processes, and to ensure sustainable land use, in order to counter overconsumption and degradation of the natural conditions of life. The attractiveness of the ES concept is based on its integrative, interdisciplinary and transdisciplinary character, as well as its linking of environmental and socio-economic elements (Müller and Burkhard 2007).

The ES concept is not, however, entirely new; the ecology movement had at an early date laid the foundations for it (e.g. Ehrlich and Ehrlich 1974; Westman 1977). The fact that nature and/or ecosystems provide free services to humankind, e.g. the decomposition of matter, the balance of water runoff, or the production of oxygen, has long been known (Graf 1984). We may recall that Bobek and Schmithüsen (1949) introduced the concept of 'potential' (▶ Chap. 2), as the 'spatial arrangement of naturally provided possibilities for development', analogous to the study of vegetation, where Tüxen saw natural vegetation as the integral that characterised the totality of growth conditions at a given site (Tüxen 1956).

In two points in particular, the ES approach differs from the concept of natural spatial potential and landscape functions, previously established particularly in the German-speaking area; this has strongly emphasised the concept of landscape ecology (Grunewald and Bastian 2010):

- First, ES evaluation is expressly anthropocentric, i.e. with regard to human quality of life.
- Second, the different functions, goods and services of nature, which often constitute 'public goods', are measured (if possible) with the aid of a single standard that integrates the interests of the ecology, of the economy, and of social sustainability. For this purpose, a monetary valuation is the proposed goal to be achieved by means of a methodological mixture of direct and indirect market evaluation (Costanza et al. 1997). However, there are still serious points of criticism with regard to a market-like evaluation of non-market related assets (e.g. Spangenberg and Settele 2010; Schröter et al. 2014), as a result of which there has recently been a tendency to move away from the concept of evaluating ES primarily or even exclusively in monetary terms, and towards using a broader spectrum of indicators instead (UNEP-WCMC 2011).

In the business community, too, there is growing realisation that scarcity of natural resources, reduced biodiversity and the degradation of ES not only bear a growing level of risk for companies, investors, banks and insurance companies, but also that solving these problems may open up opportunities of great financial significance. Leading companies are increasingly realising that the maintenance and protection of nature is not merely a marginal issue, nor is it something that can be dealt with by the commitment of volunteers. Rather, biodiversity and ES must be firmly rooted in their business models and core strategies, as a key precondition for ensuring sustainable growth and success (BESWS 2010).

Seeing nature as a productive force–along with capital and labour–makes the ES approach relevant for the public, government decision-makers and administrative bodies (◘ Fig. 1.1). Accordingly, the economic sciences have for some years made an effort to develop methods to permit ecosystems and the changes in them to be evaluated economically. In resource economics, the concepts of 'external effects' and 'economic total value' have been created for this purpose. Especially, if the efforts for so-called 'total environmental economic calculation' have to date met with little success, especially in Germany, Schweppe-Kraft (2010) has nonetheless summed up the situation as follows:

"The economic evaluation of ecosystem services, including existence, option and bequest values, conceptually permits the complete ascertainment of the effects of land-use and biotope changes upon societal welfare".

Ecosystem Services (ES)

describe the services rendered by nature and used by humankind. According to the MEA 2005, these are supporting services (such as soil formation, photosynthesis), and following provisioning services (such as food), regulation services (such as erosion control) and cultural services (such as landscape aesthetics as basis of recreation and tourism). We recommend a trinomial classification, with provisioning, regulation and sociocultural services; ► Chap. 3.2), since these correspond with sustainability categories. Effects vitally necessary for human well-being are based on these services, including provision with food, protection from natural hazards, with the supply of clean water. Societal value creation is to be weighted by means of the ES concept, and evaluated in part–but not entirely–monetarily (cost-benefit calculations), so as to promote commitment to the preservation of nature for economic reasons as well (Jessel et al. 2009).

At the latest since the study by Costanza et al. (1997), in which worldwide ES were calculated, their significance for humankind is no longer deniable. The extent of the dependence of humankind

◘ **Fig. 1.1** Nature conservation and economy–a new alliance?

on ES has been shown dramatically by the example of pollination by wild bees, upon which 15 to 30 % of US food production, with a total value of $30 billion, depends (Kremen 2005; EASAC 2009).

Nonetheless, ascertaining the scope of ES and the societal negotiation of its value priority, including an economic assessment of its value, still faces numerous challenges. Numerous ES, such as the beneficial effects of biological diversity, have been little investigated to date (Mosbrugger and Hofer 2008). In particular, there has been a lack of quantitative systemic understanding, i.e. of comprehensive knowledge of the process interconnections.

ES is such a current, exciting, complex, integrative, and open-end issue that numerous scientists and practitioners worldwide have been involved with it. ◘ Figure 1.2 shows that the number of scientific papers on ES in recent years has risen exponentially. There have been numerous attempts to index, quantify, and map ES. However, recent meta-analyses in that area (e.g., Elsasser and Meyerhoff 2007; Goldman et al. 2008; Feld et al. 2009; Jacobsen and Hanley 2009; Seppelt et al. 2011; Brouwer et al. 2013) have shown that there is as yet no comprehensive, generally accepted methodological system.

The ES Concept has Received Considerable Recognition from the Scientific Community and from Political Decision-Makers

However, what that means locally is usually not clear; in any case, such labels as 'the economisation of conservation', or 'improvement of the quality of life' are misleading and short-sighted. In this context, we will repeatedly be dealing with complex, ambiguous terms, such as ecosystem, service, capital, landscape, environment, function, space, time or value, from a number of perspectives. The following quote from the Senate Commission on Future Directions in Geoscience of the German Research Foundation (DFG; cf. DFG 2011) shows some of the stumbling blocks: "One important challenge is to quantitatively ascertain the biogeochemical turnover processes which drive global material cycles. These processes are called 'ecosystem functions' and 'ecosystem services'. They are of great significance for humankind and for climate change".

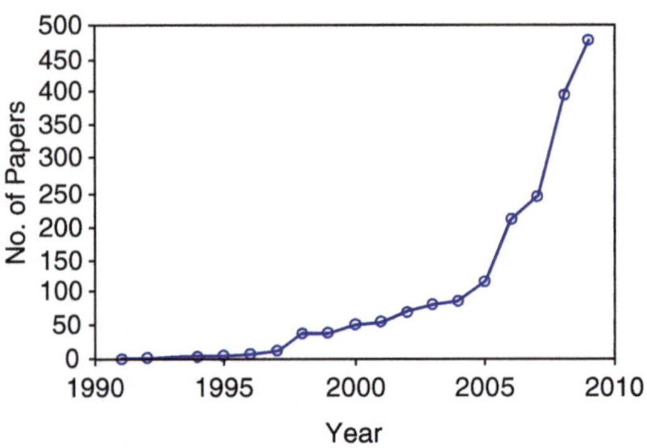

Fig. 1.2 Growth in number of papers on ecosystem services since 1990. The graph is based on searching ISI web of science using the terms *ecological*or *ecosystem service(s)*. Most ES-related papers were published in the journal *Ecological Economics*. (Source: Peterson 2010)

How Endangered are our Ecosystems? Why are ES and Biodiversity Often Mentioned in the Same Breath?

The approximately 1300 participants in the international study MEA (2001–2005) reached the conclusion that a sufficient supply of ES for future generations cannot be assured, because ecosystems are being changed, damaged, and transformed. In a survey by the Austrian Forum for Science and the Environment, experts gave a largely negative assessment of the development of our habitats ('Die Presse', print version, 1.4.2010). Human use of nature is causing a death of species at 100 to 1000 times the natural rate (Rockström et al. 2009).

In the EU and Germany, biodiversity goals, i.e. the goal of stopping the decline of biodiversity, have not been achieved to date, which has had a negative effect on such services provided by ecosystems as pollination services. Investigations have shown that without new policy approaches, the loss of biodiversity will continue (PBL 2010). Everybody realises that 'in a prosperous world with approximately 7 billion people (2011), a powerful advance of innovation must be initiated in order to secure ES and to make a resource-saving development possible' (WBGU 2011).

The Term Ecosystem

goes back to the British biologist and plant ecologist Arthur George Tansley, who introduced it as a fundamental principle in the ecology (Tansley 1935). An ecosystem contains the structure of interrelationships of living beings to one another and to their inorganic environment. In the less abstract sense, an ecosystem is characterised by its long-term relationship (biocenosis) and its habitat (biotope) (Ellenberg et al. 1992). Since Tansley, an international interdisciplinary and transdisciplinary ecosystem research community has emerged and has attempted to develop and apply holistic and systemic concepts. Ecosystem research is a conceptual approach with which particularly natural scientists identify, since analytic models of the structure and dynamics of spatial segments can be processed.

Calls for Valuation of ES

'We need strong awareness of the value of ecosystems and their services. Moral appeals and alarmism are of little help to nature. What can help counter species loss is effective management of well networked protected areas, and new land-use models with synergy effects. Primarily however, the economic value of 'green infrastructure' must at long last be accorded recognition'. (Beate Jessel, President of the German Federal Agency for the Conservation of Nature, in: *umwelt aktuell*, April 2010).

'The loss of natural capital (including ecosystems, biodiversity and natural resources) has direct and widespread negative effects on financial performance. Climate change and the financial crisis suggest that significant systemic risk requires coordinated policy intervention. The financial markets do not yet understand that many companies face specific risks from disruptions of vital ecosystems through their supply chains, and that they need to plan for the impact of new regulation'. (Colin Melvin, Hermes Equity Ownership Services Ltd., in: Demystifying Materiality: Hardwiring Biodiversity and Ecosystem Services into Finance, UNEP-CEObriefing, October 2010).

'Maybe the ecology movement shouldn't always just appealed to people's consciences, but rather view the issue from the point of view of the market economy' (Ebert 2011).

Biodiversity or Biological Diversity,

according to the Convention on Biological Diversity (CBD), means 'the variability among living organisms from all sources including, *inter alia,* terrestrial, marine and other aquatic ecosystems and the ecological complexes of which they are part; this includes diversity within species, between species and of ecosystems' (CBD 2010). According to this definition, which is binding under international law, biodiversity consists of the diversity of species, the diversity of ecosystems, and genetic diversity.

Since the diversity of ecosystems, biotic associations and landscapes are part of biodiversity, ES and biodiversity are often mentioned in the same breath (e.g. Ridder 2008; TEEB 2009). Biodiversity particularly supports the 'functioning of ecosystems'; however, it can also be defined as an ES in its own right–the ecosystem service of providing biodiversity. While both concepts overlap, they are in no sense identical. Undoubtedly, the continued loss of biodiversity will also have an impact on ES; however, generally, no simple, linear relationship can be assumed (Giller and O'Donnovan 2002; IEEP 2009; Trepl 2012). For many ES evaluated, maximum possible biodiversity is not necessary; rather, sometimes a lower number of species is favourable or sufficient, and sometimes a higher number.

Jessel (2011) describes the differences between the terms biodiversity and ES as follows:

- ES is broader than biodiversity (e.g. sociocultural ES).
- The perspectives are fundamentally different: ES focuses on properties of ecosystems for the purpose of maintaining their services, whereas biodiversity focuses on the number and characteristics of the biotic components of nature.
- The ES approach is more strongly anthropocentrically oriented.
- The protection of biodiversity implicitly presupposes the preservation of variety in all its components, and is hence fundamentally statically oriented; for ES, by contrast, not all components of the ecosystem are in all cases necessary in order to maintain services.

"A planning guide for the loss of biological diversity is difficult to define, due to the large number of species, the extreme difference of their significance for the functioning of the ecosystem, and our huge knowledge gaps" (WBGU 2011). The worldwide network of protected areas, which has been considerably expanded in recent decades (■ Fig. 1.3), and which is to be expanded further from the current 12 % of the world's land area (+ 10 % of the maritime

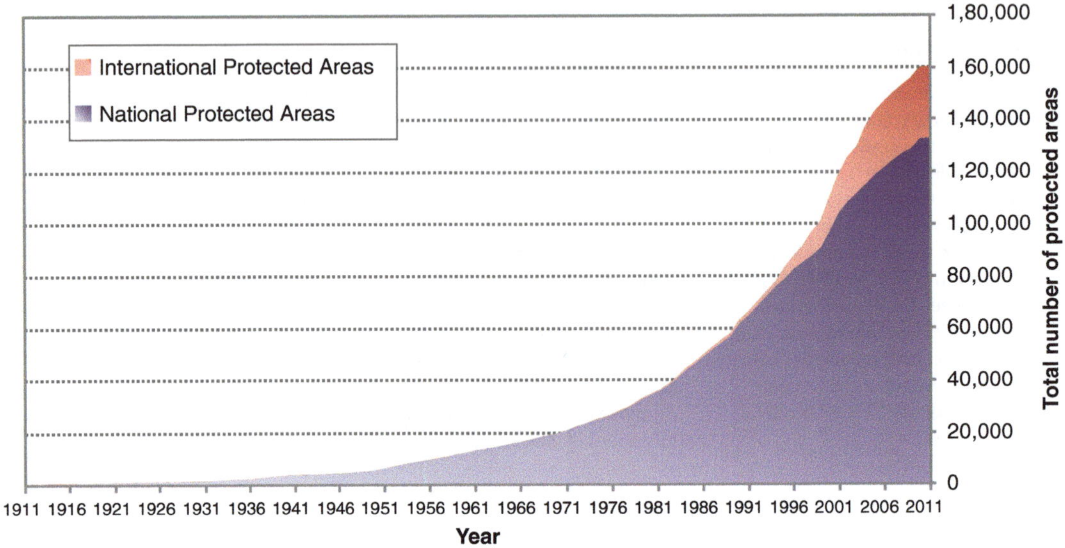

Fig. 1.3 Growth in number of nationally and internationally designated protected areas (1911–2011). © IUCN and UNEP-WCMC. The World Database on Protected Areas (WDPA): February 2012 (▶ http://www.wdpa.org/Statistics.aspx, assessed on 9.1.2014)

area) to 17 % by 2020 (CBD 2010), will certainly have a positive effect on ES and biodiversity, but has not been able to halt, much less reverse, the rate of loss with respect to most biodiversity parameters, especially species development and habitat conditions (Fig. 1.4). For this reason, the focus will in future be increasingly upon issues of 'sustainable land use' on more or less intensively managed areas of land.

Clearly, even though humankind has since time immemorial used the services of the ecosystems and landscapes, and experts are increasingly conscious of the value of the natural processes in ecosystems, society is still far from any general acceptance of these facts, and from any action to be derived from them.

Political Backgrounds and Stipulations

The preservation and improvement of ES are required not only by the EU's targets for sustainable growth and for climate protection and climate adaptation, but also by economic, spatial and social contexts; moreover, they serve to protect Europe's cultural heritage. The great political relevance of ES is shown, e.g. by the fact that in June 2010, the Intergovernmental Science-Policy Platform on Biodiversity and Ecosystem Services (IPBES) of the UN was established, as analogous to the Intergovernmental Panel on Climate Change (IPCC). Building upon the above-mentioned MEA 2005, the European Commission funded an international project for assessing 'The Economics of Ecosystems and Biodiversity' (TEEB 2009), which recommended taking into account the economic value of biodiversity in decision-making processes, accounting and reporting, in order to ensure the sustainable use and preservation of ES. This recommendation was proclaimed at the 10th Conference of the Parties to the CBD in Nagoya, Japan, in the autumn of 2010, as a key point for the strategic plan for the coming decade.

The EU biodiversity target for 2020 is, among other things, to contribute to Action 5 of the Biodiversity Strategy, 'Improving knowledge of ecosystems and their services in the EU':

"Member States, with the assistance of the Commission, will map and assess the state of ecosystems and their services in their national territory

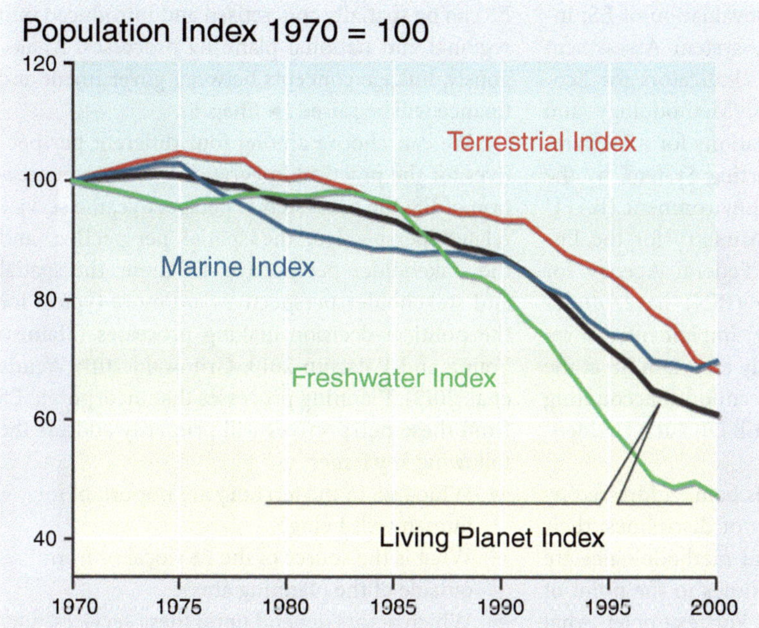

Fig. 1.4 Global Living Planet Index 1970–2000, aggregated by terrestrial, freshwater, and marine indices. © World Wide Fund for Nature and UNEP World Conservation Monitoring Centre

by 2014, assess the economic value of such services, and promote the integration of these values into accounting and reporting systems at EU and national level by 2020." (EC 2011)

Behind the targets set by various policy levels is the fact that in view of the virtually unhampered worldwide loss of biodiversity, and the growing anthropogenic load upon ecosystems, it is ever more urgent to control the multifarious and increasing demands on our limited resources and to ensure sustainable land use (see above). In future, ES and biodiversity are to be taken into account in all decision-making processes (EC 2011). However, do we have available to us the necessary knowledge and an appropriate methodological toolkit? To date there is a lack of indicators and instruments for the integration of ES that could be broadly applied at the national and regional levels, so that ES are still difficult to appropriately be taken into account in political decision-making processes. Both the monetary and the spatially explicit analysis of potential and existing ES have proven to be very time-consuming, high-effort tasks (Kienast 2010). Many practical problems, such as how to perform a comprehensive and full coverage calculation of ES, remain unsolved.

The goal of taking the role of ecosystems and ES into account in all future decision-making processes is, however, also viewed critically, due to the fear that it will come into conflict with the goals of deregulation and the simplification of decision-making. For example, certain interest groups want no additional restrictions on business and transport infrastructures, e.g. in the context of the expansion of long-distance power lines needed for renewable energies. The Federal Council (the upper house of the German Parliament) has therefore made an appeal not to call into question the foundations of the harmony of ecology, economy and social interests as the worldwide approved goals of a sustainability strategy (Bundesrat 2011). It is therefore important to ascertain the resilience of the ES concept as a pillar of policy-making, and to provide substantiation for the advantages of the integration of ES assessment in decision-making.

A number of projects in various countries, including in Europe, are currently involved in the process of the inventory and evaluation of ES, including the UK National Ecosystem Assessment (UKNEA 2011), or the study 'Indicators for Ecosystem Services: Systematics, Methodology and Implementation Recommendations for a Welfare-Related Environmental Reporting System', by the Swiss Federal Office for the Environment (BAFU 2011). The German Federal Ministry for the Environment (BMU) and the Federal Agency for the Conservation of Nature (BfN) have, under the heading 'TEEB Germany', implemented a research project to systematically ascertain ES at the national level and develop an economic accounting system for it, to an extent (TEEB DE 2012; Hedden-Dunkhorst et al. 2014).

The complex issue of ES is being addressed by scientists from a wide variety of disciplines; their approaches, terminologies and methodologies are accordingly variegated, sometimes to the point of causing misunderstandings. For example: what does the 'service capacity of nature', what does 'natural capital' mean? What is the difference between potential, functional and service approaches? Which services of nature should we analyse, and how should we evaluate them? Can all services really be quantified or even monetarised?

In the following chapters, these problems will be raised and discussed. The concept of ES will be explained, terminology elucidated, categories presented and methodological framework for analysis and evaluation of ES in its facets shown, and represented on the basis of case studies and applicability. The primary point is to gain a better understanding of systemic context and dynamics between natural capital, ecosystem structures and processes, ecosystem services/welfare effects of various scales, and in the context of multiple drivers. For that purpose, approaches of complexity research, the work of various levels and scales, and the approximation of various perspectives will be used. ◘ Figure 1.5 shows the various levels that are to be addressed and discussed in the following chapters. The conceptual considerations are compiled at the end in a Guide for analysis and evaluation of ES (▸ Chap. 7.1).

The focus will be on the central European area, and the existing system of ecological spatial planning in Germany. Among the questions this raises will be that of how the societal or macroeconomic use of measures for the improvement of quantifiable ES can be spatially concretised and introduced into regional and national planning processes. Analogously, linkage concepts between government and finance will be raised (▸ Chap. 5).

We can choose among four different perspectives for the practical integration and communication of ES: the ecosystemic perspective, the service related perspective, the spatial perspective, and the stakeholder perspective. Of these, the spatial and stakeholder perspectives are more typical for the political decision-making processes (Haines-Young and Potschin 2010; Grünwald 2011; Wende et al. 2012). Planning processes that incorporate ES from these perspectives will primarily address the following key issues:

- Which ES in the territory are important for human well-being?
- What is the source of the ES (local or from outside of the planning area)?
- Which actors depend upon these services, and with what kind of capacity (local or from outside of the planning area)?
- Which value and which priority does each service have (is replacement or exchange possible; reference of the service from elsewhere)?
- How can management and other activities improve the services (especially positive or negative effects upon other services)?
- Who benefits from management and activity options and measures?

The Task of Analysis and Evaluation of ES

The concept of ecosystem services (ES) is increasingly determining the debate over the issues of biodiversity and sustainable land use. ES can be ascertained with the methods of a variety of scientific disciplines in order to develop the term into a usable evaluations standard for policy-makers. For dealing with the problem of ES, a well-founded and broadly accepted conceptual framework will be necessary. Clear terminology is especially important. Currently, the main issue is the development of methods for the ascertainment and evaluation of the endangerment, and procedures for preservation/restoration of ES, and also the demonstration of 'public acceptance' for the concept of ES, with its possibilities and limitations, and, if necessary, its integration into planning and decision-making processes.

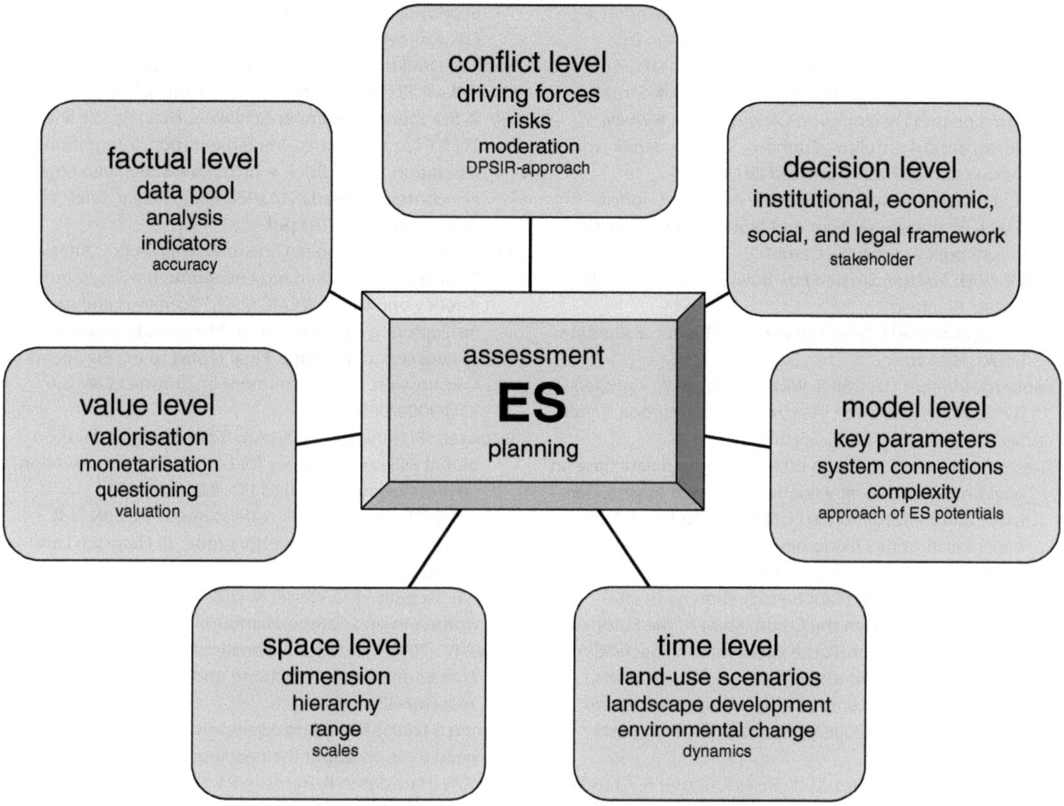

Fig. 1.5 Scheme of an approach to ES (multi-level approach), including both multiple socio-economic and scientific methods, numerous interdisciplinary technical terms (incomplete selection) © Karsten Grunewald

Central to the ES approach is the issue: what are human use demands with regard to the services which nature provides, and how can these demands be made visible and integrated into rational activity? Will we be better able, with the ES concept, to communicate the importance of nature for humankind, and to better take it into account, when balancing it against other goals? How can ES be secured, developed further, and protected from derogation?

References

BAFU–Bundesamt für Umwelt (2011) Indikatoren für Ökosystemleistungen. Systematik, Methodik und Umsetzungsempfehlungen für eine wohlfahrtsbezogene Umweltberichterstattung. Herausgegeben vom Bundesamt für Umwelt BAFU, Bern

BESWS–Biodiversity and Ecosystem Service Work Stream (2010) Demystifying materiality: hardwiring biodiversity and ecosystem services into finance. UNEP FI CEO Briefing, Genève

BfN–Bundesamt für Naturschutz (2007) Die Lage der biologischen Vielfalt 2. Globaler Ausblick. Naturschutz und biologische Vielfalt, Heft 44, Bonn

Bobek H, Schmithüsen J (1949) Die Landschaft im logischen System der Geographie. Erdkunde 3:112–120

Brouwer R, Brander L, Kuik O, Papyrakis E, Bateman I (2013) A synthesis of approaches to assess and value ecosystem services in the EU in the context of TEEB. TEEB follow-up study for Europe. University Amsterdam

Bundesrat (2011) Lebensversicherung und Naturkapital: Eine Biodiversitätsstrategie der EU für das Jahr 2020. Drucksache 309/11 vom 25.5.11, Berlin

CBD–Convention on Biological Biodiversity (2010) Global Biodiversity Outlook 3. CBD Secretariat, Montreal

Costanza R, d'Arge R, de Groot R, Farber S, Grasso M, Hannon B, Limburg K, Naeem S, O'Neill R, Paruelo J et al (1997) The value of the world's ecosystem services and natural capital. Nature 387:253–260

Daily G (ed) (1997) Nature's services: societal dependence on natural ecosystems. Island Press, Washington DC

DFG–Deutsche Forschungsgemeinschaft (2011) DFG-Senatskommission für Zukunftsaufgaben, 10.4–Struktur, Funktion und Dynamik von Ökosystemen. ►www.sk-zag.de/10.4_Struktur_Funktion_und_Dynamik_von_Oekosystemen.html. Accessed 29 Dec 2011

EASAC–European Academies Science Advisory Council (2009) Ecosystem services and biodiversity in Europe. EASAC policy report 09, Cardiff

Ebert V (2011) Machen Sie sich frei. Rowohlt Taschenbuch Verlag, Reinbek

Ehrlich PR, Ehrlich AH (1974) The end of affluence. Ballantine Books, New York

Ellenberg H, Weber HE, Düll R, Wirth V, Werner W, Paulißen D (1992) Zeigerwerte von Pflanzen in Mitteleuropa, 3. Aufl. Scripta Geobotanica 18, Göttingen

Elsasser P, Meyerhoff J (2007) A bibliography and data base on environmental benefit valuation studies in Austria, Germany and Switzerland, Part I: Forestry studies. Arbeitsbericht Institut für Ökonomie, 2007/01, Hamburg

European Commission (EC) (2011) Our life insurance, our natural capital: an EU biodiversity strategy to 2020. Communication from the Commission to the European Parliament, the Council, the Economic and Social Committee and the Committee of the Regions, Brussels, 3.5.2011. ► http://ec.europa.eu/environment/nature/biodiversity/comm2006/pdf/2020/1_EN_ACT_part1_v7%5B1%5D.pdf

Feld CK, da Silva PM, Sousa JP, de Bello F, Bugter R, Grandin U, Hering D, Lavorel S, Mountford O, Pardo I, Pärtel M, Römbke J, Sandin L, Jones KB, Harrison P (2009) Indicators of biodiversity and ecosystem services: A synthesis across ecosystems and spatial scales. Oikos 118:1862–1871

Giller PS, O'Donnovan G (2002) Biodiversity and ecosystem function: Do species matter? Biol Environ Proc Royal Irish Acad 102B:128–138

Goldman RL, Tallis H, Kareiva P, Daily GC (2008) Field evidence that ecosystem service projects support biodiversity and diversify options. Proc Natl Acad Sci USA 105:9445–9448

Graf D (1984) Gratisleistungen und Gratiseffekte. In: Graf D (ed) Ökonomie und Ökologie der Naturnutzung. Gustav Fischer, Jena, pp 42–45

de Groot RS (1992) Functions of nature: evaluation of nature in environmental planning, management and decision making. Wolters-Noordhoff, Groningen

Grünwald A (2011) Zukunft Landschaftsplan. Perspektiven einer methodischen Weiterentwicklung unter Anwendung des Konzepts der Ökosystemdienstleistungen. Masterarbeit, TU Dresden

Grunewald K, Bastian O (2010) Ökosystemdienstleistungen analysieren–begrifflicher und konzeptioneller Rahmen aus landschaftsökologischer Sicht. GEOÖKO 31:50–82

Haines-Young R, Potschin M (2010) Proposal for a common international classification of ecosystem goods and services (CICES) for integrated environmental and economic accounting (V1). Report to the European Environment Agency

Hedden-Dunkhorst B, Braat L, Wittmer H, Hundorf J, Hendriks K, Rizzotti N, Grunewald K, Panis J, Erhard M, Singh R, Bradburne R, Kümper-Schlake L, Hu L, Rančić IP (2014) TEEB Country Studies–Implementation, Integration and Utilisation. Policy Brief. ► http://www.teebweb.org/wp-content/uploads/2013/12/TEEB__Policy_Brief_Vilm_Workshop_May_2013.pdf

IEEP–Institute for European Environmental Policy, Alterra, Ecologic, PBL–Netherland Environmental Assessment Agency und UNEP-WCMC (2009) Scenarios and models for exploring future trends of biodiversity and ecosystem services change. Final report to the European Commission, DG Environment on Contract ENV.G.1/ETU/2008/0090r

Jacobsen JB, Hanley N (2009) Are there income effects on global willingness to pay for biodiversity conservation? Environ Resource Econ 43:137–160

Jessel B (2011) Ökosystemdienstleistungen. In: BBN (ed) Frischer Wind und weite Horizonte. Jb Natursch Landschaftspfl 58:72–87

Jessel B, Tschimpke O, Waiser M (2009) Produktivkraft Natur. Hoffmann und Campe, Hamburg

Kienast F (2010) Landschaftsdienstleistungen: ein taugliches Konzept für Forschung und Praxis? Forum für Wissen:7–12

Kremen C (2005) Managing ecosystem services: what do we need to know about their ecology? Ecol Lett 8:468–479

Luck GW, Harrington R, Harrison PA, Kremen C, Berry PM, Bugter R, Dawson TP, de Bello F, Dia S, Feld CK, Haslett JR, Hering D, Kontogianni A, Lavorel S, Rounsevell M, Samways MJ, Sandin L, Settele J, Sykes MT, Van de Hove S, Vandewalle M, Zobel M (2009) Quantifying the contribution of organisms to the provision of ecosystem services. Bioscience 59:223–235

MEA–Millennium Ecosystem Assessment (2005) Ecosystem and human well-being: scenarios, vol 2. Island Press, Washington

Mosbrugger V, Hofer H (eds) (2008) Biodiversitätsforschung in der Leibniz-Gemeinschaft: Eine nationale Aufgabe. Leibniz-Gemeinschaft, Bonn, 48 pp

Müller F, Burkhard B (2007) An ecosystem based framework to link landscape structures, functions and services. In: Mander Ü, Wiggering H, Helming K (eds) Multifunctional land use–meeting future demands for landscape goods and services. Springer, Berlin, pp 37–64

PBL–Netherlands Environmental Assessment Agency (2010) Rethinking global biodiversity strategies. Exploring structural changes in production and consumption to reduce biodiversity loss. PBL, Biltvoven

Peterson G (2010) Growth of ecosystem services concept. ► http://rs.resalliance.org/2010/01/21/growth-of-ecosystem-services-concept/. Accessed 10 April 2012

Ridder B (2008) Questioning the ecosystem services argument for biodiversity argumentation. Biodivers Conserv 17:781–790

Rockström J, Steffen W, Noone K, Persson Å, Chapin III FS, Lambin E, Lenton TM, Scheffer M, Folke C, Schellnhuber H, Nykvist B, De Wit CA, Hughes T, van der Leeuw S, Rodhe H, Sörlin S, Snyder PK, Costanza R, Svedin U, Falkenmark M, Karlberg L, Corell RW, Fabry VJ, Hansen J, Walker BH, Liverman D, Richardson K, Crutzen C, Foley J (2009) A safe operating space for humanity. Nature 461: 472–475

Schröter M, van der Zanden EH, van Oudenhoven APE, Remme RP, Serna-Chavez HM, de Groot RS, Opdam P (2014) Ecosystem services as a contested concept: a synthesis of critique and counter-arguments. Conserv Lett. doi:10.1111/conl.12091

Schweppe-Kraft B (2010) Ökosystemdienstleistungen: ein Ansatz zur ökonomischen Bewertung von Natur. Local land & soil news 34/35 II/10, The Bulletin of the European Land and Soil Alliance (ELSA) e. V., pp 11–14

Seppelt R, Dormann CF, Eppink FV, Lautenbach S, Schmidt S (2011) A quantitative review of ecosystem service studies: approaches, shortcomings and the road ahead. J Appl Ecol 48:630–636

Spangenberg JH, Settele J (2010) Precisely incorrect? Monetising the value of ecosystem services. Ecol Complex 7:327–337

Tansley AG (1935) The use and abuse of vegetational concepts and terms. Ecology 16:284–307

TEEB DE–Naturkapital Deutschland (2012) Der Wert der Natur für Wirtschaft und Gesellschaft–Eine Einführung. München: Ifu-Plan; Leipzig: Helmholtz-Zentrum für Umweltforschung–UFZ; Bonn: Bundesamt für Naturschutz

TEEB–The Economics of Ecosystems and Biodiversity (2009) An interim report. Europ. Comm., Brussels (▶ www.teebweb.org)

Trepl L (2012) Biodiversitätsbasierte Ökosystemdienstleistungen. ▶ www.scilogs.de/chrono/blog/landschaft-oekologie/biodiversitat-und-aussterben/2012-02-20/biodiversit-tsbasierte-kosystemdienstleitungen. Accessed 22 Feb 2012

Tüxen R (1956) Die heutige potentielle natürliche Vegetation als Gegenstand der Vegetationskartierung. Angewandte Pflanzensoziologie 13:5–42

UKNEA–UK National Ecosystem Assessment (2011) Synthesis of the key findings. Information Press, Oxford

UNEP-WCMC–World Conservation Monitoring Centre of the United Nations Environment Programme (2011) Developing ecosystem service indicators: experiences and lessons learned from bus-global assessments and other initiatives. Secretariat of the Convention on Biological Diversity, CBD Technical Series 58

WBGU–Wissenschaftlicher Beirat der Bundesregierung. Globale Umweltveränderungen (2011) Welt im Wandel: Gesellschaftsvertrag für eine Große Transformation, Berlin

Wende W, Wojtkiewicz W, Marschall I, Heiland S, Lipp T, Reinke M, Schaal P, Schmidt C (2012) Putting the plan into practice: implementation of proposals for measures of local landscape plans. Landsc Res 37:483–500 doi:10.1080/01426397.2011.592575

Westman W (1977) How much are nature's services worth? Science 197:960–964

Development and Fundamentals of the ES Approach

K. Grunewald, O. Bastian (eds.), *Ecosystem Services – Concept, Methods and Case Studies*,
DOI 10.1007/978-3-662-44143-5_2, © Springer-Verlag Berlin Heidelberg 2015

2.1 Key Terms

K. Grunewald and O. Bastian

"We know the price of everything and the value of nothing (adapted from Oscar Wilde)."

Despite, or perhaps because of the wide distribution and almost inflationary use of the ES term there is no question that a clear and uncontroversial, universally accepted definition does not exist. For example, what distinguishes the service of an arable land from the service of a natural ecosystem? What are the limitations of what we may call 'service'? What is an ecosystem property of the underlying service? What do we mean by a potential and what is meant by a function?

In the context of an integrative ES concept, it is important to create a concept system, similarly understood and accepted by economics, ecology, and sociology, by scientists, practitioners, and policymakers. That this only partly succeeded is partly due to the distinct subject-specific names (delimitation of a field of knowledge through technical terms). On the other hand, there are differences between regional common definitions and their contents. An example of this is the concept of function, which is described in German as the service of the ecosystem for humans (Bastian and Schreiber 1994), but is mostly used in English as 'functioning' of the ecosystem (see below).

As the environmental debate presently as far is largely determined by climate change and energy policy issues in Central Europe, the concept of sustainable development is overlayed by the **ecosystem service** term (definition, ▸ Chap. 1). The term ecosystem service was introduced by Ehrlich and Ehrlich (1981) respectively Ehrlich and Mooney (1983). Probably in the knowledge of the, inter alia, by Neef (1966) and Haase (1978) developed approach of 'natural potential', van der Maarel and colleagues designed in the Netherlands a 'global-ecological model' (van der Maarel and Dauvellier 1978), which later was further developed to the ES concept (Albert et al. 2012) by de Groot (1992) and working groups in the USA (Daily 1997).

The ES concept has the political agenda to foster awareness in society in regards to the position and importance of the environment. This corresponds to the choice of the metaphorical term 'service', which is subject to natural and legal persons in the national economy as well as in the daily linguistic usage, which provides the services ('Nature as a service provider'). Services or goods always have a specific purpose, usually according to an individual's requirement. Nature exerts negative effects on humans (so-called disservices), for example volcanic eruptions, earthquakes, floods, or avalanches.

In the current scientific literature, mainly the following definitions of ES are cited: ES are the conditions and processes through which natural ecosystems and the species they represent, sustain and fill human life (Daily 1997); advantage or benefit of ecological systems for humans (Costanza et al. 1997b; MEA 2005) or direct and indirect contributions of ecosystems to human well-being (de Groot et al. 2010).

Other authors explicitly differentiate between ES and the benefits from these, for example, Boyd and Banzhaf (2007): 'Benefits = the welfare the services generate'. According to Boyd and Banzhaf (2007) ES are 'ecological components' in physical terms (not monetarily measurable). The authors argue things or characteristics as well as end products of nature (i.e. in fact 'goods') that are consumed directly or that can be enjoyed and produce human well-being. They complain that many of the services mentioned by Daily (1997) or the Millennium Ecosystem Assessment (MEA 2005) are ecosystem processes in fact. The simultaneous use of the terms functions and services, without clarity to define both and to distinguish from each other, is not uncommon (e.g. Vejre 2009; Willemen et al. 2008).

The definitions have in common that ES are always defined by the societal view on the ecosystem, biophysical processes, and functions (Fisher et al. 2009). However, the authors provide different views on how functions and ES can be differentiated analytically and how they can be heuristically distinguished between ES and the benefits or the value of ES (Boyd and Banzhaf 2007; Wallace 2007; Costanza 2008; Fisher et al. 2009; Loft and Lux 2010). Following and in particular in the context of the case studies (▸ Sect. 6.1), common features and definitional boundaries of the ES term and its content will be further specified in relation to biodiversity and sustainability.

- **Ecosystem (Nature, Resources and Landscape)**

With focus on 'nature', the **ecosystem term** (definition, ► Chap. 1) has been established in the international ES debate. The term 'nature services', proposed by Westman (1977), did not gain acceptance.

An important basis for the ecosystem concept was developed in the context of a major research project carried out in West Germany in the 1960s under the influence of a crisis, which was in relation to the forest dieback in the Central German mountain ranges. The 'Solling Project' carried out in Lower Saxony under the leadership of Heinz Ellenberg (1973) examined the structures, functions, and processes of a Central European beech forest. Since then the ecosystem has been understood as an interactive structure between organisms and the environment, which is open to other systems, but differs from these in terms of its own structures and its own composition (Haber 2004; Nentwig et al. 2004; Steinhardt et al. 2011). Therefore, structures and processes of the earth system at different levels of scale play the main role in the ecosystem approach.

However, the *resource term* in the strictest sense includes raw materials and energy, whereas in a broader sense the natural basis of human life, such as air, water, soil, flora, fauna, and the interactions between them. The latter correspond to the (environmental) protection goods of nature protection in accordance with § § 1 and 10 of the German Federal Nature Conservation Act (BNatSchG 2009). Natural resources are divided into renewable and non-renewable. There is a differentiation from the ES concept that focuses in general only on renewable resources (MEA 2005; ► Chap. 1).

As a metaphor for biotic and abiotic components of the earth, the term *natural capital* is used. In a broader sense, ecosystems, biodiversity, and natural resources are included therein (BESWS 2010). Thereby the connection between nature and economy and the generation of values for human society due to the condition and processes of nature should be expressed. Natural capital provides as capital in kind a stock for services (Common and Stagl 2005). ES can be regarded as components of natural capital. The latter can be partially replaced by work performance (e.g. water treatment), which is associated with economic costs.

One difficulty is to separate methodologically explicitly between services of nature (ecosystem processes, natural capital) and activities of humans (means of production, technological processes, human capital). Therefore Matzdorf and Lorenz (2010) use the term 'environmental services', as the realisation of benefits (e.g. crops, biomass) of cultural shaped ecosystems (arable land, grassland) in addition to the ecological processes human work and artificial matter input (farming, fertilisation, maintenance, etc.) is required.

From landscape ecology and landscape planning the term *landscape services* was introduced into the discussion (Termorshuizen and Opdam 2009; Grunewald and Bastian 2010; Kienast 2010; Hermann et al. 2011; Albert et al. 2012), among others, to assess better the spatial relationships of ES or cultural landscapes with their characteristic elements (► Sect. 3.4). In this regard, the question of the usefulness of a further term is not entirely unjustified, especially as the decades-long and sometimes controversial running discussions on landscape cannot be overlooked and today there remains no unanimity regarding the content and application of the landscape concept, but there are quite different patterns of interpretation. Thus, landscape can be understood as a territorial entity a 'manageable space', which can be seen as positivistic (landscape as an ecosystem complex; e.g. Neef 1967), and constructivistic (as an aesthetic phenomenon or even mental construct; Leibenath and Gailing 2012) or as a space of action (Blotevogel 1995, Kirchhoff et al. 2012).

According to the MEA 2005, a landscape is typically composed of a number of different ecosystems, each generate a whole bundle of different ES. Therefore, it is justified to identify landscapes with similar or alike overall character (or to use as reference units) in order to interpret their conditions for effective and at the same time sustainable use by society (Bernhardt et al. 1986; Hein et al. 2006; TEEB 2010).

Also, new terms such as 'Green or Blue infrastructure' finally mean properties, functions, or services that are provided through a network of suitable ecosystems, with a particular focus on the connectivity.

■ Potentials of Nature and Ecosystems

The 'geographical concept of potential' was introduced in the German literature by Bobek and Schmithüsen as early as 1949, initially as a 'spatial arrangement of naturally provided possibilities for development'. The technical literature moreover contains such concepts as 'natural potential' (Langer 1970; Buchwald 1973) and 'natural performance power' (Buchwald 1973); Lüttig and Pfeiffer (1974) drew 'maps of nature potentials' (for related attempts Durwen 1995 and Leser 1997). In botany, the term 'potential' appeared in the form of 'potential natural vegetation', which was an integral used to indicate the totality of growth conditions at a given site (Tüxen 1956).

By making natural landscape potentials scientific categories and having them ascertained according to specific parameters of natural processes, they can be distinguished from natural resources, which represent an economic category (Mannsfeld 1983). Haase (1973, 1978) offered a way out of this hardly manageable complexity by suggesting that instead of a summary energy standard for a theoretically conceivable overall potential specific factors (properties, indicators) should be addressed in a particular case, and so-called partial natural spatial potentials defined with a clear focus on more specific socio-economic or societal goals and basic functions. These would include for example biotic yield potentials and regulatory potentials, water supply and disposal potentials, and construction and recreational potentials. The 'concept of potential' assesses nature's gifts from the point of view of the potential user, by means of a primarily scientific mode of operation. It elaborates the service capacities of an ecosystem or physical landscape as a field of options available to society for use, and also to take into account resilience, which limits or may even exclude certain intended uses (Grunewald and Bastian 2010).

Parallel to that, van der Maarel (1978) and Lahaye et al. (1979) in the Netherlands addressed 'landscape potencies', which might contribute to the fulfilment of certain societal needs (!). The term 'potential' is also found e. g. in Bierhals (1988), Finke (1994) and Durwen (1995), while e.g. Marks et al. (1992) and Leser (1997) prefer the terms 'service capability' or 'capacity' of the landscape balance. The international preferred term is 'capacity' of ecosystems (to sustain a specific function) (e.g. Führer 2000; Burkhard et al. 2012).

'Land-use suitability' on the other hand, focuses more on a certain use claim, which is considered primarily in societal, less in scientific terms. To determine land-use suitability, reference to the type of land use is definitely necessary (Niemann 1982). According to Messerli (1986), land use represents a 'decisive hinge position between societal and natural processes (◘ Fig. 2.4, ► Chap. 6), … by mediating as a link between processes in the socio-economic and natural systems. It enables the transfer of processes of an economic, social and cultural nature, which are describable in factual dimensions, to spatial dimensions, thus making them relevant ecologically, and in a reversed direction, of ecological, aesthetic and emotional information to society'. Land-use suitability can be seen 'potentially' ('use possibility'), e.g. the suitability of a field or a landscape for maize cultivation (without having maize actually being cultivated there at present), or an existing maize field can be assessed as to whether it is really suitable for such use, e.g. maize cultivation might involve intolerable risks.

This is illustrated in ◘ Fig. 2.1 exemplarily. Thanks to the fertile loess soil, the Lommatzscher Pflege landscape in Saxony not only has the potential for productive agriculture, but that potential has in fact long been used, so that it fulfils a societal function (or provides ES). However, the increasing intensification, particularly the expansion of rapeseed and maize cultivation, is giving rise to conflicts, e.g. with regard to protection of the soil and water (erosion, eutrophication), species and biotope protection (reduction of biodiversity), and the value of the landscape as an experience (monotony). The hill fortification of Zschaitz, which was already settled during the early Iron Age (800–500 B.C.), and refortified once more during the tenth century AD, is not suitable for agriculture–although it is at present so used–since soil removal is severely damaging this nationally significant archaeological site. Since, unlike the crops produced in this area, ideational or scientific values (or services) have no market, it is difficult to gain acceptance for any restriction of agricultural use.

Fig. 2.1 The increasingly intensive use (= social function) of the fertile loess soils of Lommatzscher Pflege landscape in Saxony (high production potential) leads to an impairment of archaeological sites on the plateau of the hill fortification of Zschaitz by soil erosion © Olaf Bastian

Functions

While potentials describe the possibility of the use of nature, the reality of the use of nature is expressed in the functional concept. According to this functional-spatial viewpoint, every part of the earth's surface fulfils societal functions. The Latin term 'function' *(fungi)* generally means 'carrying out' 'managing' or 'task' or 'activity' (Brockhaus Encyclopaedia 1996).

Thus, Speidel (1966) described the multifariousness of the functions of the forest, which benefit humankind, and which go far beyond wood production. Niemann later designed a methodology for ascertaining the degree of functional performance of landscape elements and units (Niemann 1977, 1982). Preobrazhenski (1980) referred to the natural functions of landscape, De Groot (1992) generally to 'functions of nature'. In spatial and regional planning, functions are defined as 'tasks which an area is to fulfil for the needs of life of the people' (ARL 1995). According to Wiggering et al. (2003), the determination of the multiple ecological, social, and economic functions of the landscape (multifunctionality) in their regional differentiation is the prerequisite for sustainable land use. The protection of efficacy and functionality is today provided by, e. g. the German Federal Conservation Law and the Federal Soil Protection Act.

However, the term 'function' is not used uniformly in the literature, frequently leading to terminological uncertainties and misunderstandings (Jax 2005). Thus, a purely ecological interpretation is common, in the sense of ecosystemic 'functioning' or the 'manner of function', as a scientifically determined organisation of structural-procedural contexts (e.g. food chains and nutrient cycles; cf. Forman and Godron 1986, where function is 'the interactions among the spatial elements, that is, the flows of energy, materials, and species among the component ecosystem'). In the TEEB study (TEEB 2009), functions are also regarded as purely ecological phenomena. According to

Costanza et al. (1997b), and in the MEA (2005), functions can support ecosystem services (ES). For Boyd and Banzhaf (2007), functions are 'intermediate products' of ES. Eliáš (1983) distinguished between two basic groups of functions: ecological functions (important for the existence of the ecosystems, regardless of concrete societal use claims), and social functions (which reflect societal needs).

Additional imprecisions of definition appear in the widespread blurring of the difference between function and potential. Thus, Marks et al. (1992) refer to the 'functions and potentials of the landscape balance' without providing any logical, conclusive differentiation between the two. De Groot et al. (2002) see 'ecosystem functions' as 'the capacity of natural processes and components to provide goods and services which directly and/or indirectly satisfy human needs'.

Petry (2001) sees the distinction between functions and potentials as a discussion within German-speaking, geographically oriented landscape ecology, which, while highlighting theoretical differences in meaning, causes more confusion than clarity at the international level, and with regard to application. Mannsfeld too (in Bastian and Schreiber 1994) noted: "A juxtaposition of the concept of natural landscape potentials as a structural aspect, and the performancev possibilities of the ecosystemic functional viewpoint based on the gifts of nature, … shows that a sharp separation of the two approaches is neither useful nor appropriate." Here, however, the objection is that it is not at all inconsequential whether one refers to the capacity of ability to render socially utilisable services (the potential concept), or of its actual realisation, or the actual rendering of such a service (the function concept).

The difference between potential and function can be illustrated as follows, using an example: An undeveloped South Sea island might have a high recreational potential; however, its recreational function will only be fulfilled if it is actually discovered and visited by tourists.

◘ Figure 2.2 shows a coastal section (ecosystem and landscape) in Mecklenburg-Western Pomerania. The recreational potential (possibility) is used by many tourists (realisation of the recreational function), and contributes to the well-being of the visitors (beneficial relevance of ES).

Another example illustrates ◘ Fig. 2.3: Due to centuries of withdrawal of fallen conifer needles as straw for cattle stables (straw use: a function and ES), the forest soils in question have been degraded, accompanied by a reduction of its biotic yield potential. Such forest forms have now become rare, and represent not only a habitat for animal and plant species in decline but also a valuable cultural-historical relict of past methods of economic use– with a potential for environmental education and tourism that has hardly been utilised to date.

■ Governance of ES

Spatial distributions and socio-economic aspects are of particular interest for benefits and welfare effects of ecosystems in the sense of the ES approach. This is reflected in ◘ Fig. 2.4 on the one hand by the change of land use and on the other by the delta of the incentive structures originated from the social side. Conceptually, the ecosystem structures and processes are related to ecology, the benefits and values to social and economic sciences. ES should be bridging both (for more details see ► Sect. 3.1).

The control and regulating system for ES is not only dominated by the State, so that the term *governance* comes into play. Governance refers not only to the structure and process organisation of government, administration, and community but also by private or public organisations (Ostrom 2011). Governance processes take place at several levels and need to be coordinated through the institutions acting in accordance to the principles of (1) accountability, (2) responsibility, (3) openness and transparency of structures and processes, and (4) fairness (Ostrom 2011; ► Sect. 5.4).

Ecosystem Services (ES)

ES has become established as a conceptual framework on the international stage. In German-speaking countries the conceptual system is oriented to functions and 'objects of protection' of nature so far (BNatSchG 2009), so it should be adjusted and further developed. Although the distinction between functions, services and benefits, is to be regarded as important especially for the economic evaluation, often no consistent classifications can be made, because smooth transitions, overlaps, and different interpretations of these terms exist.

Fig. 2.2 Many visitors use the recovery potential of the Baltic Sea beach in Kühlungsborn–the potential has turned to ES. The visitors have benefits (recreation, health). The potential remains depending on the ecosystem structures and processes. © Karsten Grunewald

ES generate human well-being in combination with the means of production and human capital. The largest welfare effect results from the optimum interaction between them. Individual ES can be replaced by technology and labour up to a certain extent. At a complete loss, the welfare effect is equal to zero and human existence cannot be maintained. Changes in the natural capital of any kind lead to changes of costs or benefits for ensuring human well-being.

2.2 ES in Retrospect

K. Mannsfeld and K. Grunewald

Scientific-Historical Roots

Currently, the concept of ecosystem services is one of the central themes in the scientific and environmental policy debates over the goal of preserving our natural resources. If, as stated above, this term is meant to encompass the benefits that society draws from the functions and capabilities of the ecosystem, then it is important to consider the lengthy evolution of the basic concepts behind this modern terminology for a fundamental societal goal. First empirically and then increasingly systematically, humankind has experienced the benefits, potentials, and also the risks and hazards associated with the use of nature, and, with increasing knowledge, has begun to put these insights to use.

A holistic view of our ambient spatial structures as a synthesis of natural and societal processes is indispensable in order to fully grasp the entire context of ecosystem services. The earliest signs for such a view can possibly be attributed to Alexander von Humboldt (1769–1859), who, by means of observation and measurement, sought to determine the 'Totalcharakter' (translating roughly as total character) of the region of the earth, and who therefore, in his later works, observed that only research that keeps the balance between specialisation and integration in nature as a whole could guarantee the desirable conditions for human life.

◘ **Fig. 2.3** Bizarre pines in the protected area Königsbrücke Heath, Saxony: Straw use has reduced the biotic yield potential but formed the potential for environmental education and tourism. © Olaf Bastian

Hence, Humboldt's basic concept of the character of nature as a whole with reference to societal and natural-scientific aspects is still a fundamental and challenging question in the present day (Neef 1971).

Only shortly thereafter, Ernst Haeckel (1866), who approached the issue from the biological point of view, coined the term 'ecology' to describe this 'interaction' between the animate and inanimate elements in nature; later, with Troll's (1939) landscape ecology, the term would very consciously incorporate the inseparable links between the biological and the geological components of our environment, by encompassing anthropogenic effect factors, and thus describing and emphasising the systemic context, which the theory of landscape ecology saw in the effective connection between nature, technology, and society (Neef 1967, p. 41). Neef describes this complex as follows:

"Hence, landscape ecology, although oriented toward the natural-scientific order of matter, must incorporate all factors which stem from the work of humankind and which will impact the natural balance."

In the decades after Humboldt's death, the analyses and interpretations of his 'total character of spatial phenomena on the Earth's surface' began to increasingly–albeit hesitantly–consider the factor humankind, and, conversely, recognise the positive and negative effects of natural factors on human desires for utilisation. However, it was a lengthy process for the research-historical unilateralism, which only considered anthropogenic effects in the landscape when they were clearly dependent on the balance of nature to be overcome, especially in geological and biological sciences. One milestone in overcoming this deterministic view with regard to the anthropogenic component in the real environment was the influence of late nineteenth-century economists on the theoretical conceptualisation of the footprint of humankind in nature and environment. They pointed out a problem in the then-accepted views of the relationships between humankind and nature, and should therefore be seen as 'contributors' to today's modern ES concepts. Specifically, they emphasised labour processes as the key factor in the interaction between humankind and nature, by which the necessary conditions for human existence were generated and upheld–entirely on the basis of natural and environmental conditions. In this respect, we should mention not only Adam Smith, Johann Heinrich von Thünen and others, but also Karl Marx in particular.

Marx used the term 'metabolism between society and nature' to describe the category under which he subsumed the role of humankind in withdrawing those materials from the landscape which were needed for its economic activity, so as to fulfil the necessities of life. He wrote:

"Labour is, in the first place, a process in which both man [sic] and Nature participate, and in which man of his own accord starts, regulates, and controls the material re-actions between himself and Nature. He opposes himself to Nature as one

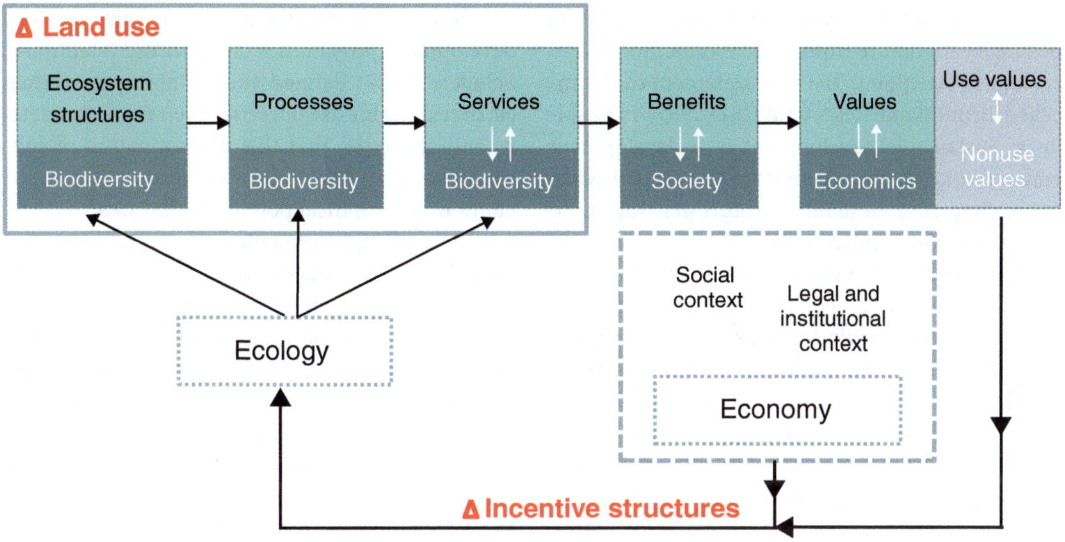

■ **Fig. 2.4** Identification and evaluation of ES as well as integration into instruments and incentive structures. © Ring 2010 based on Brouwer et al. 2011

of her own forces." (Marx 1867; ► https://www.marxists.org/archive/marx/works/1867-c1/ch07.htm)

In this context, he also pointed to the so-called 'free services' of nature, which positively affected the process of this metabolism. He noted that, as a result of the effects of natural forces–i.e. with no labour effort–such services of nature as photosynthesis, pollination, groundwater recharging, etc. positively accompany this metabolism, and thus substitute for human activity.

We can credit Carl Ritter (1779–1859; quoted in Leser and Schneider-Sliwa 1999), with calling upon the predominant specialised research activities in the geographic disciplines not to neglect the practical interests of their results. Later, Alfred Hettner (1859–1941) raised the postulate of a 'practical geography' (Hettner 1927), the core statement of which was to evaluate and predict the effects of human impacts and changes on the basis of knowledge of the causal contexts of natural processes. From that, he drew the conclusion that such an evaluation should primarily be derived from the given state of the natural systems in the cultural landscape, and that scientifically grounded proposals for improving utilisation should include concepts to preserve and protect the forces of nature. His conceptual proximity to the instrument of compensation/offsetting the impacts of human use of natural resources–which is still in use today–or the environmental impact assessment can hardly be overlooked.

The key realisation upon which this history-of-science oriented reflection is based is that if Marx's metabolic process becomes critical, which is the case today on both local and global levels, the effects caused by use processes must be ascertained systematically and according to a number of different standards. Otherwise, given the continued overtaxing of nature's 'free services' the healthy development of ecosystems, i.e. a development subjected to only low levels of disturbance and detrimental interference, can no longer be guaranteed. In this respect, it is no coincidence that the ES concept and its numerous predecessors (see below) have placed the preservation of the precious forces of nature at the centre of their considerations.

With reference to the global character of the growing imbalance between availability of natural resources and the degree of utilisation and the resulting destruction of landscape structures and their ecosystems, the report prepared by the World Commission on Environment and Development (WCED) at the end of the twentieth century gave a stern warning for humankind to reconsider its dealings with nature from an economic, social and ecological viewpoint. The core statement of the so-called Brundtland Report (WCED 1987) is as follows: Sustainable development is a development which meets the requirements of the present without endangering the ability of future generations to meet their own requirements.

This basic statement of sustainable development has proven to be of great relevance with regard to setting goals for a permanent environmentally appropriate economic and social order. On the other hand, there has to this day been no feasible methodological concept following up on this sustainability triad; it is largely a regulatory idea, a guiding concept characterised by the ethical principle of generational justice. Nonetheless, today the term carries significant meaning whenever policy-makers, business leaders or academics employ it to identify the linkage between economic development and ecological carrying capacity as a major goal of today's societal policy, so as to be able to leave a liveable and usable environment to future generations. Indisputably, the ES approach, which is currently being widely discussed, is viewed as a fundamentally suitable instrument for the implementation of the idea of sustainability.

▪ The Substantive and Methodological Precursors of ES

Especially, the German geographic community has, by way of a number of small steps, begun to approach the question of the extent to which it is necessary and possible to refer to the service capacity of a natural abundance (natural balance) which functions in a manner appropriate to the ecosystem (▶ Sect. 2.1). One early source is an essay by Schmithüsen (1942) on site ecology and its importance for the cultural landscape, in which he explains that people use the service possibilities existing in the natural plan of a landscape to secure their livelihoods, by drafting a 'cultural service plan' of natural and labour processes for distinguishable spatial structures. A few years later, Bobek and Schmithüsen (1949) designated 'regional nature' (*Landesnatur*; a term meaning the totality of naturally provided interactive contexts) in the cultural landscape as a range of potentials, and hence a spatial pattern of arrangements for naturally provided development possibilities (societal use intentions). Schultze (1957) defined the suitability of certain earth regions for use purposes even more concretely, and suggested that this determination of suitability be reformulated into a determination of the cultural-geographical potential of an area.

The growing exploitation of natural resources, with the well-known consequences for the condition of 'protected goods', as we would call them today, confronted society and hence a number of scientific disciplines with the task of seeking answers and proposing solutions as to how to ascertain the service capacity of natural systems and how to preserve and secure them over the long term. Within the geographic community which, as we know, has to deal with hybrid material systems in the cultural landscape surrounding us (abiotic, biotic and societal/cultural components), Neef (1966) presented an initial study for the evaluation of the potentials of natural systems, the essence of which involve the idea of making all aspects of natural factors comparable with the anthropogenic creations in the cultural landscape, and similarly capable of valuation, by defining their various elements in terms of energy content. He entitled this study in which he describes the use of this energy content concept for the elucidation of the relationships between naturally related and economic components of societal activity in the natural environment 'Questions of regional economic potentials', clearly highlighting what he believed was involved. He saw it as an important part of this concept and also an absolute necessity to transfer natural scientific findings into societally familiar, i.e. primarily economic, categories if utility, sustainability, resilience and protection of natural resources were to be considered as societal activities at all.

The epistemological phenomenon which he describes as the 'transformation problem' became part of the application-oriented foundations of East German landscape research. Neef saw his proposal as an important bridge towards objectifying the various processes of nature and society, and the transition from one causal area to another, and towards making the metabolism between human society and nature, which had up to that time been described only as a fairly general phenomenon, usable for such purposes as balancing-of-interests decisions (Neef 1969). Over 45 years later, the German Federal Government's Advisory Council on Global Change (WBGU 2011) has now used the term transformation research in a study titled *Zukunftsprojekt Erde*[The earth as a project for the future], albeit with a more specialised meaning–and without mentioning the preceding ideas.

However, the proposed exclusive use of an energy scale (Neef 1966) lead to methodological difficulties of implementation, especially with regard to the specific-use demands of society upon the natural-spatial service capacity. The later proposals by von Haase (1973, 1978) provided a way out: Instead of energy as the standard of measurement for service capacity, a thorough analysis of the characteristics of the 'Naturkapital' (natural capital) was to be employed in order to evaluate the fulfilment of basic societal functions. Only when the scientific and social goals were clearly defined did material and energetic properties of the services of nature become 'potentials', since they referred to the specific distribution of such service possibilities in the spatial context, 'natural-spatial potentials' (▶ Sect. 2.1). Thus, the concept is able to illustrate not only the actual degree of tolerance towards societal utilisation, but also the resilience, especially under the conditions of realistic multiple utilisations. The spatially differentiated service capacity of nature suitable for societal development processes has been defined as the natural-spatial potential. Due to the different demands placed upon this capacity by society, it is, for methodological reasons, structured into a number of sub-potentials (partial natural spatial potentials), including for example:

- The Biotic Yield Potential, or the capacity to produce organic substances and to regenerate the conditions for such production (site fertility).
- The Biotic Regulation Potential, or the capacity to sustain biological processes and to regulate them once again after disturbances (the biodiversity aspect).
- The Recreation Potential, or the capacity of nature to contribute to the recreation and health of people by psychological and physical effects.

These brief examples describing the properties of potentials indeed show that the occasionally uttered opinion that the concept of natural spatial potential puts too much emphasis on its natural-scientific elements and fails to sufficiently capture societal or economic aspects is unfounded. A broad range of methodological procedures have been developed by von Haase (1991), Jäger et al. (1977), Mannsfeld (1983), and others through which the advantages and disadvantages of potential utilisation interests can be clearly fleshed out on the basis of an initially unbiased and value-neutral analysis of space. The potential approach was at an early stage also adopted into the system of landscape management and landscape planning (Langer 1970; Buchwald 1973; Lüttig and Pfeiffer 1974).

Complementary to the derivation of suitability for utilisation potentially provided by the abundance of nature, a functional-spatial paradigm began to be developed, according to which certain landscape spaces are to fulfill societal functions. This involves not so much the functioning of ecosystems as the scientifically determined organisation of structural procedural contexts (Forman and Godron 1986). The German Federal Conservation of Nature Law underscores in § 1 Item 5 the mandate to preserve the service provision and functionality of landscapes (BNatSchG 2009).

Particularly Niemann (1977), also van der Maarel (1978), Bastian (1991), de Groot (1992), Marks et al. (1992), Durwen (1995), Willemen et al. (2008) and others have addressed this functional approach thoroughly and in great depth. As a result, these and other authors have developed often closely corresponding categorisations into main and partial functions, for example production (economic) functions, regulatory (ecological) functions, and habitat (social) functions–a structuring that clearly reflects a proximity to the three pillars of the sustainability thought discussed earlier. The more recently introduced suggestion to follow a transparent action plan that seeks to secure ES at the interface between conservation of nature and societal/economic goals is based, similarly to the landscape function concept, on the economic, ecological, and sociocultural services provided by ecosystems, and this pragmatic subdivision reflects a great conceptual proximity to fundamental concepts that were already conceived two or three decades prior.

The concept of landscape functions was widely accepted in West German landscape planning during the 1980s (e.g. Langer et al. 1985), since it had proven itself advantageous in communications with political decision-makers (Albert et al. 2012). However, landscape functions in general only see those aspects of the landscape that are ignored by the commercial markets, and hence need to be managed by public planning (von Haaren 2004; Albert et al. 2012).

The knowledge gained from landscape ecological studies about the natural processes are generally not suited for incorporation into economic calculations, due to which reason they are not usually considered in spatial planning decisions. Hence, a correct handling of the transition of natural quanta into economic data (the transformation problem) remains indispensable. Neef, in an essay published in 1969, wrote the following in this regard:

"The role of natural functions in economic contexts, and the feedback effects of societal impacts into the natural balance can only be properly understood if both are placed into a relationship with one another. In order to derive a foundation for the evaluation of natural potentials, it is necessary to juxtapose the potential quanta to the effort of societal labour that needs to be performed (Neef 1969)".

With the results of a large-scale exemplary project involving the regions north of Dresden (Mannsfeld 1971) an attempt was made to find an economic standard for the implementation of these basic concepts. A method was developed to indirectly ascertain the suitability of a natural space for agricultural utilisation in terms of service capacity and functionality by analysing its lack of suitability in that regard. For this purpose, agricultural sites were evaluated according to their deficits in terms of a defined optimum (e.g. fertile loess soil), and the costs of upgrading them to a higher level of yield were ascertained. By multiplying the theoretical costs for drainage or irrigation, removal of stones, fertilisation, deep-loosening of the soil, and humus enrichment with the area shares of the respective sites, a standard of comparison could be obtained that is also expressive in financial terms. Even though this process is quite involved, the approach itself might serve as a real conceptual model for how the 'welfare services' of natural potentials could be directly translated into monetary quantities. However, the example also shows that technological and human labour inputs are made to optimise ES, and have to be treated methodologically separately (in the sense of 'environmental services' ►Sect. 2.1 and 4.2). Haber (2011) noted in this regard:

"A wheat field–or, for that matter, any other field upon which corn, potatoes, or beets are grown–is not an ecosystem from the 'natural-ecological' point of view, for no such thing could exist in nature. It is an 'artificial' food production system created by people; it would not and could not exist without them, so that it does not really fit into the concept of ecosystem services. Of course, it is put together of natural components, [...]."

Ultimately, ecological economics claims to have decisively contributed to the development of the ES approach (Røpke 2004, 2005), with its roots especially in the American scientific debate, which should be more broadly considered with respect to its cultural historical development framework and its applicability to central Europe. The arguments are made that the independent evolution of ecology and economy and their growing apart into two special and independent sciences is an ill development that should be overcome (cf. the *Full World Model of the Ecological Economic System* of Costanza 1997a), and that economics must learn from ecology the limitations that human economic activity must take into account in a non-growing biophysical world. Jetzkowitz (2011) notes in this context that ecological economics, which sees nature as the ultimate limiting characteristic for economic processes, must be distinguished from neoclassical environmental economics. The latter, he argues, while it does depend on the ES concept, nonetheless attempts to economise nature as much as possible.

The ES concept found its way into the international political limelight by way of the Millennium Ecosystem Assessment (Synthesis Report, MEA 2005). The Report was drafted between 2001 and 2005 under the auspices of the UN, and coordinated by the UN Economic Program (UNEP). It pursues the goal of identifying the consequences from changes in the ecosystems for human well-being and to thus create the scientific basis for the necessary activities for the sustainable use of the ecosystems. For this purpose, no primary knowledge was developed; rather, the report depended on the existing scientific literature, relevant data and models, and knowledge from the private sector, local communities and indigenous peoples. The acute deterioration of no less than 15 of the 24 ES examined worldwide, and the resulting negative effects on future human well-being, were a key message of the report.

Based on the idea that the perception and appreciation of nature changes when it is also perceived from an economic point of view, novel conceptual approaches have emerged in the environmental and conservationist community which are further discussed in ▶ Chap. 5. In this context we would like to refer to the study *The Economics of Ecosystems and Biodiversity* (TEEB) carried out in 2007 with the support of the German Federal Ministry of the Environment. This was an important international initiative designed to direct attention towards the global utility of ecosystems and to highlight the growing costs generated by the loss of biodiversity and ES. In the reports, expert knowledge from the scientific community, and business and policy-makers were brought together in order to develop concepts for sustainable solutions (reports and information under ▶ www.teebweb.org).

Even if one were to accept the premise that the current ES approach places particular emphasis on the preservation of biodiversity and the economic valuation of the services of ecosystems, the relevant literature on the procedural foundations (including Costanza 1991; de Groot 1992; Daily 1997) contained no reference to such conceptual models as 'natural spatial potential' or 'landscape functions'. Recently, Gomez-Baggethun et al. (2010) argued that the source of the ES concept is to be found in the late 1970s, and they refer primarily to Dutch, American and Spanish authors. However, it would seem appropriate to also identify the other intellectual and methodological forerunners (see above), the credo of whose research was that the intensive use of regenerable, natural resources often has negative effects to such an extent that they could no longer–without naivety–be seen as free services; rather, they should be accepted as natural assets of limited self-generating capacity. Precisely this realisation has opened the way to a systematic review of the service capacity of nature, in order to be able to target and implement a sustainable and optimal use of ES.

For today's ES concept and the problems of its implementation–such as the repeatedly mentioned transformation from the material level to the values level, etc.–a look back at older methodological approaches may provide necessary indicators. Ecosystems and their services for society consist not only of elements, such as sinks, regulators, or processes, which function as a system, they also have a spatial reference. A sufficient accounting for these factors–as in Neef's (1963) 'theory of geographical dimensions'–in the form of standardisation is still absent, for small-scale overviews, while they may be appropriate for awakening consciousness with regard to the actual problems, are little suited for providing concrete proof of spatial service availability. A methodological differentiation of the ES approach into local/regional and global scopes of standards is hardly apparent, although it would be necessary, e.g. for the categories of ecological planning. But also the already discussed concepts of natural spatial potentials and landscape functions currently lack any adequate 'knobs' to properly handle and incorporate the continually emerging use of nature-dependent changes in ecosystem quality–e.g. in climate change. The ES concept, even in its international dimension, will only be successful in preventing further overuse and even destruction of the free services of nature if such and similar fundamental methodological questions are settled and taken into account.

2.3 Values and Services of Nature for Humans

K. Grunewald and O. Bastian

'Better living and increase our wealth', are important drivers of human existence. Nature provides the basis for that and is considered to be a significant growth generator, which brings prosperity to the extent that its elements are protected and developed (Jessel et al. 2009). Human societies require resources from nature. The regional value creation in agriculture, forestry and fisheries, measures of landscape management, or tourism in national parks and other protected areas are solid economic variables of society. However, renewable resources such as food, wood or fiber are counted as ecosystem services and goods but not fossil row materials such as oil or coal (► Sect. 2.1).

There is also the benefit that individuals or society derive from a variety of indirect (supporting) services of nature: maintenance of soil fertility, biodiversity, clean air, fixation of carbon by photosynthesis, groundwater recharge, nitrogen fixation, aesthetically attractive landscapes and the innovation potential of nature for technical innovations or pharmaceutical development.

▪ Values and Value Shift

An effective, yet democratically legitimised sustainability policy must be accepted by the majority of people to give consent and enable people to participate. One aspect of the eco-ethical value level is that 'nature is full of values', which are subsequently discovered and recognised by valuing people (Ott 2010). Accordingly, basic values are assessments that are shared by the vast majority of the population. This includes in principle the benefit foundations of ES (health, food provisioning, security, etc.). Basic values can be associated with standards and–important for ES–collective goods.

Values

are in the sense of Emanuel Kant, what is highly valued, what is respected, what is dear to us. Societies are always also a community of values, i.e. a society without value orientations is not conceivable. People feel bound by and to values, but at the same time they are not unfree to act or transform their value systems (apparent paradox according to Joas 1997). Values affect wishes, interests, and preferences. However, a value is not a standard or rule. Value orientations describe relatively stable preferences with respect to different values of individual persons (Häcker and Stapf 1994). They are always linked to a cultural and social context and need negotiations and disputations in pluralistic societies.

Banzhaf and Boyd (2012) refer to a fundamental difference between ES (per se) and the values they present. ES are biophysical qualities and quantities that are directly related to market goods and services. It is initially not an evaluation in the strictest sense. This occurs only when a stronger relation to a benefit is given, especially on monetary values.

Following value concepts are to be distinguished with reference to biodiversity (Potthast 2007):

- *Exchange value*: economic; measures the value of an object against which you can trade on the market; measuring value is usually the price, which does not necessarily say something about the 'real' value.
- *Value in use*: instrumental; 'useful for something…'; biodiversity is valuable because of its function as a resource for human economic purposes; substitutability as an essential feature of the value in use; monetised utility values are substantial arguments–but more important than their absolute level is their distribution.
- *Intrinsic value*: inherent; biodiversity has intrinsic value for me when I appreciate them for their own sake, not for their use of sake, simply because they exist (existence value), because it has for me biographical or cultural significance (reminder value, home), because it is unique and special (character), because it allows experiences for me (e.g. wilderness), because it is to be preserved for posterity ('heritage') (bequest value); intrinsic values evades a monetisation in principle, but they are communicable, that means, they are comprehensible for others, they need to be weighed with each other and against each other; and the value lies in the specific relationship.

While an anthropocentric basic position, according to persuasive environmental and nature conservation rationales finally must always refer to human interests, needs, etc. ('nature conservation is human protection'), represents a natural or physiographic focused position the view that some or all of the natural beings are to be protected for their own sake (ethical naturalism, according to which the various natural objects or in the sense of this position 'natural subjects' have an intrinsic value,

Value of City Trees

Even if they are planted by people and their habitats are not very close to nature–city trees provide numerous services and benefits. They sequester CO_2 and produce O_2, improve the urban climate, generate biomass, serve wildlife (e.g. birds, bats, insects) as habitat and enhance the scenery of cities. Furthermore, urban districts with lots of trees increase the value of residential property. Trees contribute to the natural experience of the urban population and generate emotions, for example by flowers and leaves sprouting in spring or by colourful fall foliage. In this case, not everything is perceived as a whole positive by the population (e.g. falling leaves or bird droppings under the trees).

The example of city trees also shows that it is neither possible nor useful to calculate all these services in euros. However, a quantification–Where city green is missing, how does it evolve (see, e.g. city of Berlin; Hermsmeier and Marrach 2012)–and in individual cases also monetisation, can raise awareness in dealing with nature.

which is in doubt validity also independent of human interests). The pathocentrism (Teutsch 1985) recognises only pain-sensitive beings (humans and higher animals) have this intrinsic value. The biocentrism, however, gives it to all living things, while holism to all of nature involved (even inanimate and system wholes), so the landscape. The German Federal Nature Conservation Act (BNatSchG 2009, § 1) expressly refers to the "protection of nature and landscape due to their inherent value and for future generations".

Focusing too much on one of the two basic positions is not helpful, since one would be a 'natural-forgotten anthropocentrism', the other a 'forgotten human environmentalism'. It is finally the secondary sources from which comes the moral justification for the protection of nature (and landscape), if more from religious or secular reasons. A serious respect to anthropocentric nature conservation obligations would have approximately the same results as the strictest compliance of biocentric or theological rules–nature would just as well be preserved (Ott 2010).

Ott (2010) proposes pragmatic solutions to the 'self-value problem'. He argues finally that not too much is to be focused on the question "whether all the grains of sand, every drop of water, all the blades of grass, all soil bacteria, all blueberries, all squid, etc. have a self-value". All this distracts from the actual environmental challenges (climate change, water supply, agriculture and fisheries, protection of forests and wetlands, wildlife conservation, and ecological urban redevelopment).

Value is generated in various relationships and interactions between humans and nature, which happen less and less successfully in our mechanised and increasingly urban world. Not all of these relationships and interactions are adequately addressed as 'utilisation'. Subjective, qualitative value judgments are sometimes more meaningful than (often supposedly) objective, quantitative values (► Chap. 4).

Since the beginning of modern times attitudes and positions designed for individual utility maximisation have prevailed. We have got used in today's industrial society that the basic resources for life, the daily supply of food, water, and all the necessities of civilisation have become a matter of course (Haber 2011). With the advent of industrial mass production 'good life' was increasingly equated with material wealth (WBGU 2011). Rational cost-benefit calculations provide in industrial societies such as Germany action-bearing patterns of interpretation. Thus, money and monetary values are of paramount importance in our economically driven world. Money is among other things a symbolic medium, it is also for nature conservation. But it must not be an end in itself but only a means to an end in the sense of social benefit.

Appreciations for ES (Ott 2010):

- can be evaluated by surveys (e.g. the acceptance of nature conservation),
- articulate themselves within value judgments ("I like mountains rather than the sea."),
- are of different intensities (e.g. different feelings at the sight of a sunset or a spider–graded joy, happiness, aversion, indifference, etc.).

2

Commons Don't Simply Exist; They are Created

Silke Helfrich (in Ostrom 2011), writes:
'Resources are free. They know neither property rights nor borders. Resources do not know if we need them to live or if we don't. We, however, are tied in one way or another to these things: to limits, to ownership and–above all–to resources themselves. The old German word for commons, 'Allmende', preserves this nexus for us, because, 'Allmende' derives from words for 'all' and 'community', according to language historians. Thus the term itself conveys the core challenge of the commons debate: Everyone who belongs to a particular community and collectively uses its resources needs to agree on how to share. But to agree on usage rules for resources and monitor their compliance is anything but 'child's play'.

However, according to a survey of the Emnid Institute on behalf of the Bertelsmann Foundation published in autumn 2010 (Bertelsmann Foundation 2010) a value shift is to be noted. Thus, growth and material prosperity isn't (no longer) everything: a prosperity which is bought by damaging the environment or high national debt, more than 80 %, are principle that Germans do not accept. Nine out of ten people are demanding a new economic order which takes the social balance and the careful handling of our livelihoods and finances stronger into account.

Conclusion

A mere economic efficiency focused policy seems to have lost its appeal and plausibility. However, the perception of the problem leads not automatically to "right choices" of people, for example to environmentally friendly actions (Kuckartz 2010). This is partly caused due to lack of long-term orientation and loss aversion (WBGU 2011), on the other hand, goods and services of nature are mostly 'public goods' characterised by non-excludability and non-rivalry (▶ Table 3.5). The recreation effect of landscapes, biodiversity as gene pool or 'intrinsic value' are examples of public goods in the sense of the ES concept. In case of such goods and services it is best for the individual, if he profits, but is not involved in providing ('free riders'). Those who wish to participate in principle to the provision fear that due to the 'free rider' only a minimal level of provision at high individual costs comes about, so it is not worth participating in the provision ('the realists'). In addition, one does not want to be the fool who participates in the provision, and the other free benefit ('those who do not want to be exploited').

The uncompensated impact of economic decisions on unaffiliated market participants is referred as an external effect in economics. External means that the effects (side effects) of a behaviour are not adequately reflected in the market. They are not included in the calculation or decision of a causer. In economic terms they are a cause of market failure and thus government intervention can become necessary. Negative externalities are also referred to as external or social costs, positive externalities as external benefits or social income (Mankiw 2004).

■ Implementation of Biodiversity and ES

The aim of the ES concept is to prioritise and to assess services of nature, in particular also in monetary terms (cost-benefit calculation) to advocate for economic reasons the conservation of nature (Jessel et al. 2009). As attractive and new this approach may be, it must be complemented by the ethical principles of protecting nature for its own sake, laid down in the German Federal Nature Conservation Act (BNatSchG 2009, § 1).

ES are supported by markets if their base products are manufactured (free market, such as purely agricultural products). Many public goods and services are, however, exploited, overused and destroyed because market mechanisms do not work ('market failure'). Here, market-based instruments can help–if they are placed correctly–to create incentives for codes of behaviour. Currently, there are two main economic instruments for environmental policy in order to maintain and enhance ES and thus biodiversity:

1. Positive incentives aimed to make nature and environmental protection lucrative in financial

Can Biodiversity and Cultural Landscape be Profitable?

Nature and landscape are scarce resources, as well as public goods, which in many cases are beyond conventional market mechanisms (allocation of public goods, market failure). Maintenance of a stone wall landscape in the Ore Mountains or the old wine cultural landscape in the Elbe valley near Dresden is labour and cost-intensive. Who would want to process the steep slopes mechanically and effectively, would have to destroy the dry stone walls. But so the charm of the landscape would be lost. A simple profit and loss account almost always ends up in the red. It is the duty of society/the state to preserve such heritage in financially reasonable limits. The return is in an identity-creating landscape with a high feel good factor for locals and visitors. This requires an intersubjective agreement on drafts for future, the resilience of ecosystems but also about the cost.

terms (example: remuneration of special environmental services from agriculture);
2. Negative price signals in form of price control (user fees, environmental taxes, compensation) or quantity control (example: CO_2-emission certificates), which increase the costs of environmentally damaging behaviour.

Ideally, a 'common good market' is developing, e. g. for landscape maintenance. In this context the institutional side is important. In general, the State must act as buyers of services of general interest, and it must also be a 'provider' of the service, for example, in the form of a farm or landscape maintenance associations (► Sect. 6.5).

Biodiversity and many ES have not been calculated in conventional economic evaluations and were generally accepted as free of charge. However, these services have a high value but which can be determined often only indirectly, as it is only inadequately reflected in markets and prices. Scientists make every effort to assign a real value to natural capital. Examples are illustrated in Jessel et al. (2009) or TEEB (2010). In all questions concerning the methodological and assessment approaches, the attempt is to recognise ES and not only to address it, but to quantify and evaluate them in a holistic manner, so that they are comparable with economic goods.

Economists use the concept of 'total economic value' to determine the economic value of ecosystems and biodiversity (► Sect. 4.2.2). It includes both the 'potential' and 'real use values' as well as the so-called nonuse values. This means that the economic evaluation of natural captures are not only the direct benefits of nature. The economic value of a good or service results from the appreciation by individuals and the scarcity of the resource and must not necessarily be monetary. Empirical studies show that the nonuse values often account for the largest part of the human appreciation for threatened ecosystems. For this, however, no tax is paid as a general rule, which means, users of nature are not adequately involved in the coverage of costs. The ES concept is seen as a promising way to achieve substantial improvements regarding this dilemma.

Potthast (2007) points out the limits of the commodification of biodiversity (► Sect. 4.2). He calls:

- Moral limitations: What commodification? Clarification of the normative presuppositions, man and nature images.
- Methodological limitations: determining the 'right' monetary value for nature as home or for feelings. What can be 'measured' and how does it correlate with actions?
- Empirical limitations: feasibility and limits of determination of **all** partial sums of total economic value. Commodification and monetisation are always partial and task-/interest-related.
- Strategic limitations: complete substitutability (by money) is objectively and strategically wrong; option of complete renunciation always threatens, if the price is too low or high.
- Political limitations: a rationally well-founded price of nature does not provide the security of conservation. Economic rationality is not regularly correlated with appropriate political rational decisions, which is the case individually and socially.

Table 2.1 Overview of concepts for welfare and sustainability measures. (Source: WBGU 2011)

Type of the measurement concept	Name of the index/indicator	Economical dimension	Social dimension	Ecological dimension
Extensions of GDP: monetised indicators/indices	Measure of economic welfare	+	+	+
	Index of sustainable economic welfare (ISEW)	+	+	+
	Genuine progress indicator (GPI)	+	+	+
	Full costs of goods and services (FCGS)	+	–	+
	National welfare index + (NWI)	+	+	+
Extensions of GDP: integrated environmental and economic accounting/satellite systems	Integrated environmental and economic accounting/UN system of environmental and economic accounting (SEEA)	+	–	+
Non-monetised indicators/ indices	Ecological footprint	–	–	+
	Living planet Index	–	–	+
Composite indicators/indices (integration of monetised and non-monetised values)	Human development index (HDI)	+	+	–
	Index of economic well-being	+	+	+
	Happy planet index[a]	–	+	+
	KfW-Nachhaltigkeitsindikator (sustainability indicator)	+	+	+
	Sustainable development indica-tors (Eurostat)	+	+	+
	Index of economic freedom	+	+	–
	Environmental sustainability index (ESI)	+	–	+
	Environmental performance index (EPI)	–	–	–
	Gross national happiness[a] (GNH, Bhutan)	+	+	+
	Canadian index of well-being[a] (CIW)	–	–	–
	Corruption perception index (CPI)	–	+	–
	National accounts of well-being[a]	–	+	–

[a]Index encloses subjective indicators

■ **Welfare and Sustainability Measurement**
The search for alternatives to gross domestic product (GDP) as a welfare indicator is an expression of change in values. Current concepts are presented in ◘ Table 2.1; in each case the sustainability dimensions are assigned.

The GDP per capita is a measure for all economic activities transacted on markets and in monetary terms. Goods and services which have no market prices or real traded–as most ES–are not recorded in GDP. A rising GDP leads not automatically to an increase in subjective well-being (Inglehart 2008).

The debate on the latest indicators shows on the one hand that measure for welfare and sustainability are necessary beyond the GDP and will be developed (◘ Table 2.1). On the other hand, the political decision on which an alternative is given preference is still under discussion. This depends on the orientation of the goals as well as on data availability and data quality. Thus, for example, the introduction of an 'ecosystem index' by the German Federal Statistical Office failed, because this was scientifically not tenable (Radermacher 2008). Here too, new, especially methodological impulses are expected regarding the ES concept.

Implementation of ES

It is the intention to give environmental policy better strategic orientation, focusing on the benefit of the people. ES deliver such benefits, e.g. recreation service for health and well-being or protection against floods for the security needs. Economic arguments are intended to supplement the classical ethical rationales for conservation, without replacing them. In addition to economic values (based on efficiency and cost-effectiveness) there are always environmental values (based on ecological sustainability/load capacity), and sociocultural values (based on justice and perception as well as ethical considerations) necessary. Decisions on land use, for example in connection with the energy turnaround in Germany, with all its normative questions are concerned with genuinely ethical and legal dimensions and determine much about the future structure and function of ecosystems, the existence and spread of animal and plant species as well as the life chances of people. All this poses great challenges for the analysis of ES and their complex, integrative evaluation.

References

Albert C, von Haaren C, Galler C (2012) Ökosystemdienstleistungen. Alter Wein in neuen Schläuchen oder ein Impuls für die Landschaftsplanung? Naturschutz und Landschaftsplanung 44:142–148

ARL–Akademie für Raumforschung und Landesplanung (ed) (1995) Handbuch der Raumordnung. Hannover

Banzhaf HS, Boyd J (2012) The architecture and measurement of an ecosystem services index. Sustainability 4:430–461

Bastian O (1991) Biotische Komponenten in der Landschaftsforschung und -planung. Probleme ihrer Erfassung und Bewertung. Habilitations-Schrift Martin-Luther-Universität Halle, Wittenberg, 214 p

Bastian O, Schreiber KF (eds) (1994) Analyse und ökologische Bewertung der Landschaft. G Fischer, Jena. (2., erheblich veränderte Aufl. 1999. Spektrum Akademischer Verlag, Heidelberg)

Bernhardt A, Haase G, Mannsfeld K, Richter H, Schmidt R (1986) Naturräume der sächsischen Bezirke. Sächsische Heimatblätter 4/5

Bertelsmann-Stiftung (2010) Bürger wollen kein Wachstum um jeden Preis. ▶ www.bertelsmann-stiftung.de/cps/rde/xchg/SID-1CE81901-A2FDE973/bst/hs.xsl/nachrichten_102799.htm. Accessed 5 Jan 2011

BESWS–Biodiversity and Ecosystem Service Work Stream (2010) Demystifying materiality: hardwiring biodiversity and ecosystem services into finance. UNEP FI CEO Briefing, Genf

Bierhals E (1978) Ökologischer Datenbedarf für die Landschaftsplanung. Landschaft und Stadt 10:30–36

Blotevogel HH (1995) Raum. In: Akademie für Raumforschung und Landesplanung (ed) Handwörterbuch der Raumordnung. ARL, Hannover, pp 733–740

BNatSchG–Bundesnaturschutzgesetz (2009) Gesetz über Naturschutz und Landschaftspflege. BGBl. I, p 2542

Bobek H, Schmithüsen J (1949) Die Landschaft im logischen System der Geographie. Erdkunde 3:112–120

Boyd J, Banzhaf S (2007) What are ecosystem services? The need for standardized environmental accounting units. Ecol Econ 63:616–626

Brockhaus Enzyklopädie (1996) 20., 2nd edn., Gütersloh

Brouwer R, Oosterhuis FH, Ansink JH, Barton DN, Lienhoop N (2011) POLICYMIX WP4: Guidelines for estimating costs and benefits of policy instruments for biodiversity conservation. POLICYMIX Technical Brief 6. ▶ http://policymix.nina.no

Buchwald K (1973) Landschaftsplanung und Ausführung landschaftspflegerischer Maßnahmen. In: Buchwald K, Engelhardt W (eds) Landschaftspflege und Naturschutz in der Praxis. BLV-Buchverlag, München

Burkhard B, Kroll F, Nedkov S, Müller F (2012) Mapping supply, demand and budgets of ecosystem services. Ecol Indic 21:17–29

Common M, Stagl S (2005) Ecological economics. An introduction. Cambridge University Press, Cambridge

Costanza R (ed) (1991) Ecological economics: the science and management of sustainability. Columbia University Press, New York

Costanza R (2008) Ecosystem services: multiple classification systems are needed. Biol Conserv 141:350–352

Costanza R, Cumberland JC, Daly HE, Goodland R, Norgaard R (1997a) An introduction to ecological economics. St. Lucie Press, Boca Raton

Costanza R, d'Arge R, de Groot RS, Farber S, Grasso M, Hannon B, Limburg K, Naeem S, O'Neill R, Paruelo J et al. (1997b) The value of the world's ecosystem services and natural capital. Nature 387:253–260

Daily G (ed) (1997) Nature's services: societal dependence on natural ecosystems. Island Press, Washington DC

de Groot, RS (1992) Functions of nature: evaluation of nature in environmental planning, management and decision making. Wolters-Noordhoff, Groningen

de Groot RS, Wilson M, Boumans R (2002) A typology for description, classification and valuation of ecosystem functions, goods and services. Ecol Econ 41:393–408

de Groot R, Fisher B, Christie M, Aronson J, Braat L, Gowdy J, Haines-Young R, Maltby E, Neuville A, Polasky S, Portela R, Ring I (2010) Integrating the Ecological and Economic Dimensions in Biodiversity and Ecosystem Service Valuation. In: Kumar P (ed) The Economics of Ecosystems and Biodiversity: Ecological and Economic Foundations. Earthscan, London, Washington, S 9–40

Durwen KJ (1995) Naturraum-Potential und Landschaftsplanung. (Landschaftsökologie und Vegetationskunde als Grundlage der Landnutzung). Nürtinger Hochschulschriften 13:45–82

Ehrlich PR, Ehrlich AH (1981) Extinction: The causes and consequences of the disappearance of species. Random House, New York

Ehrlich PR, Mooney HA (1983) Extinction, substitution, and the ecosystem services. BioScience 33:248–254

Eliáš P (1983) Ecological and social functions of vegetation. Ekológia (ČSSR) 2:93–104

Ellenberg H (1973) Die Ökosysteme der Erde: Versuch einer Klassifikation der Ökosysteme nach funktionalen Gesichtspunkten. In: Ellenberg H (ed) Ökosystemforschung. Springer, Berlin, S 235–265

Ellenberg H, Zeller O (1951) Die Pflanzenstandortkarte am Beispiel des Kreises Leonberg. Forschungs- u. Sitzungsbericht der Akademie für Raumforschung u. Landesplanung II, Hannover, S 11–49

Finke L (1994) Landschaftsökologie. Das geographische Seminar, Braunschweig

Fisher B, Turner RK, Morlin P (2009) Defining and classifying ecosystem services for decision making. Ecol Econ 68:643–653

Forman RTT, Godron M (1986) Landscape ecology. Wiley, New York

Fry G (2000) The landscape character of Norway–landscape values today and tomorrow. In: Pedroli B (ed) Landscape-our Home. Lebensraum Landschaft, Indigo Zeist, S 93–100

Führer E (2000) Forest functions, ecosystem stability and management. For Ecol Manage 132:29–38

Gomez-Baggethun E, de Groot R, Lomas PL, Montes C (2010) The history of ecosystem services in economic theory and practice: from early notions to markets and payment schemes. Ecol Econ 69:1209–1218

Grunewald K, Bastian O (2010) Ökosystemdienstleistungen analysieren–begrifflicher und konzeptioneller Rahmen aus landschaftsökologischer Sicht. GEOÖKO 31:50–82

Haase G (1973) Zur Ausgliederung von Raumeinheiten der chorischen und der regionischen Dimension–dargestellt an Beispielen aus der Bodengeographie. Petermanns Geogr Mitt 117:81–90

Haase G (1978) Zur Ableitung und Kennzeichnung von Naturraumpotenzialen. Petermanns Geogr Mitt 22:113–125

Haase G (1991) Naturraumerkundung und Bewertung des Naturraumpotentials. Schriftenreihe des dt. Rates für Landespflege 59, Hannover, pp 923–940

Haaren C von (ed) (2004) Landschaftsplanung. UTB, Eugen Ulmer, Stuttgart

Haber W (2004) The ecosystem–power of a metaphysical construct. Schriftenreihe Landschaftsökologie Weihenstephan 13:25–48

Haber W (2011) Umweltpolitikberatung–eine persönliche Bilanz. Studienarchiv Umweltgeschichte 16:15–25. ► www.iugr.net

Haeckel E (1866) Generelle Morphologie der Organismen. G. Reimer, Berlin, 2 vol (Anatomie der Organismen und Entwicklungsgeschichte der Organismen)

Häcker H, Stapf KH (1994) Dorsch: Psychologisches Wörterbuch. Hans Huber, Bern

Hein L, van Koppen K, de Groot RS, van Ierland EC (2006) Spatial scales, stakeholders and the valuation of ecosystem services. Ecol Econ 57:209–228

Hermsmeier L, Marrach K (2012) Schauen Sie mal, wie grün Ihr Bezirk ist. Neue Zahlen vom Berliner Senat: In Steglitz-Zehlendorf stehen die meisten Straßenbäume. BZ–Berliner Zeitung vom 25. Mai 2012

Hermann A, Schleifer S, Wrbka T (2011) The concept of ecosystem services regarding landscape research: a review. Living Rev Landscape Res 5. ► http://landscaperesearch.livingreviews.org/Articles/lrlr-2011-1/. Accessed 1. April 2011

Hettner A (1927) Die Geographie–ihre Geschichte, ihr Wesen, ihre Methoden. Hirt-Verlag, Breslau, 466 S

Inglehart R (2008) Changing values among western publics from 1970 to 2006. West European Politics 31:130–146

Jax K (2005) Function and "functioning" in ecology: what does it mean? OIKOS 111:641–648

Jäger KD, Mannsfeld K, Haase G (1977) Bestimmung von partiellen und komplexen Potentialeigenschaften für chorische Naturräume. (Methoden und Beispielsuntersuchungen). F/E-Bericht, Institut f. Geographie u. Geoökologie, Leipzig, Sächsische Akademie der Wissenschaften, Leipzig, 125 p

Jessel B, Tschimpke O, Waiser M (2009) Produktivkraft Natur. Hoffmann und Campe, Hamburg
Jetzkowitz J (2011) Ökosystemdienstleistungen in soziologischer Perspektive. In: Groß M (ed) Handbuch Umweltsoziologie. VS Verlag für Sozialwissenschaften, Wiesbaden, S 303–324
Joas H (1997) Die Entstehung der Werte. Suhrkamp, Frankfurt a. M.
Kienast F (2010) Landschaftsdienstleistungen: ein taugliches Konzept für Forschung und Praxis? Forum für Wissen:7–12
Kirchhoff T, Trepl L, Vicenzotti V (2012) What is landscape ecology? An analysis and evaluation of six different conceptions. Landscape Res. doi:10.1080/01426397.2011.640751
Kuckartz U (2010) Nicht hier, nicht jetzt, nicht ich–Über die symbolische Bearbeitung eines ernsten Problems. In: Welzer H (ed) KlimaKulturen: soziale Wirklichkeiten im Klimawandel. Campus, Frankfurt a. M., pp 144–160
Lahaye P, Harms B, Stortelder A, Vos W (1979) Grundlagen für die Anwendung landschaftsökologischer Erkenntnisse in der Raumplanung. Verh Ges Ökol 7:79–84
Langer H (1970) Zum Problem der ökologischen Landschaftsgliederung. Quaest Geobiol 7:77–95
Langer H, von Haaren C, Hoppenstedt A (1985) Ökologische Landschaftsfunktionen als Planungsgrundlage–ein Verfahrensansatz zur räumlichen Erfassung. Landschaft + Stadt 17:1–9
Leibenath M, Gailing L (2012) Semantische Annäherung an die Worte "Landschaft" und "Kulturlandschaft". In: Schenk W, Kühn M, Leibenath M, Tzschaschel S (eds) Suburbane Räume als Kulturlandschaften. ARL, Hannover, pp 58–79
Leser H (1997) Landschaftsökologie. 4. edn. Eugen Ulmer, Stuttgart, 644 p
Leser H, Schneider-Sliwa R (1999) Geographie: Eine Einführung. Westermann, Braunschweig
Loft L, Lux A (2010) Ecosystem Services–Ökonomische Analyse ihres Verlusts, ihre Bewertung und Steuerung. BiK-F Knowledge Flow Paper 10
Lüttig G, Pfeiffer D (1974) Die Karte des Naturraumpotentials. Ein neues Ausdrucksmittel geowissenschaftlicher Forschung für Landesplanung und Raumordnung. N Arch f Nds 23:3–13
Maarel E van der (1978) Ecological principles for physical planning. In: Holdgate W, Woodman MJ (eds) The breakdown and restoration of ecosystems. Conf Ser I Ecology 3, Plenum Press, New York, pp 413–450
Maarel E van der, Dauvellier PJ (1978) Naar een globaal ecologisch model voor de ruimlijke entwikkeling van Niederland. Studierapp. Rijksplanologische Dienst, Den Haag, 9
Mankiw NG (2004) Grundzüge der Volkswirtschaftslehre, 3. edn. Schäffer Poeschel, Stuttgart
Mannsfeld K (1971) Landschaftsökologie und ökonomische Wertung der Westlausitzer Platte. Dissertation, TU Dresden, Fakultät Bau-, Wasser-und Forstwesen, Dresden
Mannsfeld K (1981) Landeskulturelle Auswirkungen moderner Agrarproduktion an Beispielen aus dem Westlausitzer Hügelland. Wiss. Abhandl. Geogr. Gesellschaft DDR 15
Mannsfeld K (1983) Landschaftsanalyse und Ableitung von Naturraumpotentialen. Abhandl. Sächsische Akademie der Wissenschaften, Leipzig, math.-nat. Kl., vol 35, Akademie-Verlag, Berlin
Marks R, Müller MJ, Leser H, Klink HJ (eds) (1992) Anleitung zur Bewertung des Leistungsvermögens des Landschaftshaushaltes, 2. edn. Forsch. zur Deutschen Landeskunde, vol 229, Trier
Matzdorf B, Lorenz J (2010) How cost-effective are result-oriented agri-environmental measures? An empirical analysis in Germany. Land Use Policy 27:535–544
MEA–Millennium Ecosystem Assessment (2005) Ecosystem and human well-being: Scenarios, vol 2. Island Press, Washington
Messerli P (1986) Modelle und Methoden zur Analyse der Mensch-Umwelt-Beziehungen im alpinen Lebens- und Erholungsraum. Erkenntnisse und Folgerungen aus dem schweizerischen MAB-Programm Nr. 25, Bern
Neef E (1963) Dimensionen geographischer Betrachtung. Forschungen und Fortschritte 37:361–363
Neef E (1966) Zur Frage des gebietswirtschaftlichen Potentials. Forschungen und Fortschritte 40:65–70
Neef E (1967) Die theoretischen Grundlagen der Landschaftslehre. Haack, Gotha
Neef E (1969) Der Stoffwechsel zwischen Gesellschaft und Natur als geographisches Problem. Geogr Rundsch 21:453–459
Neef E (1971) Über das Weiterwirken der Ideen Alexander von Humboldt in der Geographie. Acta Historica Leopoldina 6:17–29
Nentwig W, Bacher S, Brandl R (2004) Ökologie kompakt (3. edn. 2011). Spektrum Akad. Verlag, Heidelberg
Niemann E (1977) Eine Methode zur Erarbeitung der Funktionsleistungsgrade von Landschaftselementen. Arch Naturschutz Landschaftsforschung 17:119–158
Niemann E (1982) Methodik zur Bestimmung der Eignung, Leistung und Belastbarkeit von Landschaftselementen und Landschaftseinheiten. Wissenschaftliche Mitteilung d. Instituts f. Geographie u. Geoökologie, Akademie der Wissenschaften der DDR, Leipzig, Sonderheft 2
Ostrom E (2011) Was mehr wird, wenn wir teilen. Vom gesellschaftlichen Wert der Gemeingüter. Oekom, München
Ott K (2010) Umweltethik zur Einführung. Junius, Hamburg
Petry D (2001) Landschaftsfunktionen und planerische Umweltvorsorge auf regionaler Ebene. Eine landschaftsökologische Verfahrensentwicklung am Beispiel des Regierungsbezirkes Dessau. UFZ-Bericht Nr. 10/2001, UFZ-Umweltforschungszentrum Leipzig-Halle
Potthast T (2007) Biodiversität–Schlüsselbegriff des Naturschutzes im 21. Jahrhundert? Naturschutz und Biologische Vielfalt 48, Bundesamt für Naturschutz, Bonn
Preobraženskij VS (1980) Issledovanie landšaftnyh system dlja celej ochrany prirody–struktura, dinamika i razvitie

landšaftov. Inst. Geografii AN SSSR (Akadademie der Wissenschaften der UdSSR), Moskau
Radermacher W (2008) Beyond GDP–A Ecosystem services as part of environmental economic accounting. ► www.uni-kiel.de/ecology/users/fmueller/salzau2008/Abstracts_Salzau2008.pdf. Accessed 1 April 2011
Ring I (2010) Die Ökonomie von Ökosystemen und Biodiversität–die TEEB-Initiative. Vortrag auf dem 5. Dresdener Landschaftskolloquium, Wert und Potenziale sächsischer Landschaften, Dresden. ► www.ioer.de/aktuelles/veranstaltungen/19-november–2010–5-dresdner-landschaftskolloquium/. Accessed 19. Nov 2010
Røpke I (2004) The early history of modern ecological economics. Ecol Econ 50:293–314
Røpke I (2005) Trends in the development of ecological economics from the late 1980s to the early 2000s. Ecol Econ 55:262–290
Schmithüsen J (1942) Vegetationsforschung und ökologische Standortslehre in ihrer Bedeutung für die Geographie in der Kulturlandschaft. Z Ges Erdkunde, Berlin, pp 113–157
Schultze HJ (1957) Die wissenschaftliche Erfassung und Bewertung von Erdräumen als Problem der Geographie. Die Erde 88:241–298
Speidel G (1966) Zur Bewertung von Wohlfahrtswirkungen des Waldes. Allg Forstzeitschr 21:383–386
Steinhardt U, Blumenstein O, Barsch H (2011) Lehrbuch der Landschaftsökologie, 2 edn., Elsevier, Spektrum Akademischer Verlag, Heidelberg
TEEB–The Economics of Ecosystems and Biodiversity (2010) The economics of ecosystems and biodiversity: ecological and economic foundations. Kumar P (ed). Earthscan, London
Termorshuizen JW, Opdam P (2009) Landscape services as a bridge between landscape ecology and sustainable development. Landscape Ecol 24:1037–1052
Teutsch GM (1985) Lexikon der Umweltethik. Vandenhoeck und Ruprecht, Düsseldorf
Troll C (1939) Luftbildplan und ökologische Bodenforschung. Z Ges Erdkunde, Berlin 7/8:241–298
Tüxen R (1956) Die heutige potentielle natürliche Vegetation als Gegenstand der Vegetationskartierung. Angew Pflanzensoziol (Stolzenau) 13:5–42
Vejre H (2009) Quantification and aggregation of landscape functions/services in periurban landscapes. Proc Europ IALE Conf, Salzburg. Breuste J, Kozová M, Finka M (eds), pp 430–432
Wallace KJ (2007) Classification of ecosystem services: problems and solutions. Biol Conserv 139:235–246
WBGU–Wissenschaftlicher Beirat der Bundesregierung. Globale Umweltveränderungen (2011) Welt im Wandel: Gesellschaftsvertrag für eine Große Transformation. Berlin
WCED–World Commission on Environment and Development (1987) Our common future. University Press, Oxford
Westman W (1977) How much are nature's services worth. Science 197:960–964
Wiggering H, Müller K, Werner A, Helming K (2003) The concept of multifunctionality in sustainable land development. In: Helming K, Wiggering H (eds) Sustainable development of multifunctional landscapes. Springer, Berlin, pp 3–18
Willemen L, Verburg PH, Hein L, van Mensvoort MEF (2008) Spatial characterization of landscape functions. Landsc Urb Plan 88:34–43

Conceptual Framework

K. Grunewald, O. Bastian (eds.), *Ecosystem Services – Concept, Methods and Case Studies*,
DOI 10.1007/978-3-662-44143-5_3, © Springer-Verlag Berlin Heidelberg 2015

3.1 Properties, Potentials and Services of Ecosystems

O. Bastian and K. Grunewald

3.1.1 The Cascade Model in the TEEB Study

Anyone wants to analyze and evaluate ecosystem services is going to be challenged finding the suitable methodology. Due to the complexity of the issue 'services of nature for society', this is no simple feat. Generally valid methodical requirements refer to the scientific foundations, intersubjective comprehensibility, and communicability. The interdisciplinary approach is not only the terminology, but rather the diversity of methods and the different perspectives and approaches (▶ Fig. 1.5), which must be geared to specific, quite concrete questions. It is necessary to distinguish between general principles and concepts on the one side and specific investigation methods on the other side.

In view of the complexity and multidisciplinarity of the problem area ecosystem service (ES) it is no surprise that different scientific-theoretical approaches and practical methods have emerged over time, which complement each other, or which are partially congruent or divergent. This is reflected in the classification of ES (▶ Sect. 3.2), but also in the different theoretical-methodical concepts.

The cascade model of Haines-Young and Potschin (2009) and Maltby (2009) is a frequently cited framework and was also adopted by TEEB (2010) (◘ Fig. 3.1). The graph presents the chain from the ecosystems to human well-being. The ES mediate between the structures, processes and functions (functioning) of ecosystems and the benefits and values belonging to the human well-being. In the real world, however, the relation is not so simple as the graph might communicate. Nevertheless, the general structure proposed by the scheme is widely accepted among experts.

3.1.2 The EPPS Framework

Based on this scheme (◘ Fig. 3.1) and taking the knowledge of various schools of landscape ecology and the international scientific discussions into account, we consider the framework depicted in ◘ Fig. 3.2 appropriate for ES issues. According to this, the 'functions' in the sense of ecosystem integrity are directly attributed to the left pillar ('properties of ecosystems'), while the societal functions are subsumed in the ES. This better corresponds with the German understanding of the term 'function' (▶ Sect. 2.1). In the cascade model of Haines-Young and Potschin (2009) (◘ Fig. 3.1), functions represent their own intermediate step between the structure and processes on the one side and the ES on the other side. This subgroup of ecosystem processes is essential for and directly contributes to the generation of ES (Albert et al. 2012). The potentials of an ecosystem (or a landscape) show its performance and possible utilisation and, thus, they are a logical intermediate step between the properties (structure and processes) and the ES themselves (real use of nature and landscape, or demand). This conceptual concept is called EPPS framework (derived from ***E**cosystem **P**roperties, **P**otentials and **S**ervices*, cp. Grunewald and Bastian 2010; Bastian et al. 2012b).

The basic elements of the EPPS framework will be explained in the following section.

Ecosystem Properties

On the left side of the EPPS framework (◘ Fig. 3.2) are the properties of ecosystems–individual objects, parts of objects and even entire ecosystem complexes–and the structures and processes (e.g. soil qualities, nutrient cycles, biological diversity), which form the basis for all ES and, moreover, for the existence of humans and of human society in general. According to van Oudenhoven et al. (2012), ecosystem properties are the set of ecological conditions, structures, and processes that determine whether an ES can be supplied. Since this ecological endowment is, first of all, scientifically based, it has to be assigned mainly to the factual level. The (scientific) analysis of ecosystem properties is the research starting point as it enables an understanding of the functional principles of nature.

It is the matter of the performance basis, i.e. those components of nature that provide services, e.g. the particular components of specifications of ecosystems, which ensure primary production, flood regulation or aesthetic values. As a component

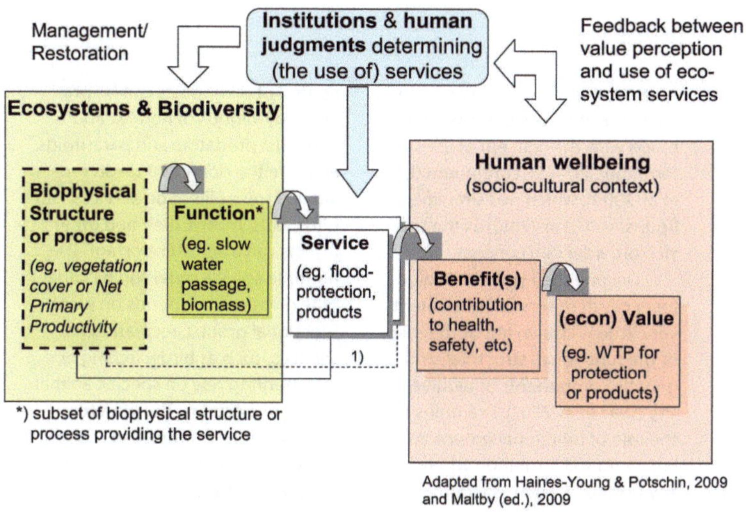

Fig. 3.1 The chain from ecosystem structures and processes to human well-being (from TEEB 2010; ► www.teebweb.org/EcologicalandEconomicFoundationDraftChapters/tabid/29426/Default.aspx, Chap. 1, p. 11)

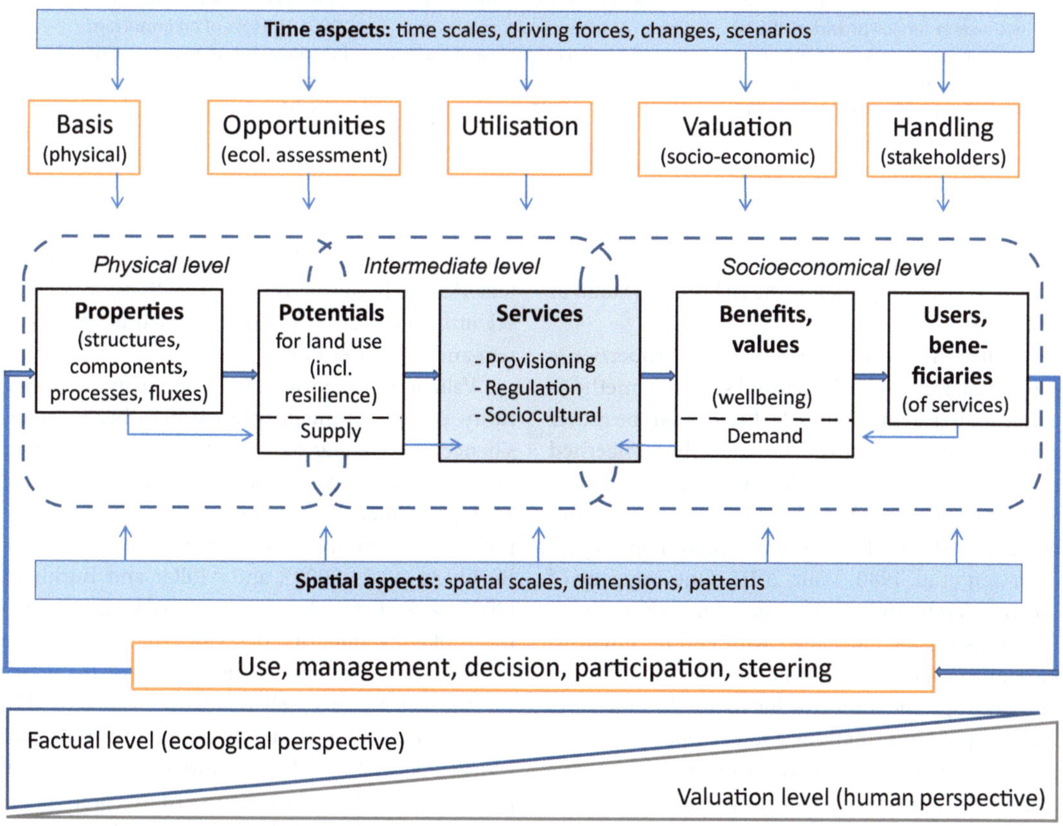

Fig. 3.2 Conceptual framework for the analysis and evaluation of ES with a particular focus on space and time aspects (from Grunewald and Bastian 2010, modified)

Functional Traits

Sometimes only specific parts of ecosystems, single species, individuals or parts of them (roots or leafs of plants) are relevant to ES. The issue that functional groups, populations, communities and different genotypes or species may contribute to service provision to different degrees at different times or in different places has been discussed for several years (de Bello et al. 2010). This involves the concepts of functional traits (Lavorel et al. 1998) and service providing units (SPU–Luck et al. 2003; Harrington et al. 2010; Haslett et al. 2010): a SPU is 'the collection of organisms and their characteristics necessary to deliver a given ecosystem service at the level required by service beneficiaries'. Kremen (2005) emphasised the importance of key ecosystem service providers (ESPs) and functional groups of species (e.g. population abundance and spatiotemporal variation in group membership) for service provision. Later, the SPU concept was combined with the concept of ESPs to form the SPU–ESP continuum (Luck et al. 2009), which was simplified by Rounsevell et al. (2010) as the service provider (SP) concept.

Despite their potential value for ecosystem service assessments, very little is known about the role of the functional, structural, and genetic components of biodiversity (Diaz et al. 2007). Examples for the role of functional groups are known in soil formation where key taxa exist, such as legumes, which are able to fix atmospheric nitrogen and build up nitrogen stores in the soil. An other example is deep-rooted species that can relocate nutrient elements from the parent material to the surface layers. At a finer scale, sequestration of carbon in stable aggregates depends on the activity of the soil fauna: in many managed systems, control of plant pests can be provided by various generalist and specialist predators and parasitoids. Bees are the dominant taxon providing crop pollination services, but birds, bats, moths, flies, and other insects can also be important. The multiple service provision by subalpine grasslands depends on plant functional groups; recreational services, such as birdwatching or duckhunting rely on specific animal taxonomic groups. The literature also mentions examples of ESs provided by such single species as the Eurasian jay (*Garrulus glandarius*), which ensures oak seed dispersal, or the Eurasian wildcat (*Felis silvestris*), whose presence makes it a flagship species in terms of recreational/touristic value (Vandewalle et al. 2008; Haines-Young and Potschin 2009). The loss of an important functional group may cause drastic changes in the functioning of ecosystems.

of nature, this basis for services is materially manifested and can, in principle, be measured (Staub et al. 2011).

Hence, the analysis of ecosystem properties is predominantly driven by natural scientific methods using analytical indicators. Indicators can be rather easily analyzed and they illustrate the concerned problem especially well. Without them it is almost impossible to decipher the complicated network of relationships of ecosystems (and landscapes) (Durwen et al. 1980; Walz 2011). One category of indicators is bioindicators: organisms, whose living functions can be correlated with certain environmental factors in such a manner that they can be used as a specific indicator for them. As indicators may simplify informations and present them comprehensively, they enable decision-makers to give convincing reasons for their decisions.

There is now extensive experience in the field of analyzing ecosystems, their structures, processes and changes (e.g. in the framework of the 'Ecosystem Research Germany'–Fränzle 1998), as well as scientific literature (e.g. Leser and Klink 1988; Bastian and Schreiber 1999).

Valueless categories like complexity, diversity, rarity, ecosystem integrity, ecosystem health or resilience also belong to the category of 'ecosystem properties' (de Groot et al. 2002). The concept of 'ecological integrity' as a precondition for the supply of ES is applied by the assessment method of Burkhard et al. (2009), and Müller and Burkhard (2007, ► Sect. 4.1). According to Barkmann (2001) the 'ecological integrity' describes the maintenance of those structures and processes that are necessary for the ecosystems' self-regulation capacity. The ecological integrity is mainly based on variables of energy and matter balance, as well as on structural properties of whole ecosystems. These components are similar to those defined in other ES-studies as supporting services (e.g. in MEA 2005).

Biomass Potentials

The concept of potentials will be described in Sect. 4.4.2 using the example of the 'energetic use of biomass' as a presently widely discussed topic (utilisation of the biotic productivity, the so-called 'biotic yield potential' to produce energetically usable biomass, ► Sect. 4.4.2). The land potential for bioenergy in Germany is c. 3–4 million ha (SRU 2007), including energy crops and biomass from landscape management, the application of which can be honored by a so-called landscape management bonus under the Renewable-Energies-Act–EEG 2009). In the future, regionally energetic use of biomass from landscape management measures shall make a tangible contribution to satisfy our energy needs.

By order of the Saxon State Office for Environment, Agriculture and Geology (LfULG), the consulting firm Bosch & Partner has calculated the biomass potential of the Free State of Saxony (Peters 2009). For this purpose, databases for the relevant area types like grassland, water margins, and roadside greenery were established (◘ Fig. 3.3). This potential was regionalised with the aid of a geographic information system (GIS). Thus, biomass potentials of c. 204,000 ha with c. 667,500 t annual yield are available in the Free State of Saxony, an amount sufficient for workable realisation concepts.

This example shows how the feasibility of natural resources usage may be analyzed and evaluated. This provides an important basis for planning purposes. On this basis, the existing but still almost not used potential may be used properly–in this case for bioenergy production–if there are appropriate framework conditions (e.g. technology, logistics, remuneration). Not only would the energy sector benefit from it, but also the socio-economic significance and the social standing of nature conservation.

Ecosystem Potentials–The Capacity or Supply Side

Depending on their properties, ecosystems are able to supply services; they have particular potentials or capacities for that. Potentials (► Chap. 2) have consciously been included as the second, so as to distinguish between the possibility of use and an actual use, which is the expression of the real service (Bastian et al. 2012a). Potentials can be regarded and quantified as stocks of ES, while the services themselves represent the actual flows (Haines-Young et al. 2012).

In terms of ecosystem potentials, various preconditions need to be considered, e.g. the ecological carrying capacity and the *resilience*, which is defined as 'the capacity of a system to absorb and utilise or even benefit from perturbations and changes that attain it, and so to persist without a qualitative change in the system' (Holling in Ring et al. 2010).

This is closely related to the *ecological stability*, i.e. the persistence of an ecological system and its capacity to return to the initial situation after changes. Within the 'stability', we can distinguish between constancy and cyclicity (without extraneous factors), as well as between resistance and elasticity (with extraneous factors). In this regard, the *carrying capacity*, meaning the range of a possible use should be mentioned. It indicates to which extent particular utilisations may be tolerated. For example, high (natural) soil fertility allows the assumption of a high potential for farming, though, this alone is not sufficient if, for example, risk factors like high erosion disposition may damage the topsoil at some point, which eventually causes the loss of the usability for farming.

The assessment of ecosystem potentials also pursues the goal of ascertaining the potential use of particular services, and is more normative than a mere accounting of ecosystem properties. It constitutes an important basis for planning, e.g. for the implementation of sustainable land-use systems: the suitability of an ecosystem to carry different forms of land use can be established, the available but still unused potentials can be put to actual use, and risks can be estimated.

Ecosystem Services

Only human needs or demands actually convert a potential into a real service. ES, the third pillar of the framework (◘ Fig. 3.2), reflect an even stronger human perspective (value level), since the services (and goods) are in fact currently valued, demanded, or used. In other words, the status of an ES is influ-

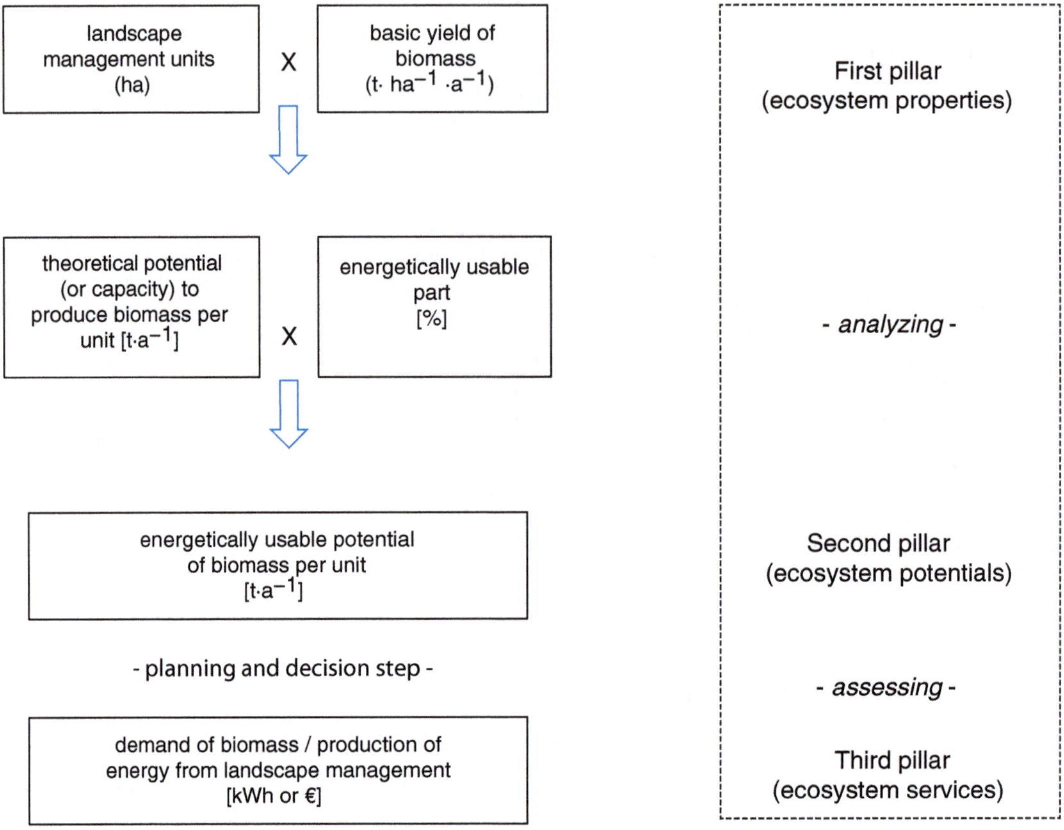

Fig. 3.3 Algorithm for the calculation of the biomass potential for energetic use. (According to Peters 2009)

enced not only by its provision of a certain service, but also by human needs and the desired level of provision for this service by society, which connects inseparably supply and demand of ES (Burkhard et al. 2012; Syrbe and Walz 2012).

We regard services and (societal) functions as synonyms. The term 'function' stands for a benefit-oriented view, not for the functioning of ecosystems in the sense of processes, cycles, etc. We prefer a tripartite classification of functions (Bastian and Schreiber 1999) or ES (Grunewald and Bastian 2010): provisioning, regulation and sociocultural services (► Sect. 3.2).

The analysis of ES always involves a valuation step, e.g. scientific findings (facts) are transformed into human driven value categories. The decisive factor is the combination of the various causal areas in the relationship between society and nature, one example being economic valuation (e.g. Costanza et al. 1997; Spangenberg and Settele 2010).

Intact ecosystems provide a wide variety of ES that are characterised by complex interrelations (trade-offs, see below). Some ES are strictly related or occur in bundles and, therefore, are influenced positively or negatively if a particular ES is enhanced (e.g. the maximisation of the yield of an arable field at the expense of regulation ES, like carbon sequestration, or habitat services). The manner of connections and interrelations between single ES is still an issue with significant knowledge gaps (MEA 2005).

Although the EPPS framework focuses on the benefits produced, it also implicitly includes negative social or economic effects of ecosystems (and landscapes) to human well-being, so-called 'disservices' (Lyytimäki and Sipilä 2009; Dunn 2010).

Classification of ES-Related Values (▶ Sects. 2.3 and ▶ 4.2)

ES-related values can be classified into two categories: use values and nonuse values. Use values refer to the present, future, or potential use of an ES. They encompass direct and indirect use, option, and quasi-option values.

Values for hunting, fishery, and medical plants are examples for use values. All provisioning and some sociocultural services (e.g. recreation) provide direct use values. Indirect use values refer especially to the positive effects of ecosystems. Examples are the values of pollination and decomposition of toxic substances.

Option values und quasi-option values are connected with information and uncertainty. As humans are not sure what are their future demands, circumstances of life and then available information, they evaluate the option of a possible future use, and they take the expected information growth into account.

Society attributes nonuse values to the mere existence of an ecosystem regardless of the use of its services (existence values). Altruistic values (benefits of existence for other people) and bequest values (benefits for the well-being of future generations) also belong to this heading. It is often difficult to differentiate the single categories of nonuse values, both conceptually and empirically (Hein et al. 2006).

Quasi-option values represent the value of irreversible decisions, until new information is not available, which may indicate today still unknown values of ecosystems. Quasi-option values, too, are difficult to assess in practice (Hein et al. 2006). They are strongly corresponding with the concept of natural potentials.

As previously stated, the term ES is only justified if ecosystems and their processes generate a benefit for humans. Status and value of ES are determined by the demand depending on the societal conditions. The actual land use reflects such a demand. For the application of the ES concept, the demand side plays a crucial role. Nevertheless, in contrast to 'ecological' assessments and plannings (e.g. within landscape planning, cp. Wende et al. 2011a; Albert et al. 2012), spatially precise representations of the demand or the comparison of supply and demand are still rarely implemented (▶ Sect. 5.3). The demand for services, however, is the basis of an appropriate spatial planning. To analyze the demand, information about the actual, intended or desired use of ES is needed, e.g. through socio-economic modelling, statistics, or questionnaires (Burkhard et al. 2012). Suitable data is often only available to a limited extent. They must be specifically collected, which mostly entails a significant amount of work.

Both sections ▶ Chaps. 4 and ▶ 5 and the case studies give an overview of common methodological approaches for assessing ES (▶ Chap. 6).

Benefits, Values and Welfare

Through the link 'ecosystem services', human beings benefit from ecosystems. That means, ecosystems yield benefits and values (fourth pillar of the EPPS framework), which contribute to human well-being. The benefit is the sociocultural or economic welfare gain provided through the ES, such as health, employment, and income. Moreover, the benefits of ES must have a direct relationship to human wellbeing (Fisher and Turner (2008). Value is most commonly defined as the contribution of ES to goals, objectives or conditions that are specified by a user (van Oudenhoven et al. 2012). Actors in society can attach a value to these benefits. Monetary value can help to internalise so-called externalities (impacts and side effects) in economic valuation procedures so that they can be better taken into account in decision-making processes at all levels. It should be noted that not all dimensions of human well-being can be expressed in monetary terms, e.g. cultural and spiritual values.

For human well-being factors like health, prevention of psychological damages, aesthetic pleasure, recreation, food supply, and economic prosperity are crucial. They are influenced positively by ES. For the Millennium Ecosystem Assessment (MEA 2005) and several other authors (e.g. Costanza et al. 1997; Wallace 2007) ES and benefits are identical.

In order to measure benefits and values, an evaluation step is necessary. Generally, an evaluation is a relation between an evaluating subject and an object of evaluation, or the degree of fulfilment in comparison with predetermined objectives. This relation has two dimensions:

- Factual dimension: facts on the object to be evaluated for the reflection of the reality
- Value dimension: value system or basic values as a normative basis for the value judgment (Bechmann 1989, 1995)

The evaluation shows the extent to which the present state differs from the desired or planned one (Auhagen 1998). The literature often uses the term 'evaluation' ambiguously (Wiegleb 1997), e.g. in the sense of basic assessment (scaling), judgment, ranking (relative comparison), or plan/actual comparisons (= evaluation sensu stricto).

An *evaluation* is the crucial step to process analytical data concerning decision-making and action, i.e. to convert scientific parameters into socio-political categories.

An *evaluation sensu stricto* indicates the extent and the manner of necessary measures. It provides the norms and orientations for the concrete action, which is always a decision between several options. If an evaluation shall be generally valid, the consensus of the human society is necessary; it is a matter of conventions and, thus, depending on the situation and time. Therefore, evaluation can never be objective. The skill of evaluation is the combination of facts and standards of value with sensible judgement. Evaluations are always based on the competence of the evaluating subject. On no account does subjectivity mean arbitrariness or irrationality since an evaluation is or should be also comprehended by other subjects (intersubjectivity). Necessary preconditions for this are disclosed facts and standards of value that are combined in a systematical manner, i.e. using well-defined assessment procedures (Bechmann 1995; Bastian and Steinhardt 2002).

There are quite different motivations to valuate ES. These motivations heavily depend on moral, aesthetic, and other cultural perspectives (Hein et al. 2006).

It is often neglected that scientific findings are in principle free of value. That means that there is no logical conclusion on the desired situation (normative consideration) from being (actual state, descriptive consideration). In other words: it is not possible to derive value judgments from ecological findings or to answer respective questions such as 'Which nature we want to protect?' or 'How nature shall be protected?' Things are not valuable per se, but because we appreciate them and decide so.

Already Hume (1740) referred in his 'A Treatise of Human Nature' to the problem of the dichotomy between what is and what ought to be. As a term for the derivation of norms from nature, Moore introduced the term 'naturalistic fallacy' in his 'Principa Ethica' in 1903 (see Erdmann et al. 2002). Terms like naturalness, rarity, etc. don't necessarily prejudge a value decision. The protection of rare species must be justified because not all rare things are per se worthy of protection. A near-natural vegetation is not generally desirable, e.g. from the farmer's point of view if he looks at his weedy arable field. However, from a nature conservation point of view, a near-natural vegetation can also be undesired if, for instance, a colourful flowering meadow owing its existence to human influences shall be conserved and not become fallow-field, shrubland or forest.

The sense of formalised evaluation algorithms is to rationalise the (landscape) planning process and to increase the acceptance of the results by society.

For the analysis of benefits and values in the ES context, monetary valuation is often regarded as the method of choice. The sole orientation to the monetary valuation of ES, however, is increasingly regarded critical (Spangenberg and Settele 2010). On the other side, studies on the implementation of measures and their financial consequences (e.g. Lütz and Bastian 2000; von Haaren and Bathke 2008; Grossmann et al. 2010), have shown that a monetary valuation of services may provide incentives for alterations in existing management rules or decision support for certain problem solutions. Monetary values served to internalise so-called externalities (external influences, impacts) in economic valuation methods in order to take them better into account in decision processes at all levels (► Sect. 4.2).

In addition to the economic evaluation, other approaches must also be observed to show the importance of ES. Other dimension of human well-being that cannot be expressed in monetary values, e.g. cultural and spiritual values, should also be integrated. Participative methods have a great significance, i.e. the participation of stakeholders. The preferences for certain ES are negotiated within society. As a basis, adequate background knowledge is indispensable, which entails ecological as well as economic information (▶ Sect. 4.3).

In principle we distinguish between three types of methods for the evaluation of ES (▶ Sect. 4.1, ▶ Sect. 4.2 and ▶ Sect. 4.3): quantitative expert methods (mainly ecologically or physically based), economic/monetary methods and participative, scenario-based methods. Complex methods as combinations of these three methods are discussed in ▶ Sect. 4.4.

Beneficiaries of ES/Actors

An ecosystem service is only a service if there is a human benefit. Without human beneficiaries, there are no ES (Fisher et al. 2009). Accordingly, a disservice only exists if humans suffer harm. The stakeholders, providers, users or beneficiaries of ecosystems and their services (pillar 5 of the EPPS framework) can be single persons, groups, or society as a whole. Not only do they depend or benefit from ecosystems, they in turn react upon ecosystems through land use, management, decision, regulation, etc. (▶ Chap. 5).

The identification of beneficiaries of ES helps to develop environmental-political steering instruments to set incentives in a targeted manner for a more careful management of ecosystems and the services they deliver. The key question is: Who benefits where from which ES? The following cases can be distinguished (Kettunen et al. 2009):

- Local public benefits: a site's role in supporting local identity, local recreation, local nonmarket forest products and the local 'brand', etc.
- Local private benefits: a site's support to natural water purification resulting in lower pretreatment costs to the local water supply company, etc.
- Local public sector benefits: a site's abilities to mitigate floods resulting to lower public investment in flood control and/or flood damage, etc.
- Regional and cross-border benefits: regulation of climate and floods, mitigation of wild fires, provisioning and purification of water in transnational river basins, etc.
- International/global public benefits: a site's provision of habitat for a migratory species at some point in its annual cycle, regulation of climate (carbon capture and storage), maintenance of global species and genetic diversity, etc.
- International private benefits: new pharmaceutical or medicinal product derived via bioprospecting, etc.

Trade-Offs, Limit Values, Driving Forces and Scenarios

Other very important points of view regarding ES are related e.g. to the so-called *trade-offs*. They describe the multiple interactions and linkages among services; this means that management aimed at providing a single service (e.g. food, fibre, water) often reduces biodiversity and the provision of other services (Ring et al. 2010). Some ES co-vary positively but others negatively. For example, the increase of provisioning ES may reduce many regulation ES. Thus, the growth of agricultural production may reduce carbon storage in the soils, water regulation and/or sociocultural ES. The TEEB study (TEEB 2009) distinguishes between: 1. Temporal trade-offs: Benefits now–costs later, 2. Spatial trade-offs: benefits here–costs there, 3. Beneficiary trade-offs: Some win–others lose, 4. Service trade-offs: Enhancing one ES–reduces another.

All pillars or categories of the framework can or should be analyzed and differentiated in terms of space (e.g. scale, dimension, patterns) and time (e.g. driving forces, changes, scenarios) aspects (▶ Sect. 3.3).

Ecosystems can go through fairly big changes: If *critical thresholds* or *limit values* are exceeded, substantial changes cannot be excluded, e.g. the eutrophication of lakes, the degradation of farmland, the collapse of fish stocks or coral reefs.

Ecosystem changes can be triggered by various, partly superposed *driving forces*. Artner et al. (2005) distinguished between fixed factors or drivers, e.g. the ongoing globalisation, the demographic change and variable factors like the economic development, the societal governance, leisure behaviour, the traffic volume, the consumption of resources and the structural development.

The status of ES can be predicted or analyzed under the assumption of different *scenarios*. In contrast to a prognosis, a scenario is no forecast and not correlated with a statement on the probability of occurrence. Instead it represents a possible development under defined, predictable conditions. A set of scenarios can be used to simulate possible long-term effects and consequences of decisions (Dunlop et al. 2002) (► Sect. 4.3). Scenarios inform the decision-maker about possible welfare gains and losses. Not only do the changes in ecosystems and ES have to be considered, but also the **variability of values.** Value orientations are subject to cycles and trends (one of the best examples are fashion trends). The future development of societal values depends on many factors. As the value scales, e.g. the value of money, may change, monetary valuations of future states are subject to considerable uncertainties (see the discounting of ES, ► Sect. 4.2).

3.1.3 The Application of the EPPS Framework–The Example 'Mountain Meadow'

Finally, the application of the EPPS framework will be demonstrated with an example, the ecosystem (type) 'mountain meadow'.

Mountain meadows are species-rich, extensively used meadows of fresh to medium moist sites of mountains above c. 500 m a.s.l. Depending on the geographical situation, nutrient content, moisture balance of soils, type and intensity of use or management, e.g. cutting frequency and fertilisation, mountain meadows occur in different specifications.

For the capacity of mountain meadows to deliver ES, particular characteristics, combinations of them or parts of ecosystems (functional traits–see above) are crucial, e.g. nutrient and water balance, the combination of species and usage intensity. Mountain meadows have the potential to deliver manifold ES of all three classes–provisioning, regulation and sociocultural services, among them:

- Provisioning services: provision of fodder plants for livestock, biochemical/pharmaceutical substances (spignel plants–*Meum athamanticum*–and other herbs), drinking water
- Regulation services: cold air production, water retention and flood prevention, erosion control, habitat services
- Sociocultural services: aesthetic values (e.g. scenery), recreation and eco-tourism, culture-historical aspects

Not all of these potentials are really used. There is almost no demand for the biomass from species-rich but low yielding meadows since the current dairy cattle farming trimmed for high-performance has no use for it. The energetic use of scrap materials from landscaping is not very advanced either. Until a market or customers for such materials will come into existence, no benefit or value in an economic sense can be attributed. The situation with biodiversity and aesthetic values is quite different, although a quantification or even monetisation is anything but easy. Irrespective of this, colourful flowering meadows contribute to human well-being because of their beauty and if their occurrence is related to the attractiveness of holiday regions, economic values can be derived, for instance, in the form of the number of tourists traveling there just because of these attractive mountain meadows. In this case, tourists and touristic enterprises can be regarded as beneficiaries, with regard to the maintenance of biodiversity the whole society or even the European Community (in the case of Natura 2000, ► Sect. 6.6.1).

Mountain meadows seem to be natural, but they represent ecosystems created by humans through regular cutting. Hence, an adequate usage or management must be ensured so that the mountain meadows as such and the related/relevant ES are maintained. This requires human labour, e.g. of agricultural enterprises, landscape management associations, or nature conservation organisations. The ones ensuring the ongoing existence of the meadows and the provision of ES with their activities are not always identical with the beneficiaries. However, as society is interested in, for example, the conservation of biodiversity, which is reflected in many laws, contracts, conventions and strategies at different levels, the expense is remunerated in monetary terms (► Sect. 6.2). Simultaneously, society ensures for necessary legal instruments in the form of protected areas (nature reserves, Natura 2000, etc.).

All these levels, starting from the ecosystem 'mountain meadow' (physical level, factual level) over the ES (intermediate level) to the benefits and beneficiaries (socio-economic level) are subjected to manifold space and time aspects (▶ Sect. 3.3). Thus, at the ecosystem level, the size of the mountain meadow or its arrangement in the biotope mosaic is important, so that the requirements of particular species are met. As a rule, a large mountain meadow delivers more services than a small one, if the other properties are more or less identical; a big flowering meadow has a higher aesthetic effect as a smaller one. Also, benefits and beneficiaries are subject of strong spatial relationships. Thus, the local landscape management association ensures the maintenance of the mountain meadow, and the travelling tourists benefit from its aesthetic values. The effect of the 'conservation of biodiversity' is difficult to narrow down in terms of its effective radius, but it may refer–as with Natura 2000–to the whole EU and even other countries.

In terms of time aspects, first of all the changes to which the ecosystems are subjected should be regarded, this is especially the case with mountain meadows due to improper or missing usage or management. Over time attitudes and value systems of people may change.

Changes are triggered by driving forces: globalisation and the Common Agricultural Policy (CAP) of the EU, but also technological progress reducing the attractiveness of mountain meadows for agriculture. Demographic change goes hand in hand with a shortage of personnel in voluntary nature conservation, i.e. less actors are available who will take care of the mountain meadows (Wende et al. 2012). Climate change, too, will doubtless have some measure of impact on such sensible ecosystems.

3.2 Classification of ES

O. Bastian, K. Grunewald and R.-U. Syrbe

3.2.1 Introduction

In view of the diversity and complexity of ecosystems and the services they supply, it is difficult to develop a classification of ES which is clear, widely accepted, and meets broad requirements. With respect to the classification of ecosystem and landscape functions, potentials and services, there are numerous proposals, classification systems and partly divergent opinions. Depending on the goals of the assessment, spatial scales and specific decision-making context, they all show both strengths and weaknesses.

For the past decades science has been trying to determine a way of classifying ecosystem functions (and services). In 1977, Niemann distinguished four groups of functions: production, landscape-shaping (ecological), human-ecological, and aesthetic ones. Van der Maarel and Dauvellier (1978) declared production, carrier, information, regulation and reservoir functions as societal functions of the physical landscape. Bastian and Schreiber (1999) divided landscape functions into three groups: so-called production functions (economic functions), regulation functions (ecological functions) and habitat functions (sociocultural functions). Each group was again classified into main-functions and sub-functions.

De Groot et al. (1992, 2002) defined regulation, production, habitat, and information functions (or services). The TEEB study also identifies the habitat services as a separate category to stress the importance of ecosystems to provide habitat for migratory species and gene-pool 'protectors' (TEEB 2010). Using the definition of Costanza et al. (1997), the Millennium Ecosystem Assessment (MEA 2005) provided a simple typology of services that has been widely taken-up in the international research and policy literature:

- Provisioning services, e.g. food, drinking water, timber
- Regulating services, e.g. flood protection, air pollution control
- Cultural services, e.g. recreation services
- Supporting services: all processes that ensure necessary preconditions for the existence of ecosystems, e.g. nutrient cycle.

The ES classification systems outlined above shows numerous commonalities, mainly in the three classes provisioning, regulating and cultural services. There is disagreement about the assignment of phenomena, which are the basis for the services of the three other classes. This applies to the supporting services (or basic services, ecosystem integrity–e.g. Müller and Burkhard 2007). We consider supporting services an intermediate (analytical) stage. They are a prerequisite for defining the other three groups of services, but they are more related to the first pillar of our EPPS framework (► Sect. 3.1), that of ecosystem properties. Other authors (e.g. Pfisterer et al. 2005, Burkhard et al. 2009, Hein et al. 2006, OECD 2008, Haines-Young and Potschin 2010) also suggest treating them differently from the other ES, which provide their benefits directly to humans. Due to thematic overlaps with regulating ES there is a high risk of double-counting (Hein et al. 2006, Burkhard et al. 2009, see Box p. 51).

The breakdown into productive (economic), regulating (ecological), and societal functions or services (Bastian and Schreiber 1999, Bastian et al. 2012b) has the advantage that it can be linked to both fundamental concepts of sustainability and risk using the established ecological, economic, and social development categories. We adjust the supporting services–depending on the respective situation–to the regulative services or the ecological processes (e.g. nutrient cycles, food chains).

Ultimately, the classification depends on the respective researcher. As a rule, three or four groups with a total of 15 to 30 functions or services are distinguished. For useful results, they must be further specified. Moreover, information on suitable indicators that describe these ES is necessary. In this respect there are still severe deficits in the literature (Jessel et al. 2009, TEEB 2009).

Below we present an overview of ES supplied by terrestrial and aquatic ecosystems based on current knowledge (e.g. Costanza et al. 1997, de Groot et al. 2002, Müller and Burkhard 2007, Vandewalle et al. 2008) and on our own experiences and reflections (◘ Tab. 3.1–3.3). We classify 30 ES according to three main categories: provisioning, regulation and sociocultural services -each with subdivisions. Furthermore, we provide a short definition and description with examples and mention selected indicators for the analysis or the assessment of the ES with no claim to completeness.

3.2.2 Provisioning Services

Ecosystems may provide many goods and services from oxygen and water to food and energy to medicinal and genetic resources, and materials for clothing and shelter. As a rule, these goods and services refer to renewable biotic resources, i.e. the products of living plants and animals. Abiotic resources (raw materials near the earth's surface), wind and solar energy cannot be assigned to particular ecosystems; hence, they are not, in our view, to be considered ecosystem goods and services. Especially in ecosystems strongly modified by humans (e.g. farmland) it is difficult to differentiate between the natural and human inputs in labour, material and energy to a service or a good (◘ Table 3.1).

3.2.3 Regulation Services

The biosphere and its ecosystems are the main preconditions for human life. Processes like energy transformation mainly from solar radiation into biomass, storage and transfer of mineral material and energy in food chains, bio-geochemical cycles, mineralisation of organic matter in soils and climate regulation are essential for life on earth. On the other hand, these processes are influenced and enabled by the interaction of abiotic factors with living organisms. The existence and functioning of–particularly natural and semi-natural–ecosystems must be ensured so that people will be able to continue benefiting from these processes in the future. Due to the 'merely' indirect benefits of regulation services (◘ Tab. 3.2), they are often overlooked and not sufficiently considered until they are damaged or lost, although they are the basis for human life on earth (De Groot et al. 2002).

3.2.4 Sociocultural Services

Especially natural and semi-natural ecosystems provide manifold opportunities for enjoyment, inspiration, intellectual enrichment, aesthetic delight and recreation. Such 'psychological-social' services are no less important to people than regulation and provisioning services; however, they

Table 3.1 Provisioning services

Code/Name of the ecosystem services	Definition/Description	Examples	Selected indicators
I Food (provision of plant and animal materials)			
P.1 Food and forage plants	Cultivated plants as food/forage for humans and animals	Cereals, vegetables, fruits, edible oil, hay	Harvested yields (dt ha^{-1}), contribution margin (€ ha^{-1})
P.2 Livestock	Slaughter and productive livestock	Cattle, pigs, horses, poultry	Stock density (livestock units per ha), contribution margin (€ ha^{-1})
P.3 Wild fruits and game	Edible plants and animals from the wilderness	Berries, mushrooms, game	Shooting quota (animals per ha), yields (€ ha^{-1})
P.4 Wild fish	Fishes and seafood caught in waters	Eels, herrings, shrimps, shells	Catch quota and numbers, harvest amounts (t ha^{-1}), revenues (€ ha^{-1})
P.5 Aquaculture	Fishes, shells or algae growing in ponds or farming installations	Carps, shrimps, oysters	Produced amounts (t ha^{-1}), revenues (€ ha^{-1})
II Renewable raw materials			
P.6 Wood and tree products	Raw materials from trees in forests, plantations or agro-forest systems	Timber, cellulose, resin, natural rubber	Stock, growth, yields (m^3 ha^{-1}, t ha^{-1}), revenues (€ ha^{-1})
P.7 Vegetable fibres	Fibres from herbaceous plants (from nature or cultivated)	Cotton, hemp, flax, sisal	Yields (t ha^{-1}), revenues (€ ha^{-1})
P.8 Regrowing energy sources	Biomass from energy crops and wastes	Fire wood, charcoal, maize, rape, dung, liquid manure	Yields (t ha^{-1}), energy amount (MJ ha^{-1})
P.9 Other natural materials	Materials for industry, crafts, decoration, arts, souvenirs	Leather, flavorings, pearls, feathers, ornamental fishes	Sold units (e.g. furs per year), revenues (€ ha^{-1})
III Other renewable natural resources			
P.10 Genetic resources	Genes und genetical information for breeding and biotechnology	Seeds, resistance genes	Number of species
P.11 Biochemicals, natural medicine	Raw materials for medicine, cosmetics and others to enhance health and well-being	Etheric oils, tees, *Echinacea*, garlic, food supplements, leeches, natural crop protection products	Yields, amounts of active substance (kg ha^{-1}), revenues (€ ha^{-1})
P.12 Freshwater	Clean water in ground- and surface waters, precipitation and in the underground for private, industrial and agricultural use	Rain, spring and fountain discharge, bank filtrate	Raw water, drinking water (Tm^3 a^{-1}), revenues (€ ha^{-1})

3

Tab. 3.2 Regulation services

Code/Name of the ecosystem services	Definition/Description	Examples	Selected indicators
I Climatologic and air hygienic services			
R.1 Air quality regulation	Air cleaning, gas exchange	Filter effects (fine dust, aerosols), oxygen production	Proportion of forests (%), leaf area index
R.2 Climate regulation	Impacts on the maintenance of natural climatic processes and on reducing the risks of extreme weather events	Cold air production, humification, reducing temperature by the vegetation, weakening of extreme temperatures and storms	Proportion of forests and open areas (%), slope (°), albedo
R.3 Carbon sequestration	Removing carbon dioxide from the atmosphere and relocation into sinks	Photosynthesis, fixation in the vegetation cover and in soils	Proportion of vegetation areas (%), soil forms (e.g. peat)
R.4 Noise protection	Reducing noise immissions by vegetation and surface forms	Noise protection effects of vegetation	Vitality, layering and density of vegetation
II Hydrological services			
R.5 Water regulation	Balancing impacts on the water level of watercourses and the height, duration, delay and avoiding floods, droughts and (forest) fires, protection against tidal flooding (e.g. by coral reefs, mangroves), water as transport medium, water power	Natural irrigation, soil storage, leaching/ groundwater recharge	Slope (°), land use (land cover) (%), soil types
R.6 Water purification	Filter effects, storage of nutrients, decomposition of wastes	Nitrogen retention, denitrification, self-purification of rivers and lakes	Land cover (%), soil type, water structure and stream margins
III Pedological services			
R.7 Erosion protection	Effects of vegetation on soil erosion, sedimentation, capping and silting	Protection against landslides and avalanches, breaking winds	Slope (°), soil types, land use, permanent land cover, slope protection forests, crop spectrum
R.8 Maintenance of soil fertility	Regeneration of soil quality by the edaphon (soil organisms), soil generation (pedogenesis) and nutrient cycles	Nitrogen fixation, waste decomposition, humus formation and accumulation	Crop diversity, soil types, removal of harvest remnants and wood
IV Biological services (habitat functions)			
R.9 Regulation of pests and diseases	Mitigating influences on pests and the spread of epidemics	Songbirds, lacewings, ladybirds, parasitic wasps, tics (*Encephalitis*)	Biocides applied, naturalness and vitality of the vegetation, proportion of (semi-) natural vegetation areas (%), species spectrum (parasites, predators, pests)

■ Tab. 3.2 Continued

Code/Name of the ecosystem services	Definition/Description	Examples	Selected indicators
R.10 Pollination	Spread of pollens and seeds of wild and domestic plants	Honey and wild bees, bumblebees, butterflies, syrphid flies	Proportion of (semi-) natural vegetation areas (%), biocide application, proportion of flowering plants, genetically modified organisms
R.11 Maintenance of biodiversity	Conservation of wild species and breeds of cultivated plants and livestock	Refuge and reproduction habitats of wild plants and animals, partial habitats of migrating species, nursery spaces (e.g. spawning grounds for fishes), cattle breeds	Natural/semi-natural vegetation (proportion %), naturalness structural diversity, biotope compound, number of species, rarity, endangering

are often neglected or not fully appreciated. One reason is the difficulty of valuating them economically, especially in monetary terms. A second group includes information services, i.e. the contribution of ecosystems to knowledge and education (■ Tab. 3.3).

3.2.5 Additional Classification Aspects

Classification systems that combine both ecosystem processes and the results of these processes cause redundancy (Box 'The problem of double-counting'). Hence, it should be strictly distinguished between the benefit people enjoy (or the so-called 'final services') on the one hand and the mechanisms that give rise to that benefit, the so-called 'intermediate services', on the other hand. Any classification system containing both ecosystem processes and the outcomes of those processes within the same set will produce redundancy (Wallace 2008).

The literature also raises the question whether ES are delivered only by natural or semi-natural ecosystems or if they can also be delivered by cultivated areas (Cowling et al. 2008). This may cause astonishment, as even an intensively used arable field may represent a habitat for several plant and animal species. Arable land has a better infiltration rate and, hence, groundwater recharge compared to forests! Biodiversity in cities may be high. Of course, there are methodical specifications regarding the ES of highly modified or man-made ecosystems (e.g. urban ES, ▸ Sect. 6.3).

Hermann et al. (2011) present a classification that distinguishes between two main groups, namely the active and passive functions. Whereas the passive functions are divided into 'regulating and life sustaining functions' of the natural systems (environmental regulation, habitat protection, biomass generation) and the 'potentials' (biomass, raw material production and provision of territory for the different land uses and provision of information and aesthetics), the active functions are the services provided by human activities and artificial territories (settlements, infrastructure networks, recreation sites and agricultural surfaces, etc.). Apart from the fact that it is difficult in practice to draw a sharp distinction between cultivated and natural ecosystems, most of the ES of Central European cultural landscapes would be excluded by such a narrow ES concept. Instead, we

Tab. 3.3 Sociocultural services

Code/Name of the ecosystem services	Definition/Description	Examples	Selected indicators
I Psychological-social goods and services			
C.1 Ethical, spiritual, religious values	Possibility to live in harmony with nature, Integrity of Creation, freedom of choice, fairness, generational equity	Bioproducts, sacred places	Natural/semi-natural vegetation (%), extinct/threatened, genetically modified organisms, biocide application
C.2 Aesthetic values	Diversity, beauty, singularity, naturalness of nature and landscape	Flowering mountain meadows, harmonious landscape	Land use, vegetation types, crop diversity, relief diversity/slopes
C.3 Identification	Possibility for personal bonds and sense of home in a landscape	Natural and cultural heritage, places of memory, traditional knowledge	Natural and cultural monuments, historical landscape elements, architectural styles, persistence/continuity of landscape
C.4 Opportunities for recreation and (eco)tourism	Conditions for sports, recreation and leisure activities in nature and landscape	Accessibility, security, stimuli	Level of accessibility, carrying capacity, snow cover, number and area of waters, attractive species, number of visitors
II Information services			
C.5 Education and training values, scientific insights	Opportunities to gain knowledge about natural interrelations, processes and genesis, scientific research and technological innovations	Natural soil profiles, functioning ecosystems, rare species, traditional land knowledge	Natural and cultural monuments, land-use forms, naturalness
C.6 Mental, spiritual and artistic inspiration	Stimulating fantasy and inventiveness, inspiration in architecture, painting, photography, musics, dance, fashion, folklore	Impressive landscapes, mounts, rivers, cliffs, old trees	Natural and cultural monuments, diversity of the land
C.7 Environmental indication	Gaining knowledge of environmental conditions, changes and threats by visually perceptible structures, processes and species	Indication with lichens (air quality), indicator plants (site conditions)	Species spectrum (ecological groups), number of lichen species, indicator organisms, naturalness

should follow an ES definition which does not distinguish between both ecosystem types (Loft and Lux 2010).

It is not possible to simply dismiss the problem that some ES are not only resulting from ecosystem effects, but that natural effects interfere with human influences. Thus, Boyd and Banzhaf (2007) pointed out that conventional agriculture requires various inputs (soil quality, fertiliser application, human labour) that influence the yield. This, however, makes the identification and assessment of ES difficult, as too many non-natural factors are effective. In contrast, the amount of harvestable end-products of nonactively cultivated ecosystems may

The Problem of Double-Counting

A clear distinction between the ecological phenomena on one hand and their direct and indirect contributions to human well-being on the other hand is necessary to avoid double-counting, especially when considering the various ES as a whole or a sum, e.g. as the Total Economic Value. Double-counting may arise due to the fact that certain ES serve as prerequisite for others and become a part of them (Boyd and Banzhaf 2007, Balmford et al. 2008, Wallace 2008).

Intermediate ES are basing on complex interactions between ecosystem structures and processes and contribute to final ES, which directly provide human benefits and well-being (Fisher et al. 2009). For example, clean drinking water (e.g. from a lake) is traded on markets and is included as a product in the calculation of welfare, but not the upstream process of natural water filtering. This process can be described as an intermediate service; its indirect value is included in the value of the drinking water (cp. Wallace 2007). The focus of ES classifications on final ES does not mean to abstain from a comprehensive consideration and appreciation of ecosystem structures, processes and cause-effect interrelationships.

But the regulative services are often also included in other ES, e.g. pollination, which is important for the maintenance of fruit-growing in the ES 'food provision'. Mäler et al. (2008) classified only provisioning and cultural ES to final ES but regulative and supporting ES to the 'final' ES as both provisioning and cultural ES would influence human well-being directly, whilst the other two would do so only indirectly.

After Costanza (2008) all ES are 'only' means to achieve human well-being. Ecosystem processes may appear also as ES (the roles of process or service are not mutually exclusive), hence, on a case-by-case basis the same ES may be intermediate or final. Thus, the lake with drinking-water quality mentioned above may be regarded as a final product of nature, and a direct societal benefit can be attributed if the lake serves as water reservoir (final ES). In another context, the same lake may be regarded as a supplier of an intermediate ES, i.e. if the direct benefit consists of fishing for leisure and the special water quality only ensures the fish stock in the lake enabling fishing in the first place. In this case, the indirect benefit of the special water quality is included in the direct benefit, which delivers the fish stock for the angler (Boyd and Banzhaf 2007, Loft and Lux 2010).

Generally, there are quite controversial expert opinions with respect to a clear classification of intermediate and final ES, and agreement may hardly be achieved. Finally, the respective context and pragmatic points of view are crucial.

be a measure for the assessment of ES. Also, the example of sport fishing mentioned above shows the difficulty while analyzing and evaluating ES. Often, they can deliver benefits only through interactions with other goods and services since the recreation value arising through sport fishing consists of natural conditions (landscape, lake, fishes) and artificial goods (fishing rods, boat, etc.). In other words: without technical tools like the fishing rod the recreation value would not come into effect (Boyd and Banzhaf 2007; Loft and Lux 2010).

For cases where for the provision of benefits for humans not only ecosystem processes are necessary, but also human impacts, Matzdorf et al. (2010) suggested the term 'environmental services'. For example, the maintenance of semi-natural meadows with their ES relies on regular mowing. Conscious exclusion of permitted actions, such as the application of fertilisers can be regarded as a human performance, too. Species-rich grassland may be regarded as a final environmental good, for its production both human and ecosystem impacts are necessary. Hence, the evaluation, especially in monetary terms, the anthropogenic part (human performance = private costs) must be subtracted. This means that agriculture delivers environmental services but no ES. Instead it uses them for the production of demanded environmental goods (► Sect. 6.2.4).

The method developed for a welfare-oriented perspective of the Swiss environmental reporting, defines altogether 23 final ES (Final Ecosystem Goods and Services–FEGS) in the benefit categories 'health', 'safety', 'natural diversity' and 'economic performances' (BAFU 2011). The attribute 'service type' indicates whether the provided service:

1. Is a directly usable ES
2. Is an input factor for the production of market goods by the economy

3

Table 3.4 Ecosystem services classified according to their spatial characteristics (adapted from Costanza 2008)

ES group	Examples
1. Global non-proximal (does not depend on proximity)	Climate regulation Carbon sequestration, carbon storage Cultural/existence value
2. Local proximal (depends on proximity)	Disturbance regulation/storm protection Waste treatment Pollination Biological control Habitat/refugia
3. Directional flow related: flow from point of production to point of use	Water regulation/flood protection Water supply Sediment regulation/erosion control Nutrient cycle regulation
4. In situ (point of use)	Soil formation Local food production
5. User movement related: flow of people to unique natural features	Genetic resources Recreation potential Cultural/aesthetic values

3. Is provided by a natural/healthy habitat (ecosystem)
4. Contributes to final ES as an intermediate ES

(1) Directly usable ES cause discrete benefits to humans (e.g. recreation service). Input factors (2) are final services of the ecosphere; they are integrated in a market product (e.g. timber growth), they belong to the benefit category 'economic performance'. The performance type (3) natural/healthy habitat (ecosystem) contains welfare contributions of the environment, which–in contrast to the classical ES–are not 'produced' by ecosystems but they rather represent qualities of the habitat enabling humans' health life (e.g. air quality). The (4) intermediate ES are considered only exceptionally (e.g. CO_2 sequestration), namely if the resulting ES occur only with long delay, and therefore cannot be measured at the moment, yet.

The classification of ES according to spatial characteristics is another possibility. This can be useful if they are used as a basis for decisions on different scales, or if Service Providing Area and Service Benefiting Area are not congruent (Fisher et al. 2009, ► Sect. 3.3).

Tab. 3.5 Classification of ecosystem services according to their excludability and rivalness (after Costanza 2008)

	Excludable	Nonexcludable
Rival	Market goods and services (most provisioning services)	Open access resources (some provisioning services)
Nonrival	Club goods (some recreation services)	Public goods and services (most regulatory and cultural services)

Costanza (2008) grouped ES into five categories according to their spatial characteristics (■ Table 3.4). For example, he classified carbon sequestration (CO_2 and other greenhouse gases; an intermediate input to climate regulation) as *global: non-proximal* since the spatial location of carbon sequestration does not matter. *Local proximal* services, however, are dependent on the spatial proximity of the ecosystem to the human beneficiaries. For example, 'storm protection' requires that the ecosystem performing the protecting is proximal to the human settlements being protected. *Directional flow related* services are related to the flow from upstream to downstream as is the case for water supply and water regulation.

Another way to classify ES is according to their *excludability and rivalness* status (Costanza 2008, ■ Tab. 3.5). Thus, individuals can be excluded from benefiting from *excludable* goods and services. Most privately owned, marketed goods and services are relatively easily excludable. One can prevent others from eating the tomatoes one has grown or the fish one caught unless they pay for these goods. But it is difficult or impossible to exclude other people from benefiting from many public goods like a well-regulated climate, fish in the ocean, or the beauty of a forest. Goods and services are *rival* if one person's benefiting from them interferes with or is rival with other people benefiting from them. If one person eats the tomato or the fish, another cannot. But if one person benefit from a favourable climate, other people can also do the same. There are cultural and institutional mechanisms available to enforce exclusion, while rivalness is a function of demand (How do benefits depend on other users?).

Conclusion

All attempts to develop a generally applicable **classification system** must be viewed with caution as they are not targeted to a certain extent. ES arise through complex interactions of the biotic and abiotic environment, claims on utilisation and the expectations of the users. An inappropriate classification system as a basis of assessments hardly leads to reliable results. If decisions shall be taken on an economic evaluation, the classification after the Millennium Ecosystem Assessment (MEA 2005) is less useful since multiple counting may occur. For this purpose, a classification should distinguish between intermediate and final services and benefits. Nevertheless, there is striving for internationally consistent classification systems. The European Environmental Agency (EEA) is promoting the *Common International Classification of Ecosystem Goods and Services (CICES). The goal of* CICES is, starting from the Millennium Ecosystem Assessment, to develop a new classification system, which is compatible with the already existing national accounting systems (Haines-Young and Potschin 2010).

3.3 Space and Time Aspects of ES[1]

K. Grunewald, O. Bastian and R.-U. Syrbe

3.3.1 Fundamentals, Control Scheme

"Space and time are modes in which we think, not conditions in which we exist (A. Einstein)."

There are significant deficits in knowledge and many open questions concerning spatial aspects of ES. Ecosystems and their services are always linked to space and time. This issue was addressed repeatedly in the literature, but so far relatively few operationalised and systematised in terms of conceptual and methodological aspects (e.g. Hein et al. 2006; Bastian et al. 2012a). However, there are more and more international publications that operate spatially explicit, e.g. the results of the PEER project (*PEER Research on EcoSystem Services*, ▶ www.peer.eu/projects/press/).

The term 'space' is used and considered very important and constitutive in a wide range of scientific disciplines, e.g. philosophy, mathematics and physics, but also history, medicine, theology, archaeology, education science and sociology. Of course, this term is especially important for the inter- and multidisciplinary spatial sciences, such as geography, environmental sciences, urban development and architecture, spatial planning, traffic sciences and also sociology and economics (cf. Müller 2005). According to Blotevogel (1995) we understand 'space' as a:

a. tangible physical space (pattern of different ares and cubes), which can be described objectively;
b. the natural human environment (e.g. landscape); and
c. social space (social construction of reality, spaces of collective actions, areas of spatial allocations).

Various main research questions need to be resolved in order to better integrate ES into landscape planning, management and decision-making, as identified by De Groot et al. (2010), who calls for a focus on aspects such as: 'How can ecosystem/landscape functions and services be spatially defined (mapped) and visualised?', and: 'What is the influence of scaling-issues on the economic value of ecosystem and landscape services to society?'

The arrangement patterns and spatial relationships of ecosystems are hardly ever taken into account (Blaschke 2006), and 'spatial and temporal dimensions of ecosystem service production, use, and value are not well understood' (TEEB 2010).

If the space–time dimensions of the ES concept are not well understood, the conclusion is inevitable that nature and its services cannot be integrated adequately into political decision-making processes. This is especially true of cases involving distribution options.

There are further questions, e.g.: to what extent specific methods are necessary for analyses and

1 Sect. 3.3 is in main parts based on the paper of Bastian et al. (2012a)

3

Table 3.6 Physical and social space perspective and EPPS approach as a framework methodology

Space	Pillar 1: Ecosystem properties	Pillar 2: Ecosystem potentials	Pillar 3: Ecosystem services
Terms/types/reference units	Definition, identification and delimitation of spatial units Scales, hierarchy, homogeneity/heterogeneity		
Causal relations, interdependences	Matter and energy fluxes between ecosystems, neighbourhood effects, functioning	Suitable units, risk space	Trade-offs (e.g. flows of values) between ES, overlapping, supply and demand
Complementary spatial approaches, landscape perspective	Extended perspective, e.g. by means of ethical and aesthetic aspects	Specifics of landscape units, planning alternatives	Benefit-transfer, costs of planning alternatives, assessment approaches based on cultural landscape aspects, complex and integrative approaches

evaluations in the particular scales? How can spatial approaches in the areas of nature and society be harmonised? How can we define clear space and time relations, especially with regard to distribution options?

Within the EPPS-framework (▶ Sect. 3.1) principle solutions for capturing spatially relevant aspects are offered; Table 3.6 gives an orientation for it. Appropriate representations and visualisations of spatial aspects of ES should also be considered.

Time aspects are of great relevance related to space. ES are subject to various temporal dynamics. Of particular importance are the different, for the formation of the respective services necessary time spans, the nonsimultaneity in the multifunctional use, and the temporary differences between the provision and use of services or goods (Fisher et al. 2009). The changes in individual services over time are very relevant because functional effects of interventions, plans and other policy measures can be evaluated either retrospectively or may be estimated in advance (scenarios and forecasts). Another aspect is the variability of individual and societal value systems.

Spatial Aspects of Ecosystems

The spatial reference of ES appears in many ways. The generation of ES requires ecosystems with specific (including spatial) characteristics. To be able to supply ES, special *areal requirements* (minimum areas) of the ecosystems concerned are necessary. For example, animal populations need specific minimum areas of appropriate quality for their stability and their survival; a forest must have a size of several hectares to be able to influence the microclimate in the vicinity; a body of groundwater must have a minimum size or rate of groundwater recharge in order to be able to supply usable amounts of drinking water. Sometimes only single parts of ecosystems, single (organism) species, individuals or parts of them (roots or leafs of plants) are responsible for ES generation.

Frequently, a specific *spatial composition* or pattern of several ecosystems is necessary to generate ES. Composition aspects are also manifested in the spatial congruence or divergence of ES (e.g. Anderson et al. 2009), or in mutual influences. There can be spatial concordance among different services. Some ES co-vary positively: for example, maintaining soil quality may promote nutrient cycling and primary production, enhance carbon storage and hence climate regulation, help regulate water flows and water quality and improve most provisioning services, notably food, fibre and other chemicals (Ring et al. 2010). Other services co-vary negatively (▶ Sect. 3.1).

Multiple ES can be interconnected and interlinked in 'bundles' (MEA 2005). Willemen (2010) refers to interactions between landscape functions (or ES), which can be categorised into three classes:

1. Conflicts: the combination of several landscape functions reduces the provision of services to society of a particular landscape function
2. Synergies: the combination of functions enhances a particular function
3. Compatibility: landscape functions co-exist without reducing or enhancing one another

Whether different ES co-vary positively or negatively often depends on the *configuration* of the ecosystems or landscape elements involved at a specific scale. Productive land uses require compensation areas for the maintenance of key ecosystem providers. By contrast, sensible ecosystems need buffers to shelter them from harmful side effects. Nonetheless, in places without enough space for all desired functions in a landscape to operate equally, complex structures and sophisticated sequences of different ecosystems might be able to maintain the majority of them. In practice, mainly at local levels rather than at regional scales, we are familiar with structural environmental quality standards, such as buffer stripes, habitat connection, wildlife corridors and SCA concepts, as described below. A well-known example is the zoning within large protected areas (national parks, biosphere reserves), where core zones (wilderness) are buffered by managed, near natural zones, which in turn provides a gradient to the more intensively used areas (e.g. farmland) outside the protected areas.

Spatial Aspects of ES Providers and ES Beneficiaries (Functional Connections)

In spatial analyses of ES, not only the source area of a service is interesting but also the demand area, i.e. the areas where the benefits are required and realised. Hence, we need to address both providers and beneficiaries of ES: who provides the ES? For whom are they provided or who benefits from them? Within which spatial position is the ES generated and supplied and where is it used (where are providers and beneficiaries located)? We should also consider spatial cost/benefit relationships, such as spatial, 'benefits here–costs there' trade-offs, where a service is provided in one location for the benefit of another. This creates a relation between the ES provider (the person/or group responsible for an ES or environmental responsibility) and the ES beneficiary, or between winner/s and loser/s (Ring et al. 2010).

There are often distinct spatial differences between areas where ES are generated (SPA–Service Providing Areas) and areas which benefit from the ES (SBA–Service Benefiting Areas, correspond to the SPU ▶ Sect. 3.1.2). If providing and benefiting areas (SPA and SBA) do not adjoin, there will necessarily be a space between them, the so-called Service Connecting Area (SCA) (Syrbe and Walz 2012). For instance, flood protection is provided mainly in the mountains (by water storage reservoirs) and benefits cities along the middle and lower stretches of a river. In between, the river course can alter a flood wave. The SCA should be identified to support the transmission from the SPA to the SBA, for instance by avoiding or removing barriers (e.g. in water streams or in biotope networks). Thus, a natural floodplain, which is connected with the river and not separated by dams, can be regarded as a SCA, too. It can contribute to flood mitigation in favour of downstream settlements. The identification of SP and beneficiaries helps to avoid free riders or at least to reduce their effect on ES consumption.

Fisher et al. (2009) proposed a classification scheme that describes relationships between service provision and benefit (i.e. where and by whom benefits are realised):

a. both the service provision and benefit occur at the same location (e.g. soil formation, provision of raw materials)
b. the service is provided omni-directionally and benefits the surrounding landscape (e.g. pollination, carbon sequestration)
c. specific directional benefits, e.g. down slope units benefit from services provided in uphill areas in mountains; the service provision unit could be coastal wetlands providing storm and flood protection to a coastline

An additional case could be added to these classes as the counterpart to (b):

d. the service is provided in large (hardly limited) areas and benefits small, discrete locations (e.g. a settlement).

The cases described in (b) and (c) necessarily lead to scale transfers (▶ Sect. 3.3.1 Scale and Dimension). According to such spatial characteristics, Costanza (2008) groups ES into five categories. For example, services like carbon sequestration are classified as 'global: non-proximal', since the spatial location of carbon sequestration does not matter. Nowadays, due to carbon trades spatial scales in CO_2 storage area are becoming more crucial and need to be considered on a finer scale. When one pays for CO_2 storage, e.g. by planting trees, he would like to know where the trees are planted and how much carbon will be sequestrated. 'Local proximal' services, on the other hand, are dependent on the spatial proximity of the ecosystem to the human beneficiaries. For example, 'storm protection' requires that the ecosystem doing the protecting be proximal to the human settlements being protected. 'Directional flow related' services are dependent on the flow from upstream to downstream, as is the cases of water supply and water regulation. Other services are 'in situ (point of use)' (e.g. soil formation) or 'user movement related: flow of people to unique natural feature' (e.g. recreational potential).

Aspects of Time

Ecosystems do not only need special time spans for their regeneration, they are also subject to natural fluctuations and trends, which can alter their functionality and capacity (to supply ES) periodically, episodically or permanently. The Millennium Ecosystem Assessment (MEA 2005) predicts a decline of many ES. Land use (intensification) is or will be a major reason for this (EASAC 2009). Changes in ecosystems and the ES they supply are increasingly caused by humans. The knowledge of time-dependent changes of ES are of great practical importance since it helps to evaluate practical consequences of impacts, plans and policies for humans and societies either ex-post or ex-ante (scenarios and prognoses). Not only ecosystems or ecological properties can change; so, too, can economic values and the values that different stakeholders attach to the services. For example, infrastructure and transportation costs can change, which leads to new spatial and economic relationships between SP and beneficiaries. Methods are needed to reveal natural fluctuations or changes of ecosystems more detailed in order to be able to better adapt impacts caused by human utilisations.

Systematically, the following time aspects are especially important:

1. The minimum *time requirements* for the generation of particular ES
2. The disparity in the multifunctional use requirement for adequate *temporal sequences* in the provision and utilisation of ES (e.g. concerning water sampling, flood runoff, fishing)
3. The temporal differences between supply and demand or use of goods and services, so-called time lags (e.g. between water sampling from the water bodies and water consumption, or between water accumulation in the mountains and the crisis situation in the valley; e.g. Grunewald et al. 2007).

Functional traits (or SPU, ESP–see above) may contribute to service provisions to a different degree, not only in different places, but also at different times (De Bello et al. 2010).

To consider the capacity of ecosystems to supply ES sustainably is a basic issue for the development of the ES concept and also needs to be fundamentally implemented in its methodology. Thus, it is also crucial to adjust the sequence of different land uses in an intelligent manner to minimise impacts. For instance, crop rotation can influence flood regulation. A tight crop rotation, adapted intercrops, or conservative cultivation can close critical bare fallow periods to reduce erosion and surface runoff.

One of the most important issues refers to the sometimes huge differences between the periods in which natural developments occur and the time frames of social processes (public awareness, political opinion-making, parliamentary terms, human lifetimes).

Ring et al. (2010) highlight the question of temporal trade-offs: benefits now–costs later. Such trade-offs represent the central tenet of sustainable development stipulating that it '... meet the needs of the present generation without compromising

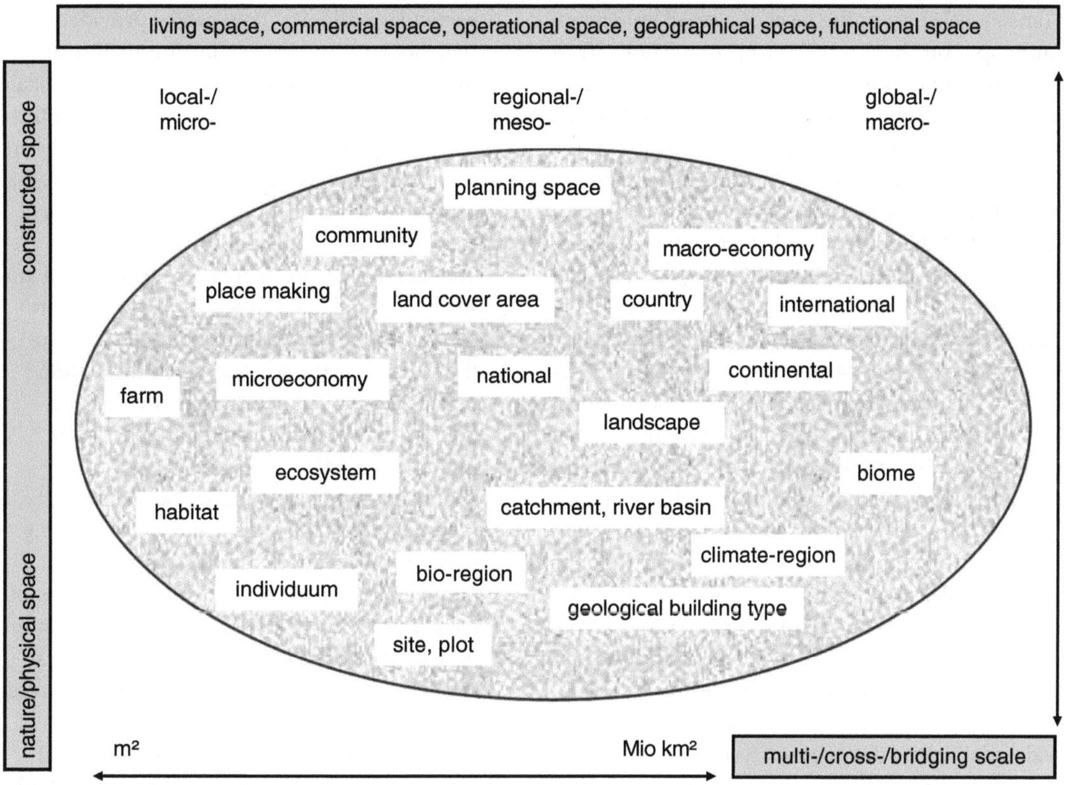

Fig. 3.4 Selected spatially relevant phenomena reflecting different scales. © Grunewald

the needs of future generations …' Therefore, even the inter-generational time lags need to be addressed (► Sect. 2.2).

Time differences between the supply of ES on the one hand and the use of goods and services on the other can usefully be expressed by the concept of natural potentials (► Chap. 2 and ► Sect. 3.1). The concept of natural potentials (see Neef 1966; Haase 1978; Mannsfeld 1979, Bastian and Steinhardt 2002; Burkhard et al. 2009; Grunewald and Bastian 2010; Bastian et al. 2012b) aims to display the service capacities of an area as a field of options available to society to use while taking different categories into account, which limit or even exclude certain intended uses, such as risks, carrying capacity, and the capacity to handle stress (increasingly summarised today in the term 'resilience'). Analogously, e.g. de Groot et al. (2002) and Willemen (2010) use the term 'capacity' and define 'ecosystem functions' (and 'landscape functions') as 'the capacity of natural processes and components to provide goods and services which directly and/or indirectly satisfy human needs', and the Millennium Ecosystem Assessment (MEA 2005) refers to 'the capacity of the natural system to sustain the flow of economic, ecological, social and cultural benefits in the future' (see also option values, ► Sect. 3.1 and ► Sect. 4.2).

Scale and Dimension

The scale dependence of ES is an additional but rather poorly investigated aspect (MEA 2005; Hein et al. 2006). Recent research emphasises that both the manner in which we are dissecting our reality and the scale of investigation influence the results significantly (Blaschke 2006). Ecological structures and processes as well as ES manifest themselves at different scales and in quite different manners at the local, the regional and the global scale (■ Fig. 3.4).

According to its original definition, ecosystems can be defined at a wide range of spatial scales

(Tansley 1935), from the level of a small ephemeral sunlit spot on the forest floor up to a whole forest ecosystem spanning several thousands of kilometres and persisting for decades or centuries (Forman and Godron 1986). The supply of ES depends on the functioning of ecosystems, which is in turn driven by ecological processes operating across a range of scales (MEA 2003; Hein et al. 2006). Hence, ES depend on several scale issues. Often, specific ES are generated and supplied at particular scales (Hein et al. 2006; Costanza 2008; Bastian et al. 2012a).

As an example, carbon sequestration and climate regulation are related more to the global scale–notwithstanding the fact that the global balance will be improved by a multitude of local measures. On the other hand, protection against floods by coastal or riparian ecosystems as well as regulation of erosion and sedimentation requires various scales. Pollination (for most plants) and regulation of pests and pathogens refer to the ecosystem level or the local scale (Hein et al. 2006).

According to various scale levels, scale-dependent process variables and magnitudes require scale-adapted methods of analysis and evaluation, which has already been addressed by the dimension theory (Neef 1963). Using this, the approaches developed at the local and regional scales can be transferred (adapted, applied and checked) to the supra-regional or even to the global context (bottom-up strategy). But the reverse approach (top-down) is possible as well. For example, the results of the Millennium Ecosystem Assessment (MEA 2005) (global scale) need to be underpinned by case studies at the local to regional levels (Neßhöver et al. 2007) (▶ Chap. 6). Due to the fact that the combination and processing of data from quite different temporal and spatial scales and the transition from one scale to another can cause problems concerning the expressiveness and interpretation of data and information (Neef 1963), the choice of a suitable dimension is essential for any conceptual and/or methodological ES framework.

It is necessary to distinguish between scales related to socio-economic and ecological issues:

- Ecological and institutional boundaries seldomly coincide and stakeholders of ES often cut across a range of institutional zones and scales (de Groot et al. 2010). Services generated at a particular ecological level can be provided to stakeholders at a range of institutional scales, from the individual and household to the local/municipal, state/provincial, national and international/global community levels. Stakeholders at a particular institutional scale can receive ES generated at a range of ecological scales (Hein et al. 2006; de Groot et al. 2010).
- The fact that ES are generated and supplied at various spatial scales has a strong impact on the value that various stakeholders attach to the services as the scale at which the system service is supplied determines which stakeholders may benefit from it and what their interests would be.
- Spatial trade-offs in terms of local costs and regional or global benefits and vice-versa (e.g. of water purification, carbon sequestration, biodiversity conservation), so-called spatial externalities (Ring et al. 2010), are also a question of scale. The costs of conserving ecosystems and biodiversity fall mostly on local land users and communities whereas the beneficiaries of conservation are not only found at the local level but also far beyond it at the national and global scales.
- There are also various scales at which decisions on natural resources and ES are made. The identification of scales and stakeholders allows an analysis of potential conflicts in environmental management, in particular between local stakeholders and those at larger scales. Considering scale issues in ecosystem management can be important as a basis for establishing compensation payments to local stakeholders who face opportunity costs of ecosystem conservation. Furthermore, they provide insight into the appropriate institutional scales for decision-making on ecosystem management (Hein et al. 2006).

There is a strong need to examine the various scales at which ES are generated and used and, subsequently, how the supply of ES affects the interests of stakeholders at various scales (Tacconi 2000; MEA 2003; Turner et al. 2003; Hein et al. 2006). Hence, the possible scale transitions of ES and the relevant traits need to be examined carefully.

Scale trade-offs are very difficult to manage (► Sect. 3.1.2 Trade-offs, Limit Values, Driving Forces and Scenarios), because they include, in both space and time, shifts of costs and benefits transcending levels of magnitude–small- and large-scale, as well as short- and long-term. Threats on biodiversity and climate, deforestation, and desertification do not imply simple transfers of costs from just one area to other regions or continents. Most likely there will be transfers to later periods and future generations. This problem can render ecosystem payment systems as well as immediate political reactions difficult or even impossible. Regarding time scales it is very important that 'analyses of the dynamics of ES supply require consideration of drivers and processes at scales relevant for the ES at stake' (de Groot et al. 2010). Due to the scale trade-off problem, the transfer of ES assessments over the different scales ('glocal valuation') needs to analyze the specific units and scales of Service Providing Areas (SPA) and Service Benefiting Areas (SBA) (► Sect. 3.3.2).

Scale issues lead to the question of reference units. Adequate spatial reference units are necessary for the sampling, analysis, and assignment of data, as well as for the assessment and modelling of ES (Bastian et al. 2006). The reference units should be related to scales that are ecologically reasonable and policy relevant and they should express the complexity of facts and relationships. Examples for ecological units are ecosystems, watersheds, landscapes and geo-chores (Haase and Mannsfeld 2002; Bastian et al. 2006; Blaschke 2006). For example, the supply of the hydrological service depends on a range of ecological processes that operate, in particular, at the scale of the watershed (de Groot et al. 2010). Examples for socio-economic reference units are: administrative units (municipality, district, state, country) and land-use units. The mismatch of administrative/socio-economic and ecological units and data is a crucial problem (e.g. population statistics on administrative units not matching catchment boundaries), which needs special attention.

Ecological reference units can be used for benefit transfers (benefit-transfer, ► Sect. 4.2; e.g. Plummer 2009): Ecological data and analyses from a particular reference unit can be transferred to a certain degree to ecologically similar and therefore comparable units (incl. the capacity to supply goods and services).

Control Scheme for ES Space and Time Considerations

In order to check and improve the given methodological ES frameworks and studies concerning the consideration of important space and time aspects, we propose the following check list (◘ Table 3.7). It can help avoid overlooking or missing important aspects, and it provides a guideline for the quality control of ES assessments as well as for the analysis of the aspects taken into consideration. The relevant issues (space, time and scale aspects) have been described above (The relevant key words are in italic). We explicitly intend to introduce the check list even into fields that have not been affected by the ES concept to date. The scheme is demonstrated by the example of the European Water Framework Directive (WFD 2000), which addresses many space and time aspects. In fact, it does not mention the ES concept and terminology, but implicitly aims to maintain and improve several ES.

3.3.2 Case Study: EU-Water Framework Directive (WFD) and ES

WFD–Contents

The application of the EU Water Framework Directive (WFD 2000) implies consideration for many space and time aspects, as we seek to demonstrate below, using the example of the Elbe River management plan (◘ Tab. 3.8). The WFD is a directive designed to harmonise the legal framework of water policy in the EU. It also aims at a stronger orientation of the water policy towards a sustainable and environment-friendly use of water. Due to the quite heterogeneous natural conditions within the EU, the WFD is confined to establishing general quality goals and to indicating methods for meeting those goals and achieving favourable water quality.

The core of this directive is the establishment of the WFD of environmental goals including sustainable land use (long-term sustainable water management basing on a high level of protection for the aquatic environment), and also the optimisation of ES (e.g. human health protection, economic consequences).

Table 3.7 General check list of space and time issues related to ES (Bastian et al. 2012a)

Pos.	Issue	Criteria, examples
1 Space aspects		
1.1	Areal requirements	Minimum area (for the supply of ES) with a special quality (structure, abiotic characteristics, biodiversity)
1.2	Spatial composition	Completeness of required partial habitats of animals, land cover diversity, patch richness, a set of ESP
1.3	Spatial configuration	Shape, core areas, buffers, land-use gradients, proximity, mesh size
1.4	General: functional connection	Supply-transfer-demand relations, transmission and transfer likelihood (e.g. habitat networks, river–floodplain relations)
2 Time aspects		
2.1	Time requirements	Minimal process time, regeneration time (of ecosystems and ES)
2.2	Temporal sequences	Natural oscillation, land-use time pattern and interferences, storage capacity for ES
2.3	Time lags	Precaution measures, risks, option values, inter-generational time lags (the present generation benefits, the next pays for environmental damages)
3 Scale and dimension		
3.1	Suitable dimension	Compatibility of scale and measures, reference units, areal and temporal resolution
3.2	Transition	Consideration of upper/ lower scale effects (up-scaling, down-scaling), analysis of transition risks, transfer offsets

Tab. 3.8 Scheme of spatial levels in the Elbe River management plan

Scale	Physical level	Institutional level
Macro	Total catchment area, watershed-related coordination units	International Commission for the Protection of the Elbe River, countries/states
Meso	Partial catchment areas, coordination and planning units	States, counties, catchment areas, area-specific panels
Micro	Small catchment areas, study areas, surface waters and groundwater bodies	Districts, municipalities, working groups and commodity teams, clearing meetings

The 'translation' of normative regulations in the WFD into numerical class limits of a 'favourable state' applies scientific methods. Socio-economic aspects are also taken into consideration by the WFD in the form of 'exceptions' from the goals, and of cost efficiency analyses.

The goals of the WFD imply mainly the following benefits, reflecting a whole bundle of ES:

- Human health protection by water-related utilisations, e.g. bathing-water quality, drinking-water quality
- Lower costs for water purification
- Maintenance of water supply
- Improvement of life quality by increasing the recreation value of surface waters
- Coping with conflicts and regional damages through the balance of interests among different social groups

The precautionary principle, information and transparency shall be considered consequently. The WFD contains mechanisms to assure that socio-economic effects are considered in decision-making processes and that cost-effective options are preferred. The implementation of the environmental goals, however, can cause additional costs but it can be profitable for some beneficiaries (e.g. landscape management companies) and–in the long run–for the whole society. According to the particular watershed, the goals depend on the difference between the actual and the target state as well as on the choice of instruments and management measures. Space-time approaches play a decisive role.

Selected Spatial and Scale Aspects of the WFD

The spatial orientation towards river basins is decisive. Until recently, Germany's water body management was organised predominantly according to political borders and administrative units. The water policy changed first in Great Britain and in France where it was oriented on watershed units. This gave the impulse for a European regulation. As the watersheds of many large European rivers (Meuse, Rhine, Elbe, Oder, Danube) exceed state borders, a common European regulation was advisable. A similar situation applies to groundwater bodies, which are also independent of political borders.

The international Elbe river basin unit contains 146,828 km^2 and it is divided into 10 coordination units. The Czech Republic is responsible for five coordination units (Upper and Middle Bohemian Labe/Elbe, Upper Vltava/Moldau, Berounka, Lower Vltava/Moldau, Ohře/Eger), while Germany is responsible for the other five coordination units (Mulde-Elbe-Black Elster, Saale, Havel, Middle Elbe/Elde, Tidal Elbe). Except for the coordination unit Lower Vltava/Moldau, minor parts of the coordination units with Czech responsibility are situated in Germany (Ohře/Eger and Lower Bohemian Labe/Elbe, Berounka, Upper Vltava/Moldau), Austria (Upper Vltava/Moldau) and Poland (Upper and Middle Bohemian Elbe). The International Commission for the Protection of the Elbe River (ICPER) has the role of a supra-national coordination agency (e.g. water monitoring, supra-regional goals and strategies).

Management plans for large-scale river basin units, e.g. the plan for the Elbe watershed in Germany, contain, due to these large dimensions, specifically for the dimensions, strongly aggregated statements. They refer to such questions as: 'Who provides the ES and who pays for them?' They also consider the specific spatial categories for ecological analyses, planning and decision-making.

As EFTEC (2010) noticed, the spatial analysis of the management plans:

- Helps better organise locally specific data on water bodies and provides a consistent basis for accounting the context-specific nature of economic values, in particular in terms of spatial variation
- Allows better representation of WFD implementation impacts (e.g. in identifying the location of improvements in environmental quality)
- Provides a basis for assessing spatial variation in economic values. This implies that more robust estimates of aggregate costs and benefits can be obtained and additionally, that the distributional impacts can also be examined.

The real planning and implementation of measures takes place at the regional and local levels within meso- and microscale spatial subunits. For this purpose, combined top-down and bottom-up approaches are necessary: supra-regional environmental goals and needs must be down-scaled to regional and local action targets. In contrast, the measures must be aggregated according to the related river basin units and coordination units.

After EFTEC (2010), a key aspect of the WFD implementation is concerned with the spatial and geographic aspects of water bodies. It is necessary to understand how the impacts of measures may vary over spatial scales. These effects will not only have an impact on the direct benefits related to the water bodies themselves but can also have indirect beneficial or detrimental impacts elsewhere. In the case of water quality, and in particular rivers, most of the relationships between ES production areas and benefit areas are 'directional' in a downstream direction (rather than 'in situ'). In some cases the beneficial effects can be spatially very remote from the area of a targeted intervention. For example, reducing diffuse pollution may enhance terrestrial biodiversity, soil quality and erosion control in addition to the water quality benefits downstream (Grunewald et al. 2005, 2008; EFTEC 2010) (◘ Table 3.7 und ◘ Tab. 3.10, line 3.2: scale transition).

Accordingly, for management purposes (assessments of the state, targeting) the Elbe river basin has been divided into 61 planning units ranging in size from 300 to 5600 km^2, 3896 surface water bodies and 327 groundwater bodies. The institutional levels and the information levels, including the accuracy of data, should be in reference to these scales (◘ Table 3.7 und ◘ Tab. 3.10, line 3.1: suitable dimension).

The chemical, biological and ecological quality of waters depends on a variety of influences. In order to assess them and to take action, an integrated approach and a broad database are the key necessities. The WFD prescribes consistent and therefore comparable criteria for the provision and updating of these data. For example, Article 10 of the WFD prescribes that the loads from point sources (especially industrial wastes and from sewage purification works) and diffuse sources (especially from agricultural land) should be considered together.

This is based on spatially-specific analyses and documentations of loads (main sources). Typical questions are: Which waters (surface waters, groundwater) are polluted by nutrients (N, P) and to which extent? What is the contribution of parts of catchment areas or of countries/states to the eutrophication of the North Sea and what are the specific potentials for reducing these loads? Such spatially relevant distribution options were traded off in the framework of the Elbe river basin Agency (Flussgebietsgemeinschaft Elbe–FGG Elbe 2009). It is obvious that the efforts to reduce N can and should be especially high in the German states of Schleswig -Holstein and Saxony, while the potentials to reduce P are especially high in Thuringia, Schleswig-Holstein, Saxony-Anhalt and Saxony (◘ Table 3.9 and ◘ Table 3.10, 1.4: functional connection).

This supra-regional distribution of nutrient reductions must be further underpinned in the water basin subunits. In terms of spatial aspects, for example, it needs to be clarified whether agro-environmental payments, e.g. for intermediate crops, or soil protection against erosion are provided for all arable fields, or if they are concentrated on focus areas. Analyses of efficiency and acceptance are necessary for this (Grunewald and Naumann 2012). It is also essential to make arrangements for cooperative efforts and to negotiate solutions between the land users (farmers) and the beneficiaries of ES (here society as a whole).

◘ Table 3.9 Expected reductions of nutrient loads of the Elbe River for the protection of the North Sea in tributary rivers, by country/ German state (reference year: 2006; measurements between 2009 and 2015; nutrient inputs into primary flowing waters, as per FGG Elbe 2009)

Country/ state	Nitrogen		Phosphorus	
	%	$t\,a^{-1}$	%	$t\,a^{-1}$
Czech Republic	5	~3 120	7	~150
Brandenburg, Berlin	0,8	~47	1.5	~8
Bavaria	3.5–7.5	~195	2–5	~3
Hamburg	10	~85	10	~3
Mecklenburg-Western Pomerania	19	~400	5	~5
Lower Saxony	2.7	~270	2.7	~12
Schleswig-Holstein	16.6	~1 650	18.7	~70
Saxony	10–11	~2 740	11–13	~75
Saxony-Anhalt	3.9	~625	13.4	~60
Thuringia	5	~600	23.6	~80

Time Aspects of the WFD

The WFD outlines several time limits for the legal implementation of the Directive itself, the analyses, the monitoring programme, the management plans and the specific programmes (time tables) for the undertaken measures. More important, it is established until when a 'favourable state' of the water(s) has to be reached. Time aspects are especially considered with respect to the practical implementation of the WFD. The clear requirements for ES providers and beneficiaries correspond to the time spans for the realisation of measures, e.g. for reducing nutrient loads or the reporting obligation of the countries/states (◘ Table 3.10, line 2.1: Time require-

Table 3.10 Check list of space and time issues exemplified by WFD (2000)

Pos.	Issue	Implementation in WFD (examples)	ES–example: Groundwater Recharge
1 Space aspects			
1.1	Areal requirements	Minimum sizes of standing waters (50 ha) and catchments (of flowing waters: 10 km^2) in the WFD taken into account; catchment alignment instead of administrative units	Mapping areas and state of groundwater bodies
1.2	Spatial composition	Combined consideration of surface and groundwater, management of entire catchments	Mapping groundwater recharge (supply) and groundwater extraction (demand), accounting balance
1.3	Spatial configuration	Configuration issues only partially implemented with mappings of the waters' structure; fish migration ability considered; confined to big- and medium-sized water bodies (i.e. two-third of streams are not considered in terms of their structure)	Hydrogeological maps, land use, etc.
1.4	General: functional connection	Orientation towards human health, quality of life, joint consideration of biological, chemical and ecological quality	Maps of groundwater protection
2 Time aspects			
2.1	Time requirements	Differentiating management measures by graduated time periods	Time aspects of groundwater flows, monitoring (water level gauge)
2.2	Temporal sequences	Targets in accordance with ecological processes are differentiated according to specific time periods; flexible management priorities	Natural conditions can vary (precipitation necessary for water infiltration, crop rotation), trends (e.g. climate change)
2.3	Time lags	Strict application of the precautionary principle, (flood) risk minimisation	e.g. water protection areas
3 Scale and dimension			
3.1	Suitable dimension	Combined top-down and bottom-up approach, planning and management regional, but measured locally	Hierarchy of catchment areas
3.2	Transition	Partly considered: influences on adjoining seas and estuaries as well as effect on climate protection goals–rather good; on floodplains and floods:poor	Many local measures can effect groundwater recharge regionally (or regarding the whole water body)

ments). The concrete, super-ordinate timetable with milestones is obligatory for all parties concerned: beginning with the transformation of the WFD into national legislation in 2003 and ending with the achievement of the 'good ecological state in river basins' in 2015, with the possibility of extending this time limit until 2021 or 2027 (WFD 2000).

It must be considered that waters need time to reach such goals after development measures (timespan until results of the measures are achieved). The temporal sequence (Table 3.10, line 2.2) of requirements refers to the duration of natural processes, as well as to the time needed to accomplish management measures. In fact, WFD aims at a 'good ecological state' of all waters by 2015. But the directive also allows exceptions: extensions of time or reduced environmental targets, if they cannot be achieved in time for objective reasons. The exceptions are designed to avoid excessively high costs of management measures. Without valid cost calculations it is difficult to justify exceptions. For the practical implementation of the WFD, the countries (in Germany also the federal states) are responsible. All countries interpret the directive independently, but they have implemented working groups to harmonise the national regulations to a certain extent.

The WFD puts an end to previous time lags, it contributes to ensuring water-related ecosystem potentials for the future. The precautionary principle is already implemented since the WFD ensures water reasonable quality. But even economic time lags (i.e. the next generation has to pay for our success now) will be avoided.

The member states of the EU were obligated to implement an appropriate water fee policy by 2010 with incentives for water users to use the resources economically. The various water users (industry, households, agriculture, etc.) are to contribute adequately to cover the costs of water ES including costs related to the environment and the resources (Article 9 WFD). The evaluation of financial disproportions (cost excessiveness) also needs the balancing of costs and benefits, i.e. typical core aspects of the ES approach are considered (▶ Sect. 4.2). The WFD also mandates that the water supply was to be organised in such a way by 2010 that all costs were covered (the cost-covering principle). The question is: 'Who pays?' Formerly, the general public paid for the protection of drinking water. Now, the waste producer has to pay but the principle of solidarity is applied. It must be noted that to date these regulations and obligations have been only partially implemented.

Control Scheme for ES Space and Time Considerations in the WFD

The check list for space and time aspects (◘ Table 3.7) was completed and exemplified by means of relevant aspects of the European Water Framework Directive. ◘ Table 3.10 shows that the directive meets most of the space and time issues concerned, e.g. the size of catchments and the differentiation of measures in terms of space and time. On the other hand, the table also reveals possible deficits, such as the incomplete consideration of spatial configuration or of scale transition aspects.

Conclusion

ES demonstrate a wide range of space, time and scale dependent relations.

In respect to the analysis and evaluation steps as well as to the supply and demand perspectives, not only the ecological aspects are concerned, but also socio-economic and cultural ones. Often, space and scale effects are related mainly to ecological phenomena. According to our concept of space, we have tried to widen this perspective and to include socio-economic aspects as well. This is in line with the UK National Ecosystem Assessment (UKNEA 2011), which notes that institutional mechanisms linking across spatial scales (from small- to large-scale in terms of area) would 'provide opportunities for stakeholder engagement and greater collaboration between actors, and for the involvement of local groups and nongovernmental organisations'. From the perspective of ecological regional development, the ES concept is of particular importance because the human-environment relationship is emphasised. Thereby the social concept of space (perception, area for interaction) can be associated with physical concepts of space (order, place, location, spatial intersections, distances, boundaries in space).

All main aspects of the ES approach can be found in the European Water Framework Directive (EU-WFD), e.g. conflict relevance, focus on problems, goal setting, environmental and economic data, quantitative and model-based approaches, integrated approach, participatory approaches, decision support systems, cost-benefit considerations, and solutions-oriented approach. Even in terms of space and time approaches, the WFD represents an enormous advance over previous approaches, simply because of clear definitions and conceptual hierarchies. Some of the special questions concerning space, time and scale relationships in ES assessments could be solved and discussed by reference to the example of the WFD and the Elbe river watershed, e.g. spatial configuration and composition (patterns), reference units, concordance of physical and socio-economic space concepts, the spatial position of services providers and service beneficiaries, service connecting areas, the role of temporal sequences (of land uses, supply and demand) and time lags (precautionary principle, intergenerational lags), the shift from one scale to another and practical consequences resulting from these factors.

In order to take space, time and scale effects into consideration adequately, a check list is useful, which we have developed and tested successfully using the example of the WFD. Such a check list can be applied to all frameworks and studies where ES are to be assessed. This check list is a flexible scheme that can be modified according to the particular situation.

Space, time and scale aspects of ES are of great practical interest, e.g. for land-use and landscape management, for spatial planning, regional development and financial policies (balancing of costs and benefits arising from ES). After EFTEC (2010), spatial analysis improves the economic valuation and it can help to 'target' policies (e.g. maximise aggregate benefits given a resource budget, or to redistribute benefits to disadvantaged groups). The example of the WFD reveals the practical relevance in many ways, e.g. the choice of relevant reference units, the spatial and temporal distribution of costs and benefits, time frames for reaching particular goals with consideration for ecological preconditions (e.g. the regeneration capacity of waters) and also of economic scales (economic carrying capacity, payments over adequately great time periods). The WFD takes ecological periods into account (development, seasonality, regeneration, matter transfers) and it gives a clear orientation in terms of time horizons, which is important for users and other stakeholders. In the WFD, such issues are better addressed than–for instance–in the EU Habitats Directive and other regulations (► Sect. 6.6.1).

3.4 Landscape Services

O. Bastian, K. Grunewald, M. Leibenath, R.-U. Syrbe, U. Walz and W. Wende

As explained in ► Sect. 3.3, the creation and also the use of ES is always tied to concrete spaces. It is manifested in spatial differentiation, and in various dimensions and scales. Critical voices have claimed that to date there has been little or no localisation, i.e. that the pattern of arrangements and relationships of ES in space has hardly been taken into account at all (Syrbe and Walz 2012), and that merely statistical information, such as land cover, has been included instead (Blaschke 2006). Moreover, it is claimed that the practical applicability and the connections of ES to the planning process have been insufficient (Termorshuizen and Opdam 2009).

One promising way to eliminate these deficits is to link ES to the landscape approach and to the definition of landscape services in order to emphasise the spatial connection and to arrive at statements, which can better be used in the planning and/or practical context (Burkhard et al. 2009; Termorshuizen and Opdam 2009; Frank et al. 2012; Schenk and Overbeck 2012).

This is true in spite of the fact that the term landscape has been highly controversial in the scientific discourse, with a broad spectrum of interpretations and substantive meanings existing, depending not only on levels of education, socialisation and professional backgrounds, but also on language and cultural area. 'The landscape' has been an object of investigation for various scientific disciplines, and also of other areas of life, such as aesthetics, painting, literature, philosophy, geography, conservation and landscape care, agriculture and silviculture, etc. A farmer, a geologist, a forester, a recreation seeker–each of them sees the landscape differently and focuses on something different (Jessel 1998).

In common parlance landscape is usually seen as a piece of land that can be perceived all at once with the naked eye. The word 'landscape' comes from the old Germanic *lantscaf*, with *scaf* having developed to '*shape*' in English and '*schaffen*' ('to create', 'to achieve;' in some dialects, 'to work') in German (Haber 2002). Hence, the landscape is literally the 'land shaped' or created by people. However, the landscape as a dimension that can be visually experienced was for centuries only to a lesser extent a consciously created object. It was merely seen as a product of the top-priority activity: the provision of the food supply. Nonetheless, even at an early date landscapes were often shaped in such a way that various positive side effects were realised. Examples include the rows of fruit trees on embankments, which are otherwise difficult to utilise, so as to provide fruit for food and at the same time shade for the peasants on their long walks to the work in the fields, or else, in the proximity of farms and villages, as planted groves which served as a windbreak and improved the microclimate. Aesthetic aspects, too, may certainly have played a part. The consciously shaped landscape, which would later also be marketed as a tourist attraction had its roots in the Enlightenment–the ideal of the English landscape garden–and culminated in park designs of major cities in the nineteenth century, such as New York's Central Park. Today, this constant is a firm part of landscape and spatial planning (Kienast 2010).

According to Leibenath and Gailing (2012), landscape can be interpreted in any of four different ways:

1. The landscape as a physical space or complex of ecosystems
2. The cultural landscape in the context of the human-environment relationship
3. The cultural landscape as a metaphor; and
4. The cultural landscape as a social construct, or as an object of communication.

Backhaus and Stremlow (2010) distinguish the following four basic disciplinary approaches to landscape:

1. The ecosystemic and geomorphological approach
2. The psychological and phenomenological approach
3. The constructivist/cultural-scientific approach
4. The political and social scientific approach

An understanding of landscape as an intermediate phenomenon between natural-scientifically ascertainable objective reality on the one hand and a mental construct on the other is expressed in such definitions as that of the Council of Europe in the European Landscape Convention (Article 1; Czybulka 2007 [Engl: ► http://www.coe.int/t/dg4/cultureheritage/heritage/Landscape/default_en.asp]): '…an area, as perceived by people, whose character is the result of the action and interaction of natural and/or human factors; or by Fry (2000): …' a physical and mental reflection of the interaction between societies and cultures and their natural environment. In this context, landscape can also be seen as a section of the earth's shell of varying orders of magnitude, prepared by natural conditions, overformed to varying degrees by human activity, perceived or felt by people as characteristic, and delimited according to rules which are to be stipulated (Bastian 2006, 2008, modified).

According to the Millennium Ecosystem Assessment (MEA 2005), a landscape is typically composed of a number of different ecosystems, each of which generates a whole package of different ES. Hence, it is certainly justified to certify landscape areas of identical or similar overall character–or to use them as units of reference–in order to interpret their characteristics for an effective but gentle use by society (Bernhardt et al. 1986; Hein et al. 2006; TEEB 2009).

Most landscape definitions fulfill the requirement of spatial reference or of spatial expanse, and of holism in accordance with Alexander von Humboldt's 'total impression of a region', or of the 'landscape-like' (Humboldt 1847, pp. 92, 97). Often, landscape and people are seen as two opposite poles, an attitude which, by the way, is promoted even by a term such as 'people and nature'. It is easy to ignore the fact that people are also part of nature (Oldemeyer 1983, in Gebhard 2000). Increasingly, however, material and intellectual aspects are being taken into account in a more balanced way, and people are being directly involved.

For the ES concept, we see the definition of landscape as a physical space or an ecosystem complex as particularly helpful. Many ES are influenced by the landscape structure and the geographic context, for instance by the arrangement of landscape elements or land-use units. Landscape structure largely determines flows and cycles of waters, nutrients and organisms. The spatial relationship between biotic factors, such as vegetation, and abiotic factors, such as soil, is decisive for the manner in which many ES are provided, so that the whole–the landscape and the ecological mosaic linked to it–is more significant than the sum of its parts (Odum 1971; Haber 2004). The matrix of the landscape determines the effectivity and significance of its biotic components to a much greater degree than would

Landscape vs. Ecosystem

In view of the multiplicity of meanings of the term 'landscape' and the difficulty in delimiting concrete landscape areas, the concept of landscape may appear as too non-concrete, fuzzy and unscientific compared with the concept of ecosystems. However, is the ecosystem paradigm really that unproblematic, by comparison? Certainly not, for it, too, is subject to the criticism that it is too diffuse and contradictory (O'Neill et al. 1986) and suffers from methodological deficits in its application in research and practice. Naveh and Lieberman (1994) raise the question of whether ecosystems could indeed be considered real existing phenomenon or whether they were not simply conceptual aids for the analysis of the flows of energy, materials and information in ecological systems.

Noss (2001) sees ecosystems as functional systems with their spatial boundaries either undefined or defined more or less arbitrarily. In his opinion ecosystems are open systems between which the exchange of materials, energies and organisms take place.

Naveh (2010) raises serious issues regarding the ecosystem paradigm with respect to their spatial aspects: first, he says, what is at issue is the assumption that interactions and feedback loops exist within ecosystemic boundaries. In reality, however, the spatial dissemination of the participating organism populations may be much broader. Second, spatial homogeneity is often assumed. This simplification overlooks some of the essential properties of the system, for precisely heterogeneity is the precondition for the lives of these organisms. Another major failing of the paradigm of 'natural' ecosystems is, he says, the common practice of categorizing human activity as an external disturbance.

Nonetheless, it may certainly be useful to generally prefer the abstract term 'ecosystem' for the ES concept, (► Chap. 1, ► Chap. 2, ► Chap. 3) since it emphasises the natural structures and processes more strongly and does a better job of creating the connections to 'ecology' as a category of sustainability, and/or as a class of functions and services.

be the case if these components were merely added together (Frank et al. 2012; Syrbe and Walz 2012).

Landscape services constitute the link between landscape and human well-being. They imply a strong spatial orientation and regional differentiation, as well as a reference to actors, planners and decision-makers. The concept of landscape services is also of particular significance inasmuch as it raises the issue of the human-environment relationship and of anthropogenic transformation more strongly, and hence links the societal concept of space–space for perception, and also space for action–with the physical concept of space.

The incorporation of landscape services as a special form of the ES approach has the following advantages:

- Landscapes as units of reference enhance the perspective beyond the services provided by ecosystems and place a greater emphasis on the aesthetic, ethical and sociocultural aspects, as well as on the anthropogenic modification (e.g. land use) and the overall character of an area (peculiarities of the landscape).
- Spatial aspects are expressed more strongly, for example the arrangement of ecosystems and land-use units in their spatial context, structural and process-determined interactions, the spatial difference of supply and demand–in the form of so-called 'service-providing areas' and 'service-benefiting areas'–or the reference to different dimensions and scales (► Sect. 3.3). Interactions between spaces and ES can be shown with reference to many functional aspects relevant for practice: the problem of the conflicting needs of upstream vs. downstream residents in watersheds, the relationship between cities and their surrounding countrysides, or the relationship between economic areas, impact areas and places used for compensation and offsetting measures, etc. To some extent, the ES generated at certain places can only be transferred to the areas of demand via specific spaces, known as 'service-connecting areas', e.g. the feeding of cold air into cities via cold-air corridors; (► Sect. 3.3).
- The emphasis on the reference to a landscape improves the interaction (or integration) of various disciplines since nature, culture, and use aspects all have to be addressed in equal measures–even though the definitions of

'landscape' differ between the various academic disciplines. Especially the physical landscape approach enhances the relevance for practical spatial planning, including landscape planning, as well as for the landscape development of management, and favours participatory approaches, which recognise the landscape as an element providing identity and as an area for action, with a connection to the actors.

Another advantage of the reference to landscapes is provided by the fact that in spite of the controversial scientific discourse on the definition of 'landscape', the sustainable use and protection of landscapes is gaining growing support worldwide, in the first European environmental report, the so-called Dobřiš Assessment of the European Environmental Agency, and in the European Landscape Convention of the Council of Europe of 2000. One of the demands is that visions, or models, for European landscapes are established, and that landscape protection be integrated into sectoral policy, e.g. in the EU's Common Agricultural Policy and its regional policy, in order to support regional identities and landscape peculiarities (Czybulka 2007).

In the Territorial Agenda of the European Union of 2007 (EU 2007, p. 7), cultural landscapes are designated as the 'foundation for environmentally and culturally oriented development … which offers development perspectives … particularly in regions that are lagging behind or undergoing structural changes'. Fürst et al. (2008) call for placing greater emphasis–once again–on seeing cultural-landscape development as a catalyst and as a vehicle, i.e. as 'the essential element for new kinds of problem-solving in regional development'. The concept of the landscape must be integrated into all relevant policy areas in this context, e.g. in connection with the Common Agricultural Policy of the EU after 2013, in Natura 2000, and with regard to issues of bio-energy.

We consider landscape services to be a special case within the overall concept of ES (analogously to Kienast 2010; Hermann et al. 2011). However, the landscape approach is broader and more complex since it includes not only ecological aspects but also to a peculiar degree aesthetic, cultural, psychological, as well as other aspects. In this case we are examining services with a specific connection to the landscape. Thus, we explicitly emphasise the analysis and evaluation of landscape services as it is usually already implied in the main focus of the work: landscape planning, landscape care, evaluation of the cultural landscape and the appearance of the landscape (cf. ► Sect. 5.3 **and particularly** ► Sect. 6.5**).**

The term 'landscape' moreover has a strong connection to planning and is especially familiar to spatial planners. Likewise, the broader public has a greater understanding of this term than of 'ecosystem'. According to Termorshuizen and Opdam (2009), landscape planners have for decades viewed landscape as a human-ecological concept and have addressed its economic, cultural and ecological values.

Rather than treating single components or protected assets as isolated from one another, landscape planning is taking the complexity of the investigated object into account, which is one of its fundamental requirements. Even during the 1970s and 1980s landscape and spatial planning assigned potential functions to the landscape, which were for the most part cartographically recorded. In that respect, landscape and spatial planning was actually very close to the concept of landscape services, even if the landscape-specific functions were not yet called 'services' (► Sect. 2.2).

The selected landscape approach (see above) not only enhances the relevance for practical spatial planning, including landscape planning (► Sect. 5.3), and for landscape development and management (► Sect. 6.5), it also favours participatory approaches, which see the landscape as an identity-providing element and as a space for action (► Sect. 4.3; Fürst and Scholles 2008). The landscape, not the ecosystem, is the space of reference for public participation; it permits a large number of local stakeholders to identify with the landscape in which they live, work and enjoy life, and to have an influence upon it, to take responsibility for it and to help shape it. By contrast, the term 'ecosystem' often means new natural, more or less untouched areas, often associated with a protected status, with recreational function, with species diversity and with undisturbed natural processes (Termorshuizen and Opdam 2009). The landscape is also a public-relations factor; it can be 'sold' as a good place where recreation can be found and where people can live and work (Wascher 2005).

Conclusion

In the final analysis, ecosystem and landscape services cannot be fundamentally distinguished. The latter emphasises spatial aspects more and is oriented towards complex approaches by reference to interfaces of ecological, economic and social aspects. Moreover, it is more oriented towards spatial planning, communications and the participation of actors and stakeholders, of 'local people'. Methods for ascertaining and evaluation are largely similar or identical; however, landscape services as a result of broader, more multidisciplinary approaches take a more comprehensive spectrum of methodologies into account. A thorough and detailed discussion of the landscape services issue is published in the Journal of *Landscape Ecology* in 2014 (Bastian O et al. 2014).

References

Albert C, von Haaren C, Galler C (2012) Ökosystemdienstleistungen. Alter Wein in neuen Schläuchen oder ein Impuls für die Landschaftsplanung? Nat Schutz Landsch Plan 44:142–148

Anderson BJ, Armsworth PR, Eigenbrod F, Thomas CD, Gillings S, Heinemeyer A, Roy DB, Gaston KJ (2009) Spatial covariance between biodiversity and other ecosystem service priorities. J Appl Ecol 46:888–896

Artner A, Frohnmeyer U, Matzdorf B, Rudolph I, Rother J, Stark G (2005) Future Landscapes. Perspektiven der Kulturlandschaft. Bundesamt für Bauwesen und Raumordnung, Bonn

Auhagen (1998) Verbal-Argumentation oder Punkte-Ökologie–Bewertungsverfahren unter der Lupe des Planers. Sächsische Akademie für Natur und Umwelt im Sächsischen Staatsministerium für Umwelt und Landesentwicklung in Zusammenarbeit mit dem Lehr- und Forschungsgebiet Landschaftsplanung der Technischen Universität Dresden (eds) Dresdner Planergespräche–Vom Leitbild zur Quantifizierung–Bewertungsprobleme und ihre Lösung in der Landschafts- und Grünordnungsplanung. Bericht zur wissenschaftlichen Arbeitstagung am 14. und 15. November 1997. Dresden, 137 S

Backhaus N, Stremlow M (2010) Handlungsraum Landschaft: Wege zur Förderung transdisziplinärer Zusammenarbeit. Nat Landsch 85:345–349

BAFU–Bundesamt für Umwelt (2011) Indikatoren für Ökosystemleistungen. Systematik, Methodik und Umsetzungsempfehlungen für eine wohlfahrtsbezogene Umweltberichterstattung. Herausgegeben vom Bundesamt für Umwelt BAFU, Bern

Balmford A, Rodrigues ASL, Walpole M, ten Brink P, Kettunen M, Braat L, de Groot R (2008) The economics of biodiversity and ecosystems: scoping the science. European Commission, Cambridge

Barkmann J (2001) Angewandte Ökosystemforschung zwischen Biodiversitäts-, Landschafts- und Ressourcenschutz. Petermanns Geographische Mitteilungen 145:16–23

Bastian O (2006) Landschaft als Basiskonzept für integrative Forschungsansätze in der Geographie. (Vortrag im Institut für Länderkunde Leipzig. Workshop "Raum–Landschaft–Region als Bezugsgrößen integrativer Projekte in der Geographie", 14./15.7.2006 in Leipzig)

Bastian O (2008) Landscape classification–between fact and fiction. In: Lechnio J, Kulczyk S, Malinowska E, Szumacher I (eds) Landscape classification. Theory and practice. Warsaw, S 13–20

Bastian O, Grunewald K, Syrbe RU (2012a) Space and time aspects of ecosystem services, using the example of the European Water Framework Directive. Int J Biodivers Sci Ecosys Services Manage 8(1–2):5–16

Bastian O, Grunewald K, Syrbe RU, Walz U, Wende W (2014) Landscape services–concept and practical relevance. Landsc Ecol 29(6):1463–1478. DOI 10.1007/s10980-014-0064-5

Bastian O, Haase D, Grunewald K (2012b) Ecosystem properties, potentials and services–the EPPS conceptual framework and an urban application example. Ecol Indic 21:7–16

Bastian O, Krönert R, Lipsky Z (2006) Landscape diagnosis on different space and time scales–a challenge for landscape planning. Landsc Ecol 21:359–374

Bastian O, Schreiber KF (eds) (1994) Analyse und ökologische Bewertung der Landschaft. Fischer, Jena, 502 S

Bastian O, Schreiber KF (eds) (1999) Analyse und ökologische Bewertung der Landschaft, 2. edn. Spektrum Akademischer Verlag, Heidelberg, Berlin, 564 pp.

Bastian O, Steinhardt U (eds) (2002) Development and perspectives of landscape ecology. Kluwer, Dordrecht, 498 pp.

Bechmann A (1989) Die Nutzwertanalyse. In: Storm P-C, Bunge T (eds) Handbuch der Umweltverträglichkeitsprüfung. Schmidt, Berlin, S 1–31

Bechmann A (1995) Anforderungen an Bewertungsverfahren im Umweltmanagement–Dargestellt am Beispiel der

Bewertung für die UVP-Dokumentation zu den 11. Pillnitzer Planergesprächen am 29./30.9.1995, pp 6–39
de Bello F, Lavorel S, Diaz S, Harrington R, Cornelissen JHC, Bardgett RD, Berg MP, Cipriotti P, Feld CK, Hering D, da Silva PM, Potts SG, Sandin L, Sousa JP, Storkey J, Wardle DA, Harrison PA (2010) Towards an assessment of multiple ecosystem processes and services via functional traits. Biodivers Conserv. doi:10.1007/s10531-010-9850-9
Bernhardt A, Haase G, Mannsfeld K, Richter H, Schmidt R (1986) Naturräume der sächsischen Bezirke. Sächsische Heimatblätter 4, 5
Blaschke T (2006) The role of the spatial dimension within the framework of sustainable landscapes and natural capital. Landsc Urban Plan 75:198–226
Blotevogel HH (1995) Raum. In: Akademie für Raumforschung und Landesplanung (ed) Handwörterbuch der Raumordnung, Hannover, pp 733–740
Boyd J, Banzhaf S (2007) What are ecosystem services? The need for standardized environmental accounting units. Ecol Econ 63:616–626
Burkhard B, Kroll F, Müller F, Windhorst W (2009) Landscapes' capacities to provide ecosystem services–a concept for land-cover based assessments. Landscape Online 15:1–22
Burkhard B, Kroll F, Nedkov S, Müller F (2012) Mapping ecosystem service supply, demand and budgets. Ecol Indic 21:17–29
Costanza R (2008) Ecosystem services: Multiple classification systems are needed. Biol Conserv 141:350–352
Costanza R, d'Arge R, de Groot R, Farber S, Grasso M, Hannon B, Limburg K, Naeem S, O'Neill R, Paruelo J, et al (1997) The value of the world's ecosystem services and natural capital. Nature 387:253–260
Cowling RM, Egoh B, Knight AT, O'Farrell PJ, Reyers B, Rouget M, Roux DJ, Welz A, Wilhelm-Rechman A (2008) An operational model for mainstreaming ecosystem services for implementation. Proc Natl Acad Sci USA 105:9483–9488
Czybulka D (2007) Die Europäische Landschaftskonvention. Zeitschrift für Europäisches Umwelt- und Planungsrecht (EurUP) 6:250–258
Díaz S, Fargione J, Chapin FS, Tilman D (2007) Biodiversity loss threatens human wellbeing. PLoS Biol 4:e277. doi:10.1371/journal.pbio.0040277
Dunlop M, Turner G, Foran B, Poldy F (2002) Decision points for land and water futures. Resource Futures Program Working Document 2002/08, CSIRO Sustainaible Ecosystems, Canberra, Australia
Dunn RR (2010) Global mapping of ecosystem disservices: the unspoken reality that nature sometimes kills us. Biotropica 42:555–557
Durwen KJ, Schreiber K-F, Thöle R (1980) Ein pragmatischer Ansatz zur Aufbereitung ökologischer Determinanten für die Raumplanung. Arb.-Ber. Landschaftsökologie, Münster 2:3–12
EASAC–European Academies Science Advisory Council (2009) Ecosystem services and biodiversity in Europe. EASAC policy report 09, London
EEG (2009) Gesetz zur Neuregelung des Rechts der Erneuerbaren Energien im Strombereich und zur Änderung damit zusammenhängender Vorschriften. Bundesgesetzblatt 2008, Teil I(49), Bonn, 21.10.2008
EFTEC (2010) Scoping study on the economic (or non-market) valuation issues and the implementation of the Water Framework Directive. Final Report for the EU Comm. DGE (ENV.D1/ETU/2009/0102rl) (Economics for the Environment Consultancy–eftec, Ozdemiroglu E, Provins A, Hime S), London
Erdmann KH, Shell C, Todt A, Küchle-Krischun J (2002) Natur und Gesellschaft: Humanwissenschaftliche Aspekte zum Naturschutz. Nat Landsch 77:101–104
EU–European Union (2007) Territorial Agenda of the European Union. ► http://www.eu-territorial-agenda.eu/Reference%20Documents/Territorial-Agenda-of-the-European-Union-Agreed-on-25-May-2007.pdf. Accessed 22 April 2014
FGG Elbe (2009) Nationales Überwachungsprogramm Elbe 2009– Deutscher Teil des internationalen Messprogramms, aufgestellt: Wassergütestelle Elbe, 1/09. ► www.fgg-elbe.de. Accessed 3 Jan 2011
Fisher B, Turner RK (2008) Ecosystem services: classification for valuation. Biol Conserv 141:1167–1169
Fisher B, Turner RK, Morlin P (2009) Defining and classifying ecosystem services for decision-making. Ecol Econ 68:643–653
Forman RTT, Godron M (1986) Landscape ecology. Wiley, New York
Frank S, Fürst C, Koschke L, Makeschin F (2012) A contribution towards a transfer of the ecosystem service concept to landscape planning using landscape metrics. Ecol Indic 21:30–38
Fränzle O (1998) Grundlagen und Entwicklung der Ökosystemforschung. Handbuch der Umweltwissenschaften, 3. Erg. Lfg. 12/98, pp 1–24
Fry G (2000) The landscape character of Norway–landscape values today and tomorrow. In: Pedroli B (ed) Landscape our Home/Lebensraum Landschaft. Indigo, Zeist, pp 93–100
Fürst D, Gailing L, Pollermann K, Röhring A (eds) (2008) Kulturlandschaft als Handlungsraum–Institutionen und Governance im Umgang mit dem regionalen Gemeinschaftsgut Kulturlandschaft. Rohn, Dortmund, 328 pp
Fürst D, Scholles F (eds) (2008) Handbuch Theorien und Methoden der Raum- und Umweltplanung, 3. edn. Rohn, Dortmund
Gebhard U (2000) Naturschutz, Naturbeziehung und psychische Entwicklung. Nat Schutz Landsch Plan 32:45–48
de Groot RS (1992) Functions of nature: evaluation of nature in environmental planning, management and decision making. Wolters-Noordhoff, Groningen
de Groot RS, Wilson M, Boumans R (2002) A typology for description, classification and valuation of ecosystem functions, goods and services. Ecol Econ 41:393–408

de Groot RS, Alkemade R, Braat L, Hein L, Willemen L (2010) Challenges in integrating the concept of ecosystem services and values in landscape planning, management and decision making. Ecol Complex 7:260–272

Grossmann M, Hartje V, Meyerhoff J (2010) Ökonomische Bewertung naturverträglicher Hochwasservorsorge an der Elbe, Abschlussbericht des F + E-Vorhabens "Naturverträgliche Hochwasservorsorge an der Elbe und Nebenflüssen und ihr volkswirtschaftlicher Nutzen. Teil: Ökonomische Bewertung naturverträglicher Hochwasservorsorge an der Elbe und ihren Nebenflüssen", BfN 89

Grunewald K, Bastian O (2010) Ökosystemdienstleistungen analysieren–begrifflicher und konzeptioneller Rahmen aus landschaftsökologischer Sicht. GEOÖKO 31:50–82

Grunewald K, Naumann S (2012) Bewertung von Ökosystemdienstleistungen im Hinblick auf die Erreichung von Umweltzielen der Wasserrahmenrichtlinie am Beispiel des Flusseinzugsgebietes der Jahna in Sachsen. Nat Landsch 1:17–23

Grunewald K, Weber C, Schröder K (2005) Schadstoffbelastung und Auennutzung nach dem Elbehochwasser 2002 in Sachsen–eine Synopse. In: Steinberg C, Calmano W, Klapper H, Wilken R-D (eds) Handbuch Angewandte Limnologie. 22. Erg.Lfg. 7/05, 42 pp

Grunewald K, Scheithauer J, Monget JM, Nikolova N (2007) Mountain water tower and ecological risk estimation of the Mesta-Nestos transboundary river basin (Bulgaria-Greece). J Mt Sci 4:209–220

Grunewald K, Dehnert J et al (2008) Nährstoffmodellierung zur Aufstellung der Maßnahmenprogramme nach WRRL in Sachsen. Wasser + Abfall 3:15–19

Haaren C von, Bathke M (2008) Integrated landscape planning and remuneration of agri-environmental services, Results of a case study in the Fuhrberg region of Germany. J Environ Manage 89:209–221. doi:10.1016/j.jenvman.2007.01.058

Haase G (1978) Zur Ableitung und Kennzeichnung von Naturraumpotenzialen. Petermanns Geogr Mitt 22:113–125

Haase G, Mannsfeld K (2002) Naturraumeinheiten, Landschaftsfunktionen und Leitbilder am Beispiel von Sachsen. Deutsche Akademie für Landeskunde, Flensburg

Haber W (2002) Kulturlandschaft zwischen Bild und Wirklichkeit. In: Schweizerische Akademie der Geistes- und Sozialwissenschaften, Bern, Akademievorträge, Heft IX, 19 pp

Haber W (2004) Über den Umgang mit Biodiversität. In: Bayerische Akademie für Naturschutz und Landschaftspflege (ANL): Schwerpunkte: Leitbilddiskussion: Traditionen und Trends/Biodiversität als Aufgabe/Biografie: Alwin Seifert. Berichte der ANL, 28:25–43; Laufen/Salzach (Bayerische Akademie für Naturschutz und Landschaftspflege)

Haines-Young RH, Potschin MB (2009) Methodologies for defining and assessing ecosystem services. Final Report, JNCC, Project Code C08–0170-0062, 69 pp

Haines-Young RH, Potschin MB (2010) Proposal for a Common International Classification of Ecosystem Goods and Services (CICES) for Integrated Environmental and Economic Accounting (V1). Report to the European Environment Agency

Haines-Young RH, Potschin M, Kienast F (2012) Indicators of ecosystem service potential at European scales: mapping marginal changes and trade-offs. Ecol Indicators 21:39–53

Harrington R, Anton C, Dawson TP, de Bello F, Feld CK, Haslett JR, Kluvánkova-Oravská T, Kontogianni A, Lavorel S, Luck GW, Rounsevell MDA, Samways MJ, Settele J, Skourtos M, Spangenberg JH, Vandewalle M, Zobel M, Harrison PA (2010) Ecosystem services and biodiversity conservation: concepts and a glossary. Biodiv Conserv. doi:10.1007/s10531-010-9834–9 19, 10: 2843–2853

Haslett J, Berry P, Bela G, Jongman RG, Pataki G, Samways M, Zobel M (2010) Changing conservation strategies in Europe: a framework integrating ecosystem services and dynamics. Biodivers Conserv 19:2963–2977. doi:10.1007/s10531-009-9743-y

Hein L, van Koppen K, de Groot RS, van Ierland EC (2006) Spatial scales, stakeholders and the valuation of ecosystem services. Ecol Econ 57:209–228

Hermann A, Schleifer S, Wrbka T (2011) The concept of ecosystem services regarding landscape research: a review. Living Rev Landscape Res 5. ► http://landscapeesearch.livingreviews.org/Articles/lrlr-2011-1/, 01.04.2011

Humboldt A von (1847) Kosmos. Entwurf einer physischen Weltbeschreibung, Bd 2. Cotta, Stuttgart, Tübingen, 544 pp

Jessel B (1998) Landschaften als Gegenstand von Planung. Schmidt, Berlin

Jessel B, Tschimpke O, Waiser M (2009) Produktivkraft Natur. Hoffmann und Campe, Hamburg

Kettunen M, Bassi S, Gantioler S, ten Brink P (2009) Assessing socioeconomic benefits of Natura 2000–a toolkit for practitioners. Institute for European Environmental Policy, London, Brussels

Kienast F (2010) Landschaftsdienstleistungen: ein taugliches Konzept für Forschung und Praxis? Forum Wiss 2010:7–12

Kontogianni A, Luck GW, Skourtos M (2010) Valuing ecosystem services on the basis of service-providing units: a potential approach to address the 'endpoint problem' and improve stated preference methods. Ecol Econ 69:1479–1487. doi:10.1016/j.ecolecon.2010.02.019

Kremen C (2005) Managing ecosystem services: what do we need to know about their ecology? Ecol Lett 8:468–479

Lavorel S, McIntyre S, Landsberg J, Forbes TDA (1997) Plant functional classifications: from general groups to specific groups based on response to disturbance. Trends Ecol Evol 12:474–478

Lavorel S, Touzard B, Lebreton JD, Clement B (1998) Identifying functional groups for response to disturbance in an abandoned pasture. Acta Oecol 19:227–240

Leibenath M, Gailing L (2012) Semantische Annäherung an die Worte "Landschaft" und "Kulturlandschaft". In: Schenk W, Kühn M, Leibenath M, Tzschaschel S (eds) Suburbane Räume als Kulturlandschaften. ARL, Hannover, pp 58–79
Leser H, Klink HJ (Hrsg) (1988) Handbuch und Kartieranleitung Geoökologische Karte 1: 25000 (KA GÖK 25). Forsch z. dt. Landeskunde 228
Loft L, Lux A (2010) Ecosystem Services–eine Einführung. BiKF–Biodiversität und Klima. Knowledge Flow Paper 6
Luck GW, Daily GC, Ehrlich PR (2003) Population diversity and ecosystem services. Trends Ecol Evol 18:331–336
Luck GW, Harrington R, Harrison PA, Kremen C, Berry PM, Bugter R, Dawson TP, de Bello F, Diaz S, Feld CK, Haslett JR, Hering D, Kontogianni A, Lavorel S, Rounsevell M, Samways MJ, Sandin L, Settele J, Sykes MT, van den Hove S, Vandewalle M, Zobel M (2009) Quantifying the contribution of organisms to the provision of ecosystem services. BioScience 59:223–235
Lütz M, Bastian O (2000) Vom Landschaftsplan zum Bewirtschaftungsentwurf. Z Kulturtech Landesentwick 41:259–266
Lyytimäki J, Sipilä M (2009) Hopping on one leg–The challenge of ecosystem disservices for urban green management. Urban For Urban Green 8:309–315
Maarel E van der, Dauvellier PJ (1978) Naar een globaal ecologisch model voor de ruimlijke entwikkeling van Niederland. Studierapp. Rijksplanologische Dienst, Den Haag, 9 p
Mäler KG, Aniyar S, Jansson A (2008) Accounting for ecosystem services as a way to understand the requirements for sustainable development. Proc Natl Acad Sci USA 105:9501–9506
Maltby E (Hrsg) (2009) Functional assessment of wetlands: towards evaluation of ecosystem services. Woodhead Publishing, Cambridge
Mannsfeld K (1979) Die Beurteilung von Naturraumpotentialen als Aufgabe der geographischen Landschaftsforschung. Petermanns Geogr Mitt 123:2–6
Matzdorf B, Reutter M, Hübner C (2010) Gutachten-Vorstudie: Bewertung der Ökosystemdienstleistungen von HNV-Grünland (High Nature Value Grassland)–Abschlussbericht im Auftrag des Bundesamtes für Naturschutz, Bonn (z2.zalf.de/oa/46452aad -17c1-45ed -82b6-fd9ba62fc9c2.pdf)
MEA–Millennium Ecosystem Assessment (2003) Ecosystems and human wellbeing: a framework for assessment. Island Press, Washington DC
MEA–Millennium Ecosystem Assessment (2005) Ecosystems and human wellbeing. Synthesis. Island Press, Washington DC
Müller B (2005) Stichwort "Raumwissenschaft". In: Handwörterbuch der Raumordnung. 4. Aufl. ARL, Hannover, S 906–911
Müller F, Burkhard B (2007) An ecosystem based framework to link landscape structures, functions and services. In: Mander Ü, Wiggering H, Helming K (eds) Multifunctional land use–meeting future demands for landscape goods and services. Springer, Berlin, Heidelberg, pp 37–64
Naveh Z (2010) Ecosystem and landscapes–a critical comparative appraisal. J Landsc Ecol 3:64–81
Naveh Z, Lieberman AS (1994) Landscape ecology. Theory and application. 2nd edn. Springer, Berlin
Neef E (1963) Dimensionen geographischer Betrachtungen. Forsch Fortsch 37:361–363
Neef E (1966) Zur Frage des gebietswirtschaftlichen Potentials. Forsch Fortschr 40:65–70
Neßhöver C, Beck S, Born W, Dziock S, Görg C, Hansjürgens B, Jax K, Köck W, Rauschmeyer F, Ring I, Schmidt-Loske K, Unnerstall H, Wittmer H, Henle K (2007) Das Millennium Ecosystem Assessment–eine deutsche Perspektive. Nat Landsch 82:262–267
Niemann E (1977) Eine Methode zur Erarbeitung der Funktionsleistungsgrade von Landschaftselementen. Arch Nat Schutz Landsch Forsch 17:119–158
Noss RF (2001) Beyond Kyoto: forest management in a time of rapid climate change. Conserv Biol 15:578–590
Odum EP (1971) Fundamentals of ecology, 3. edn. Saunders, Philadelphia, 574 pp
OECD (2008) Strategic environmental assessment and ecosystem services. DAC Network on Environment and Development Co-operation (ENVIRONET), Paris
O'Neill RV (2001) It is time to bury the ecosystem concept? Ecology 82:3275–3284
O'Neill RV, DeAngelis DL, Waide JB, Allen TFH (1986) A hierarchical concept of ecosystems. Princeton University Press, Princeton
van Oudenhoven APE, Petz K, Alkemade R, Hein L, de Groot RS (2012) Framework for systematic indicator selection to assess effects of land management on ecosystem services. Ecol Indicators 21:110–122
Peters W (2009) Biomassepotenziale aus der Landschaftspflege in Sachsen. FuE-Bericht im Auftrag des Sächsischen Landesamtes für Umwelt. Landwirtschaft und Geologie (unpubl.), Bosch & Partner, Berlin
Pfisterer AB, Balvanera P, Buchmann N, He JS, Nakashizuka T, Raffaelli D, Schmid B (2005) The role of biodiversity for ecosystem services: current knowledge. Institute of Environmental Sciences, University of Zurich
Plummer M (2009) Assessing benefit transfer for the valuation of ecosystem services. Front Ecol Environ 7:38–45
Ring I, Hansjürgens B, Elmqvist T, Wittmer H, Sukhdev P (2010) Challenges in framing the economics of ecosystems and biodiversity: the TEEB initiative. Curr Opin Environ Sustain 2:15–26
Rounsevell DA, Dawson TP, Harrison PA (2010) A conceptual framework to assess the effects of environmental change on ecosystem services. Biodivers Conserv 19:2823–2842
Schenk W, Overbeck G (2012) Suburbane Räume als Kulturlandschaften–Einführung. In: Schenk W, Kühn M, Leibenath M, Tzschaschel S (eds) Suburbane Räume als Kulturlandschaften. ARL, Hannover, 1–12 pp

Spangenberg JH, Settele J (2010) Precisely incorrect? Monetising the value of ecosystem services. Ecol Complex 7:327–337

SRU (2007) Klimaschutz durch Biomasse. Sondergutachten des Sachverständigenrates für Umweltfragen (SRU). Schmidt, Berlin

Staub C, Ott W et al (2011) Indikatoren für Ökosystemleistungen: Systematik, Methodik und Umsetzungsempfehlungen für eine wohlfahrtsbezogene Umweltberichterstattung. Bundesamt für Umwelt, Bern. Umwelt-Wissen 1102, 106 S

Syrbe R-U, Walz U (2012) Spatial indicators for the assessment of ecosystem services: Providing, benefiting and connecting areas and landscape metrics. Ecol Indic 21:80–88

Tacconi L (2000) Decentralization, forests and livelihoods: theory and narrative. Global Environ Chang 17:338–348

Tansley AG (1935) The use and abuse of vegetational concepts and terms. Ecology 16:284–307

TEEB–The economics of Ecosystems and Biodiversity (2009) An interim report. Europ. Comm., Brussels. ► www.teebweb.org. Assessed 20 June 2013

TEEB (2010) The economics of ecosystems and biodiversity: ecological and economic foundations. Kumar P (ed) Earthscan, London, Washington

Termorshuizen JW, Opdam P (2009) Landscape services as a bridge between landscape ecology and sustainable development. Ecol Soc 16(4):17. ► http://dx.doi.org/10.5751/ES -04191–160417

Territoriale Agenda der Europäischen Union (2007) Für ein wettbewerbsfähigeres nachhaltiges Europa der vielfältigen Regionen. Angenommen anlässlich des Informellen Ministertreffens zur Stadtentwicklung und zum territorialen Zusammenhalt in Leipzig am 24./25. Mai 2007. ► http://www.bmvbs.de/SharedDocs/DE/Artikel/SW/territoriale-agenda-der-europaeischen-union.html. Accessed 10 July 2012

UKNEA–UK National Ecosystem Assessment (2011) Synthesis of the key findings. Information Press, Oxford

Vandewalle M, Sykes MT, Harrison PA, Luck GW, Berry P, Bugter R, Dawson TP, Feld CK, Harrington R, Haslett JR, Hering D, Jones KB, Jongman R, Lavorel S, Martins da Silva P, Moora M, Paterson J, Rounsevell MDA, Sandin L, Settele J, Sousa JP, Zobel M (2008) Review paper on concepts of dynamic ecosystems and their services. RUBICODE Deliverable D2.1. ► www.rubicode.net/rubicode/RUBICODE_e-conference_report.pdf. Accessed 5 Feb 2009

Wallace KJ (2007) Classification of ecosystem services: problems and solutions. Biol Conserv 139:235–246

Wallace KJ (2008) Ecosystem services: multiple classifications or confusion? Biol Conserv 141:353–354

Walz U (2011) Landscape structure, landscape metrics and biodiversity. Living Rev Landsc Res 5:3–23

Wascher DM (Hrsg) (2005) European landscape character areas–typologies, cartography and indicators for the assessment of sustainable landscapes. Final Project Report. Alterra Report 1254

Wende W, Kästner A, Bastian O, Walz U, Blum A, Oertel H (2011) Analyse des ehrenamtlichen und privaten Naturschutzes in Sachsen. Abschlussbericht im Auftrag des Sächsischen Staatsministeriums für Umwelt und Landwirtschaft. ► www.umwelt.sachsen.de/umwelt/download/Abschlussbericht_Ehrenamt_01_09_2011_mit_Anlagen.pdf. Accessed 4 May 2012

Wende W, Wojtkiewicz W, Marschall I, Heiland S, Lipp T, Reinke M, Schaal P, Schmidt C (2012) Putting the plan into practice: implementation of proposals for measures of local landscape plans. Landscape Res 37: 483–500 doi:10.1080/01426397.2011.592575

Wiegleb G (1997) Leitbildmethode und naturschutzfachliche Bewertung. Z Ökol Nat Schutz 6:43–62

Willemen L (2010) Mapping and modeling multifunctional landscapes. PhD- Thesis, Wageningen, 152 pp

WFD–Water Framework Directive (2000) Directive 2000/60/EG of the European Parliament and of the Council establishing a framework for the Community action in the field of water policy.- published in the *Official Journal* (OJ L 327) on 22 December 2000

Ascertainment and Assessment of ES

K. Grunewald, O. Bastian (eds.), *Ecosystem Services – Concept, Methods and Case Studies*,
DOI 10.1007/978-3-662-44143-5_4, © Springer-Verlag Berlin Heidelberg 2015

4.1 Indicators and Quantification Approaches[1]

B. Burkhard, F. Müller

4.1.1 Introduction

4

The need for applications and tools of the–frequently mainly conceptually used–ecosystem service (ES) ideas has become more and more obvious during the last years (Daily et al. 2009). Practical applications are necessary to further develop and improve the conceptual base of ES on the one hand. On the other, tools for environmental and resource management are needed in order to further establish ES in decision-making processes (Kienast et al. 2009). The recognition and the appropriate quantification of ES are fundamentals for their valuation, independently whether the valuation is conducted with biophysical, social or economic methods. Their application and integration is one of the biggest challenges of contemporary ES science (Wallace 2007).

The supply of ES is based on geo-biophysical structures and processes, which are changing in intensity as well as in spatial and temporal distribution. Anthropogenic impacts, especially land-use and land-cover changes or climatic variations are among the major factors determining the qualities and quantities of ES supply. Land-use patterns and changes in land cover can be surveyed, spatially analysed and regionally assessed. They deliver direct measures for human activities (Riitters et al. 2000) and clearly demonstrate the relations between ES supply and demand (Burkhard et al. 2012). Spatially explicit identification and mapping of ES distributions and the analysis of their spatio-temporal dynamics therefore enable the aggregation of highly complex information. The respective ES visualisations can support decision-makers in the environmental sector by providing powerful tools to support sustainable landscape planning and ES trade-off assessments (Swetnam et al. 2010). Spatially explicit ES quantification and mapping have therefore been named as one of the key requirements for the implementation of the ES concept in environmental institutions and decision-making processes (Daily and Matson 2008).

One key problem of each ES quantification is, besides the difficult and comprehensive data acquisition, the selection of an ES categorisation system which is appropriate for the specific study region and the particular research question. Most of the currently available ES classification systems (e.g. de Groot et al. 2010a; Wallace 2007) distinguish the three classes with regulating ES, provisioning ES and cultural ES. Some authors additionally include *habitat services* (de Groot et al. 2010a; TEEB 2010). Habitat services are, however, often assigned to the category of ecosystem functions, which in the Millennium Ecosystem Assessment (MEA 2005a) were called *supporting ecosystem services*. Many ecosystem functions or habitat properties do not deliver direct or final ES. Therefore, the distinction between ecosystem functions and ES has become more common and accepted. This distinction also proved to be advantageous for the avoidance of double counting of closely correlating functions and services, for example in monetary valuations.

Numerous methods and tools for the characterisation of ecosystem functions and services in landscapes have been developed especially within the last 10 years. Additionally, existing methods and data collection programmes are ready to be integrated in the ES concept due to their thematic diversity (e.g. monitoring within the long-term ecological research (LTER) network; Müller et al. 2010). They include measurements, monitoring programmes, mapping activities, expert interviews, statistical analyses, model applications or transfer-functions (de Groot et al. 2010b). Natural structures and processes (e.g. flows of energy, matter and water) are central in biophysical assessments. These approaches are different from monetary valuations, where the actual assessment of values is carried out by monetisation. Monetary ES approaches such as cost-benefit analyses (CBA) or willingness-to-pay (WTP) surveys are applicable and well-established concepts (Farber et al. 2002). However, results are often disappointing especially for nonmarket goods and services such as many regulating ES, ecosystem functions or biodiversity characteristics (Ludwig 2000; Spangenberg and Settele 2010).

1 Section 4.1 is in main parts based on the paper of Burkhard et al. (2012).

Suitable ES indicators are needed for all quantification approaches. These indicators have to be quantifiable, sensitive for land-use changes, temporally and spatially explicit and scalable (van Oudenhoven et al. 2012). Indicators are tools for communication, enabling the reduction of information about highly complex human-environmental systems. After Wiggering and Müller (2004), indicators in general are variables delivering aggregated information about certain phenomena. They are selected to support specific management purposes by providing integrating synoptic values, depicting not directly accessible qualities, quantities, states or interactions (Dale and Beyeler 2001; Turnhout et al. 2007; Niemeijer and de Groot 2008).

4.1.2 Ecosystem Service Supply and Demand Assessment at the Landscape Scale–the 'Matrix'

Different landscapes can be characterised by different ecosystem structures, functions and consequently by varying capacities to supply ES (Burkhard et al. 2009), depending on the natural settings as well as human activities (e.g. land use) within the research area. Different land-use patterns, heterogeneous population distributions as well as multiple ecological and socio-economic conditions cause varying demands for ES (► Fig. 3.2).

In this chapter, a method for the assessment of different land-cover types' capacities to support ecosystem functions (assessed based on the ecological integrity concept and respective indicators for ecosystem structures and processes; for detailed information see Müller 2005; Burkhard et al. 2009, 2012), to supply multiple ES and to identify demands for ES will be shortly introduced. The method has been applied in different case studies, for example for the assessment of ES in boreal forest landscapes in northern Finland (Vihervaara et al. 2010), in urban–rural regions in central eastern Germany (Kroll et al. 2012) or for the calculation of flood regulation capacities in a Bulgarian mountainous region (Nedkov and Burkhard 2012).

The approach is based on an *assessment matrix*, which links relative and mainly non-monetary ES supply capacities or ES demand intensities to different geospatial units (e.g. different land-cover types). Based on this interrelation analysis, resulting ecosystem function and ES scores can be visualised in maps. Differentiations between ES supply and demand but also between ES potential and *de facto* flows (ES actually used by humans) are needed (see below). Supply and demand of/for different ecosystem goods and services are often spatially and temporally decoupled and managed by transport, trade and storage opportunities in today's globalised world. Nevertheless, calculations of these two variables deliver data that are highly relevant for ES budget assessments for specific spatial or temporal units. Self-sufficiency rates and ES flows within and between regions can be calculated on this basis. Ecosystem functions and several regulating ES such as nutrient regulation, erosion control and natural hazard protection are exceptions. They are normally not transportable and therefore, a physical connection between the service providing unit (SPU) and service benefiting/demand area (SBA) must exist (Nedkov and Burkhard 2012; Syrbe and Walz 2012; ► Sect. 3.3).

Such information, especially in a regionalised form, and the related ecological and socio-economic data are highly relevant for environmental management and for ES-based landscape planning. Thus, requests for appropriate tools are numerous (Kienast et al. 2009). When assessing the *potential* of a landscape, a land-use type or an ecosystem, usually the (hypothetical) maximum of ES supply under the given conditions is being assessed. Often it is not considered whether there is a human use of these ES or not. *Flows* of ES on the contrary describe the capacity of a defined spatial unit to supply a specific ES set (ES bundle) actually used by humans within a given time period (after Burkhard et al. 2012; see Box). This distinction becomes relevant for certain ES, for example when assessing protected ecosystems. These systems undoubtedly supply numerous goods and services. However, e.g. in the case of core zones in national parks, where any human activity may be prohibited, many of these ES (e.g. timber, game) cannot be used. Of course, ecosystem functions, such as nutrient cycling or biodiversity, take place anyway. They provide positive effects on ecological integrity within the protected area itself, but often also on adjacent ecosystems.

Conceptual Background for ES Supply and Demand (after Burkhard et al. 2012)

- *ES supply* refers to the capacity of a particular area to provide a specific bundle of ecosystem goods and services within a given time period. For detailed analyses, a differentiation between ES potentials and actual ES flows is needed.
- *ES demand* is the sum of all ecosystem goods and services currently consumed or used in a particular area over a given time period. Up to now, demands are assessed not considering where ecosystem services actually are provided. These detailed provision patterns are part of the
- *ES footprint* which (closely related to the ecological footprint concept; Rees 1992) calculates the area needed to generate particular ecosystem goods and services demanded by humans in a certain area and a certain time. Different aspects of ecosystem service generation are considered (production capacities, waste absorption, etc.) for assessing the ES footprint.

For many regulating ES, it can be assumed that ES potentials and flows are comparable (▶ Sect. 2.1).

Ecosystem functions, ES supply, ES demand and ES budgets in different land-use types can be assessed by the help of *ES matrices*. The first matrix in ◘ Fig. 4.1 contains ecosystem functions (ecological integrity) and ES on the x-axis. The geospatial units (here CORINE land-cover types; EEA 1994) are placed on the y-axis (after Burkhard et al. 2009, 2012). All relevant ES capacity scores are entered, using a relative scale between 0 (equivalent to no relevant capacity to support the respective ecosystem function or to supply the respective ES), 1 (low relevant capacity), 2 (relevant capacity), 3 (medium relevant capacity), 4 (high relevant capacity) and 5 (maximum capacity in the study area) at the intersections. Based on the 44 different CORINE land-cover classes and 39 ecosystem functions and services, altogether 1716 capacity scores have to be given (◘ Fig. 4.1). Due to this high number of scores needed and the related high assessment efforts, existing databases or expert evaluations need to be harnessed. These data can successively be checked and replaced by more exact information resulting from modelling, measurement, monitoring or in-depth interviews (Burkhard et al. 2009).

The matrix in ◘ Fig. 4.1 shows clear patterns of ES capacity distributions across the different land-cover types. Especially, the forest land-cover types (including broad-leaved, coniferous and mixed forests) show high scores for a multitude of ES. Such multifunctionality is typical for forest ecosystems. Also the other generally more natural land-cover types such as natural grasslands, wetlands and water bodies are characterised by high ES capacities. Strongly anthropogenically influenced ecosystems, such as urban fabrics, industrial or commercial units and transport units (in the upper part of the matrix), show comparably low ES capacities. Of course, these areas also supply ES, but in comparison with the other land-cover types, their ES supply is rather low (▶ Sect. 6.4).

The whole ES concept is a highly anthropocentric approach. Fisher et al. (2009) defined that only those services with a clear benefit to human societies can be denoted as ES. Services without direct human benefits should be termed as ecosystem functions or intermediate services. Thus, a societal demand should be identifiable for all individual ES. Data about actual anthropogenic uses of each ES are needed for their assessment (see definitions in Box 1). Major parts of this information can be derived from statistics, modelling, ecological and socio-economic monitoring or from interviews. ◘ Figure 4.2 shows a respective matrix, which, comparable to the ES supply matrix (◘ Fig. 4.1), provides exemplary information about the ES demands within the different CORINE land-cover classes. The y-axis contains regulating, provisioning and cultural ES. The ecological integrity variables are not relevant here because they (per definition) do not provide direct benefits to humans. The scores were given in a similar manner as in the ES supply matrix; 0 (light pink) denotes no relevant human demand within the particular land-cover type and 5 (dark red) illustrates maximum demand.

◘ Figure 4.2 clearly shows that the overall highest demands for manifold ES are located within the

Ecological integrity

	Land-cover type	1 Exergy capture	2 Entropy production	3 Storage capacity	4 Cycling & Nutrient loss reduction	5 Biotic water flows	6 Metabolic efficiency	7 Heterogeneity	8 Biodiversity
1	Continuous urban fabric	0	0	0	0	0	0	0	0
2	Discontinuous urban fabric	1	1	1	1	1	1	1	1
3	Industrial or commercial units	0	0	0	0	0	0	1	1
4	Road and rail networks	0	0	0	0	0	0	2	2
5	Port areas	0	0	0	0	0	0	1	1
6	Airports	1	1	0	2	1	1	1	1
7	Mineral extraction sites	0	0	0	0	0	0	2	2
8	Dump sites	0	0	5	0	0	0	2	1
9	Construction sites	0	0	0	0	0	0	2	1
10	Green urban areas	4	3	2	3	2	1	3	3
11	Sport and leisure facilities	4	3	2	3	2	1	2	2
12	Nonirrigated arable land	5	4	4	1	3	4	3	2
13	Permanently irrigated land	5	4	3	1	5	2	3	2
14	Ricefields	5	4	3	1	5	1	3	2
15	Vineyards	3	2	2	0	3	1	3	2
16	Fruit trees and berries	3	2	3	2	4	2	4	3
17	Olive groves	3	2	3	1	3	2	3	2
18	Pastures	5	5	4	2	4	5	2	2
19	Annual and permanent crops	4	3	3	2	3	2	2	2
20	Complex cultivation patterns	4	3	3	1	3	2	4	3
21	Agriculture & Natural vegetation	3	2	3	2	3	2	3	3
22	Agro-forestry areas	4	3	4	4	4	3	4	4
23	Broad-leaved forest	5	4	5	5	5	4	3	4
24	Coniferous forest	5	4	5	5	4	4	3	4
25	Mixed forest	5	4	5	5	5	4	3	5
26	Natural grassland	4	3	5	5	4	4	3	5
27	Moors and heathland	4	3	5	5	4	5	3	4
28	Sclerophyllous vegetation	3	2	2	4	2	3	3	4
29	Transitional woodland shrub	3	2	2	4	2	3	3	4
30	Beaches, dunes and sand plains	1	0	1	0	1	1	3	3
31	Bare rock	0	0	0	0	0	0	3	3
32	Sparsely vegetated areas	1	1	1	1	1	0	2	3
33	Burnt areas	0	0	3	0	0	0	2	1
34	Glaciers and perpetual snow	0	0	0	0	0	0	2	1
35	Inland marshes	4	3	5	3	4	4	3	2
36	Peatbogs	4	3	5	5	4	4	3	4
37	Salt marshes	3	2	5	3	4	3	2	3
38	Salines	0	0	0	0	0	0	1	1
39	Intertidal flats	1	1	1	4	0	2	2	3
40	Water courses	3	1	1	3	0	3	4	4
41	Water bodies	4	2	4	3	0	4	4	4
42	Coastal lagoons	5	4	4	3	0	5	4	4
43	Estuaries	5	4	2	3	0	5	3	3
44	Sea and ocean	3	2	1	4	0	3	2	2

Regulating services

	Land-cover type	9 Global climate regulation	10 Local climate regulation	11 Air quality Regulation	12 Water flow regulation	13 Water purification	14 Nutrient regulation	15 Erosion regulation	16 Natural hazard regulation	17 Pollination *	18 Pest and disease control *	19 Regulation of waste *
1	Continuous urban fabric	0	0	0	0	0	0	0	0	0	0	0
2	Discontinuous urban fabric	0	0	0	0	0	0	0	0	0	0	0
3	Industrial or commercial units	0	0	0	0	0	0	0	0	0	0	0
4	Road and rail networks	0	0	0	0	0	0	0	0	0	0	0
5	Port areas	0	0	0	0	0	0	0	3	0	0	0
6	Airports	0	0	0	0	0	0	0	0	0	0	0
7	Mineral extraction sites	0	0	0	0	0	0	0	0	0	0	0
8	Dump sites	0	0	0	0	0	0	0	0	0	0	0
9	Construction sites	0	0	0	0	0	0	0	0	0	0	0
10	Green urban areas	1	2	1	2	1	1	2	0	1	1	1
11	Sport and leisure facilities	1	1	1	2	1	1	1	0	1	1	1
12	Nonirrigated arable land	1	2	0	1	0	0	0	1	0	2	2
13	Permanently irrigated land	1	3	0	0	0	0	0	1	0	1	2
14	Ricefields	0	2	0	2	0	0	0	0	0	1	1
15	Vineyards	1	1	0	1	0	0	0	0	0	1	1
16	Fruit trees and berries	2	2	2	2	1	1	2	2	5	3	2
17	Olive groves	1	1	1	1	1	1	1	0	0	3	2
18	Pastures	1	1	0	1	0	0	4	1	0	2	4
19	Annual and permanent crops	1	2	1	1	0	0	1	1	0	2	2
20	Complex cultivation patterns	1	2	0	1	0	0	0	1	0	3	2
21	Agriculture & Natural vegetation	2	3	1	2	1	0	3	1	0	3	2
22	Agro-forestry areas	1	2	1	1	1	1	2	1	3	3	3
23	Broad-leaved forest	4	5	5	2	5	5	5	3	5	4	4
24	Coniferous forest	4	5	5	2	5	5	5	3	5	4	4
25	Mixed forest	4	5	5	2	5	5	5	3	5	5	5
26	Natural grassland	3	2	0	1	5	5	5	1	0	1	2
27	Moors and heathland	3	4	0	2	4	3	0	2	2	2	3
28	Sclerophyllous vegetation	1	2	0	1	0	0	0	1	2	2	3
29	Transitional woodland shrub	0	1	0	0	0	0	0	0	2	2	3
30	Beaches, dunes and sand plains	0	0	0	1	0	0	0	5	0	1	1
31	Bare rock	0	0	0	1	1	0	0	1	0	0	0
32	Sparsely vegetated areas	0	1	0	1	0	0	0	1	0	1	1
33	Burnt areas	0	1	0	0	0	0	0	0	0	0	1
34	Glaciers and perpetual snow	3	3	0	4	0	0	0	0	0	1	1
35	Inland marshes	2	2	0	2	0	4	0	4	0	2	3
36	Peatbogs	5	4	0	3	4	3	0	3	2	3	4
37	Salt marshes	0	1	0	0	0	2	0	5	0	2	2
38	Salines	0	2	0	0	0	0	0	0	0	1	1
39	Intertidal flats	0	1	0	0	0	1	0	5	0	2	3
40	Water courses	0	1	0	1	3	3	0	2	0	3	5
41	Water bodies	1	2	0	2	0	1	0	1	0	3	5
42	Coastal lagoons	0	1	0	0	0	0	0	4	0	3	5
43	Estuaries	0	0	0	0	3	3	0	3	0	4	5
44	Sea and ocean	5	3	0	0	0	5	0	0	0	3	5

Provisioning services

	Land-cover type	20 Crops	21 Biomass for energy	22 Fodder **	23 Livestock (domestic)	24 Fibre	25 Timber	26 Wood Fuel	27 Fish, seafood & edible algae	28 Aquaculture	29 Wild foods & resources	30 Biochemicals & Medicine	31 Freshwater	32 Mineral resources ***	33 Abiotic energy sources ***
1	Continuous urban fabric	0	1	0	0	0	0	0	0	0	0	0	0	0	1
2	Discontinuous urban fabric	1	1	1	0	0	0	0	0	0	1	0	0	0	1
3	Industrial or commercial units	0	1	0	0	0	0	0	0	0	0	0	0	0	1
4	Road and rail networks	0	0	0	0	0	0	0	0	0	0	0	0	0	0
5	Port areas	0	0	0	0	0	0	0	0	0	0	0	0	0	0
6	Airports	0	0	1	0	0	0	0	0	0	0	0	0	0	0
7	Mineral extraction sites	0	0	0	0	0	0	0	0	0	0	0	0	5	5
8	Dump sites	0	1	0	0	0	0	0	0	0	0	0	0	0	0
9	Construction sites	0	0	0	0	0	0	0	0	0	0	0	0	1	0
10	Green urban areas	0	0	0	0	0	0	1	0	0	1	0	0	0	0
11	Sport and leisure facilities	0	0	0	0	0	0	0	0	0	0	0	0	0	0
12	Nonirrigated arable land	5	2	3	0	5	0	0	0	0	0	1	0	0	0
13	Permanently irrigated land	5	1	2	0	2	0	0	0	0	0	1	0	0	0
14	Ricefields	5	0	2	0	0	0	0	0	0	0	0	0	0	0
15	Vineyards	4	1	0	0	0	0	1	0	0	0	0	0	0	0
16	Fruit trees and berries	5	1	0	0	0	4	4	0	0	0	0	0	0	0
17	Olive groves	4	1	0	0	0	4	4	0	0	0	0	0	0	0
18	Pastures	0	1	5	5	0	0	0	0	0	0	0	0	0	0
19	Annual and permanent crops	5	1	5	5	5	0	0	0	0	0	1	0	0	0
20	Complex cultivation patterns	4	1	3	0	4	0	0	0	0	0	2	0	0	0
21	Agriculture & Natural vegetation	3	2	2	3	4	3	3	0	0	3	1	0	0	1
22	Agro-forestry areas	3	2	2	3	2	3	3	0	0	0	0	0	0	0
23	Broad-leaved forest	0	1	1	0	0	5	5	0	0	5	5	0	0	0
24	Coniferous forest	0	1	1	0	0	5	5	0	0	5	5	0	0	0
25	Mixed forest	0	1	1	0	0	5	5	0	0	5	5	0	0	0
26	Natural grassland	0	0	1	3	0	0	0	0	0	5	0	0	0	0
27	Moors and heathland	0	2	0	2	0	0	2	0	0	1	0	0	0	0
28	Sclerophyllous vegetation	0	0	0	2	0	0	2	0	0	1	3	0	0	0
29	Transitional woodland shrub	0	1	0	2	1	0	2	0	0	1	0	0	0	0
30	Beaches, dunes and sand plains	0	1	0	0	0	0	0	0	0	0	0	0	1	0
31	Bare rock	0	0	0	0	0	0	0	0	0	0	0	0	1	0
32	Sparsely vegetated areas	0	0	0	0	0	0	0	0	0	0	0	0	0	0
33	Burnt areas	0	0	0	0	0	0	0	0	0	0	0	0	0	0
34	Glaciers and perpetual snow	0	0	0	0	0	0	0	0	0	0	0	5	0	0
35	Inland marshes	0	0	5	2	0	0	0	0	0	0	0	0	0	0
36	Peatbogs	0	2	0	0	0	0	0	0	0	0	0	0	0	0
37	Salt marshes	0	0	0	2	0	0	0	0	0	0	0	0	0	0
38	Salines	0	0	0	0	0	0	0	0	0	0	0	0	0	0
39	Intertidal flats	0	1	0	0	0	0	0	0	0	0	0	0	0	0
40	Water courses	0	3	0	0	0	0	0	3	0	4	0	5	0	5
41	Water bodies	0	0	0	0	0	0	0	3	5	4	0	5	0	5
42	Coastal lagoons	0	1	0	0	0	0	0	4	5	4	0	0	0	0
43	Estuaries	0	2	0	0	0	0	0	5	5	4	0	0	0	0
44	Sea and ocean	0	3	1	0	0	0	0	5	5	0	0	0	1	3

Cultural services

	Land-cover type	34 Recreation & Tourism	35 Landscape aesthetics & Inspiration	36 Knowledge systems	37 Religious & Spiritual experience	38 Cultural heritage & Cultural diversity	39 Natural heritage & Natural diversity
1	Continuous urban fabric	3	3	1	5	1	0
2	Discontinuous urban fabric	3	2	1	5	1	0
3	Industrial or commercial units	0	1	0	0	1	0
4	Road and rail networks	0	0	0	0	1	0
5	Port areas	1	2	0	0	1	0
6	Airports	0	0	0	0	1	0
7	Mineral extraction sites	0	0	1	0	0	0
8	Dump sites	0	0	0	0	0	0
9	Construction sites	0	0	0	0	0	0
10	Green urban areas	3	3	1	0	2	0
11	Sport and leisure facilities	5	1	0	0	1	0
12	Nonirrigated arable land	1	1	2	0	3	0
13	Permanently irrigated land	1	1	2	0	3	0
14	Ricefields	1	1	2	0	4	0
15	Vineyards	5	2	2	0	5	0
16	Fruit trees and berries	5	2	2	0	4	0
17	Olive groves	5	2	2	0	4	0
18	Pastures	3	2	2	0	3	0
19	Annual and permanent crops	1	1	2	0	2	0
20	Complex cultivation patterns	2	2	2	0	3	0
21	Agriculture & Natural vegetation	2	2	3	0	2	3
22	Agro-forestry areas	3	2	2	0	3	0
23	Broad-leaved forest	5	5	5	3	3	5
24	Coniferous forest	5	5	5	3	4	5
25	Mixed forest	5	5	5	3	4	5
26	Natural grassland	3	4	5	1	4	3
27	Moors and heathland	5	4	5	1	2	5
28	Sclerophyllous vegetation	2	3	4	1	2	4
29	Transitional woodland shrub	2	3	4	1	2	2
30	Beaches, dunes and sand plains	5	4	4	0	2	2
31	Bare rock	4	3	4	0	2	0
32	Sparsely vegetated areas	0	1	4	0	2	0
33	Burnt areas	0	0	5	0	0	0
34	Glaciers and perpetual snow	5	5	3	2	0	0
35	Inland marshes	0	2	4	0	2	0
36	Peatbogs	4	2	3	0	2	4
37	Salt marshes	3	2	3	0	2	0
38	Salines	2	2	3	0	2	0
39	Intertidal flats	4	2	3	0	2	0
40	Water courses	4	4	4	0	3	5
41	Water bodies	5	4	4	0	3	4
42	Coastal lagoons	4	4	4	0	2	4
43	Estuaries	4	5	4	0	2	3
44	Sea and ocean	4	5	4	0	4	2

* These ecosystem services are named because they can be of high importance in some ecosystems although the potential of double-counting must be noted.
** Potential double-counting when fodder is used for feeding on the same farm.
*** These services are often not counted as ecosystem services; but they can be of high importance for policy decisions, land-use management strategies and scenarios.

Fig. 4.1 Land-cover types (*y-axis*) and ecosystem functions and services (*x-axis*) illustrating the capacities of different land-cover types to support ecosystem functions and to supply ES on a scale from 0 (no relevant capacity; *pink*) to 5 (maximum relevant capacity; *dark green*); exemplarily assessed for a central European 'normal landscape' (after Burkhard et al. 2009, 2012)

highly human-modified land-cover types in the upper part of the matrix. Urban areas as well as industrial and commercial areas are the land-cover types with the highest demand scores. It also becomes obvious that in the more natural land-cover types (lower part of the matrix), generally lower demands for ES can be found. This can of course be justified by the lower population numbers and related lower consumption rates in these areas. Agrarian land-cover types show high demands for regulating ES (e.g. nutrient regulation, water purification, erosion control). Similarly to the ES supply matrix, ES demand maps can also be compiled based on the ES demand matrix.

Taking the information from the ES supply and demand matrices as starting points, sources and sinks for individual ES can be identified. As both components–supply and demand–were normalised to the same relative units (0–5), ES budgets can be calculated by subtracting the ES demand scores from the ES supply scores. And also the resulting ES budget scores can be illustrated in a matrix and

Regulating services

	Land-cover type	9 Global climate regulation	10 Local climate regulation	11 Air quality regulation	12 Water flow regulation	13 Water purification	14 Nutrient regulation	15 Erosion regulation	16 Natural hazard regulation	17 Pollination	18 Pest and disease control	19 Regulation of waste
1	Continuous urban fabric	3	5	5	4	1	1	1	5	3	5	3
2	Discontinuous urban fabric	3	5	5	5	2	2	1	4	4	4	2
3	Industrial or commercial units	5	1	5	4	3	3	1	5	4	3	4
4	Road and rail networks	4	2	4	4	0	0	3	4	1	2	0
5	Port areas	3	2	2	5	3	0	4	5	1	4	3
6	Airports	5	2	4	1	2	1	1	5	0	5	1
7	Mineral extraction sites	0	0	0	2	0	0	4	3	0	0	3
8	Dump sites	2	2	3	0	2	0	0	5	0	3	5
9	Construction sites	0	2	1	2	2	2	2	3	0	1	2
10	Green urban areas	0	2	1	0	0	0	0	2	2	3	0
11	Sport and leisure facilities	0	2	3	0	1	0	0	3	0	3	0
12	Nonirrigated arable land	2	2	1	2	0	3	2	3	3	3	2
13	Permanently irrigated land	2	2	1	2	5	3	2	3	3	3	2
14	Ricefields	4	3	1	5	5	3	5	3	1	3	2
15	Vineyards	2	5	1	0	4	3	5	3	2	3	1
16	Fruit trees and berries	1	2	1	0	2	3	1	3	5	3	1
17	Olive groves	1	2	1	0	2	2	0	3	2	3	1
18	Pastures	3	1	0	1	2	1	0	2	0	1	3
19	Annual and permanent crops	1	1	1	1	2	5	1	2	2	3	1
20	Complex cultivation patterns	1	1	1	1	2	5	1	2	3	3	2
21	Agriculture & natural vegetation	2	1	1	0	2	3	1	1	2	3	1
22	Agro-forestry areas	1	1	1	0	2	3	0	1	2	1	0
23	Broad-leaved forest	0	0	0	0	0	0	0	0	0	0	0
24	Coniferous forest	0	0	0	0	0	0	0	0	0	0	0
25	Mixed forest	0	0	0	0	0	0	0	0	0	0	0
26	Natural grassland	0	0	0	0	0	0	0	0	0	0	0
27	Moors and heathland	0	0	0	0	0	0	0	0	0	0	0
28	Sclerophyllous vegetation	0	0	0	0	0	0	0	0	0	0	0
29	Transitional woodland shrub	0	0	0	0	0	0	0	0	0	0	0
30	Beaches, dunes and sand plains	0	0	0	0	0	0	0	0	0	0	0
31	Bare rock	0	0	0	0	0	0	0	0	0	0	0
32	Sparsely vegetated areas	0	0	0	0	0	0	0	0	0	0	0
33	Burnt areas	0	0	0	0	0	0	0	0	0	0	0
34	Glaciers and perpetual snow	0	0	0	0	0	0	0	0	0	0	0
35	Inland marshes	0	0	0	0	0	0	0	0	0	0	0
36	Peatbogs	0	0	0	0	0	0	0	0	0	0	0
37	Salt marshes	0	0	0	0	0	0	0	0	0	0	0
38	Salines	0	0	0	0	0	0	0	0	0	0	0
39	Intertidal flats	0	0	0	0	0	0	0	0	0	0	0
40	Water courses	0	0	0	0	0	0	0	0	0	0	0
41	Water bodies	0	0	0	0	0	0	0	0	0	0	0
42	Coastal lagoons	0	0	0	0	0	0	0	0	0	0	0
43	Estuaries	0	0	0	0	0	0	0	0	0	0	0
44	Sea and ocean	0	0	0	0	0	0	0	0	0	0	0

Provisioning services

	Land-cover type	20 Crops	21 Biomass for energy	22 Fodder	23 Livestock	24 Fibre	25 Timber	26 Wood fuel	27 Fish, seafood & edible algae	28 Aquaculture	29 Wild foods & resources	30 Biochemicals / medicine	31 Freshwater	32 Mineral resources	33 Abiotic energy sources
1	Continuous urban fabric	5	5	1	5	3	3	2	5	5	5	5	5	4	2
2	Discontinuous urban fabric	4	4	2	4	3	3	3	4	4	4	5	5	3	3
3	Industrial or commercial units	5	5	5	5	5	5	5	4	4	4	5	5	5	5
4	Road and rail networks	0	4	0	0	0	2	0	0	0	0	0	1	2	0
5	Port areas	2	5	2	2	2	5	2	2	2	1	1	3	3	1
6	Airports	2	5	0	2	1	1	0	1	1	1	1	3	2	0
7	Mineral extraction sites	0	3	0	0	1	2	0	0	0	0	0	2	1	0
8	Dump sites	0	1	0	0	0	0	0	0	0	0	0	2	0	0
9	Construction sites	0	4	0	0	4	4	0	0	0	0	0	2	4	0
10	Green urban areas	1	1	0	1	0	0	0	0	0	0	0	2	0	0
11	Sport and leisure facilities	2	3	1	2	1	1	1	2	2	2	3	3	1	1
12	Nonirrigated arable land	1	1	0	0	0	0	0	0	0	0	1	0	0	1
13	Permanently irrigated land	1	2	0	0	0	0	0	0	0	0	1	5	0	1
14	Ricefields	1	2	0	0	0	0	0	0	0	0	1	5	0	0
15	Vineyards	1	2	0	0	1	1	0	0	0	0	2	4	0	0
16	Fruit trees and berries	1	2	0	0	1	1	0	0	0	0	2	3	0	0
17	Olive groves	1	1	0	0	1	0	0	0	0	0	2	1	0	0
18	Pastures	0	1	3	1	0	1	0	0	0	0	1	2	0	2
19	Annual and permanent crops	1	2	0	0	0	0	0	0	0	0	1	1	0	1
20	Complex cultivation patterns	1	2	0	0	0	0	0	0	0	0	1	1	0	0
21	Agriculture & natural vegetation	1	2	0	0	0	0	0	0	0	0	1	2	0	2
22	Agro-forestry areas	1	1	0	0	0	0	0	0	0	0	1	2	0	0
23	Broad-leaved forest	0	0	0	0	0	1	1	0	0	1	0	0	0	0
24	Coniferous forest	0	0	0	0	0	1	1	0	0	1	0	0	0	0
25	Mixed forest	0	0	0	0	0	1	1	0	0	1	0	0	0	0
26	Natural grassland	0	0	0	0	0	0	0	0	0	0	0	0	0	2
27	Moors and heathland	0	0	0	0	0	0	0	0	0	0	0	0	0	2
28	Sclerophyllous vegetation	0	0	0	0	0	0	0	0	0	0	0	0	0	1
29	Transitional woodland shrub	0	0	0	0	0	0	0	0	0	0	0	0	0	0
30	Beaches, dunes and sand plains	0	0	0	0	0	0	0	0	0	0	0	0	0	0
31	Bare rock	0	0	0	0	0	0	0	0	0	0	0	0	0	0
32	Sparsely vegetated areas	0	0	0	0	0	0	0	0	0	0	0	0	0	1
33	Burnt areas	0	0	0	0	0	0	0	0	0	0	0	0	0	0
34	Glaciers and perpetual snow	0	0	0	0	0	0	0	0	0	0	0	0	0	0
35	Inland marshes	0	0	0	0	0	0	0	0	0	0	0	0	0	0
36	Peatbogs	0	0	0	0	0	0	0	0	0	0	0	0	0	0
37	Salt marshes	0	0	0	0	0	0	0	0	0	0	0	0	0	0
38	Salines	0	0	0	0	0	0	0	0	0	0	0	0	0	0
39	Intertidal flats	0	0	0	0	0	0	0	0	0	0	0	0	0	0
40	Water courses	0	0	0	0	0	0	0	1	0	0	0	0	0	3
41	Water bodies	0	0	0	0	0	0	0	1	0	0	0	0	0	0
42	Coastal lagoons	0	0	0	0	0	0	0	1	0	0	0	0	0	0
43	Estuaries	0	0	0	0	0	0	0	1	0	0	0	0	0	0
44	Sea and ocean	0	0	0	0	0	0	0	1	0	0	0	0	0	4

Cultural services

	Land-cover type	34 Recreation& tourism	35 Landscape aesthetics	36 Knowledge systems	37 Religious experience	38 Cultural heritage & diversity	39 Natural heritage & diversity
1	Continuous urban fabric	4	4	3	4	4	2
2	Discontinuous urban fabric	4	4	3	3	2	3
3	Industrial or commercial units	1	1	4	1	3	1
4	Road and rail networks	2	2	1	1	1	0
5	Port areas	2	2	2	1	2	1
6	Airports	1	1	1	1	1	0
7	Mineral extraction sites	0	0	0	0	0	0
8	Dump sites	0	0	0	0	0	0
9	Construction sites	0	0	0	0	0	0
10	Green urban areas	4	4	2	0	2	1
11	Sport and leisure facilities	3	3	1	0	2	0
12	Nonirrigated arable land	0	0	1	0	1	0
13	Permanently irrigated land	0	0	1	0	1	0
14	Ricefields	0	0	2	0	3	0
15	Vineyards	0	0	2	0	3	0
16	Fruit trees and berries	0	0	2	0	2	0
17	Olive groves	0	0	2	0	2	0
18	Pastures	0	0	1	0	1	0
19	Annual and permanent crops	0	0	1	0	1	0
20	Complex cultivation patterns	0	0	1	0	1	0
21	Agriculture & natural vegetation	0	0	1	0	0	0
22	Agro-forestry areas	0	0	0	0	0	0
23	Broad-leaved forest	0	0	0	0	0	0
24	Coniferous forest	0	0	0	0	0	0
25	Mixed forest	0	0	0	0	0	0
26	Natural grassland	0	0	0	0	0	0
27	Moors and heathland	0	0	0	0	0	0
28	Sclerophyllous vegetation	0	0	0	0	0	0
29	Transitional woodland shrub	0	0	0	0	0	0
30	Beaches, dunes and sand plains	1	1	0	0	0	1
31	Bare rock	0	0	0	0	0	0
32	Sparsely vegetated areas	0	0	0	0	0	0
33	Burnt areas	0	0	0	0	0	0
34	Glaciers and perpetual snow	0	0	0	0	0	0
35	Inland marshes	0	0	0	0	0	0
36	Peatbogs	0	0	0	0	0	0
37	Salt marshes	0	0	0	0	0	0
38	Salines	0	0	1	0	1	0
39	Intertidal flats	0	0	0	0	0	0
40	Water courses	0	0	0	0	0	0
41	Water bodies	0	0	0	0	0	0
42	Coastal lagoons	0	0	0	0	0	0
43	Estuaries	0	0	0	0	0	0
44	Sea and ocean	0	0	0	0	0	0

Fig. 4.2 Demand for ecosystem services (*x-axis*) within different land-cover types (*y-axis*) on a scale from 0 (no relevant demand; *light pink*) to 5 (maximum demand; *dark red*); exemplarily assessed for a central European 'normal landscape' (after Burkhard et al. 2012)

in maps. Figure 4.3 shows the ES budget matrix for the different CORINE land-cover types. Each field in the ES budget matrix was calculated based on the scores in the ES supply matrix (Fig. 4.1) and the ES demand matrix (Fig. 4.2). Therefore, the assessment scale ranges from −5 = demand clearly exceeds supply (undersupply), via 0 = demand = supply (neutral budget), to +5 = supply clearly exceeds demand (oversupply). Empty fields indicate land-cover types with neither a relevant ES supply nor a relevant demand for ES.

Figure 4.3 shows a clear pattern of ES undersupply in the regions with high anthropogenic

		Regulating services	9 Global climate regulation	10 ocal climate regulation	11 Air quality regulation	12 Water flow regulation	13 Water purification	14 Nutrient regulation	15 Erosion regulation	16 Natural hazard regulation	17 Pollination	18 Pest and disease control	19 Regulation of waste	Provisioning services	20 Crops	21 Biomass for energy	22 Fodder	23 Livestock	24 Fibre	25 Timber	26 Wood fuel	27 Capture fisheries	28 Aquaculture	29 Wild foods & resources	30 Biochemicals / medicine	31 Freshwater	32 Mineral resources	33 Abiotic energy sources	Cultural services	34 Recreation& tourism	35 Landscape aesthetics	36 Knowledge systems	37 Religious experience	38 Cultural heritage & diversity	39 Natural heritage & diversity
1	Continuous urban fabric		-3	-5	-5	-4	-1	-1	-1	-5	-3	-5	-3		-5	-4	-1	-5	-3	-3	-2	-5	-5	-5	-5	-5	-4	-1		-1	-1	-2	1	-3	-2
2	Discontinuous urban fabric		-3	-5	-5	-5	-2	-2	-1	-4	-4	-4	-2		-3	-3	-1	-4	-3	-3	-3	-4	-4	-3	-5	-5	-3	-2		-1	-2	-2	2	-1	-3
3	Industrial or commercial units		-5	-1	-5	-4	-3	-3	-1	-5	-4	-3	-4		-5	-4	-5	-5	-5	-5	-5	-4	-4	-4	-5	-5	-5	-4		-1	0	-4	-1	-2	-1
4	Road and rail networks		-4	-2	-4	-4	0	0	-3	-4	-1	-2				-4				-2						-1	-2			-2	-2	-1	-1	0	
5	Port areas		-3	-2	-2	-5	-3		-4	-2	-1	-4	-3		-2	-5	-2	-2	-2	-5	-2	-2	-2	-1	-1	-3	-3	-1		-1	0	-2	-1	-1	-1
6	Airports		-5	-2	-4	-1	-2	-1	-1	-5		-5	-1		-2	-5	1	-2	-1	-1	0	-1	-1	-1	-1	-3	-2			-1	-1	-1	-1	0	
7	Mineral extraction sites					-2			-4	-3			-3			-3			-1	-2						-2	4	5				1			
8	Dump sites		-2	-2	-3		-2			-5		-3	-5			0										-2									
9	Construction sites			-2	-1	-2	-2	-2	-2	-3		-1	-2			-4			-4	-4						-2	-3								
10	Green urban areas		1	0	0	2	1	1	2	-2	-1	-2	1		-1	-1		-1			1			1		-2				-1	-1	-1		0	-1
11	Sport and leisure facilities		1	-1	-2	2	0	1	1	-3	1	-2	1		-2	-3	-1	-2	-1	-1	-1	-2	-2	-2	-3	-3	-1	-1		2	-2	-1		-1	
12	Nonirrigated arable land		-1	0	-1	-1		-3	-2	-2	-3	-1	0		4	1	3		5						0			-1		1	1	1		2	
13	Permanently irrigated land		-1	1	-1	-2	-5	-3	-2	-2	-3	-2	0		4	-1	2		2						0	-5		-1		1	1	1		2	
14	Ricefields		-4	-1	-1	-3	-5	-3	-5	-3	-1	-2	-1		4	-2	2								-1	-5				1	1	0		1	
15	Vineyards		-1	-4	-1	1	-4	-3	-5	-3	-2	-2	0		3	-1			-1	-1	1				-2	-4				5	2	0		2	
16	Fruit trees and berries		1	0	1	2	-1	-2	1	-1	0	0	1		4	-1			-1	3	4				-2	-3				5	2	0		2	
17	Olive groves		0	-1	0	1	-1	-1	1	-3	-2	0	1		3	0			-1	4	4				-2	-1				5	2	0		2	
18	Pastures		-2	0		0	-2	-1	4	-1		1	1			0	2	4		-1					-1	-2		-2		3	2	1		2	
19	Annual and permanent crops		0	1	0	0	-2	-5	0	-1	-2	-1	1		4	-1	5	5	5						0	-1		-1		1	1	1		1	
20	Complex cultivation patterns		0	1	-1	0	-2	-5	-1	-1	-3	0	0		3	-1	3		4						1	-1				2	2	1		2	
21	Agriculture & natural vegetation		0	2	0	2	-1	-3	2	0	-2	0	1		2	0	2	3	4	3	3			3	0	-2		-1		2	2	2		2	3
22	Agro-forestry areas		0	1	0	1	-1	-2	2	0	1	2	3		2	1	2	3	2	3	3				-1	-2				3	2	2		3	
23	Broad-leaved forest		4	5	5	2	5	5	5	3	5	4	4			1	1			4	4			4	5					5	5	5	3	3	5
24	Coniferous forest		4	5	5	2	5	5	5	3	5	4	4			1	1			4	4			4	5					5	5	5	3	4	5
25	Mixed forest		4	5	5	2	5	5	5	3	5	5	5			1	1			4	4			4	5					5	5	5	3	4	5
26	Natural grassland		3	2		1	5	5	5	1		1	2				1	3						5				-2		3	4	5	1	4	3
27	Moors and heathland		3	4		2	4	3		2	2	2	3			2		2			2			1				-2		5	4	5	1	2	5
28	Sclerophyllous vegetation		1	2		1				1	2	2	3					2			2			1	3			-1		2	3	4	1	2	4
29	Transitional woodland shrub			1							2	2	3			1		2	1		2			1						2	3	4	1	2	2
30	Beaches, dunes and sand plains					1				5		1	1			1											1			4	3	4		2	1
31	Bare rock					1	1			1																	1			4	3	4		2	
32	Sparsely vegetated areas			1		1				1		1	1															-1			1	4		2	
33	Burnt areas			1									1																			5			
34	Glaciers and perpetual snow		3	3		4						1	1													5				5	5	3	2		
35	Inland marshes		2	2		2		4		4		2	3				5	2													2	4		2	
36	Peatbogs		5	4		3	4	3		3	2	3	4			2														4	2	3		2	4
37	Salt marshes			1				2		5		2	2					2												3	2	3		2	
38	Salines			2								1	1																	2	2	2		1	
39	Intertidal flats			1				1		5		2	3			1														4	2	3		2	
40	Water courses			1		1	3	3		2		3	5			3						2		4		5		2		4	4	4		3	5
41	Water bodies		1	2		2		1		1		3	5									2	5	4		5		5		5	4	4		3	4
42	Coastal lagoons			1						4		3	5			1						3	5	4						4	4	4		2	4
43	Estuaries						3	3		3		4	5			2						4	5	4						4	5	4		2	3
44	Sea and ocean		5	3				5				3	5			3	1					4	5				1	-1		4	5	4		4	2

Fig. 4.3 Ecosystem service supply-demand matrix showing budgets in the different land-cover types; based on matrices in Figs. 4.1 and 4.2. Scale from −5 (*dark red*) = demand clearly exceeds supply = undersupply; via 0 (*pink*) = demand = supply = neutral budget; to 5 (*dark green*) = supply clearly exceeds demand = oversupply. Empty fields indicate land cover types with neither a relevant ES supply nor a relevant demand for ES (after Burkhard et al. 2012)

influences, especially in the urbanised areas and the industrial and commercial units. The more natural land-cover types, particularly the forests, show characteristic patterns where the ES supply often exceeds the demand. More detailed information about the locations of actual ES supply (SPUs) and related flows to areas of ES demand (SBAs) could be integrated in ecosystem service footprint calculations (see Box 1). No experience with this approach is available up to now. Highly complex import and export balances would be needed, for which data on required scales are not easily available.

The following case study application from the central eastern German region Leipzig-Halle shows how empirical ES quantifications can be transferred to the relative 0–5 scale, and how the results can be illustrated in spatially explicit ES maps. The study took place as a part of the EU project PLUREL (*Peri-urban Land Use Relationships*, ► www.plurel.net/). More detailed information about the different ES quantification methods and the map compilation can be found in Kroll et al. (2012) and in Burkhard et al. (2009, 2012). The following maps from the Leipzig-Halle case study region include CORINE land-cover maps for the years 1990 and 2006 and spatial distributions of the provisioning ES 'energy' supply, demand and supply–demand budgets (◘ Figs. 4.4 and 4.5). The quantifications for the ES 'energy' refer to final energy units in gigajoule per hectare per year. Lignite as the major energy source in this region was included within the provisioning ES category. We are aware that current ecosystem functions are not involved in the generation of lignite and that the integration of natural resources is seen critical by many authors. We are following the CICES system (► http://cices.eu/) here, which includes abiotic outputs from natural systems in their accompanying ES classification. Moreover, open-pit lignite mining has enormous impacts on ecosystem structures and processes in the study area's landscapes. Thus, this ES is of high relevance for landscape planning and therefore cannot be neglected.

The energy supply map from the year 1990 (◘ Fig. 4.4, top right) shows that the large lignite open-pit mines were the only regional energy source at this time with a final energy contribution of 20,000 GJ ha^{-1} $year^{-1}$. In the year 2007 (◘ Fig. 4.5, top right), a clear reduction of the open-pit mine areas and their energetic outputs are visible. New energy sources such as wind power, biomass, solar energy or waterpower were developed, resulting in a more heterogeneous distribution of energy supply in the region.

The demands for the energy provisioning ES (◘ Figs. 4.4 and 4.5, bottom left) show a clear sink function of the industrial and commercial units and the urban areas. The pit mines themselves also have a high demand for energy. The demand for energy was generally decreasing by 20 % between 1990 and 2007, mainly due to the decline of energy-intensive industrial activities and energy saving measures. The ES supply–demand budget maps (◘ Figs. 4.4 and 4.5, bottom right) illustrate the abovementioned source-sink functions of the rural and urban areas. Based on such information and data, decisions for regional ES provision and landscape planning can be supported.

4.1.3 Conclusions and Outlook

The high applicability of the ES matrix approach presented here arises from its potential for visualisation and from the comparison of the effects of different land-use activities on ecosystem functions and services. Thereby, assessments of trade-offs between different land-use types are possible. Various ecosystem functions and services can be displayed and huge amounts of data resulting for example from expert interviews, statistics, measurements and modelling can be integrated. The normalisation to the standardised relative 0–5 scale integrates different biophysical dimensions (e.g. Joule, tons, diversity indices) or economic units (e.g. Euro, Yuan) and makes them (to a certain degree) comparable.

The application of freely available spatial data such as CORINE enables the coverage of large landscape units with a unified land-cover classification system in almost all European countries. Issues with the land-cover classification system, the spatial data resolution and generalisation problems lead to uncertainties of the assessments. Further data with higher spatio-temporal or thematic resolution can, like in the ES assessments, easily be integrated.

The matrix approach is also linked with technical and thematic uncertainties, especially if the majority of the ES scores are based on expert opinions. The uncertainties are based upon the selection of a suitable and representative case study area, the selection of relevant land-cover classes (matrix *y*-axis), spatial and geo-biophysical data acquisition, the selection of relevant ecosystem functions and services (matrix *x*-axis) and related indicators, the indicator quantification in the matrices based on

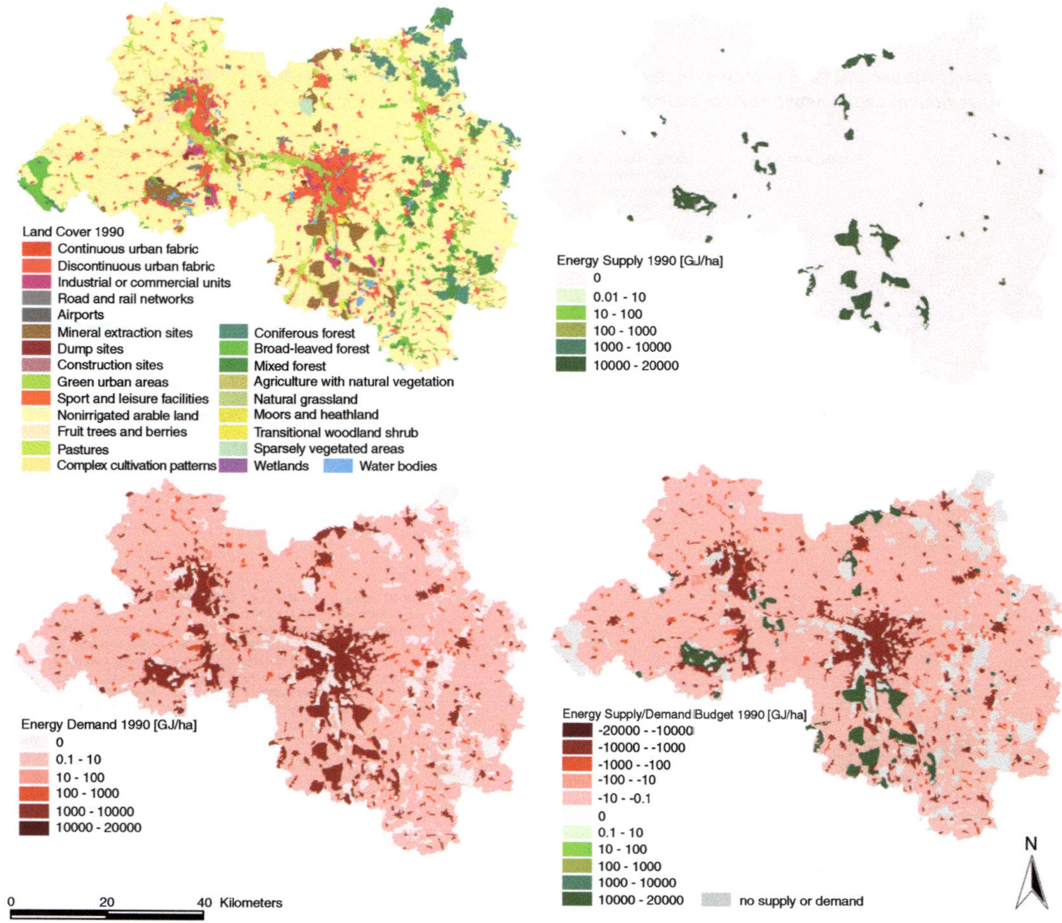

Fig. 4.4 CORINE land-cover map 1990 (*top left*); energy supply map (*top right*), energy demand map (*bottom left*) and energy budget map (*bottom right*) for the Leipzig-Halle region in the year 1990 (energy data in GJ ha^{-1} a^{-1}; after Burkhard et al. 2009, 2012)

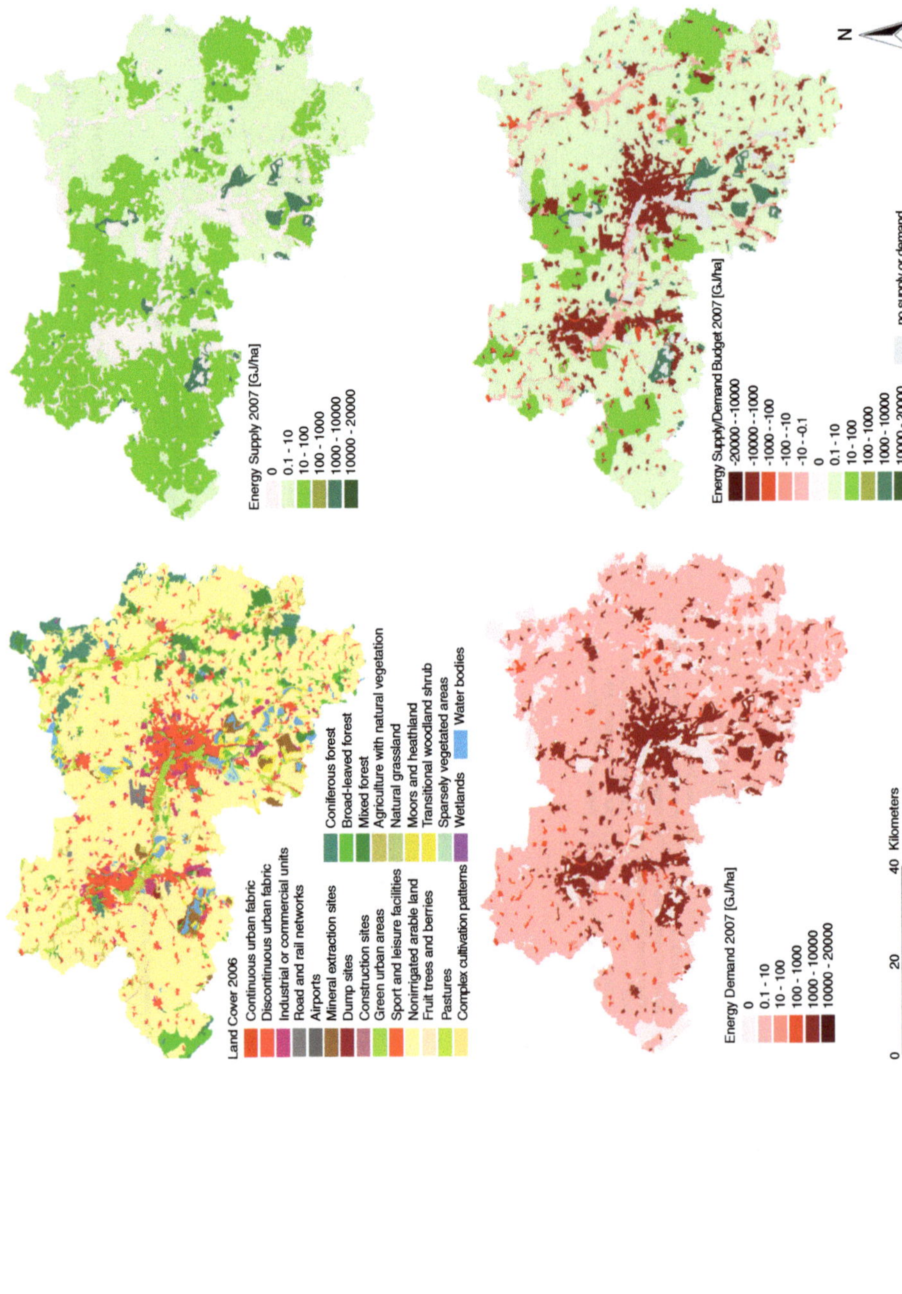

Fig. 4.5 CORINE land-cover map 2006 (*top left*); energy supply map (*top right*), energy demand map (*bottom left*) and energy budget map (*bottom right*) for the Leipzig-Halle region in the year 2007 (energy data in GJ ha^{-1} a^{-1}; after Burkhard et al. 2012)

the 0–5 scale, the linkage of the assessment values with the spatial units (map compilation) and the interpretation of the results by the end user. A detailed discussion of the different sources of uncertainties can be found in Hou et al. (2013).

Further developmental steps are needed to tackle these problems. One key issue is the inclusion of additional ES in the quantitative classifications, as shown in the energy budget case study example. Direct measurements, official statistics, simulation models or specific surveys, for example in the class of cultural ES, are needed to fill these data gaps. Moroever, regional geological, geomorphological, pedological, climatic and geobotanical site conditions as well as additional human system inputs (e.g. fertiliser, energy, materials) strongly influence ES potentials and flows. These effects should be integrated in future assessments (besides land-cover and land-use intensity) in order to minimise the assessments' uncertainties. Thereby, more exact ES scores (0–5) can be provided for example to actors in participatory processes.

Nevertheless, there are limits of intersubjectivity in such an optimisation. Related to the high amount of data needed to derive the different ES matrices, it will probably not be possible to completely abdicate from expert opinions. This statement can of course be interpreted as a critical argument. But it can also be seen positively because expert-based approaches have the advantage of relatively rapidly delivering target-oriented results which immediately can be applicable in decision-making processes.

One major demand from environmental planning is to make predictions about potential future developments' effects. Therefore, one key step in the future improvement of the matrix approach is the coupling with computer models (▶ Sect. 4.4.3). This would enable assessments of scenarios and their spatial specifications regarding the supply and demand of ES. This would seriously increase the applicability of the ES concept in practice. Due to the enormous complexity of such efforts, only common, transdisciplinary and cross-regional efforts will lead to positive outcomes.

4.2 Approaches to the Economic Valuation of Natural Assets

B. Schweppe-Kraft, K. Grunewald

4.2.1 Principles of Economic Valuation

"It is not with money that things are really purchased. (John Stuart Mill 1848)"

Economic science is, briefly put, the art of the rational and economical use of scarce resources for the fulfillment of human values and needs. Since ecosystem services are limited and their use is often at least partially mutually exclusive (trade-offs), rules are needed to make rational choices between alternatives that affect ES more or less strongly. Here, economic science seeks to maximise the general welfare, taking into account intergenerational welfare, distribution and consensual ethical rules.

Ecosystem services become economic goods, or obtain economic value, by providing benefits, and by being scarce. Not only such goods as food, water and recreational opportunities provide benefits; so, too, do the nonmaterial assets that are part of human preference and thus relevant as benefits. The right of species to exist and the value we ascribe to that right are–besides other more direct benefits–of economic importance, as soon as they become a part of individual preference. Thus, the habitat function of an ecosystem for wild species may constitute a sociocultural ES in this sense.

Scarcity means that the provision or maintenance of an ES is associated with costs (Baumgärtner 2002). An example are the costs of measures to maintain 'healthy' landscapes that provide sufficient opportunities for recreation, fertile soils, fresh water, etc. (▶ details in Sect. 6.5). Almost 50 % of the biological diversity in Germany relies on traditional or nonintensive forms of land use that are usually not economically competitive on the world market. The resources for conserving such anthropogenic biotopes and habitats are scarce. Costs can arise even if no money is paid, for example from the limitation of agricultural and forestry use in protected areas. These so-called opportunity costs are, generally speaking, benefits which the society or the individual must do without, in favour of other goals or benefits.

Ecosystems continually provide people with services. They are similar in this respect to the human-made productive assets that are used to provide us with goods and commodities. Such assets are the basis of our welfare, unless they are consumed or destroyed. The same holds true for natural assets as well: 'We must live from the interest, and should not consume [natural capital]' (Hampicke and Wätzold 2009). Destroyed or degraded ecosystems are restorable, if at all, only after a long period of time. The costs of restoration generally exceed the cost of maintenance many times over. The genetic information lost by species extinction is irreversible. Nonetheless, the economic value of the depreciation of natural capital is not easy to determine.

Unlike buildings, industrial plants or machinery, natural capital usually provides us with a number of different benefits simultaneously, each of which has to be evaluated separately. These generally include so-called public goods, such as air quality regulation, recreation in the open countryside, etc. One of the characteristics of public goods is that they cannot be privately appropriated. Therefore, there are no functioning markets which could lead to an optimum level of supply based on individual supply and demand. Market prices can be interpreted as values in the sense of willingness-to-pay and as costs, expressing scarcity. All this is lacking in the absence of markets.

In addition, each single ecosystem is embedded in a tight network of ecological dependencies with other natural assets. In such a situation, the assessment of physical changes can already be a problem, long before we arrive at the point of valuation. Moreover, there are also creeping impacts which occur later, and when they occur, then sometimes in an erratic and irreversible way. Which means, that methods, like the discounting method, are required to compare current and future costs and the difficult problem of valuing nonmarginal changes has to be solved.

If economists valuate goods or services, they as a rule assign them instrumental value, based on their usefulness for achieving a defined objective. This means that both economic valuation and the ES concept approach the issue from an anthropocentric perspective (Hampicke 1991). In addition, economic valuation is based on 'methodological subjectivism' (Baumgärtner 2002). All valuations must (or at least should, see below) build on the preferences of each individual citizen.

Economic assessments are always focused on choices between alternatives. Ecosystem services, like any other goods and services assessed in an economic cost-benefit-analysis, are not evaluated in isolation, but always in terms of their relative advantage in comparison with other goods, which, due resource scarcity, must be dispensed with. The relative advantage of one asset compared with others is its economic valuation, which, for practical reasons, is not expressed in terms of specific goods (e.g. 'How many glasses of beer is something worth to me?'), but rather in terms of the maximum amount of income which one will forego, or the maximum willingness-to-pay/ minimum willingness-to-accept, of individuals. All methods of economic evaluation, including the market-based and cost-based methods, try in principle to value (real) income changes and willingness-to-pay more or less accurately, or at least to find plausible proxies for such valuations.

Economic valuation, must, in accordance with its own principles and methodological standards, always focus on specific alternatives, e.g. restoration or no restoration of an alluvial floodplain; maintaining a grassland or converting it into farmland; urban living conditions with or without an adjacent park, etc. Economic valuations of ES are often part of a so-called cost-benefit analysis, which attempts, as far as possible, to evaluate all the economic impacts of the implementation and of the nonimplementation of a project or programme, or of various project or programme alternatives. To this end, all relevant effects of the various alternatives must first be predicted. As regards public goods, such as recreation, urban living conditions or urban climate, this encompasses an assessment of the number of persons who will benefit or suffer disadvantages due to a change with respect to these goods. Moreover, all costs, savings, income increases and income declines must be determined, including all costs and benefits measured in income equivalents (willingness to pay or to accept) which will result from the changes in public goods.

Discounting Future Costs and Benefits

The future development of costs and benefits can vary significantly between different project alternatives. Dike-shifting involves high investment costs; the future benefits include flood damage avoidance, reduced nutrient concentration in the water and restored habitats. No dike-shifting means more financial scope for consumption today, but higher damage cost, higher spending on prevention of nutrient loads and less benefit from additional biodiversity in subsequent years.

In order to make differences in temporal cost-benefit distributions comparable, all future values are discounted to their present value and then summed up (the discounted cash-flow method, illustrated by the example of nature conservation; see Herrmann et al. 2012).

The discounting of future values is justified by the consideration that (a) investments help increase production; and (b) people are willing to forego consumption today to save and invest in order to ensure a higher level of supply in the future. The model of discounting is thus fundamentally based on the assumption of future growth. If the availability of goods and services is to increase in the future, it makes sense to rate the same quantity of goods higher in the present than in the future, when the quantity and quality of available goods and services will have risen, due to investment and growth. A no-growth perspective, however, does not per se mean that any calculation based on discounting would be obsolete. In such a case, additional sustainability criteria for each of the different periods could act as limits showing where discounting is still feasible and where it is not. Nevertheless, a generally accepted method for such a case does not exist yet.

The choice of the interest rate depends, among other factors, on the type of investment that constitutes the basis for comparison. Private investments in innovative goods can achieve a very high return on capital. The rate of return of saving deposits marks the lower limit of interest rates for private investments. A prerequisite for the operation of private markets are complementary products provided by the public sector, such as infrastructure, education, jurisdiction, social security, etc. If all these costs were attributed to private market activities, the real value of the return of investments could be reduced further.

The German Federal Environment Agency suggests using interest rates of between 3 and 1.5 % in cost-benefit analyses, the latter figure for cross-generational considerations of over 20 years (UBA 2007).

Some authors (Baumgärtner et al. 2013) propose working with different interest rates, arguing that environmental goods and ecosystem services should be discounted at lower interest rates than other goods. The underlying assumption is that the supply of environmental goods and ES will deteriorate, making them more valuable per unit, or that consumer demand for environmental goods will increase with growing incomes.

However, it should be noted that the tendency to support low interest rates for environmental and growth-critical reasons, can also have negative results for environmental and natural assets in the context of concrete decisions. In the abovementioned example of dike shifting, a low discount rate leads to high values for all future benefits, such as avoided flood damage, extended habitat areas, reduced maintenance costs, or additional opportunities for recreation. But a low discount rate also means that the time of taking action, e.g. making an investment in natural capital, becomes ever more irrelevant to the value of its outcomes. At a discount rate of 3 %, the net present value (NPV) of an infinite constant stream of benefits to begin immediately is 80 % higher than one that is to start in 20 years. At an interest rate of 1 %, the value of the stream of benefits beginning today would only be 20 % higher than one which were to start in 20 years. Hence, a low interest rate can also be taken as a reason for reluctance to initiate environmental projects.

Conclusion: It is the state of the art to use different discount rates and different costing/calculation periods, and to compare the different outcomes with a critical view of the underlying assumptions.

The final step in a cost-benefit analysis, as in any economic evaluation, is the aggregation of individual values to a total value. This is done by adding all positive and negative income effects (costs and benefits) including the observed income equivalents (willingness to pay). This means that, for example, the social value of the preservation of the recreational function of a landscape and of the habitat function of its ecosystems for flora and fauna is nothing but the sum of individual willingnesses to forego income in favour of the maintenance of these functions. The social value of a land development project, e.g. an industrial plant, would result from the net income growth caused by the new plant, minus the willingness to pay for the lost recreation and conservation functions, minus the agricultural land rent (which is usually included in the price paid for the land by the new owner), minus all other external costs not included in the price, such as increased flood damage or flood regulation costs caused by the additional water run-off due to imperviousness of the land surface.

The process of evaluation and aggregation is somewhat similar to an election (Osborne and Turner 2007), but with some differences:

- The individual can only vote in accordance with the scope of his own interests (How often does he really use a recreational area? What is the share of the income generated that accrues to him?).
- The strength of a vote can differ (a greater or lesser increase in individual incomes or of income equivalents measured by willingness-to-pay).
- The individual is not directly asked to vote; rather, his 'vote' is ascertained from the extent (positive or negative) of the net income effect accruing to him.
- The net income effect does not have to be investigated for each person individually, it is sufficient if the sum is known.
- Representative sampling methods are applied to determine the benefits of public goods (▶ Sect. 4.2.3).

Economic valuation methods differ from the 'one man, one vote' rule, inasmuch as every individual valuation of public goods is in fact tied to the amount of individual earnings, i.e. valuation results can depend on income distribution. Normally, it is not the purpose of a cost-benefit analysis to examine the fairness of distribution. In industrialised countries, this is no problem, for income distribution is as a rule irrelevant to the results of a cost-benefit analysis. Different weightings for individual willingness to pay in order to compensate for income disparities usually affect the overall results only slightly. This may be different if the effects of an international scope are assessed. Ignoring income inequalities on an international scale can easily result in ethically unacceptable valuation approaches.

The abovementioned principles of economic valuation:

- Are based on individual preferences
- Assess values as relative advantages, expressed in terms of changes in income or income equivalents (willingness-to-pay)
- Involve the formation of a social value by simple aggregation of individual values

They do not mean that economic valuation completely denies the notion of values that are not simply individual, but which rather have supra-individual worth, such as divine commandments, animal rights, or the notion of binding rules for a harmonious human-nature relationship. Cost-benefit analysis accepts such values, but treats them as individual ones, assuming that they are solely valid for the person that proclaims them. A person who assumes, for example, that animal rights should be ranked higher than the pursuit of any additional welfare gains, cannot demand that all economic advantages measured in a cost-benefit analysis be set to zero. He can, however, demand that his own individual foreseeable future income growth be assessed as his willingness-to-pay against e.g. any further species extinction.

Accordingly, individuals and their choices based on individual preferences tied to their economic limits (income) on the one hand constitute elementary declarative units. That means that the economic value is determined by the subjective evaluation of individuals ascertained by means of a survey of representative samples. In the strict sense, expert judgments can only be integrated into cost-benefit analyses if they can be interpreted as approximations to the preferences of individuals which cannot be measured directly. In this view, the economic value assigned to an ES is not a quality that is inherent to that object (e.g. an ecosystem), but rather a value which depends on the overall context, not only the economic context.

The valuation of the ES 'fresh drinking water', can, for example, depend on the following aspects (Baumgärtner 2002): How much clean water is there in total? How is the supply of clean drinking water distributed in space and time? How is the access to this resource regulated? What competing demands for water exist, besides its use in households? What kind of institutional restrictions exist? What kind of alternatives are there to water use in various use areas, and what would they cost? How much would it cost to import clean water from other regions? How much does technical water purification cost?

The failure of the market, private production and private consumption to generate socially-acceptable or optimal results–i.e. a market failure–is, according to economic doctrine, the occasion for an economic evaluation. This may be the case if:

- Production and consumption cause losses of benefits or price increases for others (so-called negative external effects). Examples: intensifying agriculture by removing hedgerows impairs the recreational capacity of a landscape; diking along a river can prevent flooding of areas behind the dike, but increases the flood risk upstream and downstream.
- Public goods are involved, i.e. those which benefit a large number of people without or with only limited possibilities of excluding anyone from those benefits. Example: recreational use of the open landscape, of public bathing waters, the existence value of species/biodiversity, or possible future pharmaceutical use of a certain kinds of species. In such cases, due to the lack of user payment, there are no incentives for market activities to maintain the provision, to prevent overexploitation, or to protect the asset from detrimental external effects.
- The costs of current activities accrue over the long term, e.g. to future generations, and therefore are not taken into account by present market participants. For example soil erosion, CO_2 emissions by intensive agricultural use of peat soils.

In the case of market failure, economic valuation has the function of informing about all costs and benefits accruing to people now and in the future, and enables decision-makers to reduce external costs and maintain provisioning with public goods to an optimal extent, thus maximising welfare under consideration of all relevant costs and benefits.

Like public surveys and public participation, cost-benefit analysis can help ascertain public opinion more precisely and make individual preferences more obvious than can be done by general elections only. In addition, it can reveal a malfunction of the democratic system, for example, the lopsided influence of powerful interest groups which are able to effect political decisions against the public interest (e.g. environmentally counter-productive subsidies; Brown et al. 1993).

Economic valuations need not necessarily be carried out with monetary units (Abeel 2010). Money can even be a hindrance. It can, for instance, promote the idea that only the world of market goods (production and consumption) really counts, whereas the actual goal is to correct the results of the market, by making it clear that the production of goods entails hidden costs that can obscure their true prices. Often, we are persuaded to produce things that we would rather do without for other, nontraded goods, e.g. for biodiversity and healthy ecosystems, if we knew enough about the issues, or if it became obvious that national income consists to a considerable degree of the costs of repair of damage to the environment and nature (Leipert 1989).

Money as a valuation unit may moreover suggest that the valuated goods will in fact be priced and thereafter traded. Nonetheless, the decision as to how to deal with market failure is up to policy makers, and is completely independent of the valuation process. Whether market failure is to be corrected by public supply, by do's and don'ts, by incentives, by taxes, duties or user fees or by the creation of markets, is a matter for public decision making. Economic valuation does not imply converting public goods into commodities to be traded on the market, either directly or indirectly.

Another misconception may be that the value of an ES that is calculated and determined for a specific social, economic or ecological environment could be transferred to other situations with no adaption, like the price of a good trade on the world market, for instance a smart phone. Such an understanding, however, would overlook the fact that many ecosystem services are tied to their point of origin, so that no distribution can take place. However, distribution in response to demand is a prerequisite for the emergence of a common price level on the market.

On the other hand, valuation in monetary terms can be highly practical. A monetary value allows a trade-off involving costs, income and various other goods, including public goods, based on the views of a representative sample of citizens. Other valuation methods, such as benefits analysis (Zangemeister 1971; Hanke et al. 1981) and similar types of so-called multi-criteria analysis (Zimmermann and Gutsche 1991), also use decision-making models based on trade-offs (▶ Sect 4.1.). However, such models often depend on the opinions of a limited selection of experts and/or 'citizen experts' (Dienel 2002), which are not representatives. Although in certain cases, expert-based models may have a high problem-solving competence, the social values upon which they often implicitly build have not been validated.

Various decision-support instruments, such as cost-benefit-analyses, expert-based multi-criteria analyses or discursive processes of active citizenship, should be used in accordance with their respective strengths and weaknesses. A representative group of citizens mixed with some experts could for instance provide useful advice for the best use of a fixed local budget for various urban green-space management measures; however, when it comes to the preparation of a concept for reducing soil erosion in a district (Grunewald and Naumann 2012), an expert-based cost-effectiveness analysis would likely be better grounds for sound decision-making. The cost-benefit analysis, after all, shows its strengths when actions are to be taken that might affect a great number of people physically and financially in very different ways. This is the case, for instance, when decision support is needed on the question as to how much money a city should spend overall on green-space management. Another example would be the design of a well-balanced programme of measures for reducing soil erosion that should also take into account other effects, e.g. upon species preservation, the landscape, or water pollution, in such a way that the costs of the measures will best be outweighed by their benefits.

Example

Grossmann et al. (2010) applied a cost-benefit analysis on proposals for a bundle of nature-based flood prevention measures by increasing the retention area through dyke-shiftings (▶ Sect. 6.6.3). They calculated the avoidance of flood damage, valuated the water purification effect of an enlarged alluvial floodplain by comparing it with the cost of alternative measures for reducing water pollution, and asked people about their willingness-to-pay for the benefit of the enhancement of conservation and recreation. The value of the ES thus assessed was three times as high as the cost of the measures.

4.2.2 The Total Economic Value

The most widely accepted approach for the economic valuation of ES is the concept of Total Economic Value (TEV, Pearce and Turner 1990) (◘ Fig. 4.6). The various benefits of ecosystems are classified as either *use values* or *nonuse values*. Use values are further subdivided into *direct* and *indirect use values* and *option values*. Nonuse values are broken down into *existence values* and *bequest values*.

▪ Direct Use Values

Direct use values accrue from the direct use of ES for consumption and production, e.g. food, firewood, medicine, timber, drinking water, cooling water, etc. The use of a landscape for recreation, leisure activities, tourism or scientific or educational purposes is also considered a direct use of ES (Baumgärtner 2002). Direct use can be consumptive–example: firewood–or nonconsumptive, as with recreation. Direct use values are linked to provisioning services and goods, as well as with some sociocultural ES, such as for recreation, cultural identity, landscape aesthetics and knowledge services.

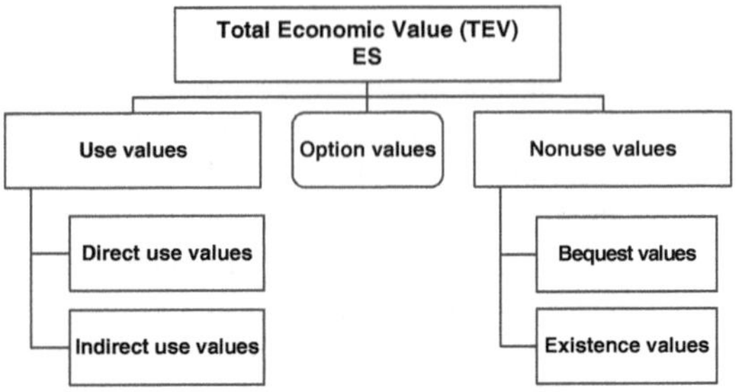

◘ **Fig. 4.6** The concept of total economic value (TEV). (Adapted from Pearce and Turner 1990; Bräuer 2002)

▪ **Indirect Use Values**

Indirect use values arise when ecosystem services interact directly or indirectly with human activities. Examples are flood control by means of water-retention measures in alluvial floodplains, the self-purification effect of water bodies, or the water-filtration capacity of soils. The so-called regulatory services generally fall into this category. The economic value of these services is measured as the change in the costs and benefits of the use that is affected by them, e.g. reduction of flood damage, benefits from additional use as a swimming location, or the decreased costs of the drinking water supply; see, by analogy, the concept of *final ecosystem services* by Boyd and Banzhaf (2007) (► Sect. 3.2).

▪ **Option Values**

Option values express the fact that there is a willingness to preserve the possibility of later use of ES, regardless of whether this will really take place or not. Option values and values to be realised in the future correspond largely to the so-called *Potentialansatz* (capacity approach) in German landscape planning (► Chap. 2 and ► Sect. 3.1). The option value can also be interpreted as an insurance premium that people are willing to pay to maintain the possibility of future use (Weitzman 2000). Option values are especially significant in the context of landscapes and ecosystems of high cultural significance and singularity, such as the Brocken peak in the Harz Mountains in Germany, or with respect to the uncertainty of a future economic use of species and their genomes (e.g. Norton 1988).

▪ **Bequest Values**

The bequest value expresses the willingness of people to forego parts of their present income in order to preserve things for future generations. This heritage can refer to sociocultural ES, but also to provisioning services.

▪ **Existence Values**

Existence value reflects the willingness-to-pay for the preservation of things regardless of whether there is any likelihood of their future use or not, just in order to preserve their existence. Such values are often ascribed to assets thought to have an intrinsic value, such as living species, e.g. in the concept of animal rights.

These different kinds of values, named above, are conclusive. Their sum is the overall economic value of an ecosystem. However, in field studies, it is often impossible to clearly separate the different values from one another.

Investigations at Natura 2000 sites have revealed that more than 50 % of their TEV were constituted by indirect use values and nonuse values (Jacobs 2004). That means that from a conservationist point of view, these values, especially the option, bequest and existence values, are the most critical ones. On the other hand, the problems of reliable evaluation increase as one moves from direct use values to nonuse values.

4.2.3 Valuation Methods and Techniques[2]

Use Values

▪ **Market Prices**

If assets provided directly by nature can also be found on markets in the same or a similar quality–e.g. mushrooms, fish, game–the market price can be used as a proxy for their value (the *market-price method*). One important precondition for the applicability of this method is that product qualities and the demand for marketed and non-marketed products are similar. This is not always the case, however. For example, experience shows that blueberries which are picked in the woods on a hike taste particularly good, this special kind of appropriation seems to give them an extraordinary quality, so that they could be rated considerably higher than purchased blueberries. On the other hand, the picking is an activity that is incidentally performed, without significant additional effort. One might also pick the berries when demand is low, and therefore have to valuate them at a price well below their market price. The same is true of self-caught fish. As an actively appropriated product, it might have a higher value than comparable market products, but it could also serve as an incidental by-product of the fishing activity itself, which is the

2 For a systematic presentation of economic valuation methods that is also addressed to noneconomists see ► www.ecosystemvaluation.org.

Fig. 4.7 The economic value of the provisioning service of a field (here cornfield near Sulingen in Lower Saxony) can be measured on the basis of the income loss resulting from abandoned agricultural use. © Burkhard Schweppe-Kraft

actual ES provided–recreational activity. If the fish is used by the family of the angler, their possibly differing preference for fish may also be important for the valuation.

The market-price method could, for example, be suitable for the valuation of the effect of an alteration in forest management on all the wild fruits to be found there, or it could be appropriate for the valuation of the improved water quality in a lake on the composition of its fish population (less biomass, but a higher proportion of game fish). In both cases, the changes on the supply side are only one side of the coin, for the extent to which the additional supply will really be used must also be assessed. Finally, the question should be answered, e.g. on the basis of surveys, to what extent the value of the products is thought to lie above or below the market price level.

- **Change of Value Added, Profits, Return on Sales Minus Cost of Production**

The majority of market goods created with the help of ecosystem services, such as drinking water, wood products, food, etc., is produced in combination with labour and capital. If the ES change, e.g. additional land used for agricultural production, causes increased sales of goods, the additional value of sales is not the only determining factor for their valuation; rather, it is the difference between the additional sales and the costs of the use of capital, precursor products, production facilities and labour power, including a normal remuneration of the labour input of the entrepreneur. The difference remaining after this calculation corresponds in the case of e.g. cropland more or less to the cost for the lease of the land being assessed, or a comparable plot. Therefore, the ground rent (lease) is often used as a proxy for the net value of the productive input of ecosystem services that are combined with certain plots of land (Hampicke et al. 1991).

Example

What loss in the value of agricultural production would result from the abandonment of this field (Fig. 4.7)? From the total loss of market revenue, one must first subtract the variable costs. In addition, adjustments with respect to labour and capital inputs will occur in the mid- or long-term and have to be considered in the evaluation. After these adjustments, the loss of ground rent (lease) remains as a permanent loss. This is determined on the basis of various favourable and unfavourable factors, such as soil fertility, water supply, climate, slope, etc. When evaluating large-scale soil loss in developing countries, one would have to assume significantly higher income losses, due to a lack of alternative employment opportunities. Nothing in the world would suffice to persuade us to do without the entirety of the agricultural land on earth–its loss would have a value of 'minus infinity' (Costanza et al. 1998).

If a corn (maize) field is converted into a species-rich damp meadow, for example due to conservation measures, a comparison of these two different provisioning services–corn and hay, respectively–would require a calculation of the difference between the proceeds from the sales of these two products, and of the above-described production costs. For the corn, this difference would be positive; for the hay, probably neutral or even negative.

For a comparison of the total economic value (TEV) of intensive–e.g. corn–and extensive farming systems–e.g. a meadow–a correct valuation of the services corn and hay could be critical. The difference between the profits is often significantly less than the difference between the sales proceeds, one reason being that intensive farming systems often require higher inputs. The different valuation of provisioning services, in one case on the basis of sales proceeds, in the other on the basis of sales proceeds minus costs, explains why in the study by Ryffel and Grêt-Regamey (2010), the calculated total value of species-rich grassland is less than that of intensively used grassland, while in the study by Matzdorf et al. (2010), the species-rich grassland comparatively outperforms the farmland (► Sect. 6.2.4).

An assessment of provisioning services on the basis of sales proceeds would mean that not only ES would be evaluated, but the value added by labour and capital, too, would be included. A correct application of the cost-benefit analysis must always subtract the costs necessary for production from the value created, to calculate the net yield. In the case of provisioning services, this means the respective earnings minus the wages for the work of the contractor plus the rent paid for the land (see environmental services ► Sect. 2.1).

Implicitly, the above calculation of provisioning services involving profits or rents is based on the assumption that the labour thus 'freed' and–at least in the medium to long term, even the capital thus 'freed'–will find uses elsewhere, and will there generate added value that corresponds to the costs. Cost-benefit analyses carried out in industrialised countries are, due to the flexibility of the markets for labour and capital, generally based on this simplifying assumption. Deviations should be clearly identified and explained. In many regions in developing countries, however, the necessary alternative opportunities are not available, particularly for the factor labour. If the destruction of the services of an ecosystem, e.g. the loss of soil fertility, or overfishing, drives the people who had depended on these services into long-term unemployed, the cost-benefit analysis would have to include as the value of the supply service concerned not only the lost profits, but the entire value, including labour and possibly capital costs. In industrialised countries like Germany, adjustment problems and deadlines are more likely to be the factors to be taken into account with respect to the factor capital.

Therefore, when determining the cost of a change in agricultural production or the abandonment of agricultural use the calculations for the short or medium term are often based on contribution margins. A contribution margin is the market revenue minus the variable costs. As the term implies, the contribution margin per hectare states the contribution that the production on one hectare of land makes to cover the fixed costs of a business, for example, to the interest payments due on the loan for stables (see case study in ► Sect. 6.2.3). A contribution-margin calculation assumes that unused capital is inflexible, i.e. it cannot be used elsewhere just as profitably. In the short term, such a method of calculation is justified; in the medium term however, adaptation possibilities have to be assumed. After the technical depreciation period of the capital involved–at the latest–it is advisable to shift to such values as lease or long-term profit outlook for

Fig. 4.8 Fruit growing areas are particularly dependent on pollination services. © Burkhard Schweppe-Kraft

the calculation of production losses. The correct handling of the costs of capital can be crucial for the actual calculated results. For example, in a case of the rewetting and use abandonment of previously farmed peat soils, Röder and Grützmacher (2012) calculated costs of € 40/t of saved CO_2 emissions, on the basis of contribution margins. If only the lease costs of, say, € 250/ha were used in the calculation, a much more favourable value of around € 9/t of CO_2 would result. Assuming a 20-year adjustment period with adaptation rates at a consistent level and a calculated interest rate of 3 %, costs of over € 17/t would result. Calculation examples from studies based on all three types of calculations can be found in the literature. This shows that major methodological differences occur not only in the evaluation of ES generally, but also that great tension is possible simply with the very conventional cost calculations, which are based on different, and often highly questionable, assumptions.

The example of using land-lease as an approximation for the long-term value of the agricultural production function of an ecosystem (provisioning services) again shows dramatically that economic valuations generally apply only to relatively small changes: The higher total value of grassland compared to farmland, which can be calculated on the basis of the study by Matzdorf et al. (2010) (▶ Sect. 6.2.4), applies only to the case of the current distribution between grassland and farmland. If, due to the currently high total economic value (TEV) of grassland, ever more farmland were to be transformed into meadowland, the supply of the various public and private goods produced using these land areas would gradually increase so greatly that the prices and the willingness-to-pay for any additional margins of these goods would fall. The total economic value per unit of converted farmland could pull even with the TEV per additional unit of grassland, and then even exceed it. This could in fact be accomplished relatively quickly, for example in the case of the species-protection function/service. For the preservation of biodiversity often optimally requires a mix of grassland and farmland, and not a grassland monoculture.

This also shows why the value of the sum of all ES cannot be calculated from the value to be set for a relatively small change to be assessed. Multiplying the total stock of farmland in the industrialised countries by the respective lease values per hectare, the result is by no means the value that society would be willing to pay for the preservation of the agricultural production output of these areas; the true figures would be significantly higher. With the increasing loss of production areas, prices would rise to an extreme degree, and the social upheaval thus provoked would have uncontrollable consequences.

Example

Within the EU, the service pollination is estimated at a value of some € 14 billion (Gallai et al. 2009). This is the value of agricultural products which are highly dependent on insect pollination. This knowledge does not help much for concrete valuations. In assessing the changes in pollinator populations in specific growing regions, the decisive factor is whether the populations there already constitute a limiting factor for production, or whether they are extant in abundance. So far for example, we know relatively little about how flower strips within fruit-growing areas impact on the net yields (Fig. 4.8).

■ Change in Production Costs

The cost of production method also ascertains the change in the difference between the sales proceeds and costs of production, but it does so for the special case that product quantities and revenues remain constant, and that only the costs of production change. The typical example of this case is the reduced effort required to provide clean drinking water if a farm field, which generates pollution is replaced by grassland. Another example would be an increased use of fertilisers to compensate for reduced soil fertility, which has resulted, for example, from intensive use, or soil erosion caused by the removal of hedgerows and other small structures.

In these cases, the production cost method was used directly to valuate the supply capacity of ecosystems (water supply, agricultural production), and also indirectly to assess the impact of regulatory services (reduction of soil pollution, and of soil erosion by small structures) upon the respective provisioning service.

- **Damage Costs, Mitigation Costs, Adjustment, Repair, Replacement Costs**

Many regulating services influence the effects of natural hazards (flooding, avalanches and mudflows, storm damage, etc.) and anthropogenically induced risks (climate change, air pollution, urban climate stress). For the evaluation, the damage and damage prevention costs and the adaptation, repair, replacement or avoidance cost can often be used. Here, the extent to which damages (including medical expenses), or the cost of prevention and repair (rehabilitation) can be changed by ecosystems and ES is examined. Examples include the prevention of flood damage through restoration of floodplains, or avoidance costs for the treatment of respiratory diseases caused by the dust-filtration effect of urban green spaces.

It is a general economic principle that a goal should be achieved at minimum cost. If a damaged item is of lower value than the cost of its repair, it is more beneficial to all concerned to monetarily compensate the aggrieved person than have the damage repaired. This principle applies not only to the compensation for damage to passenger cars, but also to evaluation in the determination of total economic value (TEV). The same applies if the damage-avoidance costs are higher than the damage. Here, too, it is cheaper to pay the lower insurance compensation for a damaged asset than the higher cost of completely avoiding the potential cause of damage. Such situations are referred to as the *least-cost principle*.

Often, only a portion of the value of an ecosystem services can be quantified by damage or repair costs, just as medical costs often reflect only the cost of treatment, but not the physical or mental suffering of the patient. If, due to increased use intensification in an area, there are no more skylarks or partridges there, the cost of resettlement or avoidance of that loss may be significantly less than its ethical and aesthetic significance. Other methods, such as willingness-to-pay analyses, should be used if damage or avoidance costs can measure only part of the total economic value of a service.

Example

During the mid-1990s, Pimentel et al. (1995) assessed the on-site and off-site costs of erosion in the USA, and arrived at a figure of about \$100/ha/yr. If this order of magnitude of replacement and damage costs is compared with the cost of erosion-mitigation measures, a very positive cost-benefit ratio of 1:5 results; the soil erosion hazards due to water and wind are thus reduced from 17 $t/ha^{-1}\,a^{-1}$ to 1 $t/ha^{-1}\,a^{-1}$. Using an analogous approach for a loess-covered, predominately agricultural area in Saxony, Grunewald and Naumann (2012) ascertained a cost-benefit ratio of approximately 1:2 (► Sect. 6.6.2).

- **Alternative Costs**

Closely connected with the above methods is the so-called alternative-cost approach. This method often valuates not the costs in fact incurred, but rather those of theoretically possible options which might be used in order to achieve a goal in an alternative manner. An example might be the evaluation of the additional self-cleaning capacity of a renaturated water body, using the two potential alternatives of, on the one hand, the measures necessary to reduce pollutant input from agriculture, and on the other, the building of additional wastewater treatment capacity to achieve the same water-quality effect. The erosion protection provided by hedgerows and small structures could, for example, be valuated not only via the production-cost method, as above, but also on the basis of the cost of soil conservation measures on the field which are equally effective.

Whether or not a corresponding alternative-cost approach is permissible depends on whether the social goals are formulated in a sufficiently binding manner or not. Strictly speaking, the alternative-cost approach only leads to correct results if the objectives are formulated in such a binding manner that the necessary measures for their alternate achievement will actually be implemented in the not-too-distant future. An example of such a binding social goal is the EU Water Framework Directive (WFD), which mandates the attainment of a certain level of water quality (▶ Sects. 3.3.2 and 6.6.2). If farmland is converted to grassland, the nutrient input into the groundwater and the surface waters is reduced, and the specified goals of the WFD become more attainable. A corresponding contribution to the reduced water pollution can be achieved by various measures in farming, or by improvements in the treatment stages. Under the least-cost principle, an alternative measure, which allows both similar relief at the lowest cost and at the same time has a realistic chance of implementation should be selected as the value of reduction of nutrient immissions due to conversion into grassland. Matzdorf et al. (2010) used a value of between € 40/ha and € 120/ha for the valuation of the reduced nutrient inputs through the preservation of grassland, based on the evaluation of data of cost-effective measures to reduce nitrogen emissions by Osterburg et al. (2007) (▶ Sect. 6.2.4).

Measures for rewetting and restoring formerly farmed peat soils halt the mineralisation of organic soil components, and thus lead to a significant reduction of greenhouse-gas emissions. The evaluation of this regulatory service 'rewetted peat soils' is possible both on the basis of damage costs and on the basis of alternative cost. In accordance with the Stern Report, the methodological convention of the German Federal Environment Agency (UBA 2007) suggests a preliminary cost estimate of approximately € 70/t of CO_2, based on a combined damage-/mitigation-cost analysis. In case of the use of wind power, 1 t of avoided CO_2 emissions costs approximately € 40; on the European carbon market, a ton of CO_2 cost € 6–7 in early April 2012. Which of the above values is to be used for the valuation of the CO_2 emissions saved by rewetting will depend on how future developments are to be assessed (▶ Sect. 6.6.4).

It can be assumed that the required reduction of CO_2 emissions cannot be implemented solely using the current favourable measures that enable the current low prices on the carbon market. Achieving the goal at these costs is thus unrealistic. Measures in the cost category of CO_2 avoidance through wind power would seem, for example, to be more realistic. If we assume, moreover, that the goal of limiting the temperature increase to 2°C will fail to be attained by a wide margin, which seems increasingly likely, even the € 70 damage costs would have to be considered too low. The example shows that even with realistic assumptions, there can be very widely divergent evaluation approaches. Evaluations should therefore always disclose the assumptions upon which they are based, and whenever possible, alternative calculations under different assumptions should be undertaken.

Example

At the beginning of the 1990s, the city of New York was forced to take action, since it no longer met the established drinking-water quality standards. A water filtration and treatment plant was to be built for $ 6–8 million, and operating costs of about $ 300 million per year would have been added. As an alternative, the issue of improving the ecological functions of ecosystems in the Catskill Mountains, the drinking-water catchment area for the city, was examined. This cost was estimated at a one-time investment of € 1–1.5 billion. Faced with a balancing of interests between the cost of improving the ecosystems on the one hand and the development of purification technology as a substitute for the reduced ES of degraded ecosystems on the other, the decision was made in favour of the ES option (Chichilnisky and Heal 1998).

▪ Real Estate Prices–Hedonic Pricing

The evaluation approaches presented above have, under the MEA (2005a) system and the ES classification (▶ Sect. 3.2), respectively, been oriented primarily towards provisioning and regulating services. The *hedonic pricing method* is oriented towards the sociocultural services recreation and aesthetics, or beyond that and in more general terms, towards the subjectively evaluated welfare functions of green elements and green spaces in the residential environment.

Under the hedonic pricing method, the goal is to ascertain the effect of near-residential green spaces on real-estate prices by statistical analysis. Hoffmann and Gruehn (2010) come to the conclusion that in densely populated inner-city districts, the green features of the residential environment accounts for 36 % of the property value. In less densely populated, smaller towns, the effect is less (▶ Sect. 6.4).

The hedonic pricing method covers only that portion of the use of urban green spaces that accrues indirectly to the property owners. Any benefits above this portion would have to be ascertained by other methods, by carrying out an additional willingness-to-pay analysis, or on the basis of the statistical data estimates of a demand function, similarly to a travel-cost analysis.

▪ The Travel-Cost Approach

The term *travel-cost analysis* covers a whole package of different methodological options, which are primarily used for the evaluation of recreation areas. Here, the relationships between the number of trips to a region or a certain type of area and the amount of the cost per trip are analysed statistically. In the newer versions of the method–also the quality of the area for recreation (e.g. landscape, landscape diversity, facilities with recreational infrastrucure) are taken into account. On this basis, a demand function for recreation in the area or area type in question is assessed. Based on a comparison of the behaviour of visitors with high- and low-access costs, respectively, it is possible to deduce that the willingness-to-pay for the first visit undertaken within a given monitoring period to a particular area or type of area is higher than for later visits. Visitors with low access costs do not need to exercise this higher willingness-to-pay for the first visit in real terms, and thus realise a so-called consumer surplus. The sum of all consumer surpluses yields the total net benefits of recreation in the assessed areas. The consumer surplus constitutes the willingness-to-pay that an individual has for a recreational activity, minus its actual cost.

In some proposed methods and evaluation studies (Ewers and Schulz 1982; UBA 2007; to some extent too, Getzner et al. 2011), the actual costs of a recreational activity are regarded as its benefits. Certainly, assuming rational behaviour, the benefits must generally be at least as high as the cost paid for them; however, as discussed above in connection with the costs for the production of agricultural products, the purpose of a cost-benefit analysis is to ascertain the difference, or the ratio of costs to benefits, for each alternative. With such a difference ascertainment, the result of a recreational activity the benefits of which are just as high as the costs, would always be neutral; the net benefit, i.e. the difference between benefits and costs, would always be zero. This result would emerge in all studied alternatives, regardless of whether the recreation areas were of average quality, are actually upgraded, or would be devalued by impacts. For it we dispense with the counterbalancing of the costs, and show the cost only in their indicator function for the minimum benefit, we will arrive at completely nonsensical evaluation results when comparing options. For example, if the construction of a bypass road were to lead to an increase in the expense of money or travel-time to be paid by the inhabitants for access to their recreation areas, this would not be recorded as an obstacle to their recreation, but rather as an increase in their recreational benefits. Hence, the simple calculation of cost is unsuitable for the evaluation of recreational benefits. The goal must be to calculate the consumer surplus, the difference between the benefits (or willingness-to-pay) and the costs.

Under the travel-cost method, which uses this approach, willingness-to-pay is derived from the observed actual behaviour of a large number of different recreation-seekers, using statistical methods. This, like the land-price method, is one of the so-called revealed-preference methods, based on an investigation of factually evident preferences, in contrast to the *stated-preference methods*, in which the preferences are directly queried.

Example

In the Eibenstock-Carlsfeld region in the western Ore Mountains of Saxony, a survey was carried out via interviews among visitors and tourist-service providers on their appreciation of the landscape scenery (Grunewald et al. 2012). The questions concerned the qualitative landscape characteristics and preferences, travel expenses and willingness-to-pay for the maintenance and appearance of the landscape. For this purpose, the monetisation approaches of the travel-cost and willingness-to-pay methods were used. The study comprised face-to-face interviews with 95 summer and 105 winter tourists; travel costs were recorded for a total of 584 individuals. The goal was the analysis and monetary valuation of sociocultural ecosystem services related to landscape aesthetics, in order to provide a foundation for the improved landscape planning and management.

The tourists' aesthetic perception of the landscape elements in the region is influenced primarily by visible, near-natural landscape elements, such as the forest and water bodies, and by their harmonic composition. An undisturbed landscape was the principal reason for travelling to the region and spending vacations there. Altogether, tourists paid about € 5.5 million per year in travel costs (extrapolated to the total number of tourists visiting the region), they are willing to pay € 170,000 per year in addition for the protection and management of ecosystems. The results show that the visitors valued public goods and services highly, a factor which will have to be considered more strongly in future planning (Grunewald et al. 2012).

- **Hunting Leases, Fishing Licences, etc.**

For some recreational activities, such as hunting or fishing, there are prices to be paid in the form of fishing licences and hunting leases. These, unlike such expenses as those for fishing equipment or the fuel used to reach a fishing spot, are an expense associated with no real costs, or only minimal ones. A payment that is not remuneration for any labour or capital cost is referred to as a 'surplus.' Even the rent for agricultural land is such a 'surplus.' By paying for a hunting lease or fishing licence, the sportsman shows that his benefit from the fishing or hunting activity is at least equal in value to that payment. In this case, as with the land-price method, this share of the benefits accrues not to him, the user, but rather to the owners of the land leased. The benefits that can be calculated from fishing or hunting leases is the lower limit of the actual benefits from that activity.

If we also wish to ascertain the net benefits to the anglers and hunters over and above this minimum, it would be necessary to apply other methods, such as the travel-cost approach or contingent valuation. It is important in cases of changes in the conditions for recreational use, to always also ascertain the possibilities of substitution. Generally, there are also other places where recreational activities may be carried out. In such cases, the increase in travel costs to remaining alternative fishing or hunting areas would be a first rough measure for the welfare loss caused by the degradation or the loss of another area. With a more precise travel-cost analysis, it would be possible to capture also the 'consumer surplus' over and above simple cost effects.

- **Admission Prices**

A method for calculating leisure and recreational use which was in the past particularly common is the *admission-price method*. Here, the recreational opportunity to be valuated–from city parks to national parks–is compared with similar recreational activities for which a price of admission is charged. One problem with this method is that people who spend time in fee-based recreational facilities, such as former horticultural exhibitions or amusement parks, may have different preferences from those of people who use free leisure facilities, such as urban forests or natural parks, so that it is difficult to find truly comparable situations. For example, admission-charging swimming pools and guarded beaches often have a distinctly different character than free swimming spots. Moreover, the price of admission reflects the lowest level of willingness-to-pay among those who avail themselves of the service; some visitors would be willing to pay a higher ticket price. Because of these problems, a valuation based on admission prices should also be supplemented by some other alternative valuation method, such as travel-cost or willingness-to-pay analysis.

- **The Willingness-to-Pay Analysis (Contingent Valuation), Choice Analysis**

In addition to, or as an alternative to the above methods, any direct or indirect use value can theoretically be assessed on the basis of direct interviews using contingent valuation or the choice analysis. These valuation techniques are used for the ascertainment of both use and nonuse values (see below). Applied to the same evaluation object, travel-cost and willingness-to-pay analyses often provide relatively similar results (Löwenstein

1994; Luttmann and Schroeder 1995; Whitehead et al. 1995). In cases where specialised knowledge is required for an evaluation, e.g. for the evaluation of changes in soil fertility, erosion, effects on water quality, flood damage, etc., complementary expert-based methods should also be used, in addition to the willingness-to-pay analysis, in which, since it is a representative approach, largely nonexperts are interviewed.

Methods for the Detection of Nonuse Values

■ **Contingent Valuation, Choice Analysis**

Preferences for nonuse values, such as the desire to preserve species and habitats as a 'value in and of itself' (existence value), or so that they can be used and experienced by future generations (bequest value), can, like option values, currently only be ascertained by direct, representative surveys. The main methods for this are the willingness-to-pay analysis and the choice analysis.

The willingness-to-pay analysis asks how much money or income an individual would be willing to do without, as a maximum, in the form of a generally mandatory landscape-maintenance tax, so that nature might be preserved, or a specific conservation programme might be implemented. In a choice analysis, the respondents are presented with different options about the future, which they can, by means of various procedures, either accept or reject. Each option here describes various conditions related to the natural environment, and an income-relevant quantum, such as a surcharge or deduction for income tax purposes. By means of statistical analysis, willingness-to-pay with respect to the various parameters can be derived from the various 'decisions' thus made.

There is an extensive body of scientific literature on the validity of stated preference methods and the possibilities for improving and securing their validity (e.g. Hoevenagel 1994; Marggraf et al. 2005).

A number of results regarding willingness-to-pay for conservation measures in Germany are now available (◘ Fig. 4.9; ► Sect. 6.6.1). They involve extensive activities, such as national programs for the conservation of biodiversity (an average of € 231 per household per year) down to such local activities as measures for the conservation of the dusky large blue butterfly on 64 ha in Landau, the Palatinate (€ 22 per household per year). The fact is that today, every household pays an average of around € 16–20 per year for conservation via public expenditures for nature conservation that are based on their tax payments.

Some authors argue that concrete locally visible measures should be queried as much as possible, this provides a more realistic assessment of willingness-to-pay (Fischer and Menzel 2005). On the other hand, results regarding smaller, more specific measures always leave the question unanswered as to how the group of those questions regarding willingness-to-pay is to be defined: only at the municipality level, or that of the district, of the entire state, or nationwide? When questioned at the local level, one has to deal with the effect that measures in sparsely populated areas tend to always obtain a lower value than measures in densely populated areas, because of the smaller population, and hence the smaller potential willingness-to-pay group. For the valuation of nature as an 'intrinsic value,' this would be a substantively unacceptable result. Moreover, it has been demonstrated that the evaluation of specific measures always includes the implicit distributional assumptions of the respondents ('If I pay for Measure A, I assume that others will pay for Measure B'; Degenhardt and Gronemann 1998). As an evaluation of ◘ Fig. 4.9 shows, a lower willingness-to-pay does tend to be expressed for special measures than for comprehensive measures; however, at the local and regional levels, the willingness-to-pay per measures unit is considerably higher. In the case of the preservation of the dusky large blue butterfly (*Glaucopsyche nausithous*) in Landau, the conversion of the willingness-to-pay results of the population to a per-ha of measure-implementation value yields € 6656/ha/yr. However, in a nationwide programme examined by Meyerhoff et al. (2012), values of only € 1000/ha for the specific grassland-part of the programme, exclusively, were obtained and 300 €/ha if the whole programme was valued.

Fig. 4.9 Willingness-to-pay for conservation programmes encompassing various spatial and substantive factors (in €/mo.). When comparing the data, one matter to consider is that no adjustment was made for inflation. (Adapted and supplemented from BfN 2012 (references other than Meyerhoff et al. 2012 see there))

Actual per ha costs of conservation measures are usually below these figures.

For concrete decisions on conservation projects or interventions at the state or federal levels, the effect due to different population densities, regional preferences or implicit distributional assumptions are not particularly helpful. Such decisions should therefore be based on willingness-to-pay analyses, with which comprehensive programmes have been evaluated. Special willingness-to-pay for individual measures within these programmes could then be roughly evaluated on a pro rata basis, for instance per area segment, or, more accurately, through more detailed expert-based scoring methods (Schweppe-Kraft 1998).

Restoration-Cost Method

A nonpreference-based method for the assessment of existence values is the restoration-cost method. It is especially applied for the evaluation of the functions or services of habitats for the preservation of biodiversity. Under this method, the costs which would accrue if one were to first destroy a habitat and then restore it, are ascertained.

If restoration is required by law, this method is only used to ascertain what a measure, such as the construction of a road, would additionally cost in the form of mandatory compensatory measures. If restoration is not required, it ascertains the costs which would be incurred if society were to recognise in the future that restoration were necessary or desirable. Under economic theory, this approach is acceptable, since international conventions and policy statements such as the European Biodiversity Strategy have made a commitment to a 'no-net-loss' strategy with respect to the conservation of biological diversity. This means that we can–hopefully–assume with a relatively high degree of probability that such a restoration will in fact occur in the future.

A particular challenge in restoration-cost methodology is the monetisation of interim losses of function. Unlike technical infrastructure, the restoration of the biodiversity of ecosystems is not completed with the conclusion of the restoration of physical initial conditions (e.g. termination of intensive use, rewetting), but rather well, beyond that, require a number of years or even centuries. A number of different methods exist for evaluating the interim loss of function (Schweppe-Kraft 1998; Dietrich et al. 2014). In the USA, a discounting procedure within the framework of the so-called habitat equivalency analysis has been widely used since about 1995 for the quantification of damages. Previously, this method had already also been proposed for use in Germany for the assessment of tree damage and damage to habitats (Buchwald 1988; Schweppe-Kraft 1996; ► Sect. 6.6.1). The restoration-cost approach is also used in the German impact-regulation system (Köppel et al. 2004).

If this method is used to assess the approximately 10 % of Germany, which are of particular significance for the conservation of biological diversity, we obtain values of between 50 cents/m^2 for farmland with endangered segetal plants and almost € 200/m^2 for intact raised bogs. The total value of this 10 % of the land area in Germany comes to approximately € 740 billion, which, at the time of calculation, equaled some 80 % of the value of German productive capital (■ Table 4.1).

Such economic valuation methods as cost-benefit analyses have the goal of evaluating the macroeconomic benefits of measures. For local decision-making, however, other quanta are often determinant, such as the effect on regional income and employment, as assessed by Job et al. (2005, 2009) for selected protected areas (■ Table 4.2).

Benefit Transfer

Here, results from other primary studies in which ES-values have already been collected are transferred to the study area and to the services to be tested. There are four stages of *benefit transfer* (Wronka 2004; TEEB 2010): direct transfer, corrected transfer, transfer of evaluation functions and meta-analysis. However, this distinction is of a more or less technical nature. Whether a direct transfer leads to acceptable results, or whether a transfer with an evaluation function is required, depends on the particular problem.

Standard values and simplified evaluation method for the transmission of the value of ES are relatively easy to determine, if the value of ecosystem services is independent of the respective location. One example of this is the value of CO_2 emissions and carbon sequestration. Both have global effects that are independent of the source. The problem in this case more likely involves the correct estimation of the physical effects, which, for example, in the case of the conversion of grassland to farmland, depends on the scope and on the share of organic matter in the soil. Standard restoration costs for the species and habitat-protection functions or services must be defined relatively independently of the location, since the place of compensation is almost always different from the place of impact. For example, nutrient inputs such as nitrates and phosphorus pollute not only the local waters, but ultimately end up in the North or Baltic Seas. Hence, for the nutrient decomposition and fixing services too, uniform values make sense. The same is true for soil erosion (► Sects. 5.3 and 6.6.2). The long-term preservation of the safety of the food supply is a global issue. Long-term shortages or surpluses can therefore also be evaluated on a global scale. The locally differentiating feature would then be the respective agricultural suitability, including soil fertility as an essential input factor.

Benefit transfer becomes more problematical if the value of the service is highly site-dependent.

4

Table 4.1 Compensation values for habitats in Germany, calculated analogously to the *Habitat Equivalancy Analysis* method, taking into account average recovery costs and times (Schweppe-Kraft 2009)

Habitat type	€ per sqm	Area ratio in %	Total value in € million
Heath	41.83	0.22	34,790
Dry and nutrient-poor grassland	8.06	0.27	8037
Molinia meadows	18.51	0.04	2591
Dump floodplain meadows and tall herb communities	6.14	0.10	2315
Extensively used hay meadows	6.14	0.48	10,991
Fens and swamps	9.80	0.03	1088
Extensively used grassland	2.66	1.19	11,897
Extensively used arable land	0.49	1.26	2318
Extensively used vineyards	13.31	0.02	982
Orchard meadows	9.75	0.93	34,125
Extensively used fish ponds	48.93	0.01	1541
Hedges, shrubberies and copses	16.28	2.00	122,100
Natural and near-natural forests	18.44	1.96	135,430
Wood-pastures	20.64	0.09	6594
Low and medium forests	4.47	0.49	8172
Natural and semi-natural forest edges	22.79	0.01	786
Natural and semi-natural forest borders	2.82	0.00	22
Raised bog, natural and near-natural	195.46	0.18	131,914
Transitional bogs and degraded raised bogs	127.42	0.21	100,023
Near-natural standing waters and streams	48.93	0.66	120,698
Total	–	**9.48**	**736,416**

Table 4.2 Economic effects of protected area tourism. (Job et al. 2005, 2009)

	Berchtesgaden National Park (2002)	Altmühltal Nature Park (2005)
Number of visitors	114,100	910,000
Average daily expenditure per capita	€ 44.27	€ 22.80
Gross sales	€ 51 million	€ 20.7 million
Income 1st and 2nd sales stages	€ 4.4 million	€ 10.3 million
Employment equivalent	206 people	483 people

Examples are the recreational performance of landscapes and the prevention of flood damage. A comparably attractive landscape will provide very different recreational services, depending on whether it is located near a metropolitan area, within a familiar tourist area, or in a sparsely populated rural area. The value of the water-retention capacity of forests or floodplains is critically dependent on how extensively and densely populated the flood-prone areas in the drainage portion of the respective watershed are (► Sect. 3.3).

In assessing the capacity of ecosystems to conserve biodiversity using contingent valuation, the question of transferability depends, among other things, on whether the the biodiversity target or programme assessed was local or regional/national in scope (see above).

4.2.4 Conclusion

Economic valuation should be viewed as one decision-supporting method among others. Its main focus of application should be in cases in which the issue is to balance environmental assets and aspects of long-term sustainability, e.g. recreation, biodiversity protection, quality of the residential environment, the self-cleaning capacity of the waterways or soil fertility, against short-term income prospects. It can be used both in decision-making with regard to projects and programmes with negative effects on ES, and for such issues as the amount of money one should invest for the restoration and maintenance of ES.

Some methods of economic valuation are not particularly controversial; for example, there is little doubt that it is useful to have a monetary estimate of the damage costs available when implementing measures that affect the risk of flooding. Nor should there be any fundamental objection against the comparison of costs for reducing the nutrient inputs in agricultural operations into the water, with equivalent measures to increase the self-purification capacity of water bodies.

However, other methods–particularly the stated-preference methods–are indeed controversial. Can we really assume that the statements made by respondents with regard to their willingness-to-pay for maintaining public assets actually reflect their real preferences? How should questions be formulated, and which assets should one ask about, so that the results will be useful in real standard decision-making situations? There is certainly still a great deal of research that needs to be done. According to the existing results, the willingness-to-pay for environmental public goods is usually much higher than what citizens would have to pay in the form of lost income for the maintenance or the provision of these goods.

To date, we are still a long way from having easily applicable valuation approaches for all ES. The criticism that economic valuations address only some aspects of problems therefore often has less to do with the concept of economic valuation. The underlying concept of 'total economic value' (TEV) is based on the preferences of the individual–which is certainly not the worst premise in a democracy–but within that limitation, it sees a very broad range of needs, desires, and motives with respect to the protection and utilisation of nature, which may well also have an altruistic or ecocentric base. If only some of the relevant aspects are to be assessed, as is often the case, this is more likely due to the lack of opportunity, or the necessary resources, to fully ascertain all the effects of the alternatives to be evaluated and assessed. Scientific/ecological impact assessment is often more problematic than economic valuation, as the case of flood protection shows.

In the development of transferable standard assessments or assessment procedures, we are still at the beginning of the development. On the one hand, more primary studies are needed in many areas on which reliable benefit transfer methods could be developed–the travel-cost analysis, which ascertains the quality of areas, has hardly been used at all in Germany; on the other, the development of standards with which those primary studies can be checked for validity is necessary.

Economic valuation is an 'art' that requires a high level of knowledge in the environmental and economic area. Not every economic valuation meets scientific standards. For the uninitiated, this is rarely visible, which can lead to an impression of arbitrariness. De Groot et al. (2002) pointed out that depending on the methods and spatial characteristics in each case, the monetary results of the evaluation of individual ES will vary widely (cf. also above, for the evaluation of agricultural supply capacity). Scientific minimum standards for evaluations could prevent apparent arbitrariness and thus facilitate the acceptance of economic evaluations–especially among those who are not supporters of the interests of the environment and nature.

One of these standards would be the requirement for a generally comprehensible, nontechnical summary, in which not only the total economic value and/or the overall cost-benefit ratio, but also the respective partial values including the explanation of the methods used and their key assumptions would be documented.

Overall, the ES studies which are now extant in large numbers, and which compare the costs and the benefits of measures for the protection of nature and biodiversity, have shown that the usefulness of such measures often significantly exceeds the associated costs. Hence, more conservation and safeguarding of ES lead to an overall gain in welfare.

A critical practice of economic valuation which discloses its assumptions and methods could help business and society find a more sustainable way to manage nature, ecosystem services and biodiversity.

4.3 Scenario-Development and Participative Methods

R.-U. Syrbe, M. Rosenberg, J. Vowinckel

4.3.1 Basics and Fields of Application

Our ecosystems underlie accelerating transitions (Bernhardt and Jäger 1985; Antrop 2005). Some of the reasons are the increased utilisation of renewable energies, globalisation, demographic change and the irresistible urban sprawl. Using scenarios, we can analyse the consequences of these changes for ecosystem services and determine how people are able to intervene in terms of control (Carpenter et al. 2006).

The development of scenarios is only one approach to investigate future trends. Other examples of methods of foresight research are Delphi studies (Dörr 2005), prognoses (Jessel 2000), forward projections (Bork and Müller 2002), the analysis of planning documents, and landscape experiments (Oppermann 2008). However, the discussion of scenarios is deemed to be the key method for argument about sustainability (Walz et al. 2007). It allows a comprehensive examination of the temporal, spatial, and dimensional aspects of ecosystem services (► Sect. 3.3.) since particularly the evaluation of intergenerational justice requires a reasonable view into the future and studies about long periods. Last but not least, the scenario method is a bridging framework for interdisciplinary collaboration on the field of social-environmental research (Santelmann et al. 2004).

Scenarios may be used 'to explore plausible futures for ecosystems and human well-being based on different assumptions about driving forces of change and their possible interactions' (MEA 2005b). A simple definition is 'scenarios are hypothetical sequences of events, constructed for the purpose of focusing attention on causal processes and decision-points' (Rotmans et al. 2000). The aim of a scenario is, therefore, to identify and to compare possible options of action. Instead of only following a single future trend, a tree of possibilities can be explored (Oppermann 2008) enabling to assess the desired and manageable ones among them.

Due to their decision-preparing function, scenarios are part of an action framework and, therefore, suitable tools:

- To draft capabilities in order to prepare for coming occurrences
- To estimate the risk potential of strategies in order to demand for action
- To draft options for action and to compare them in order to choose the most feasible
- To describe the effects of individual measures to other fields of action in order to evaluate the suitability of that measures in a broader area of consideration

Depending on the application purpose, the elaboration of scenarios can be done by experts alone (analytically) or by participation together with actors from policy, economy, NGOs, and the public. The following description of the methodical framework is restricted to the analytical version. Selected participative procedures are presented in a case study below. Both versions can be applied in two forms of expression: Either scenarios are narrated in so-called storylines (Rotmans et al. 2000) using mainly qualitative statements, or quantitative scenarios are calculated depending above all on model simulations. Analytical scenarios are often quantitative, whereby participative approaches have got predominantly a qualitative character. There is also a difference between projective and normative

Well-Known Scenarios About Environmental Issues

Environmental issues were often central for scenarios with both quantitative and qualitative approaches according to the overviews given by Alcamo (2008) and Albert (2009). The first quantitative scenarios used hydrological models (Aurada 1979). A more recent prominent example is the so-called World Water Vision (Gallopin and Rijsberman 2000). The study of Wolf and Appel-Kummer (2005), funded by the German Federal Agency for Nature Protection, addressed consequences of demographic change to nature protection. Several analyses dealt with the impacts of land-use change within rural areas (Dunlop et al. 2002; Nassauer et al. 2002; Haberl et al. 2004; Bastian et al. 2006; Bolliger et al. 2007; Lütz et al. 2007; Tappeiner 2007; Tötzer et al. 2007). But also shoreline and sea issues were central for scenarios, such as the North Sea (Burkhard and Diembeck 2006) or the Great Barrier Reef near Australia (Bohnet et al. 2008).

An increasing number of recent publications evaluate environmental scenarios using landscape functions or ecosystem services such as Dunlop et al. (2002), Nassauer et al. (2002), Fidalgo and Pinto (2005), and Seppelt and Holzkämper (2007). The Fourth Assessment Report (Pachauri and Reisinger 2008) of the Intergovernmental Panel on Climate Change (IPCC) addresses the effects of climate and socio-economic changes to a large number of ecosystem services at the global level. Examples of integrated man-environmental research through scenarios are the Millennium Ecosystem Assessment (MEA 2005b), which includes foresight and backsight analyses of 50 years, and the Global Environment Outlooks of UNEP, of which the fourth generation is available (UNEP 2007) and the fifth one is under revision (UNEP 2011).

scenarios. The former searches for the implications of assumed trends and the latter starts with (desired) future goals and explores how to act in order to meet them (Nassauer and Curry 2004).

4.3.2 Framework of Scenario Development

The methodical framework presented below is particularly designed for scenarios of landscape development that should be evaluated by ecosystem services. The framework was tested on the county of Görlitz within the Landscape Saxony 2050 research project (funded by the Saxon Department of Science and Arts). The scenario methodology consists on a combination of approved single procedures and fits them to the problems of landscape development. The methodical basis includes the works of Reibnitz (1991), Gausemeier et al. (2009) from business science and Alcamo (2008) from environmental science.

The framework uses an explorative forecast approach. This approach is open-ended, i.e. there is no direction and range of developments set from the beginning. Quantitative and qualitative approaches can dominate or be combined. The framework consists maximum of seven steps. Depending on the main question and application task, not all steps have to be run-through completely. ◘ Figure. 4.10 gives an overview of this method.

Step 1 comprises, first, the organisational preparation of scenario process, second, the formulation of a principal question and, third if necessary, a specification by core topics. The principal question defines the overall objectives. A time horizon and the delineation of the study area belong to that. If the principal question is rather complex, the object of investigation should be confined by core topics. Regarding the case study, the time horizon (2050) and the study area (Görlitz County) were fixed, but the principal question was defined rather broadly as 'How will the ecosystem services be altered due to future landscape change?' Therefore, the principal question had to be specified using the two core topics 'biodiversity' and 'renewable energy' that were treated separately. Both topics were very important in political and social debates.

Step 2 consists of the selection of driving forces and ecosystem services that should be considered. That is, the scenario expert team has to select which drivers are interesting to the principal question in respect to the core topic and the impacts they have on the ecosystem services (ES). Therefore, the selection of drivers and ES has to be done simultaneously since both depend on each other. A good selection and precise definition of driving forces is crucial for the whole scenario development because if the

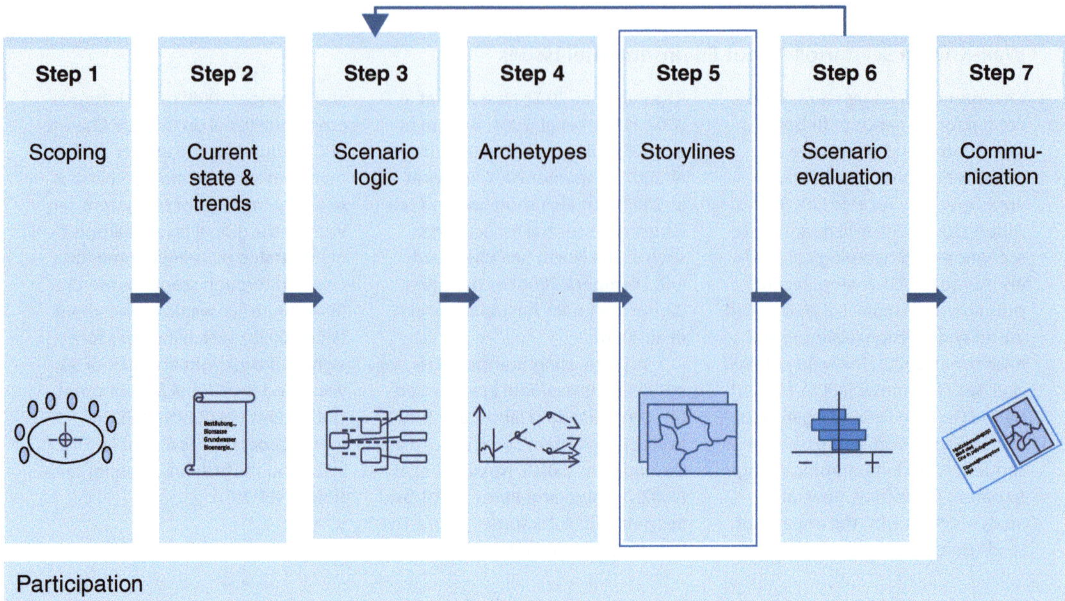

Fig. 4.10 Working steps of the described scenario framework. © IÖR/Syrbe

selection is to broad it hampers the communicability of scenarios. If the drivers are too imprecise and cannot be described by clear indicators, they will complicate the discussion as well as the quantitative processing. One bad example would be choosing 'energy and mining' as a driving force since several directions of development could be implied. On closer consideration, hundreds of driving forces can be identified. But only a small number (<10) must be considered and each of them should be describable by a single measure and a known actual value. For this, thorough investigations are necessary, which will also be useful later on.

Step 3 defines the logical scenario structure. The main purpose of scenario development is to draft different future visions. To do so, the drivers that are to be variable within the scenario process need to be chosen. A differentiation can be achieved connecting the variable drivers with diverse trends. Of course, this differentiation is only possible for a small number of drivers. Empirically it does not make sense to use and vary more than three key drivers concerning the amount of work and the straightforwardness of the whole process. The other drivers are defined to be unvaried between several trajectories. The unvaried drivers are called framework conditions and must be described as accurately as possible using also external prognoses or expertises. On the contrary, the variable drivers open up the possibility space of scenarios and, thus, are called key drivers.

Step 4 implements the abovementioned logical scenario structure. Therefore, an overview of the current situation is needed. An initial ES assessment should be made using the middle pillar of the EPPS framework (► Sect. 3.1.2) unless it already exists. The key drivers have to be connected through a small number (commonly two by four) of trends concerning their future development as it is interesting for the principal question and also relevant for altering the ES under consideration. The trends may not only be linear but can also be defined accelerating, retarding or erratic. The description does not need to be exact, but rather generic. An established way of description is using pictograms for the several trend types (■ Fig. 4.11). Not all trend combinations can be combined because contradictions are possible. An appropriate number of plausible combinations (so-called bundles) must be selected. These bundles guide the initial ways to develop and describe scenarios in detail, which will be done in step 5.

Step 5 contains the wording and specification of scenarios. The selected bundles enable to derivate several future trajectories. They receive short

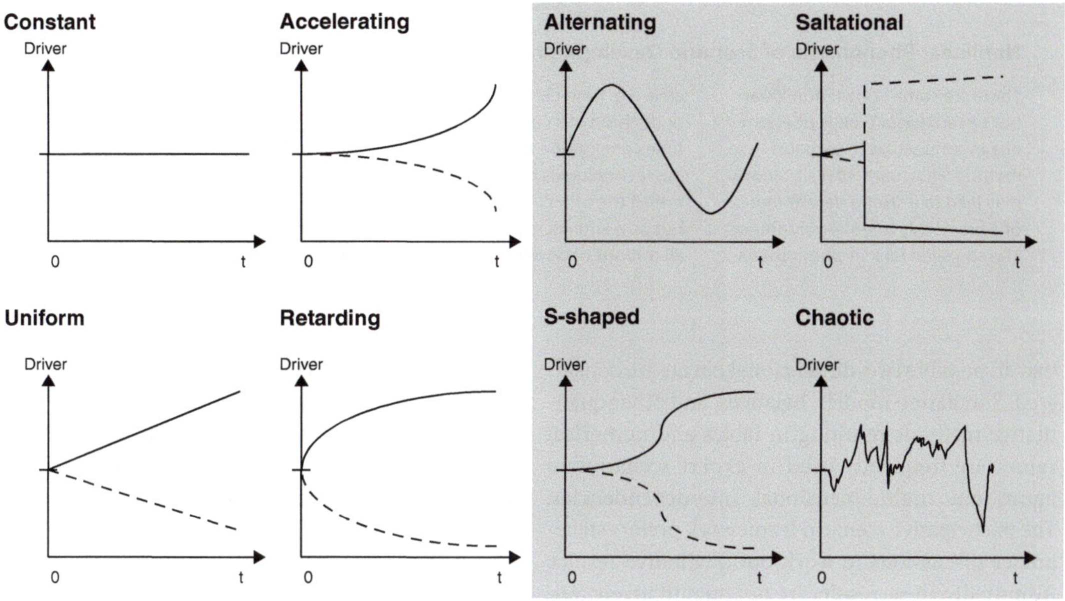

Fig. 4.11 Pictograms for the trend types of key drivers; *strait line*: positive alteration; *dashed line*: negative alteration; *white background*: basing types; *grey background*: combined types. © IÖR/Syrbe

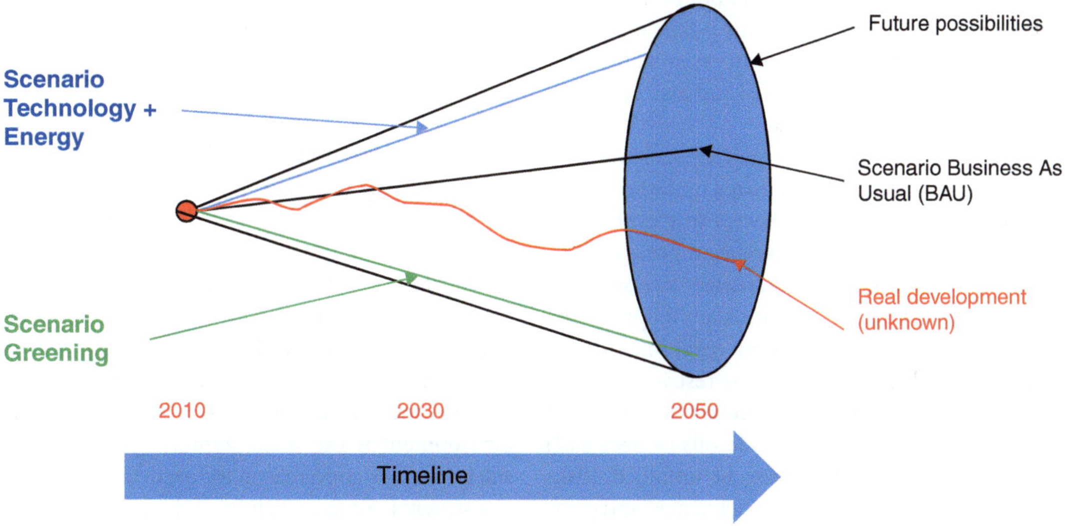

Fig. 4.12 Scenario funnel with schematically hinted trajectories. © IÖR/Syrbe

names characterising the assumable end points in future, the so-called archetypes. For instance, the archetypes of Landscape Saxony 2050 scenarios read 'Business as usual (BAU)', 'Greening' and 'Techno + Energy' Fig. 4.12. The core result of this step is a storyline that describes the future situation (sometimes also the steps towards it) and that give reasons for the most important conclusions. To achieve this goal, the interdependencies between all drivers (variable and framework conditions) have to be analyzed. The so-called cross-impact analysis can be treated with the help of a matrix to ensure

Nonlinear Phenomena of Scenario Development

There are some known nonlinear but nevertheless typical phenomena in connection to scenario method: First, particular situations may lead to a strong determination of a previously open development. The so-called lock-in-phenomena arise e.g. from exhausting resources or decision of a competition. Sometimes one option among competing technologies can win and outlive all the others. Second, a seldom but powerful incident could change all options of development. These so-called 'wild cards' should be discussed separately from the main round because many participants are not frank enough to accept them, even though their treatment may be important for taking precautions for the future.

that all possible two-dimensional effects are considered. Simulation models, balances, and other quantitative methods resulting in tables and numerical values are frequently used in expert scenarios to figure out multidimensional interdependencies. The participative scenario framework prefers stakeholder discussions to work out qualitative results. Admittedly, these results are not quantitatively representable but often more complex. A proven tool to facilitate the discussion is scenario mapping. To draw items into a map gives an overview of spatial dependencies and helps to figure out possible environmental conflicts as well as the points of interest for the actors. These maps are an essential basis for a subsequent evaluation (step 6) and instructive abstracts of scenario outcome.

Step 6 is the evaluation part of scenario outcome. Storylines, tables and maps underlie a comparing evaluation to give answers to the principal question and to ensure the scenario process quality. The evaluation can be spatial or nonspatial depending on how the scenarios are mapped. The evaluation of scenario outcome regarding ecosystem services does not need to be restricted to singular values. Rather, the future cross-impacts of the services, their so-called trade-offs (▶ Sect. 3.1.2) as well as synergies should also be unfolded. Risks and suitability areas should be delineated and compared. The main purpose of this step is to draw conclusions from scenario results for management options and possible future strategies. The aim is not only to figure out the best storyline but also the best measures that will accomplish this. It is possible that a repetition from step 4 onward is necessary to specify them anew and to rethink the scenarios therewith.

Step 7 comprises all measures of scenarios' communication and participation with the concerned actors (or customers). The participation tools are specified in the next section (▶ Sect. 4.3.3). Although it is placed as the last step, participation shall start with the beginning of a scenario development and pervade throughout the whole process. This way, the methodology can have some loops between mainly expert-oriented steps and steps with more participation. At the interface between both modes of work, data must be translated into easily comprehensible presentations, and meanings have to be quantified the other way around. Lastly, the scenario results have to be published at the end of the process to enhance public awareness and (hopefully!) application.

4.3.3 Participation and the Case Study Görlitz

'Participation' is the cooperation of actors, stakeholders or interested individuals within a scenario development or during an assessment; the concerning method is called participative. The main reason for the inclusion of decision makers by participative methods within an assessment or a scenario development is the social appreciation of the results. Another good reason for participation is that assessments are most helpful if the users take part in it (Carpenter et al. 2006). Additionally, participative scenario workshops reveal educational effects for the participants (Alcamo 2008). Therefore, it is recommendable to involve young people, particularly if long-term scenarios are being developed.

The cooperation with participants that are lay people within methodically sophisticated methods is challenging regarding the quality of communication. Experts must be able to interest people and engage them to get involved in the cause. The crucial problem is to ensure a comprehensible flow of information from scientific knowledge to messages in normal language and vice versa. Therefore, a pool of hints shall be proposed, which may be extended in several ways.

Types of Participation for Development of Scenarios or Ecosystem Services Assessment

- Workshops (with group work, presentations and perhaps stage discussion)
- Small group participation events such as world café or focus group interview
- Personal interviews (survey with prepared questions or thematic guideline)
- Public surveys (oral, by letter or on the web)
- Stalls at exhibitions, fairs or congresses
- Excursions (empirically with high motivational effect)
- Culturale events (cinema show, theatre and suchlike) with following discussion
- Teaching units in schools, other educational institutions, or outdoor
- Internet forum, blog, etc.

The participative work on scenarios, mainly using a workshop, is called a scenario exercise. It is the methodical core of the whole scenario development. The most important steps of ► Sect. 4.3.2. have to be handled therein. The scenario exercise should be combined with the working steps that are executed only among experts as well as with alternative forms of participation (► Box 'Types of Participation'), in order to minimise time exposure for the participants, to activate them without boring them, and to ensure a high degree of involvement also for those who are not keen on debates.

Elements of a Scenario Exercise

- Invitation of genuinely interested participants
- Introduction: explanation of aims and methodical steps
- Mind opener to stimulate creativity (e.g. quiz)
- Brain-storming to catch maverick ideas
- Suggestion talk(s) by experts
- Ballot about alternative proposals (e.g. by stick points)
- Plenar discussion for central decisions
- Working groups developing particular scenarios
- Breaks with social events (e.g. dinner)
- Plenar presentation of working group's results with final discussion
- Protocol shipment of the final results to all participants

The actual scenario exercise can consist of several elements (► Box 'Elements of a Scenario Exercise'). All essential information including the time frame must be communicated with the invitation beforehand to avoid the worst case: unsatisfied participants frequently discussing off-topic issues or query the meaning of the exercise in general. The first important topic on the schedule should be an introducing explanation of sense, aim, and background of the exercise, eventually completed by a short lesson on scenarios. Second, a so-called mind opener can help to get the participants in the right mood to bear creative ideas and to break away from their everyday problems, as well as to prevent them from judging prematurely. Therefore, unexpected questions, a quiz, or a flashback into the past can be recommended. These elements can also be used later to make the event less formal. The actual scenario discussion shall be done preferably in working groups. Intermediate results have to be retained periodically to ensure the progress of discussion. Spontaneous ideas should be recorded neutrally at this point and systematised only later. Because one-day workshops can be very exhausting and will only be successful for good teams, Ringland (1997) recommend two half-day rounds instead, which can be separated by an informal evening event. Graphical, textual, cinematic and interactive media help to facilitate the discussion if they are specially geared to the participants.

Some of frequently made mistakes should be mentioned. A possible participants' irritation due to incomplete information has already been noted. Additionally, frustration can arise from overloaded presentations, a boring schedule, or too slow progress in scenario elaboration. To avoid such undesirable situations, breaks should be inserted that can be used by the scenario experts to develop intermediate results further and enrich them by additional information (i.e. from simulation models) to get a faster progress and make the meeting more interesting for the participants. The hope to get quantitative data by a negotiation among actors would be mostly disappointed: data requested from participants remain often incomplete and vague; therefore, they must be completed and sophisticated by experts work. Often, a successful participation process needs more preparation time than execution time (Walz et al. 2007).

Tips for Planning a Successful Scenario Exercise

Timely invitation of participant
- Information about the venue, aims, duration, and fee as well as possible cost reimburse
- Invitation shall be motivating, provoking, exciting, or funny
- Homework (i.e. a questionnaire) can save a working step and prepare for the topic

Introduction by the scenario team
- Aims and schedule of the whole project and of the particular event
- Introduction should be short, but include organisational information (breaks, meals, etc.)
- Introduction highlights the possibilities of participation

Mind opener to activate creativity (possibilities)
- Enquiring wishes or nightmares for future
- Asking to draw an own desire scenario
- Provoking (i.e. through theses or artistic illustrations)
- 'Fairy question': 'What do you want to ask a time traveler from the future?'

Brain storm to obtain creative ideas before people hear lectures
- Ballot about drivers or evaluation criteria
- Nomination of surprising incidents to be regarded ('wild cards')
- Risks and problems for future

Key note lectures from scenario team and external experts
- Participants get comparable information as basis for discussion
- Sharing the most recent state of the art about trends and drivers
- Current state of the study area

Group work to draw particular scenarios
- Avoid strong/weak division of working groups to not confine the creativity of the weak group
- Group division should consider the interests of members
- Each group needs a moderator from the scenario team
- Job description must be prepared for groups and moderators
- Each group elects a presenter at the beginning

During two projects ('Landscape Saxony 2050' and 'LÖBESTEIN') in the East Saxon county Görlitz, Germany, additional experiences from scenario workshops were collected and will be shared below (► Box 'Experiences from Görlitz as Regional Example'). The authors developed participative scenarios about the increasing use of renewable energies and the protection of biodiversity there.

4.4 Complex Analyses, Evaluation and Modelling of ES

4.4.1 Background

K. Grunewald, G. Lupp

"To make simple things complicated, is daily routine, to make the complicate things simple, this simply is creativity. (Charles Mingus)"

Nature, our environment, and society are complex systems. Complexity means that, the reaction of a system is not predictable as a whole even if we know single reactions and interactions of its components precisely. The characteristics of complex-

Experiences from Görlitz as Regional Example

In the beginning, a world café event, where participants visited several thematic tables to discuss input variables (drivers, trends, wishes, aims, standards, values) in brief sequence, was organised.

The workshop preparation was done by Internet surveys. Online tools such as ▶ http://kwiksurveys.com/ are available that are easy to design and able to provide statistically edited results. Unfortunately, a personal email address of all participants must be known. Preconditions to use this tool are the participants' accordance and engagement. The tool worked well among the internal and external experts but not with the other participants. Therefore, survey forms (as PDF, per email of fax) were sent out in order to involve all interested actors. However, long word/excel query catalogues could not be used succesfully since they were not returned on time and fully completed except by the respective expert team.

In the workshop, statements from several experts were discussed and enriched by additional thoughts. However, the self-introduction round of participants occasionally escalated to time-consuming talks. Good experiences were made with three questions asking for one-sentence answers from all participants in the beginning (who are you, how do you feel about the topic, what is your intention). The selection of trends, drivers, and trajectories is not suitable for a full auditorium and should be implemented in other ways (see suggestions above). A good scope was to deliver several proposals that the actors could choose by participation in specific working groups or table discussions. After a certain period of difficult discussions, a change through playful insertion was appreciated. Group works with about 5 participants each were most efficient. Many participants were skilled in handling maps and used them to discuss allocation questions intensively. Therefore, well-prepared maps and drawing utensils were valuable. The moderators must keep in mind the time frame as well as those participants who don't impose themselves in discussion and activate them directly.

ity are numerous elements that interact with each other and the reaction as a whole is unpredictable (Riedel 2000). Examples for complex systems and limitations for their predictions are, for instance, weather forecasts, the prediction of market trends at the stock exchange, but also the reactions of ES. Disturbance of complex ecosystems might lead to severe and irreversible new states (SRU 2007). Land management can be considered a complex system. Land use and forestry affect nature in many ways, e.g. water cycling, soil fertility, biodiversity or regional value adding (▶ Chap. 6).

Complex or Complicated?

An airplane is a complicated thing. It consists of many different parts. However, it does not contain a real secret. This means, difficult tasks can be solved by knowledge.

A five-course meal is complex. You have to know the different ingredients. But when you prepare the different dishes, it does not necessarily means that you are getting a delicious meal. Systems with many different interactions not working on a simple 'if-then' principle are dynamic and multilayered and, thus, are complex.

The aim of the ES concept is to cope with the challenge of interactions and complexity of ecosystems and to describe impacts and consequences for human well-being. A comprehensive assessment of ES demands enormous efforts and is only partially adequate to serve as a basis for politicians and stakeholders to support decisions by involving all different demands.

By breaking down, abstracting and weighting complex issues are simplified. Therefore, just like in a caricature, they are easier to understand through simple and concise means. With simple means and a few lines, significant and striking attributes of a person or a situation can be drawn. Complex systems can only be determined by observations of patterns. They can be observed in the abiotic and biotic environment or in society (e.g. different soil substrates, routines, behaviour). ES patterns can be analysed with a matrix of supply and demand for certain land-use types (▶ Sect. 4.1) and within Contingent Economic Assessments (Examples in ▶ Sect. 4.2 and ▶ Chap. 6).

Visions and intentions like the concept of 'sustainability' and the 'ES-concept' could be seen as a tool to influence patterns and types of land uses. If new patterns occur in complex systems, a tipping point has been crossed. One of the goals of research on ES is to figure out tipping points and how they are influenced by human activities. It is one of the core challenges to determine the development of those systems (scenarios, alternatives, ▶ Sect. 4.3; modelling, ▶ Sect. 4.4.3) and forms a basis for regulation and steering (policy, incentives, planning, governance ▶ Chap. 5).

The ES concept is intended to support solving and balancing complex problems with tools and methods. It strives for integrated approaches by analysing, assessing, and weighting ES based on scientific methods by using all available information while including human needs. The ES concept requires weighting between quick and cheap assessment procedures (e.g. rough estimations based on 'rapid evaluation tools') and more detailed, elaborated, time demanding, as well as more expensive examinations (intensive assessment of all different ES aspects).

In the following section, a broad application of the ES concept will be presented using a case study on 'impacts of an increased biomass production for energy purposes'. It shows how ES can be selected and assessed, how different approaches for evaluation ES can be used, and how regional stakeholder can participate in these processes. Finally, the ES model 'InVest' is presented demonstrating its use and describing strengths and weaknesses of this model.

4.4.2 Energy Crop Production–A Complex Problem for Assessing ES

G. Lupp, O. Bastian, K. Grunewald

The increased production of biomass for energy purposes is a prime example for the increased use of ecosystems driven by strong political interest. The European Commission has set mandatory targets for all member states for the use of renewable energies. The share of renewables has to double between 2010 and 2020 according to this policy. Half of the renewables share is to be derived from biomass (Commission of the European Communities 2007). With respect to conflicts and minimising impacts, the EU commission has developed a biomass action plan and requested all member states to develop national biomass action plans. The German biomass action plan (BMELV/BMU 2009) emphasises climate protection, regional value adding, the strengthening of rural and peripheral regions, and the protection of biodiversity, soil fertility, waters as well as air quality as the core goals for biomass production using annual energy crops or woody biomass.

To achieve these goals and to minimising conflicts, stakeholders have to be included, and the acceptance for biomass has to be increased by informing and involving the lay public through adequate communication and consultation (BMELV/BMU 2009). Although ES are not explicitly mentioned in the document, ES have to be secured and enhanced in a sustainable way when energy crops or woody biomass are cultivated. This document already indicates possible methodological steps and approaches for assessing impacts of biomass production on ES.

In order to improve ES and biodiversity protection in sustainlable land-use management practices, we suggest the following steps (see also ◘ Fig. 4.13 and Lupp et al. 2011):

First, relevant economic and ecological elements, especially ES, have to be selected. In the case study food and feed production, provision of energy derived from wood and energy crops, variable cross margins for farms, biodiversity, carbon fixation, pollination, provisioning of drinking water, water discharge regulation, erosion control and outdoor recreation opportunities were chosen.

In our work, we follow the '**DPSIR**-steps' (*Driving Forces, Pressures, State, Impact, Response*) according to the OECD (2003). This approach involves a system-analysis view and describes a methodological procedure for characterising the impacts of socio-economic activities on the environment and ways to reduce or halt these impacts (BAFU/BFS 2007).

In the first step, the *Driving Forces* of an increased energy crop cultivation and timber extraction are assessed by analysing energy policies, regulations set by legal instruments and incentives

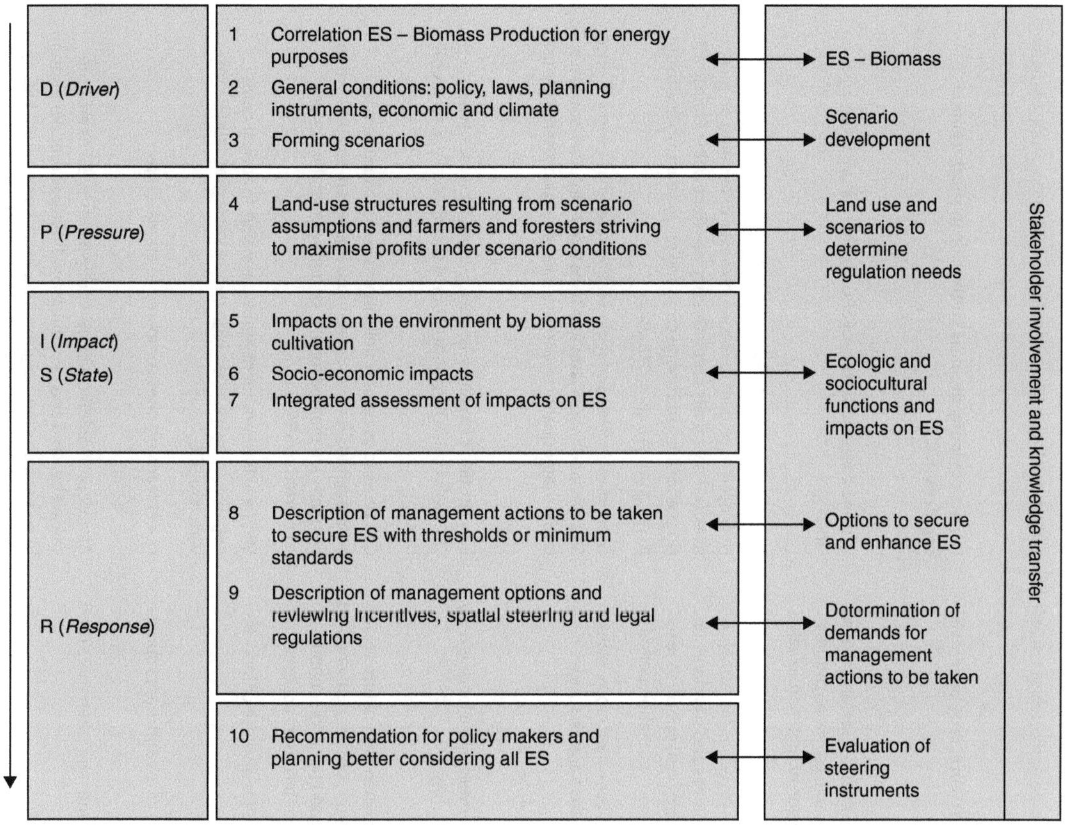

■ **Fig. 4.13** Schematic approach to assess and evaluate ES in different scenarios

as well as economic situations and climate conditions. Based on these findings, land-use scenarios are developed. By using scenarios, future land-use patterns (*State*), their impacts (*Impacts*), and *Pressures* on ES can be determined. Using this data, necessary actions to maintain or improving the provisioning of ES can be identified and possible options for improved regulations (*Responses*) can be developed (■ Fig. 4.13).

To cope with the challenges and adaptation of land management concepts, regional approaches at the landscape level seem to be among the most promising since influencing factors and the demand for specific solutions may differ (Rode and Kanning 2006). Case study regions to be selected should provide heterogeneity. Although certain factors might have global impact, different landscape units might react completely different.

To address dimension and different landscape scales, different types of units should be assessed reaching impacts on regional level down to individual land parcels. The latter is important for putting objectives into practice by farmers and foresters to carry out precise management actions to support certain species, e.g. to maintain deadwood in forests for birds and insects or provide patches for skylark (*Alauda arvensis*) in intensively managed fields as nesting habitats.

Different energy crops and the way they are cultivated lead to specific impacts on ES, some examples can be found in ■ Table 4.3. But also different natural conditions or landscape character might influence impacts on ES.

In an integrated assessment, different ES can be compared with each other. For example, so-called spider-web diagrams can be a suitable instrument to describe them (■ Fig. 4.14).

■ **Table 4.3** Impacts of energy crop cultivation on ES on the example of corn, rape and cereals compared to cultivation of woody biomass in short rotation coppices (derived from Lupp et al. 2011, modified)

ES	Factor	Impacts of annual energy crops (corn, rape or cereals)	Impact of woody biomass cultivation in short rotation coppices (SRC)
Provisioning Services			
Providing biomass for energy purposes	Land use, crop rotation	Increasing food prices, increasing prices for land tenure and land buy; crop diversity is reduced, monocultures increase, in some cases conversion of high nature value grassland into fields	In some cases cultivation of SRC on high nature value grassland
	Income of farmers, income by land tenure for land owners	Increased opportunities to gain revenues for farmers, however mainly driven by incentives and payments under the German Renewable Energy Act; option to react to different market demands every vegetation period	New options to gain revenues, however depending on incentives and payments under the German Renewable Energy act; No possibilities to change crops for 20–30 years, no flexibility to react on changing demands
Water provisioning	Groundwater level, moisture content in soils	Groundwater levels depend on cultivated crops, water quality can be affected by fertilisers and pesticides	Groundwater level and regeneration rates may be lower since SRC demand more water and have higher interception rates; Better water quality due to lower fertilising and herbicide use
Regulatory Services			
CO_2-storage	Energy input for planting, fertilizing, pesticide treatment and harvest	Substitution of fossil fuels, however quite a lot of energy input for cultivation of annual crops	Substitution of fossil fuels; energy input increases, when SRC wood chips receive additional treatments (e.g. technical drying)
	CO_2-sink, emission of greenhouse gases	Greenhouse gases are emitted when high nature value grasslands are converted, energy crops are cultivated on marshlands and drained bogs, additional greenhouse gas emissions when nitrogenous fertilisers are used	Can be used as carbon sink, humus and subsoil biomass is accumulated, however, greenhouse gases are emitted, when wetlands are drained for SRC or high-growth forests are converted to SRC
Nutrient and humus regulation	Nutrient input	Often high nutrient input and excessive use of biocides	SRC have lower nutrient input compared to annual crops; negative impacts, when high nature value grassland is converted to SRC
	Nutrient leaching	High nutrient leakage for some crops and inappropriate farming practices	Lower leaching rates compared to annual crops
Reduction of soil erosion	Vegetation cover	High risks for erosion by wind and rain, when there is no cover by crops	Almost permanent cover by woody plants reduces erosion caused by wind and rain, quick and Intensive root penetration lowers soil erosion, neighbouring fields benefit from SRC

Table 4.3 Continued

ES	Factor	Impacts of annual energy crops (corn, rape or cereals)	Impact of woody biomass cultivation in short rotation coppices (SRC)
Water retention	Water runoff	Intensive water runoff when there is no vegetation cover (relevant especially for corn)	Lower water runoff, water retention effects increase with increased height and size of SRC
Water purification	Pesticides in groundwater bodies, cost for water purification	High nutrient input in groundwater leads to increased costs for water purification, water quality might not fulfill minimum standards set by law, water has to be provided from less affected regions	Lower nutrient input in groundwater and water purification compared to annual crops
Groundwater recharge	Groundwater level	Higher groundwater recharge rates compared to perennial crops, SRC or forests	Lower groundwater recharge rates due to dense vegetation cover, higher interception rates compared to annual crops, groundwater levels might be lowered
Pest control	Resilience Amount of biocides used	Higher pest risks, greater hazards due to increased spreading of European corn borer (*Ostrinia nubilalis*) or western corn rootworm (*Diabrotica virgifera virgifera*), increased use of biocides or genetically modified plants	Higher resilience for pests, pest infestation is often less harmful for perennials, however problems with rust (*Melampsora*), beetles (*Melasoma populi*) and rodents (e.g. mice)
Pollination	Amount of pollinating insects	Corn and cereals are wind pollinated and provide no nourishment for pollinating insects, rape has only a short flowering period in spring, Intensive energy crop cultivation reduces accompanying vegetation, therefore no larger amount of permanent sources for pollinating insects	Trees are harvested before they are mature, some willow species can be used as nectar source
Biodiversity	Cultivated species, types and proveniences, cultivation patterns and forms of land use	Intensive cultivation, few different species, genetically modified species, few accompanying vegetation, harvest before maturation	Few high yield (hybrid) species, genetically modified species, few structures within a SRC, however longer rotation periods with 2–5 (20) years
	Bird species	Intensive cultivation provides little habitat for species dependent on less intensive land management practices, e.g. ortolan bunting (*Emberiza hortulana*), skylark (*Alauda arvensis*), early harvest in summer before matureness reduce feed for species depending on grains (e.g. migrating birds like geese)	Habitat and nesting for bird species depending on hedges and woodland, however mainly ubiquist species; conversion of fields reduce habitat for ground nesting birds

Table 4.3 Continued

ES	Factor	Impacts of annual energy crops (corn, rape or cereals)	Impact of woody biomass cultivation in short rotation coppices (SRC)
	Accompanying vegetation and fauna	Few accompanying vegetation, few habitats for animals (e.g. insects and small mammals), Intensive soil treatment and biocides	Accompanying vegetation and fauna increases but mainly ubiquist species; when managed with longer rotation periods, SRC can contribute to some extent for structuring landscapes and contribute to a green infrastructure; however loss of habitats, when SRC replace low impact agriculture, fallow or marginal land or high nature value grasslands
Sociocultural Services			
Ethical values	Share of food production compared to energy crop cultivation	The use of cereals and other plants suiting for human nutrition is perceived negative from an ethical perspective (increasing prices for food and starvation in developing countries)	No direct ethical conflicts, since woody biomass derived from SRC is non edible; however indirect effects: Farmland is converted for energy production and lost for food production
Identity	Cultural heritage, traditional farming practices	Traditional land-use forms and specialised crops disappear, cultivation of less productive species diminish	New landscape elements, can sometimes pick up traditional land-use forms or act as modern interpretations, willow cultivation (*Salix spec.*) used to be a common landscape element used for basket-weaving in the past, today in the context of modern, sustainable energy production
Education value	Energy crops in research projects	Only few species, research mainly on increasing yields	New land-use form, object of numerous and multidisciplinary research
Landscape aesthetics	Diversity of landscape patterns	Mainly monocultures and large unstructured fields, shortened crop rotation, however seasonality and different appearance of fields during the year; attractive yellow rape blooming in spring	Can structure large agricultural landscapes
	Visibility, vistas, viewing unique, landscape features	Visibility is reduced and vistas blocked in summer (e.g. high corn stands), distant views in winter	No visibility, vistas are blocked except a few months after initial planting and for a few months after harvesting

Based on data collected by Kort et al. 1998; McLaughlin and Walsh 1998; Börjesson 1999a, b; Heidmann et al. 2000; Liesebach and Mulsow 2003; Londo et al. 2004; Windhorst et al. 2004, Bringezu and Steger 2005; Burger 2005; NABU 2005; Rode et al. 2005; SRU 2007; Bardt 2008; Lee et al. 2008; Ericsson et al. 2009; Hillier et al. 2009; Rowe et al. 2009; Cherubini and Stromman 2010; Greiff et al. 2010

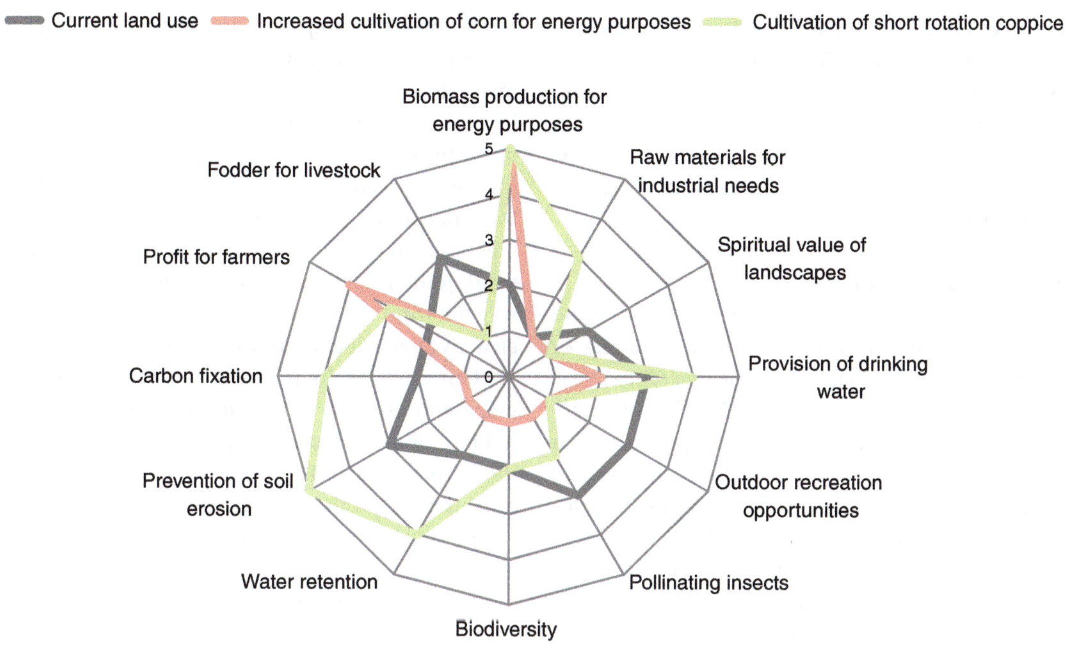

Fig. 4.14 Exemplary diagram of ES modification by energy crop cultivation

Scenarios

Scenario analyses aim to determine impacts of biomass production. Undesirable effects (Trade-offs, *disservices*) should be eliminated or at least be minimised. As demonstrated in ▶ Sect. 4.3, scenarios are suitable to evaluate the time and space aspect and to compare and weight different resulting developments or different options for action. Scenarios also provide many possibilities to involve stakeholders.

In the Moritzburg small-hilly landscape 10 km north of Dresden, expert-based scenarios were created describing impacts of different policies resulting in distinct laws and incentives like EU common agricultural payments for farmers. These assumptions lead to scenarios allowing for impact assessments for different possible developments. The three scenarios are:

- First scenario: Abandonment of livestock
- Second scenario: Biomass production for energy purposes
- Third scenario: Optimising ES from a nature conservation point of view

All three scenarios lead to different land-use patterns. In the 'Biomass' scenario, the share of corn increases. High-nature-value grassland along rivulets will be replaced by short-rotation coppice. Land use will be intensified to compensate the loss of agricultural land needed for biomass production. The third scenario 'optimising ES' will result in diversified land-use patterns.

Biophysical Approaches

To assess the impacts of an increased cultivation of energy crops on biodiversity and ES, expert-based approaches of landscape planning can be used. They are described in many methodological handbooks e.g. in Bastian and Schreiber (1999). Usually, semiquantitative assessments of the landscape functions, a subject of protection, a potential or risks, or–speaking in terms of provisioning–ES are carried out. Usually a five-step Lickert scale is used stretching from 'very good condition' to 'very bad condition'. Items evaluated are e.g. erosion sensitivity, scenic quality or biodiversity that might be affected by large scale monocultures like rape or corn (◘ Table 4.3). Often impacts on eye-catching species like skylark or lapwing are analyzed. They serve as umbrella species for certain types of habitats or groups. Choosing them helps raising awareness among different stakeholder groups and lay persons for more conceptual approaches like biodiversity or ES.

▪ **Monetary Approaches**

Many ES can have economic values, e.g. a demand for ES on markets exists or the provisioning or maintenance has costs (Baumgärtner 2002), e.g. forest growth simulators like SILVA 2.2 also integrate economic evaluations (Pretzsch 2001). For agriculture, econometric decision models exist and also provide information on economic effects of different management objectives (Kächele and Zander 1999). With these models, decisions of foresters and farmers can be described and effects of legal or regulatory frameworks can be implemented (e.g. mix of different tree species or crop rotation). The models describe developments when managers would solely act in rational profit maximising terms.

Another option is to use opportunity costs (▸ Sect. 4.1). They quantify losses, which derive from maintaining low impact practices on fields in favour of biodiversity. For example, it can be calculated how much money would be lost if a farmer does not cultivate small patches in large-scale fields to provide habitat for skylark (Brüggemann 2009).

▪ **Demand-Based Approaches**

One option to assess the demand for ES are surveys among the population. For example, the author-conducted interviews at different locations within the study area led to interesting results. The provisioning of drinking water and habitat for plants and animals is considered to be very important for the interviewees, while providing renewable energy from biomass is almost irrelevant.

▪ **Transdisciplinarity and Participation**

Transdisciplinary approaches are characterised by close cooperation between researchers and practitioners. The idea is to implement the work to solve real-life problems (Müller et al. 2000). Participation means active involvement of stakeholders and other interest groups in decision-making (UBA 2000; Förster et al. 2001). Therefore, it is useful to integrate key stakeholders in each research step, to let them participate, and to involve them actively in the project process. For example, it is possible to involve them in the scenario work, e.g. by letting them decide about key drivers (▸ Sect. 4.3). To mobilise stakeholders from different institutions, activating interviews can be conducted to see which way the wind is blowing and to produce interest in an active participation in workshops (L.I.S.T. 2011).

Our own results in the Upper Lusatian Landscape and the Ore Mountains showed that stakeholders and land users often do not decide on the basis of maximising profits, but also consider non-monetary values like traditions, attitudes, and even ethical values. They are often convinced that providing different ES for society is very important, even if they are unfamiliar with the concept of ES.

▪ **Regulation of Energy Crop Cultivation**

The cultivation of energy crops and woody biomass is mainly regulated by market prizes, incentives paid to farmers under the European Union Common Agricultural Policy and direct or indirect payments under the German renewable energy act (EEG 2008). It is therefore necessary to assess different steering instruments to see whether they can regulate energy crop production effectively and what impacts occur on ES.

It can be stated that only single ES are considered in laws and incentives in Germany, and they not as a whole. Often, no Safe Minimum Standards are defined. Laws often demand that 'deterioration has to be prevented' or 'good farming practices' have to be used. However, 'good farming practices' are more a mere code of conduct rather than safe minimum standards (Hafner 2010).

4.4.3 Application of Models of InVEST to Assess Ecosystem Services

M. Holfeld, M. Rosenberg

Models are representations of reality. They might be images, intellectual and linguistic constructs or mathematical formulas. The modelling of ecosystem services initially provides an abstract representation of ecosystems, of processes taking place and of potential changes. This is already covered by ecosystem models to a quite good extent. The challenge, however, is to incorporate the demand and benefit into the models.

In this respect, the following model approaches are currently of special relevance: *Integrated Valuation of Ecosystem Services and Tradeoffs* (InVEST, ► www.naturalcapitalproject.org), *Artificial Intelligence for Ecosystem Services* (ARIES, ► www.ariesonline.org), the BGS *ecosystem services model* (► www.bgs.ac.uk) and *Multi-scale Integrated Models of Ecosystem Services* (MIMES, ► www.uvm.edu). All these approaches aim to simplify reality so that the integrated relationships of ecosystem services can be considered.

In this section, the open source modelling approach InVEST will be introduced and experiences of its application for a case study will be discussed. According to the developers, InVEST is suitable to be used for an integrated assessment of ecosystem services at a local, regional or global scale. It has been used around the world in numerous local and national projects and studies, as well as in day-to-day decision-making processes (Daily et al. 2009; Nelson et al. 2009; Tallis and Polasky 2009; Bhagabati et al. 2012). Examples of its application include the Willamette Basin in Oregon, Oahu on Hawaii, British Columbia, California, Puget Sound in Washington State, the Eastern Arc Mountains of Tanzania, the upper Yangtze River Basin in China, Sumatra, the Amazon Basin and the Northern Andes in South America as well as Ecuador and Colombia. In the course of the realisation of the case studies, the focus is set on the identification and protection of important areas for biodiversity and ecosystem services, as well as on the demonstration of their relations.

Characterisation of the Model Approach of InVEST

InVEST was developed as a scenario tool to support decision-making in environmental planning processes. The basis of the evaluation of ecosystem services is ecological characteristics and methods of assessing economic values (Nelson et al. 2009; Tallis and Polasky 2009). InVEST is usable in combination with ArcGIS (ESRI), which provides the cartographic representation of the ecosystem services evaluation. Meanwhile, a cooperation with Idrisi is also under development (► www.clarklabs.org/about/Clark-Labs-Receives-Grant-from-Moore-Foundation.cfm).

The development and administration of the meta-model is realised by the *Natural Capital Project* with participation of several well-known American research institutions, as well as by *Nature Conservancy* and by the WWF (*World Wildlife Fund*) (Natural Capital Project 2012). Depending on the needs and expertise of the user different models with varying levels of complexity will be provided–from the simple analysis of existing relationships using a small amount of data up to a complex model, which includes different scenarios and feedback on the comprehensive analysis of ecosystem services (Nelson et al. 2008; Daily et al. 2009). However, currently only simplified procedures are offered, so that the models only require a small amount of input data.

Nevertheless, the open source model InVEST is already taking into account significant aspects of a two-dimensional modelling approach of ecosystem services. These include the spatial mapping and localisation of services and welfare effects in GIS, an integrated view of supply services, regulatory services as well as sociocultural services (TEEB 2009; Tallis et al. 2011). Furthermore, basic abiotic and biotic environmental parameters are incorporated into the assessment process. Thus, the quantification of ecosystem services within the individual models is not only based on the land use of the past, present and future, but incorporates additional parameters when necessary.

Based on the 14 models currently included, InVEST enables a biophysical and partly economic evaluation of a selection of ecosystem services of terrestrial as well as maritime systems. In ◘ Table 4.4 the seven terrestrial models for the description of services and products of land and freshwater are presented and assigned to corresponding classes of ecosystem services.

In addition to the final results of the individual models, partial results and intermediates are also taken into account. However, those partial results cannot be clearly assigned in every case to an eco-

Table 4.4 Terrestrial InVEST-models for assessment of ecosystem services (Tallis et al. 2011; date: May 2012)

InVEST-Modules	Ecosystem Services	Indicators, partial results and intermediates	► Sect. 3.2
Biodiversity	Habitat quality	– Habitat quality – Relative level of habitat degradation – Relative habitat rarity	R.11
Carbon storage and sequestration	Economic value of carbon sequestered	– Amount of carbon stored – Difference of carbon stored in future and current landscape – Volume and biomass of forest management	V.6; V.8; R.2; R.3
Reservoir hydropower production	Economic benefit of hydropower production	– Total water yield per sub-watershed – Mean water supply yield volume per sub-watershed – Total energy produced per watershed (in kWh)	V.12; R.5
Water purification: nutrient retention	Economic benefit of nutrient retention by filtration	– Total water yield per sub-watershed – Total amount of nutrient retained by each sub-watershed (in kg) – Mean amount of nutrient retained	R.5; R.6
Sediment retention	Avoiding costs of sedimentation (for dredging and water treatment)	– Total potential soil loss per sub-watershed – Mean sediment retained on each sub-watershed	R.7
Managed timber production	Net present economic value of timber production	– Volume and biomass of forest management	V.6; V.8
Crop pollination	Potential value of the pollinator supply for each agricultural land use to crop production	– Potential likely abundance of a pollinator species nesting in the landscape, given the availability of nesting sites there and of food – Relative farm value of crop production on each agricultural cell due to wild pollinators	R.10

system service as presented in ► Sect. 3.2. An assignment of models according to productive, regulatory or sociocultural ecosystem services or welfare effects will not occur. Nevertheless, each individual model and its background is explained briefly. A categorisation according to the welfare effect is not possible as some of the models do not describe a direct performance, product, or process of ecosystem services, but rather demonstrate risks–and, therefore, describe impacts on the functionality of an area at a certain land use (e.g. sediment trapping).

The programme language of all models listed is Python, which is also usable within ArcGIS. However, for calculations based on InVEST basically no knowledge of Python programming is needed, instead the usage of InVEST-models requires basic to intermediate skills in handling ArcGIS (Tallis et al. 2011). Furthermore, the computer system used has to meet certain requirements. For example, the regional and language settings need to be changed to 'English (USA)' in the system panel. This ensures that decimals are determined by a point, not a comma (as with German settings). Otherwise, incorrect results or even system crashes can be caused as the model scripts are unable to collect and process commas of the input parameters. Furthermore, a recent ArcGIS licence is required, while some models even require the ArcInfo licence level. In addition, installation and activation of the ArcGIS *Spatial Analyst* extension is required. Moreover, the model for the assessment of pollination as well as all models for assessing the maritime system require additional Python library extensions, such as *Numeric Python*, *Scientific Library for Python*, *Python for Windows* and *Matplotlib* as well as for ArcGIS 9.3, the *Geospatial Data Abstraction Library*.

Model InVEST

The scenario tool InVEST can be downloaded from the website of the *Natural Capital Project* (▶ www.naturalcapitalproject.org). The installation of the programme is very user-friendly as an entire folder structure with all scripts and training data will be unpacked–given the appropriate installation file is selected for downloading. New users of InVEST benefit from a structured data provision, as they can open the programme and test the models without a lot of prior skills or background knowledge. To apply InVEST to your own research the input data for each individual model needs to be adjusted to the specific study area according to the requirements for each model. Partly, some of these data can be found in open source databases of different state agencies or individual studies. For such data, the format needs to be adjusted in analogy to the demodata. This includes compliance with the original names of column headers and with the conventions for objects according to the instructions of the user manual, also taking into account general limitations of data management in geoinformatics. Furthermore, it needs to be considered that the computation time of the models depends on the resolution of the raster data at the beginning and at the end of the modelling process. Thus, in order to accelerate the calculation a lower spatial resolution (grid cell size) is recommended.

Continuous development of the individual models of InVEST aims to lead to a steady improvement in modelling. In this context, users need to consider the increasing demands on hardware and software. Currently, an ArcGIS 9.3 or 10 licence is required, because specific calculation algorithms of it are used in the models of InVEST.

Example of Use

While working on the project 'Landscape Saxony 2050' (▶ www.ioer.de/index.php?id=812) at the Leibniz Institute of Ecological Urban and Regional Development almost all terrestrial and one maritime model of InVEST were selected and applied to the study area–the district of Görlitz in Eastern Saxony, Germany. Those models include reservoir hydropower production, sediment retention, aesthetic quality, biodiversity–habitat quality and rarity, carbon storage and sequestration, managed timber production and crop pollination. When the simplest level of complexity in InVEST is used, most of these models are based on a matrix in which average performance parameters are assigned to the individual land use. The variables can represent both absolute values like stored carbon in tons per hectare, as well as relative values, with the highest value usually being defined as 1, while all other values are represented in their proportion to that. Depending on the programming of the individual models calculations are taking place at different levels of complexity. These calculations begin by adopting variables for land use as defined in the matrices (i.e. as in the fixation of carbon), and end with aggregated, buffered, overlaid calculations (as in biodiversity) or with neighbourhood analysis (as in aesthetics), where a decreasing influence is computed based on land use. Results are either relative values between 0 and 1, absolute values with indicators and/or economised assessments of the provided ecosystem services in the form of raster maps and tables.

In the following example, the biodiversity model and its calculation has been selected out of the mentioned InVEST models for assessments of ecosystem services, and will be processed for the district of Görlitz. This particular model has been chosen as it is characterised by high complexity, but also because of its variety of possibilities to integrate additional parameters in the calculation process, and, furthermore, because of the key significance of biodiversity as well as the possibility to represent a comprehensive topic in a highly simplified form.

Using the model biodiversity, two assessments can be carried out: habitat quality and the degree of exposure of habitat rarity. The latter describes the current decrease of the area of a habitat (in this case of land use) within a certain space compared to an earlier time. However, the actual risk or the consequences of habitat rarity are not determined or identified.

The selected area of investigation with an extent of approximately 2106 km^2 is located in the border area of Germany, the Republic of Poland and the

Attributes of LULC_value_habitat_and_sensitivity

OID	LULC	NAME	HABITAT	L_hway	L_froad	L_sroad	L_droad	L_lroad	L_urb	L_agra	L_rail
0	0	unknown	0	0	0	0	0	0	0	0	0
1	11	continuous urban fabric (high density)	0,1	0	0	0	0	0	0	0	0
2	12	continuous urban fabric (very high density)	0	0	0	0	0	0	0	0	0
3	13	discountinuous urban fabric	0,3	0,2	0,2	0,1	0	0	0	0	0,2
4	14	Industry and commercial units	0,1	0,1	0,1	0	0	0	0	0	0,1
5	15	Infrastructure	0	0	0	0	0	0	0	0	0
6	20	fill-up/pit	0,3	0,2	0,2	0	0	0	0	0	0,2
7	30	recreation	0,4	0,2	0,2	0,1	0	0	0,2	0	0,2
8	40	arable land and crop	0,4	0,2	0,2	0,1	0	0	0	0	0,2
9	41	agricultural crop land	0,2	0,2	0,2	0,1	0	0	0	0	0,2
10	42	arable crop	0,6	0,1	0,1	0	0	0	0	0	0,1
11	43	uncultivated land	0,4	0,1	0,1	0	0	0	0	0	0,1
12	50	pastures	0,8	0,3	0,3	0,2	0,1	0,1	0,5	0,2	0,3
13	60	forest area and woodland	0,9	0,1	0,1	0	0	0	0,2	0,2	0,1
14	61	broad-leaved forest	0,9	0,1	0,1	0	0	0	0,2	0,1	0,1
15	62	coniferous forest	0,8	0,1	0,1	0	0	0	0,3	0,1	0,1
16	63	mixed forest	0,9	0,1	0,1	0	0	0	0,2	0,1	0,1
17	64	clear cutting and reafforestation zone	0,7	0,1	0,1	0	0	0	0,3	0,2	0,1
18	65	transitional woodland shrub	0,8	0,1	0,1	0	0	0	0,2	0,1	0,1
19	70	water bodies	0,9	0,6	0,5	0,4	0,3	0,2	0,5	0,9	0,5
20	81	open construction sites	0	0	0	0	0	0	0	0	0
21	82	sparsely vegetated areas	0,7	0,2	0,2	0,1	0	0	0,2	0,1	0,2
22	91	moors and wetlands	0,9	0,5	0,4	0,3	0,2	0,1	0,2	0,5	0,4
23	92	arid environment	0,9	0,5	0,4	0,3	0,2	0,1	0,2	0,1	0,4
24	99	do not exist in 1992 or 2005	0	0	0	0	0	0	0	0	0

Record: 1 Show: All Selected Records (0 out of 25 Selected) Options

LULC	code of land use and land cover
HABITAT	basic habitat quality
L_hway	relative sensitivity to the threat of highway
L_froad	relative sensitivity to the threat of federal roads
L_sroad	relative sensitivity to the threat of state roads
L_droad	relative sensitivity to the threat of district roads
L_lroad	relative sensitivity to the threat of local roads
L_urb	relative sensitivity to the threat of urban areas
L_agra	relative sensitivity to the threat of agricultural areas
L_rail	relative sensitivity to the threat of railway lines

Fig. 4.15 Land-use classes, habitat value and sensitivity of habitat types to each threat (screenshot of the example of use of InVEST in the district of Görlitz)

Czech Republic. The district has a wide variety of habitats, which reach from lowlands to highlands. Open brown coal mining and recultivation have brought large-scale changes. Noteworthy are cultural and historical particularities, such as folk architecture (Umgebindehäuser) and the culture of the Sorbs, a Slavic ethnic group. Rare species such as otters, cranes, eagles and more recently even wolves, find suitable habitats here. In addition, the region is both demographically and economically affected by a strong change (▶ Sect. 4.3).

By selecting the chosen model from the toolbox of InVEST, a dialogue box opens. There, the input data and the folders for the results need to be defined. Thus, the existing paths of the sample data provided by InVEST need to be replaced with actual data of the chosen study area. The input data for the delineation of habitats are based on maps of land use and land cover, for which the habitat types and land use mapping (BTLNK) of the Free State of Saxony from 1992 and 2005 are used. These maps reflect a variety of land use classes. In order to simplify the modelling, the classes are all combined into one aggregated BTLNK mapping with 25 classes (BTLNK_25). Eventually, their contents are provided in a grid with a resolution of 20 m.

In addition, a relative habitat value (Habitat) for each land-use class needs to be defined within a spreadsheet in relation to the other classes (Fig. 4.15). Those values range from 0 (unsuitable) to 1 (perfectly suitable as Habitat). In order to define the habitat values for this case study, non-species-specific information according to Bastian and Schreiber (1999) are used. These are parameters that do not document habitat qualities of specific species or groups of species (species of open land, forest or of aquatic and wetland sites), but assign general assessments to individual habitat types with regard to their importance for species and area protection.

In addition to determining the general habitat quality of each land-use, threats that may affect this habitat quality are also determined such as highways, federal roads, state roads, district roads and local roads as well as railway lines, which were extracted from the Digital Landscape Model (ATKIS Base-DLM) on a scale of 1:25,000 for both years of the investigation and converted into a raster format. The areal threats of additionally considered urban and agricultural areas are based on the coverage of BTLNK_25.

Thus, the dimension of degradation, which is solely caused by the respective sensitivity of habitat types to each threat, has been defined between 0 and 1 (◘ Fig. 4.15) based on an evaluation of the influence of the mentioned threats on the habitat quality of the identified land-use classes. The value 1 presents the highest impact, the value 0 no or imperceptible degradation. Thus, a land use that is not displayed as Habitat (Habitat = 0) has no coefficient of degradation by threat.

Finally, the threats have been characterised based on their relative importance or weight and impact across space–range in kilometers and whether the impact of the threat decreases linearly or exponentially across space. A value of 1 is a linear decline in impact and 0 an exponential decline. The maximum range is based on the findings of Baier (2000); the remaining parameters were defined by the authors.

After completing and confirming the input data, the calculation is started. In this process, the individual steps are recorded in a separate process window. Based on the information provided by the habitat values of individual land-use classes from ◘ Fig. 4.15, a reclassification of land-use maps is taking place (H_j as general habitat quality). Simultaneously, the area sizes of individual land-use classes in the study area for the base year 1992 are compared to 2005 (the degree of hazard habitat rarity). For this application, the Eq. 4.1 is used.

$$R_j = 1 - \frac{N_j}{N_{j\,base\,year}} \quad (4.1)$$

R_j represents the degree of change of the individual land uses in the study area compared to the base year, N_j defines the area size of individual land uses in the base year and the current year. Is $N_j \geq N_{j\ base\ year}$, so $R_j \leq 0$ and the result is $R_j = 0$, otherwise there is a change in land use and R_j is greater than 0. The output of the calculation results in a grid, which values of R_j are each projected onto all present land-use areas (i.e. for the second point of time). A partial result of this calculation step is a map representing the area development of land-use classes (a so-called exposure of change in use) between a base year (here 1992) and a later point in time (in this case study 2005).

Taking into account the sensitivities of each present land use (◘ Fig. 4.15), the impact of each grid cell on its surrounding grid cells is determined within a second step by using the maximum range and impact across space for each threat and each grid cell. The individual effects of each threat on the grid cells are then summed up, which may show the impact of a threat on habitat quality. Considering the weights of the individual threats (◘ Fig. 4.15) the impacts of the threats on habitat quality are aggregated. The result of the summed degradation (D_{xj} as degradation of habitats) can be represented and compared in a raster map for the respective reference year.

In the last step, according to Eq. 4.2, the specific habitat quality (Q_{zj}) is calculated as an index by merging the aggregated degradation D_{xj} (including the half-saturation value k) with the reclassified land-use classes represented as habitat quality values (H_j). The half-saturation value is determined as half of the highest grid cell degradation value in the study area. The exponent z corresponds to the value 2.5.

$$Q_{zj} = H_j \left(1 - \left(\frac{D_{zj}^z}{D_{zj}^z + k^z} \right) \right) \quad (4.2)$$

As a first result of the modelling of the biodiversity by InVEST, the risk level of the habitats (in this case of the land-use classes) is presented in terms of their area sizes. In the context of the case study in the district of Görlitz, it was found that between 1992 and 2005 in particular the following land-use categories were affected by a strong reduction of their extent in proportion to their respective total area in the base year: reforested areas, fallow ground, mining areas as well as areas for transport and infrastructure. In addition, a decrease of the extent of meadows and pastures was discovered.

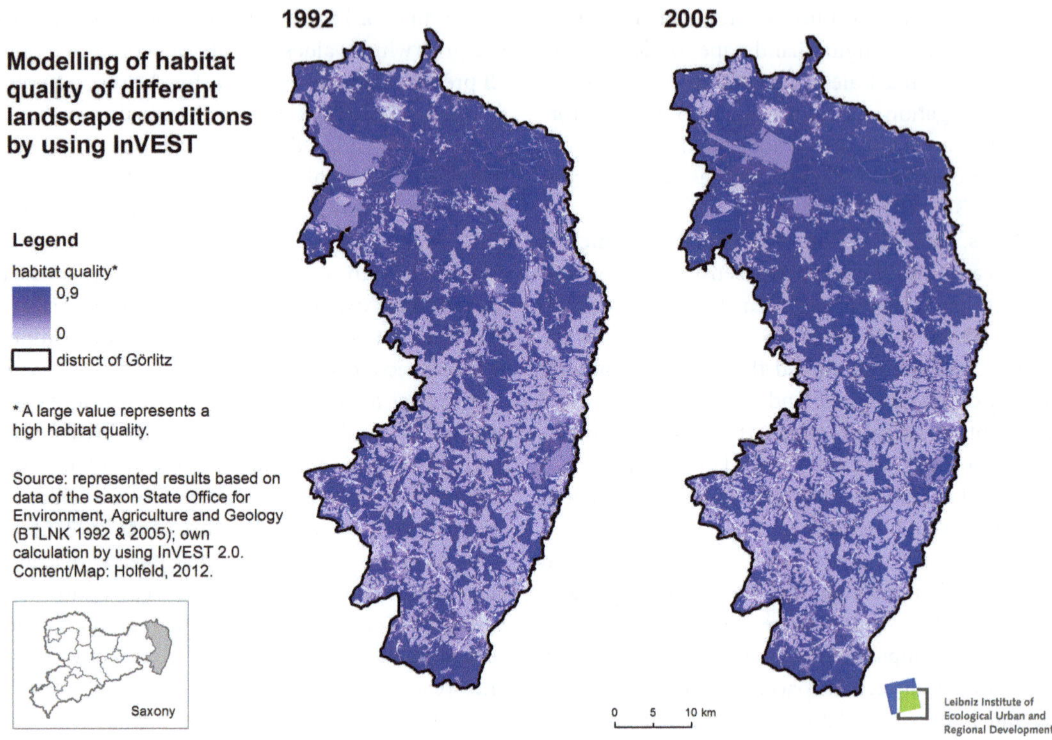

Fig. 4.16 Results of the assessment of InVEST model habitat quality in the diestrict of Görlitz for 1992 (*left*) and 2005 (*right*)

Next, the aggregated degradation as impact of threats is presented for the study area. Thus, the highest negative influences are detected at the border of urban areas and along main traffic infrastructure (highways and federal roads). The intermediate areas show no or hardly any perceptible threats. The same is found for the urban areas of the study area, which cannot be affected by any threat as they have not been assigned to the habitat function in the model.

Based on the result of the degradation, and taking into account the given habitat value of each land use, the *specific habitat quality* of each grid cell is mapped (Fig. 4.16). Thus, the highest habitat values are found mainly in the wooded north compared to the strongly agricultural influenced south of the district.

The lowest habitat values are found in the large urban areas as well as in a linear manner in the settlements along the main roads. A comparison between the base year 1992 and the year 2005 based on bluegray-scale values in the map (Fig. 4.16, left versus right) is hardly possible and also not possible as they rely on different databases. In order to compare the habitat quality of both points in time, the sum over all grid cells of a year must be calculated. The model completes this calculation automatically and writes its result into a log file, in which all input parameters are logged as well. Thus, the summed quality for 5,304,420 grid cells in the base year 1992 is 2,857,030. The total value for 2005 is 2,884,710. The habitat quality as an overall value for the district of Görlitz has improved slightly between the two assessment years, although spatially differentiated large-scale degradation in habitat quality is determined, for example, due to changes in land use. However, their scope has been fully compensated by other sub-areas within the study area. Many steps are similar to the approaches of conventional landscape planning.

Discussion

The results of the analysis of biodiversity with InVEST offer a simplified representation of the real habitat quality in the study area. Using the input data of Bastian and Schreiber (1999) average habitat values depending on the habitat type have been used for the district of Görlitz.

As an intermediate the degradation of habitats (degree of threats of habitats) was calculated, which show the downgrading of the habitat quality due to selected infrastructural threats. Within the modelling, it is basically assumed that the impacts of individual threats are adding up. In reality, however, their effect might be significantly higher (Tallis et al. 2011). Furthermore, it should be mentioned that the result is only one example out of many concerning habitat qualities, depending on the selection and consideration of individual threats as well as the considered habitats or species (Nelson et al. 2008, 2009).

Due to the manner of spatial location, the examination of habitat rarity seems hardly useful. However, the consideration of the change in land use within the biodiversity model is to be regarded as absolutely reasonable. But for this, a simple transition matrix between the different land uses would be sufficient. The current form of presentation is to be considered as very critical. Land-use types, which experience no absolute reduction or absolute increase in areal extent for the entire study area, are not assigned any degree of hazard. This includes land uses that are subject to areal change in land use in one part of the study area being fully compensated in another part.

As shown by the example of the biodiversity model, due to the low complexity of its individual models InVEST is easy to operate–as long as the user has at least basic working knowledge of geographic information systems. Through the results, some simplified relationships between land use and biodiversity or ecosystem services can be discovered (Polasky et al. 2008; Daily et al. 2009; Nelson et al. 2009; Tallis and Polasky 2009). Here the focus is more directed at the ecosystem services considering supply and demand than on biophysical processes. According to the current state of development of the models, an economic value can be assigned to an individual basis for a produced unit or for a specific process, which is used as a valuation basis for the study area. Thus, it is possible to evaluate the ecosystem services appropriately despite spatially separated locations for the demand and the provision of a service. However, the demand oriented approach is currently not available for all models contained in InVEST. Likewise, it needs to be considered if, for example, there is no water reservoir (modul: hydropower production), no service of energy can be provided.

The modelling with graded levels of complexity based existing approaches for specific modelling of landscape functions–such as SWAT or USLE (Tallis and Polasky 2009), allows to define the choice of the model complexity on the availability of data or on the user group. While simple models contribute to a better understanding of the relationships of the ecosystem services, the more complex models are intended to estimate the precisely measured services. Along with the desired development of the models, including further parameters, the demand for providing better data as well as the operability of InVEST increases (Tallis and Polasky 2009). Therefore, the provision of data and data sources in a central database would be desirable for different study areas in order to minimise the research work.

Due to the relevance to ArcGIS, results can be represented spatially in different scales (Daily et al. 2009). In order to do so it is crucial to have sufficient specific and differentiated information as input data for a certain study area. Furthermore, it has to be noted that the size of the study area depends on the considered ecosystem services (Tallis and Polasky 2009). For example, water-based services or pollination are of greater importance at a local scale (▶ Sect. 3.3) while climate-regulating processes require a global scale.

In addition to cartographic outcomes, results can be exported in a tabular form. The present results, however, are not suitable for professional use, such as in the development of detailed water and landscape plans or environmental audits as many functions and interactions are still negligible (Tallis et al. 2011). Similarly, the balance of costs and benefits of different models of InVEST is controversial even among developers, and certain ecosystem services, such as biodiversity (habitat quality), cannot be represented economically. The monetisation is furthermore criticised, because its assessment depends on spatial, temporal and sociocultural aspects that within InVEST cannot yet be considered as differentiated as their findings (Tallis and Polasky 2009). In general, average parameters are used for each evaluation of ecosystem services, which limit the validity of the result depending on the aspect to be researched and the scale of the study area.

With InVEST, the *Natural Capital Project* provides an evaluation process with great potential, even though it currently still has certain modelling weaknesses. One positive aspect is that InVEST is offered as an open source model, although its algorithms are sometimes highly complex. The open approach also allows less-experienced programmers to understand the calculation steps. The open development of the individual models ensures that both experts and laymen may submit suggestions to improve the modelling. At the same time, providing InVEST free of charge is supporting the rapid spread and development. The disadvantage of the continuous development of the models is that developers are always focussing on the latest versions of ESRI ArcGIS in order to incorporate the latest features from ArcGIS. Thus, increased system requirements of hardware are needed, but also the latest ArcGIS licences.

In conclusion, despite the identified criticism and existing weaknesses, it can be summarised that InVEST is a remarkable method to evaluate small as well as large-scale ecosystem services and to compare different regions, especially as the effort to define the input data is still small and the use of the individual models is relatively easy. However, the modelling procedures and results always need to be examined critically in order to avoid false conclusions.

Conclusion

Models provide exceptional opportunities to analyse and evaluate ecosystem services. With them, the landscape change that has already taken place as well as scenarios of future developments can be subject of an assessment. Therefore, decision-makers as well as the affected population can identify relationships and interactions of their action. Thus, the knowledge and communication of ecosystem services is strengthened. The various existing approaches to evaluate ecosystem services focus on different questions of content, spatial and/or temporal nature and still show significant deficits (Nelson et al. 2009).

With InVEST, an instrument is currently being developed, which is close to achieving the existing requirements for an evaluation of ecosystem services. In contrast to Burkhard et al. (2009) and Koschke et al. (2012), who already allow a holistic view of the ecosystem services within demarcated areas, the InVEST approach is also observing other biotic and abiotic parameters in addition to land use. However, the integration of those parameters is still at the beginning and needs further development in order to allow differentiated analyses of the ecosystem services (Nelson et al. 2009). Besides the development of computational algorithms within the models, well structured access to quantifiable data needs to be build up as the data availability is still quite limited. Simultaneously, methods are required, which allow the often individually evaluated ecosystem services to be compared and weighed up against other and to communicate their results (Holfeld et al. 2012).

4.5 Communicating ES

K. Anders

4.5.1 The Importance of Communication

In recent years, an entire new field of research, that of sustainability communication, has emerged which investigates the possibilities of communication regarding environmental issues. It encompasses a broad gradient of the issues which have been handled in various ways in various disciplines, in terms of their theoretical foundations, methodological approaches, and practical areas of application (Michelsen and Godemann 2005). In the present chapter, we will be able to examine only a few systemic decisions. The basic fact is that without appropriate communication, ecological issues will have no chance of validation in society. Only by way of communication can the relevant information in the social systems even be selected, informed and understood. Communication is therefore the key process of societal autopoiesis for social systems, i.e. it is through communication that they produce and reproduce themselves (based on Luhmann, this range of issues has e.g. been precisely circumscribed by Schack 2004).

However, how this process actually proceeds can be influenced only to a limited degree (Ziemann 2005). The feasibility of communication is widely overestimated; the definition of communication is often mechanically reduced to a more or less complicated relationship between the broadcasters in the receiver. The German phrase commonly used today, 'I'm communicating this or that,' erroneously even suggests the possibility of engaging in communication with no counterpart. However, the difficulties involved in being in control of the communications process do not imply that it is fundamentally unshapeable. Rather, one's own role as a participant in that process can certainly provide opportunities to put forward arguments, positions and assessments. In order to identify free spaces for the societal validation of ES for a number of very different fields of application–from advertising to discourse–i.e. if we are to assume that communications, in spite of its internal dynamics, is a shapeable process (Schack 2004), we will first have to take a more detailed look at the intentions connected with the term 'ecosystem services.'

4.5.2 'Ecosystem Services' as an Umbrella Term for Communicative Intent

The concept of ecosystem services is based on a very large number of different properties of ecosystems and landscapes. The initially very summary systematics of supply, regulation and sociocultural ES (► Sect. 3.2) does not follow any scientific–analytical or systematic–necessity; rather, it is designed to ensure that asymmetric processes and perspectives attained public recognition within the context of a certain topicality. A similar strategy was used several years earlier around the concept of biological diversity, in which genetic diversity, species diversity and landscape diversity were brought together without the relationship between these various levels having been clearly defined. Wilson and Piper (2010) characterised the ES use of language 'as a route to better understanding their importance and also of improving their protection.'

As a result, the term 'services' has been variously used, and the term broadly stretched. The authors of the Millennium Ecosystem Assessment admit as much: 'The condition of each category is evaluated in somewhat different ways, although in general a full assessment of any service requires consideration of stocks, flows and resilience of the service' (MEA 2005a, p. 29). While the term in such areas as supply functions has generally remained relatively closely oriented towards the usual use of the language about a service (for people; the implicit anthropomorphism is justified pragmatically), cultural services must be located more in the network of interrelationships between humankind, nature and the landscape (MEA 2005a; Freese and Anders 2010). Regulatory services, on the other hand, involve first of all the self-organisational capacity of an ecosystem; the advantages for people are thus indirect.

This leads to a difficulty of operationalisation: various processes incorporated under ES are to be found in particular landscapes and very different qualities, which resulted a problem of evaluation criteria. There are ES which can basically be provided in unlimited quantity (e.g. soil formation), while others undoubtedly violate the principles of sustainability, if their activation is not kept within limits. Often, these services are rendered at the cost of others (Trade-offs; Stallmann 2011; ► Sect. 3.1.2), resulting in requirements for a balancing of interests which have to date remained methodologically unresolved as long as the concept of planning contexts is to be used. This series of imprecisions recalls Luhmann's assessment of ecological communications in the sciences (◘ Fig. 4.17):

"The carelessness in the choice of words and the lack of awareness of theory-related decisions of great consequence are among the most notable characteristics of this literature–as if care for the environment could justify carelessness in the speech concerning it. (Luhmann 2008, p. 8)"

Whether ecosystem services have indeed become part of a discursive framework or pattern of inter-

Fig. 4.17 At the meeting of the German section of the International Association for Landscape Ecology (IALE-D) in 2010 in Nürtingen, the artists Christiane Wartenberg and Robert Lenz presented a shelf with two kinds of honey. One set of jars contained real bees' honey, labeled with the exact information regarding the place of production and also regarding the landscape development issues connected with it. Next to it was 'artificial honey'–jars with drypoint etchings of the most common terms in environmental research, from 'acceptance' to 'invasive art.' What was kept apart in this art exhibition–natural space, use, and scientific research–should also be separated more carefully in the debate over 'ecosystem services.' © Kenneth Anders

pretation, as described, e.g. by Brand and Jochum (2000), in other words, whether for example the expectation that aspects of the protection of nature and resources might better be validated has indeed been fulfilled, is a question that deserves closer examination.

The attractiveness of the concept within the environmental sciences, the business and financial world and also among policy-makers, would any case appear to be still on the increase, which should, however, not be confused with greater validation for the processes thus described. It is certainly possible, that the term 'ecosystem services' will become established without this fact having any consequences for society's relationship with the environment.

4.5.3 Government and the Market Instead of Communications?

The term communications itself is not a factor in the Millennium Ecosystem Assessment. Rather, the scientific community sees itself as a communicating actor in this context; its target system is the policy-making establishment. While the executive summary of the study for 'decision-makers' does raise the issue of participation and transparency as 'ecosystem-services'-related demands directed towards policy-makers, this is framed merely in terms of the requirements of administration, not as the constituent element of societal communications (MEA 2005b). Even a theoretically rooted concept of 'the public' is something which is not to be found in the debate around ES. Once in a while, there are merely indications about the use of publicly available information (Ruhl et al. 2007), which do correspond to basic demands for transparency in such areas as planning processes. The reason for this systematic blindness may be found in economic calculation: Unlike the 'tragedy of the commons,' the tragedy of ecosystem services is seen not as a problem of overconsumption, but rather of underproduction (Ruhl et al. 2007).

As a result, it would appear that the societal appreciation for ES will become tangible only when the market conditions for the same have been created. Communication is thus not excluded; rather, it is assumed that for ecological problems, the tool is available: the successfully established, symbolically generalised communications medium known as 'money'. That is not the place to pass judgment on the prospects for the success of this idea. However, the identification recognition of ecological processes and services, and the emergence of corresponding markets, can only be achieved through communication, in other words, the medium money cannot be transferred to ecological plans and actions merely on the basis of an assertion to that effect. The authors of the Millennium Ecosystem Assessment evidently assume that it is only necessary to convince policy-makers of the plausibility of their arguments, in order to create the necessary laws and regulations. Büscher and Japp (2010) pointed out in this context "that in the current public debates over problem solutions with respect to the 'ecological crisis', sociological arguments play no role. The salvation of the world is, as it were, to be carried out with no concept whatever of 'society'".

4.5.4 Communications Efforts as an Approach to the Shaping of Environmental Sciences

In order to be able to arrive at a statement in spite of these yet uncertain questions, let us use the term 'communications efforts' in order to do justice to the reasonable desire for the shaping of communications. 'Communications efforts' means the intent to effectualise scientific knowledge with respect to the significance of ecosystems for people outside the scientific system. Here, a changed self-consciousness is palpable in environmental research, where communications efforts have been massively enhanced in recent years. Today, we often expect that, given a general feeling of insecurity, environmental scientists should not so much bring particular ascertainments into the discourse, but should rather enter into an exchange with policy-makers regarding the weighing of ecosystemic contexts, and should assume a vanguard position in that respect. In this context, the term 'pro-active' has become fashionable.

An author such as Luhmann would doubt that this new awareness is based on any realistic analysis of the possibilities of the scientific community, for '… other functional systems must assume the task of sorting out what is useful and what is useless' (Luhmann 2008, p. 108). Precisely this step towards action is usually taken only rarely (Bechmann and Stehr 2004), which is in turn no coincidence, for research after all, due to the construct of 'consensual knowledge' (Bechmann and Stehr 2004, p. 30), is always in danger of weakening its own position as a systemic element by giving up its own medium, according to which information is selected according to the criterion of true/false. In other words, the core business of the scientific community is the question: 'Is this statement true, or is it not true?' Once one abandons the realm of this core business, one is treading on slippery ground. In order to survive in such a situation, scientists ultimately have to assume two roles: one as communicators in the sustainability discourse, and another as communicators within the scientific system. One good example is the Stern Report, *The Economics of Climate Change* (Stern 2007), in which, especially with regard to the effects of disturbed climate-regulatory functions, a political agenda ranging from the trade in emissions rights through a reduction in deforestation to targeted climate adaptation has been developed from out of the scientific community, in spite of a high degree of uncertainty.

Kuckartz and Schack (2002) pointed out that the goal of environmental communication encompasses a broad range of gradients which is not sufficiently reflected: the attempt to achieve acceptance for laws or to promote ecological products, involves very different consequences than the desired changes in behaviour, or even the claim to enable people to orient themselves amongst the complex issues of ecological action. In one case, public relations and advertising predominate; in the second, by contrast, education. This diversity also applies to communications regarding ES. In the following, we will therefore discuss several more or less established forms of scientific or planning related communications efforts with regard to their suitability to generate societal responses for certain ES.

■ **The Classical Transfer of Knowledge**

The transfer of knowledge should build an elementary bridge between the scientific community and other systems by 'publishing'–literally: 'making public'–the results of research. In other words, a communication is to be made available beyond the bounds of professional circles. In this area too, the efforts of environmental research and planning have been greatly enhanced in recent years. The goal of eliminating knowledge gaps (e.g. Schmidt et al. 2010) is appropriate, since the preparation and accessibility of sufficient information ultimately permits communication–even if such activity is in and of itself not communication. It is for precisely this reason that totalitarian systems denied the release of information, since they will be unable to control the results in the public communicative sphere. Beyond the concept of public participation in planning (Schmidt et al. 2010), it is according to the participatory intent of the authors necessary to ascertain that public opinion comes into being in the first place only through communication, and that this is precisely what the task of planning processes consist of. Communication efforts are realised through the fact of the accessibility of information; hence, it is demonstrated that certain functions of ecosystems are indispensable for human beings, or that the loss of the same would affect the general interest. Here, environmental scientists can certainly assume an active role without departing from their home turf. This includes the description of ecosystemic contexts such as soil formation, water retention or important nutrient chains (i.e. regulatory services), and also knowledge on land and water use (supply services), or descriptions of the wealth of interaction between people in the landscape (sociocultural services).

In all these cases of knowledge transfer, what is needed is not so much professional marketing strategies and campaigns, as clear statements and a generally comprehensible language based in precisely this kind of clarity. There are enough historical models for this, in which environmental scientists convey information directly, and, for good reason, do without any aggregated preparation of the same by means of 'communications profiles.' Knowledge transfer has traditionally been carried out with a high level of quality under the *Leitmotiv* of 'welfare effects' (e.g. Albert 1932; Hornsmann 1958; Altrogge 1986). The discontent around this classical role of science is often described as disillusionment regarding its societal effect. There are two variants of this; while for example, Barkmann and Schröder (2011) target the lack of the reception of scientific knowledge in society, many other authors assume that environmental knowledge is basically sufficient, but that there is a lack of corresponding behaviour resulting from it (e.g. Wehrspaun and Schoembs 2002). Indeed, the attitude of classical knowledge transfer does not ensure that the knowledge provided will also be societally used. On the other hand, the question is justified, in terms of the concept explained at the outset: To what extent is it even possible to force such an assurance?

■ **The Transfer of Knowledge and Transdisciplinary Contexts**

Beyond the 'classical' domain of knowledge transfer will–in the context of transdisciplinarity, i.e. with regard to the methods used and even with regard to the concrete research questions of a partially open process–conceptual deficits once again dominate the picture (a systematisation approach of Jenssen and Anders 2010). While knowledge transfer is correctly criticised with regard to obsolete models of the relationship between the broadcaster and the receiver (Karmanski et al. 2002), there is a lack of dialogic work methods in most research processes in which actors determinant for the landscape can weigh the relevance of the research knowledge produced and bring their own forms of knowledge–hence also their relationship to various ES–into play. Under the conditions of transdisciplinarity, knowledge transfer thus becomes an active communication task, i.e. scientists have to accept the existing heterogeneity

■ **Fig. 4.18** By means of just four positions on forest development, we can already gain a hint of the contradictions one encounters with respect to ES, if one wishes to communicate about them. In the Schorfheide-Chorin Landscape Workshop, held in the state of Brandenburg between 2006 and 2009 as part of the BMBF collaborative research project Sustainable Development of Forest Landscapes in the North German Plain (NEWAL-NET), there were over 100 such positions. Much could be gained by bringing some order into this diversity in order to create space to help enunciate aspects hitherto ignored © Kenneth Anders

of knowledge, and subject their own work to the resulting validity conflicts (■ Fig. 4.18). For reasons of quality, too, debates will become necessary, for where representatives of various disciplines and areas of practice collide, it is difficult to manage the professional standards introduced, so that valid knowledge can only be selected by means of intensive and critical discussion. With regard to ES, this means that those contradictions are invisible which emerge from the fact that landscapes are used, enjoyed and protected simultaneously. Environmental sciences can therefore not themselves per se assume the role of the advocate of various ES. The appellative stance of the Millennium Ecosystem Assessment proves ineffective in the face of the reality of such processes. Rather, environmental scientists need to clearly defined their role in communications processes, i.e. either withdraw to the relatively passive position of the 'classical scientist' (and add to that the internal dynamics of communications), or else subject themselves to the contradictions that in fact emerged from the social, economic and ecological dimensions of sustainability–in the landscape and elsewhere. The latter occurs only rarely, and is the result of an understanding that posits the knowledge is only monopolised within the scientific communities, and that nonscientific perspectives cannot claim any knowledge-related status, but are only described in terms of identity, habit, individual experience, interest or sensitivity. What then remains of communication is understood as a means for generating acceptance of consensus (critically assessed by Adomßent 2004), which again moves closer to the mechanistic understanding described at the outset.

■ Social Marketing and Considering Lifestyles with Respect to Consumer Behaviour

One approach common in Germany for raising societal awareness of sustainability issues is social marketing (e.g. Buba and Globisch 2009). The methods developed here can also be used for various ES. For example, their recognition for the area of agriculture could occur by seeing not farmers, but rather the consumers themselves, as the perpetrators of reduced biological and landscape diversity (Adomßent 2004)–at least as long as the farmers lack any possibility of financing practices for the preservation of forms of diversity on the market. Diversity is thus seen as a product to be created, and no longer as an issue existing and endangered; in that way, it can become an object of marketing.

Compared with social-scientific analyses of environmentally relevant consumer behaviours and the societal complexity upon which they are based (e.g. Brand et al. 2001), social marketing constitutes a narrowing of the perspective, with the goal of linking social-scientific research with business concepts of customer acquisition in order to ultimately effect behaviour change. This too is accompanied by a changed self-awareness on the part of the scientific community–away from critical analysis and towards 'change management' (Buba et al. 2009). First of all, social groups with certain value patterns, consumer habits and some culturally determined characteristics are identified, using a process similar to that of 'sinus-Milieus' (everyday-life worlds; cf. e.g. Theßenvitz 2009). Subsequently, the identifications obtained are used to construct target groups which are then to be won to the intended goals by means of adapted media codes; in common parlance, one might say, 'if you want to reach people, you have to go to where they're at.' This apparently simple truth becomes a distortion if one realises that communication is a process in which all participants are moving, and no one is waiting 'at' anywhere.

From the point of departure of lifestyle research, Lange (2005) described social marketing as a modest, and hence realistic, horizon of expectations, by means of which consumer behaviour could be influenced; a thorough examination of the range of possibilities available to consumerism is provided e.g. by Bilharz (2009). However, even Lange has doubts about the expectation that such consumption patterns could be permanently rooted by means of the targeted influencing of lifestyles. For lifestyles can neither be politically controlled, nor is it possible to constructively use distinction effects, e.g. for the role of eco-pioneers. Social distinction is part of social dynamics, and therefore contributes just as much to the erosion of cultural patterns as it does to their formulation. The weak correlation of lifestyle and action moreover points to the limited possibilities in our society to even practice sustainable consumer habits at all, so that the ball is now in the court of the structural-policy decision-makers. Kuckartz and Schack (2002) have confirmed empirically that the actors in environmental communication no longer even see changes in attitude and consciousness as a task to be addressed. In view of the various ES, this situation is becoming ever more acute, since not all processes compiled under its heading can be affected directly by individual consumer behaviour. Moreover, since a major share of our actions result not from lifestyle-related patterns, but rather from overall societal ones, the decision regarding the use of certain ES–especially regulatory services–can under no circumstances be left to the free market, but rather must be regulated by law (Bilharz 2009). For example, soil protection can vary obviously be better provided by legislation than by a market for intact soils.

In this respect, social marketing, too, deserves to be handled with greater care with respect to its expected effects and to the suitable fields for its application than is currently the case. The representatives of this school of thought emphasised that in addition to a designing of social groups as the object of marketing, they are expressly working towards the self-determined assumption of responsibility by these groups (Buba and Globisch 2009). However, it is doubtful that the tautology of conventional marketing can be broken by the awakening and satisfaction of needs, for the selected information and its preparations already anticipate the principles of power and validity established in the respective lifestyle circles–precisely what we have to thank for the lack of sustainability in the practice of our lives. It is conceivable that representatives of 'Lifestyle of Health and Sustainability' (LoHaS), or a 'consumer materialist' might be motivated by social marketing to make a certain decision with respect to items of purchase; however, the expectation that representatives of these target groups will as a result change their attitudes simply because we have tried to speak to them in their language, is misplaced, since just that avoids calling into question the guiding ideas and mythologies of the hitherto dominant institutional practices (Brand 2005, p. 153). Moreover, the fact is that the actors participating in communication ultimately are always open in terms of their decision-making (Ziemann 2005), and also, communication necessarily causes changes in one's own perception, as a result of which the scientists involved themselves emerge from the process with modified perspectives. In other words: if one wants to promote communication while at the same time excluding its internal dynamics, we will fail to communicate.

- **Campaigns**

In this context, efforts to generate public validation of ES by means of campaigns are conceivable. Here, the conceptual lack of clarity of the term is initially an obstacle. As Lisowski (2006) has demonstrated, at least in the European context, the linear sequence of planning, strategy and campaigns as a way of achieving democratic influence is rarely encountered; rather, campaigns develop 'evolutionarily,' along existing financial and professional spaces. Hence, certain aspects may be successful, while others fade into the background. The precondition is the existence of organisations, which represent a certain interest for the public. Their practice is also known in the area of environmental communications. Campaigns for the establishment of wilderness areas, for the preservation of endangered species and habitats, for the protection of certain landscape types, for food produced under conditions respecting the ecosystems, etc. are an everyday occurrence. They may affect decisions and help promote societal developments, as in the Stand-By Campaign (Schack 2004). Finally, Frankel (1998) demonstrated a 'greening of communications' for industrial advertising. However, it is precisely the term ES that shows us clearly that while advertising refers effectively to the respective organisations or companies that control certain landscape processes, it hardly refers at all to the ecosystems themselves (cf. the WWF Tiger campaign, described in Conta Gromberg 2006). In this respect, this form of communication suffers from an authenticity problem, since suspicion regarding motives always arises (Japp 2010). Moreover, organisations with conflicting purposes are free to promote their own respective campaigns, in which ultimately different environmental goals are pursued and addressed. Since not all functions and processes in the utilised landscape are per se mutually noncontradictory, campaigns may certainly be a possible tool for highlighting ES, but they are an unlikely tool for use in planning processes–contradictions are not considered campaign-capable.

- **Education for Sustainable Development and Education for Landscape Policy**

The goal of education for sustainable development, a transgenerational, self-organising debate, and personal skills in addressing the issue of sustainability, would appear to be close to the intent of the concept of ES, and even to offer an adequate solution to the above-described asymmetry of subsumed functions: Placing concepts in relationship with one another, permitting diversity of perspective, and acting responsibly constitute the key points within which adapted and adequately contextualised accesses to this issue could be created. What is meant here is not education for sustainable development as an 'advertisement for sustainability' (Siemer 2007), as a sub-function of social marketing, or as self-praise for environmental policy, but rather as communication. However, that would require that the autopoietic process in education itself be promoted, in other words, that its results not be prejudiced. Yet it is precisely this precept that is violated by many works purporting to promote 'education for sustainable development', instead, they rely on old concept of environmental education, albeit in new garb. For example, role-playing in which children basically provide a 'constructive solution' to a conflict have nothing to do with the purpose of the concept as described here–to promote open learning processes. The frequent restriction of the approach to questions of consumerism, too, ultimately does not result in a satisfactory proximity to the ecological aspects of the service involved. Communication of ES through education for sustainable development thus does not automatically lead to success, but rather depends on the concrete formulation of the programme. It may even cause confusion and frustration, if the individual scopes for action ultimately remain schematic which has, in personal experience, often proven to be the case.

Such approaches suffer most from their own abstraction and lack of spatial rootedness, for action always takes place in spaces of action upon which the contents are to refer in their full complexity. Scenarios which do not incorporate the logic of the locality remain ineffective. World cafés, in which the moderators stifle critical positions which stem from spatial contexts, rather than seizing upon them and using them, thereby miss their chances for success. It is not sufficient to sow a species-rich meadow or to wet a low-lying area, even if these are, beyond any doubt, good deeds. Rather, the relationship to the landscape space and the relationships existing within them is indispensable, even if the resulting balance sheet may be depressing. The logic of the school garden is useful; however, it does not yet yield any understanding of the relationship of tension between various ES.

De Haan and Kuckartz (1998) describe a 'distance gap' with respect to the perception of critical environmental situations which they interpret from various perspectives–the role of the media, interest in faraway places, or a globalised environmental consciousness. According to this thesis, environmental impacts increase with distance, while one's own surroundings remain intact. This is in fact often unwittingly reinforce by certain manners of work in education for sustainable development, due to a predominant focus on global contexts which affect humankind as a whole (cf. the development of the problem horizon in Rieß 2010, or the main syndromes of global change in de Haan and Harenberg 1999), and the corresponding environmental behaviour generally begins and ends in the perception of consumer options. In order to make use of the methods of education for sustainable development for the communication of ES and make them fertile in the participatory planning process, precisely this principle needs to be reversed. Sustainability conflicts are primarily to be found before one's own door. Such a paradigm shift would however require a critical debate, a fearless scientific description of this conflict and open questions. It seems that such precepts tend to be an exception in the present environmental communications process.

One promising path in this concept is provided by the European Landscape Convention (ELC 2000), which Germany has never signed or ratified, and which as a matter of course sees a spatial connection in education on landscape policy (as justified in a case example tested by Kulozik 2009). This approach, oriented towards the peculiarity of concrete landscapes and the changes taking place within them also promotes development of the topic of ES (▶ Sect. 3.4), since it:

1. Takes the particular landscape conditions of various processes subsumed under the heading of ES, i.e. a specific ecosystemic balance or dis-balance, as its point of departure
2. Seeks a connection with the perception of the landscape held by its own inhabitants; i.e. based on the communication process, it qualifies, processes and develops further precisely those potentials which have a prospect for gaining a response from the communicative counterpart

An orientation towards the simple and internally logically structured agenda of the landscape convention for communications regarding various ES is to be recommended, even if the demands raised herein have not yet been politically established. Such and orientation can be easily prepared by means of education about the landscape; it allows for the integration of partners such as artists, land users, conservationists, local politicians, etc., and it is evidently–like all development of the landscape–open-ended with regard to outcome. In the context of concrete landscapes, there is no need for protection against cheap arguments, since the contradictions and interdependencies of one's own space are considerably more easily recognisable than are globally conveyed contexts: behind every practical action in the landscape is an actor with societal conditions demanding a certain action. Michelsen (2002) states in this regard 'that the context of knowledge acquisition is also a decision-making factor about the relevance of knowledge for action.'

Precisely this situation makes landscape an ideal context for education. The fact that such approaches are nonetheless the exception in Germany is on the one hand due to the lack of any corresponding discursive framework–the term 'landscape' is hardly present at all in the German discourse over sustainability–and on the other, to the mistaken idea that dealing with particular landscapes will ultimately lead to a dissipation of forces, so that the overarching whole–global change–risks getting lost in the process. To this, one might respond that skill in dealing with ES can only emerge in the caring dealing with particular cases and, once it has taken shape, will always grow beyond its original dimensions.

- **Landscape Workshops–A Point of Attachment for Local Discussions, Regional Debates and Societal Discourses**

As social beings, we have various social connections. We live in a family, share in the life of a village community or an urban neighbourhood, belong to a professional grouping, and are citizens of a country. In the communications regarding ecological matters, the various levels, languages, logics and issues emerging from this situation have not been sufficiently considered to date. The oft-cited slogan 'Think globally–act locally,' which was also used for the Agenda 21 campaign, easily blurs the various communications processes which, while occurring parallel to one another, often occur without mutual reference, and with each constructing its own environment.

For the inhabitants of a major city, rural space is their nearby environment, while the inhabitants of those rural areas tend to see it as their own space which they themselves shape. Depending on the circumstances, different sustainability issues may use different symbolic places. Issues which have become established in society as a whole by way of the mass media may have been completely ignored by village communities; on the other hand, societal discourses often screen off regionally specific conditions. The limits to scientific communications efforts resulting from this situation cannot here be systematically developed, but it is certainly recommended that the level at which an ES is to be validated be precisely identified.

A local conflict, e.g. regarding a rewetting project, will have to use the scope of communications existing in a certain place; the rhetoric of climate change will seldom be of use here. On the other hand, if an international agreement on climate protection is at issue, the situation is reversed. Considerable problems may arise even at the point of transition from the space of action at the level of a cultural landscape to that of the purely local level. It is possible, by means of landscape workshops (Anders and Fischer 2010), to attempt over a lengthy period of time to continually link local, regional and societal discourses, and to thus influence them with regard to their perception of ES.

Since actors who can convincingly convey such matters as topics from the mass media into a concrete local space are few in number–generally, this is only done successfully on a temporary basis by the appearance of prominent political figures–overall societal contributions to the debate usually bypass the regions. In such cases, still there is a possibility of combining local aspects into perspectives for action at the concrete level, and to thus inject them into the debate. This approach is close to an understanding of communications science as communicating science (Ivanišin 2006), which is ultimately oriented towards the qualification of space-related discourse.

Outlook

Let us here summarise the essential statements as theses:

- Communication is a precondition for the validation of ES; however, it can only be shaped to a limited extent, i.e. the initiator of a communications process does not have sole control over its outcome.
- The term 'ecosystem services' brings together, with communicative intent, various processes of ecosystems and landscapes which have not hitherto been satisfactorily linked, a fact which has ultimately resulted in confusion in communication.
- The political sphere and the market cannot replace communication; rather, they are themselves societal subsystems, differentiated by communications. There are approaches in the environmental sciences to use the media of these systems, which requires considerable change in the self-understanding of science, but for which there has to date been no sufficient justification.

The legitimate demand to nonetheless shape communications has resulted in the formation of various schools and approaches in the context of sustainability communications:

- *Classical knowledge transfer* is today often dismissed as 'popular science.' However, the means available here permit a precise provision of scientific results for extra-scientific communication, and should therefore continue to be used.
- *Transdisciplinary knowledge transfer* is a worthwhile undertaking, but it does require that the

environmental sciences abandon, for the sake of communication, their claim to a monopoly over the concept of knowledge. Without debates, transdisciplinary processes will moreover suffer from a loss of quality due to the erosion of professional standards.

- *Social-marketing and target-group-specific communications strategies* should be critically examined with respect to the extent of their reach. Their core business is that of consumer patterns and behaviour forms which are very close to consumerism–e.g. the acceptance of laws and societal practices.
- *Campaigns* can be used effectively, but ultimately they constitute more of a service institution than an ecosystem service.
- In the context of *education for sustainable development*, global perspectives often dominate; they are important, but they should be conveyed in their own space. The communication regarding particular ES in their mutual interrelationships can be very successful in the context of landscape-policy education.
- *Local regional and societal discourses* are very difficult to link, since they constitute different environments and establish different issues. In place of the question, 'Which target groups do I want to address?' It is more promising for communication to ask, 'Which public do I want to address, i.e. within which issue contexts will I want to place a contribution which is to be communicated?'

References

Abeel K (2010) Diverse methods in cost-benefit analysis: searching for adept practices in the face of environmental and economic problems. Glossalia 2(1):13–18

Adomßent M (2004) Umweltkommunikation in der Landwirtschaft. Berliner Wissenschafts-Verlag, Berlin

Albert R (1932) Der Einfluß des Waldes auf den Stand der Gewässer und den Bodenzustand. Die Wohlfahrtswirkungen des Waldes. Vorträge und Aussprache auf der 9. Vollversammlung des Reichsforstwirtschaftsrates am 3. Februar 1932 in Berlin. Sonderdruck aus Heft 34 vom 15.3.32 der Mitteilungen des Reichsforstwirtschaftsrates

Albert C (2009) Scenarios for sustainable landscape development–a comparative analysis of six case studies. Proceedings of the IHDP open meeting, the 7th international science conference on the human dimensions of global environmental change. Bonn, Germany

Alcamo J (2008) Environmental futures: the practice of environmental scenario analysis. Elsevier, Amsterdam

Altrogge D (1986) Die Wohlfahrtswirkungen des Waldes in Zahlen. Beiträge zur Lebensqualität, Walderhaltung und Umweltschutz, Volksgesundheit, Wandern und Heimatschutz. Siegen, Heft 13

Anders K, Fischer L (2010) Landschaftswerkstatt Schlabendorfer Felder. In: Hotes S, Wolters V (eds) Wie Biodiversität in der Kulturlandschaft erhalten und nachhaltig genutzt werden kann. Fokus Biodiversität, Oekom, München, pp 262–272

Antrop M (2005) Why landscapes of the past are important for the future. Landsc Urban Plan 70:21–34

Aurada KD (1979) Ergebnisse geowissenschaftlich angewandter Systemtheorie (Vorhersage und Steuerung lang- und kurzfristiger Prozeßabläufe). Petermanns Geogr Mitt 20:409–413

BAFU–Bundesamt für Umwelt/BFS–Bundesamt für Statistik (ed) (2007) Umwelt Schweiz 2007. Bern, Neuchâtel

Baier H (2000) Die Bedeutung landschaftlicher Freiräume für Naturschutzfachplanungen. In: Bundesamt für Naturschutz (ed) Vorrangflächen, Schutzgebietssysteme und naturschutzfachliche Bewertung großer Räume in Deutschland. Schriftenreihe Landschaftspflege Naturschutz 63:101–116

Bardt H (2008) Entwicklungen und Nutzungskonkurrenz bei der Verwendung von Biomasse in Deutschland. IW-Trends, Vierteljahresschrift zur empirischen Wirtschaftsforschung. Institut der Deutschen Wirtschaft Köln, p 35

Barkmann J, Schröder K (2011) Workshop "Ökosystemdienstleistungen". Warum ein sperriges Konzept Karriere macht. Endbericht zum F & E-Vorhaben. Georg-August-Universität Göttingen

Bastian O, Schreiber K-F (1999) Analyse und ökologische Bewertung der Landschaft, 2nd edn. Spektrum Akademischer Verlag, Heidelberg

Bastian O, Lütz M, Röder M, Syrbe R-U (2006) The assessment of landscape scenarios with regard to landscape functions. In: Meyer BC (ed) Sustainable land use in intensively used regions. Landscape Europe, Wageningen. Alterra Report 1338, pp 15–22

Baumgärtner S (2002) Der ökonomische Wert der biologischen Vielfalt. In: Bayerische Akademie für Naturschutz und Landschaftspflege (ed) Grundlagen zum Verständnis der Artenvielfalt und seiner Bedeutung und der Maßnahmen, dem Artensterben entgegenzuwirken. Laufener Seminarbeiträge 2, Laufen, Salzach, pp 73–90

Baumgärtner S, Klein A, Thiel D, Winkler K (2013) Ramsey discounting of ecosystem services. European Association of Environmental and Resource Economists: 20th annual conference European Association of Environmental and Resource Economists. ► http://www.webmeets.com/EAERE/2013/prog/viewpaper.asp?pid=354. Accessed 15 May 2014

Bechmann G, Stehr N (2004) Praktische Erkenntnis: Vom Wissen zum Handeln. In: BMBF (ed) Vom Wissen zum Handeln? Die Forschung zum Globalen Wandel und ihre Umsetzung. Bonn, Berlin, pp 27–30

Bernhardt A, Jäger K-D (1985) Zur gesellschaftlichen Einflussnahme auf den Landschaftswandel in Mitteleuropa in Vergangenheit und Gegenwart. Sitzungsberichte der Sächsischen Akademie der Wissenschaften zu Leipzig. Mathematisch-Naturwissenschaftliche Klasse 117(4):5–56

BfN–Bundesamt für Naturschutz (2012) Daten zur Natur. Bundesamt für Naturschutz, Bonn

Bhagabati N, Barano T, Conte M, Ennaanay D, Hadian O, McKenzie E, Olwero N, Rosenthal A, Suparmoko, Shapiro A, Tallis H, Wolny S (2012) A green vision for Sumatra: using ecosystem services information to make recommendations for sustainable land use planning at the province and district level. A report by The Natural Capital Project, WWF-US, and WWF-Indonesia. ▶ www.ncp-dev.stanford.edu/~dataportal/pubs/Sumatra%20InVEST%20report%20combined%20Feb%202012_final.pdf. Accessed 12 July 2012

Bilharz M (2009) "Key Points" nachhaltigen Konsums. Ein strukturpolitisch fundierter Strategieansatz für Nachhaltigkeitskommunikation im Kontext aktivierender Verbraucherpolitik. Metropolis, Marburg

BMELV/BMU (2009) Nationaler Biomasseaktionsplan für Deutschland. Beitrag der Biomasse für eine nachhaltige Energieversorgung. ▶ www.bmu.de/files/pdfs/allgemein/application/pdf/broschuere_biomasseaktionsplan_anhang.pdf. Accessed 12 July 2012

Bohnet I, Bohensky E, Waterhouse J (2008) Future scenarios for the great barrier reef catchment. CSIRO Water for a Healthy Country National research flagships. ▶ www.clw.csiro.au/publications/waterforahealthycountry/2009/wfhc-future-scenarios-GBR-catchment.pdf. Accessed 12 July 2012

Bolliger J, Kienast F, Soliva R, Rutherford G (2007) Spatial sensitivity of species habitat patterns to scenarios of land use change (Switzerland). Lands Ecol 22:773–789

Börjesson P (1999a) Environmental effects of energy crop production–part I: identification and quantification. Biomass Bioenerg 16:137–154

Börjesson P (1999b) Environmental effects of energy crop production–part II: economic valuation. Biomass Bioenerg 16:155–170

Bork HR, Müller K (2002) Landschaftswandel von 500 bis 2500 n. Chr. Offenhaltung der Landschaft. Hohenheimer Umwelttagung. Günter Heimbach, Stuttgart, pp 11–26

Boyd J, Banzhaf S (2007) What are ecosystem services? The need for standardized environmental accounting units. Ecol Econ 63:616–626

Brand KW (2005) Nachhaltigkeitskommunikation: eine soziologische Perspektive. In: Michelsen G, Godemann J (eds) Handbuch Nachhaltigkeitskommunikation: Grundlagen und Praxis. Ökom, München, pp 149–159

Brand KW, Jochum G (2000) Der Deutsche Diskurs zu Nachhaltiger Entwicklung. Abschlussbericht eines DFG-Projekts zum Thema "Sustainable Development/Nachhaltige Entwicklung–Zur sozialen Konstruktion globaler Handlungskonzepte im Umweltdiskurs". Münchner Projektgruppe für Sozialforschung e. V.

Brand KW, Gugutzer R, Heimerl A, Kupfahl A (2001) Sozialwissenschaftliche Analysen zu Veränderungsmöglichkeiten nachhaltiger Konsummuster. UBA-FB 000330, München

Bräuer I (2002) Artenschutz aus volkswirtschaftlicher Sicht. Die Nutzen-Kosten-Analyse als Entscheidungshilfe. Hochschulschriften, Metropolis, Marbug

Bringezu S, Steger S (2005) Biofuels and competition for global land use. Global Issue Papers 20, Heinrich-Böll-Stiftung, Berlin

Brown K, Pearce D, Perrings C, Swanson T (1993) Economics and the conservation of global biological diversity. The global environment facility: Working Paper Nr 2, Washington

Brüggemann T (2009) Feldlerchenprojekt–1000 Fenster für die Lerche. Landesamt für Natur, Umwelt und Verbraucherschutz Nordrhein-Westfalen. Natur in NRW 3:20–21

Buba H, Globisch S (2009) Kommunikation und Social Marketing von Nachhaltigkeitskultur am Beispiel pädagogischer Initiativen. Umweltbundesamt, Dessau-Roßlau

Buba H, Globisch S, Grötzbach J (2009) Anregungen für die Nachhaltigkeitskommunikation aus kulturpolitischer Perspektive. Bausteine eines Orientierungsrahmens zu einem kulturbezogenen Konzept der Nachhaltigkeitskommunikation. Umweltbundesamt, Dessau-Roßlau

Buchwald HH (1988) Wertermittlung von Ziergehölzen–Ein neuer methodischer Vorschlag. Schriftenreihe des Hauptverbandes der landwirtschaftlichen Buchstellen und Sachverständigen e. V., Pflug und Feder, St. Augustin, Heft 122

Burger F (2005) Energiewälder und Ökologie. LWFaktuell 48:26–27

Burkhard B, Diembeck D (2006) Zukunftsszenarien für die deutsche Nordsee. Forum Geoökologie 17:27–30

Burkhard B, Kroll F, Müller F, Windhorst W (2009) Landscapes' capacities to provide ecosystem services–a concept for land-cover based assessments. Lands Online 15:1–22

Burkhard B, Kroll F, Nedkov S, Müller F (2012) Mapping supply, demand and budgets of ecosystem services. Ecol Indic 21:17–29

Büscher C, Japp KP (eds) (2010) Ökologische Aufklärung. 25 Jahre "Ökologische Kommunikation". VS Verlag für Sozialwissenschaften, Wiesbaden

Carpenter SR, Bennett EM, Peterson GD (2006) Editorial: special feature on scenarios for ecosystem services. Ecol Soc 11:32

Cherubini F, Stromman AH (2010) Life cycle assessment of bioenergy systems: state of the art and future challenges. Bioresour Technol 102:437–451

Chichilnisky G, Heal G (1998) Economic returns from the biosphere. Nature 391:629–630

Commission of the European Communities (2007) Renewable energy road map–renewable energies in the 21st

century: building a more sustainable future. Communication from the Commission to the Council and the European Parliament: Brussels
Conta Gromberg E (2006) Handbuch Sozial-Marketing. Cornelsen, Berlin
Costanza R, d'Arge R, de Groot R, Farber S, Grasso M, Hannon B, Limburg K, Naeem S, O'Neill R, Paruelo J et al (1998) The value of ecosystem services: putting the issues in perspective. Ecol Econ 25:67–72
Daily GC, Matson PA (2008) Ecosystem services: from theory to implementation. Proc Natl Acad Sci USA 105:9455–9456
Daily GC, Polasky S, Goldstein J, Kareiva PM, Mooney HA, Pejchar L, Ricketts TH, Salzman J, Shallenberger R (2009) Ecosystem services in decision making–time to deliver. Front Ecol Environ 7:21–28
Dale VH, Beyeler SC (2001) Challenges in the development and use of ecological indicators. Ecol Indic 1:3–10
de Groot RS, Wilson M, Boumans R (2002) A typology for description, classification and valuation of ecosystem functions, goods and services. Environ Econ 41:393–408
de Groot RS, Alkemade R, Braat L, Hein L, Willemen L (2010a) Challenges in integrating the concept of ecosystem services and values in landscape planning, management and decision making. Ecol Complex 7:260–272
de Groot RS, Fisher B, Christie M, Aronson J, Braat L, Haines-Young R, Gowdy J, Maltby E, Neuville A, Polasky S, Portela R, Ring I (2010b) Integrating the ecological and economic dimensions in biodiversity and ecosystem service valuation (Chap. 1). In: Kumar P (ed) The economics of ecosystems and biodiversity (TEEB): ecological and economic foundations. Earthscan, London, pp 9–40
de Haan G, Harenberg D (1999) Expertise "Förderprogramm Bildung für nachhaltige Entwicklung" im Auftrage des Bundesministeriums für Bildung. Wissenschaft Forschung und Technologie, Freie Universität Berlin
de Haan G, Kuckartz U (1998) Die Bedeutung der Umweltkommunikation im Kontext der Nachhaltigkeit. In: Umweltkommunikation und Lokale Agenda 21, Ergebnisse eines Fachgesprächs im Umweltbundesamt am 11. und 12. Dezember 1997, Berlin, pp 30–53
Degenhardt S, Gronemann S (1998) Die Zahlungsbereitschaft von Urlaubsgästen für den Naturschutz. Theorie und Empirie des Embedding-Effektes. Lang, Frankfurt a. M.
Dienel PC (2002) Die Planungszelle. Der Bürger als Chance, 5th edn. Westdeutscher Verlag, Wiesbaden
Dietrich K, Schweppe-Kraft B, Haarnacke S (2014) Methods to calculate offsets to compensate for time lags in the recovery of ecosystems and biodiversity. German Federal Agency for Nature Conservation. http://www.bfn.de/fileadmin/MDB/documents/themen/oekonomie/Foffsetting_time_lags_2014.pdf. Zugegriffen 24. Sept. 2014
Dörr H (2005) Die Zukunft der Landschaft in Mitteleuropa–Verantwortung für die Kulturlandschaft im 21. Jahrhundert–Delphi-Umfrage 2002: Dokumentation und Interpretation/Future Landscape, ein länderübergreifendes Projekt des Forschungsschwerpunktes Kulturlandschaft KLF2 im Auftrag des Österreichischen Wissenschaftsministeriums. Arp-Planning.Consulting. Research, Wien
Dunlop M, Turner G, Foran B, Poldy F (2002) Decision points for land and water futures. Resource Futures Program Working Document 2002/08, CSIRO Sustainable Ecosystems, Canberra, Australia
EEA–European Environment Agency (1994) Corine Land Cover Report–Part 2: Nomenclature. ► www.eea.europa.eu/publications/COR0-part2. Accessed 12 March 2012
EEG (2008) Erneuerbare-Energien-Gesetz (EEG). FNA: 754-22; Artikel 1 G. v. 25.10.2008 BGBl. I S 2074; zuletzt geändert durch Artikel 2 Abs. 69 G. v. 22.12.2011 BGBl. I S 3044
ELC (2000) European landscape convention. ► www.conventions.coe.int/Treaty/EN/Treaties/Html/176.htm. Accessed 12 March 2012
Ericsson K, Rosenqvist H, Nilsson LJ (2009) Energy crop production costs in the EU. Biomass Bioenerg 33:1577–1586
Ewers HJ, Schulz W (1982) Die monetären Nutzen gewässerqualitätsverbessernder Maßnahmen, dargestellt am Beispiel des Tegeler Sees in Berlin. Umweltbundesamt Erich Schmidt Verlag, Berlin, Berichte 3/82
Farber SC, Costanza R, Wilson MA (2002) Economic and ecological concepts for valuing ecosystem services. Ecol Econ 41:375–392
Fidalgo B, Pinto LM (2005) Linking landscape functions and preferences in forest landscapes–a tool for scenario building and evaluation. In: Lange E, Miller D (eds) Proceedings of our shared landscape: integrating ecological socio-economic and aesthetic aspects in landscape planning and management. A contribution from the VisuLands Project. Ascona, Switzerland, pp 34–35
Fischer A, Menzel S (2005) Die Eignung von Gütern für Zahlungsbereitschaftsanalysen. In: Marggraf R, Bräuer I, Fischer A, Menzel S, Stratmann U, Suhr A (eds) Ökonomische Bewertung bei umweltrelevanten Entscheidungen. Einsatzmöglichkeiten von Zahlungsbereitschaftsanalysen in Politik und Verwaltung. Metropolis, Marburg, pp 113–147
Fisher B, Turner RK, Morling P (2009) Defining and classifying ecosystem services for decision making. Ecol Econ 68:643–653
Förster R, Christian P, Scheringer M, Valsangiacomo A (2001) Partizipation in der transdisziplinären Forschung–Eine Positionierung und die Ankündigung des nächsten SAGUFNET-Workshops. Schweizerische Akademische Gesellschaft für Umweltforschung und Ökologie. GAIA 10:146–149
Frankel C (1998) In earth's company: business, environment, and the challenge of sustainability. New Society, Gabriola Island
Freese J, Anders K (2010) Kulturelle Dienstleistungen von Ökosystemen–was kann man sich darunter vorstellen? In: Hotes S, Wolters V (eds) Wie Biodiversität in der Kul-

turlandschaft erhalten und nachhaltig genutzt werden kann. Fokus Biodiversität. Oekom, München, pp 194–199
Gallai N, Salles JM, Settele J, Vaissière BE (2009) Economic valuation of the vulnerability of world agriculture confronted with pollinator decline. Ecol Econ 68:810–821
Gallopin GC, Rijsberman F (2000) Three global water scenarios. Int J Water 1:16
Gausemeier J, Plass C, Wenzelmann C (2009) Zukunftsorientierte Unternehmensplanung–Strategien, Geschäftsprozesse und IT-Systeme für die Produktion von morgen. München
Getzner M, Jungmeier M, Köstl T, Weiglhofer S (2011) Fließstrecken der Mur–Ermittlung der Ökosystemleistungen–Endbericht. Studie im Auftrag von: Landesumweltanwaltschaft Steiermark, Bearbeitung: E.C.O. Institut für Ökologie, Klagenfurt, 86 p
Greiff KB, Weber-Blaschke G, Faulstich M, von Haaren C (2010) Förderung eines umweltschonenden Energiepflanzenanbaus. Natschutz Landschplan 42:101–107
Grossmann M, Hartje V, Meyerhoff J (2010) Ökonomische Bewertung naturverträglicher Hochwasservorsorge an der Elbe. Bundesamt für Naturschutz, Bonn, Naturschutz und Biologische Vielfalt 89
Grunewald K, Naumann S (2012) Bewertung von Ökosystemdienstleistungen im Hinblick auf die Erreichung von Umweltzielen der Wasserrahmenrichtlinie am Beispiel des Flusseinzugsgebietes der Jahna in Sachsen. Nat Landsch 1:17–23
Grunewald K, Syrbe RU, Wachler C (2012) Analyse der ästhetischen und monetären Wertschätzung der Landschaft am Erzgebirgskamm durch den Tourismus. GEOÖKO 33:34–65
Haberl H, Fischer-Kowalski M, Krausmann F, Weisz H, Winiwarter V (2004) Progress towards sustainability? What the conceptual framework of material and energy flow accounting (MEFA) can offer. Land Use Policy 21:199–213
Hafner S (2010) Rechtliche Rahmenbedingungen für eine an den Klimawandel angepasste Landwirtschaft. UPR (Umwelt- und Planungsrecht) 30:371–377
Hampicke U (1991) Naturschutz-Ökonomie. Ulmer, Stuttgart
Hampicke U, Wätzold F (Sprecher der Initiative) (2009) Memorandum: Ökonomie für den Naturschutz–Wirtschaften im Einklang mit Schutz und Erhalt der biologischen Vielfalt. Greifswald. ► www.bfn.de/fileadmin/MDB/documents/themen/oekonomie/MemoOekNaturschutz.pdf. Accessed 12 July 2012
Hampicke U, Horlitz T, Kiemstedt H, Tampe K, Timp D, Walters M (1991) Kosten und Wertschätzung des Arten- und Biotopschutzes. Umweltbundesamt Erich Schmidt Verlag, Berlin, Berichte 3/91, 629 p
Hanke H, Boese P, Ophoff W, Rauschelbach B, Schier V (1981) Handbuch zur Ökologischen Planung, vol 1. Umweltbundesamt Erich Schmidt Verlag, Berlin, Berichte 3/81
Heidmann T, Thomsen A, Schelde K (2000) Modelling soil water dynamics in winter wheat using different estimates of canopy development. Ecol Model 129:229–243
Herrmann S, Kliebisch C, Schmitt F, Schweppe-Kraft B (2012) Naturschutz–effizient planen, managen und umsetzen. Methodenhandbuch und Ratgeber für Wirtschaftlichkeit im Naturschutz. Bundesamt für Naturschutz/Bundesverband Beruflicher Naturschutz e. V., Bonn
Hillier J, Whittaker C, Dailey G, Aylotts M, Casella E, Richter GM, Riche A, Murphy R, Taylor G, Smith P (2009) Greenhouse gas emissions from four bioenergy crops in England and Wales: integrating spatial estimates of yield and soil carbon balance in life cycle analysis. GCB Bioenergy 1:267–281
Hoevenagel R (1994) An assessment of the contingent valuation method. In: Pethig R (ed) Valuing the environment: methodological and measurement issues. Kluwer Academic, Dordrecht
Hoffmann A, Gruehn D (2010) Bedeutung von Freiräumen und Grünflächen in deutschen Groß- und Mittelstädten für den Wert von Grundstücken und Immobilien. Technische Universität, Lehrstuhl Landschaftsökologie und Landschaftsplanung, Dortmund, LLP-Report 010
Holfeld M, Stein C, Rosenberg M, Syrbe R-U, Walz U (2012) Entwicklung eines Landschaftsbarometers zur Visualisierung von Ökosystemdienstleistungen. In: Strobl J, Blaschke T, Griesebner G (eds) Angewandte Geoinformatik 2012, Beiträge zum 24. AGIT-Symposium Salzburg. Wichmann, Berlin, S 646–651
Hornsmann E (1958) Allen hilft der Wald. Seine Wohlfahrtswirkungen. BLV, München
Hou Y, Burkhard B, Müller F (2013) Uncertainties in landscape analysis and ecosystem service assessment. J Environ Manag 127:117–131
Ivanišin M (2006) Regionalentwicklung im Spannungsfeld von Nachhaltigkeit und Identität. Deutscher Universitätsverlag, Wiesbaden
Jacobs (2004) An economic assessment of the costs and benefits of Natura 2000 sites in Scotland. Final report. ► www.scotland.gov.uk/resource/doc/47251/0014580.pdf
Japp KP (2010) Risiko und Gefahr. Zum Problem authentischer Kommunikation. In: Büscher C, Japp KP (eds) Ökologische Aufklärung. 25 Jahre "Ökologische Kommunikation". VS Verlag für Sozialwissenschaften, Wiesbaden
Jenssen M, Anders K (2010) Wald und Wirtschaft. Ein systematischer Blick auf unseren Umgang mit einer nachwachsenden Ressource. In: Helmholtz-Zentrum für Umweltforschung–UFZ (ed) Nachhaltige Waldwirtschaft. Ein Förderschwerpunkt des BMBF in der Bilanz. Leipzig
Jessel B (2000) Von der "Vorhersage" zum Erkenntnisgewinn. Aufgaben und Leistungsfähigkeit von Prognosen in der Umweltplanung. Natschutz Landschplan 32:197–203
Job H, Harrer B, Metzler D, Hajizadeh-Alamdary D (2005) Ökonomische Effekte von Großschutzgebieten. Untersuchung der Bedeutung von Großschutzgebieten für den Tourismus und die wirtschaftliche Entwicklung der Region. Bundesamt für Naturschutz, Bonn, Bad Godesberg, BfN-Skripten 135

Job H, Woltering M, Harrer B (2009) Regionalökonomische Effekte des Tourismus in deutschen Nationalparken. Bundesamt für Naturschutz, Bonn, Bad Godesberg, Naturschutz und Biologische Vielfalt 76

Kächele H, Zander P (1999) Der Einsatz des Entscheidungshilfesystems MODAM zur Reduzierung von Konflikten zwischen Naturschutz und Landwirtschaft am Beispiel des Nationalparks "Unteres Odertal". Schriften der Gesellschaft für Wirtschaft- und Sozialwissenschaften des Landbaus, Agrarwirtschaft in der Informationsgesellschaft, Bd. 35. Münster-Hiltrup, pp 191–198

Karmanski A, Jacob K, Zieschank R (2002) Integration des sozialwissenschaftlichen Wissens in die Umweltkommunikation: Verbesserung des Wissenstransfers zwischen den Sozialwissenschaften und den umweltpolitischen Akteuren. Forschungsbericht. UNESCO-Verbindungsstelle im Umweltbundesamt, Berlin

Kienast F, Bolliger J, Potschin M, de Groot RS, Verburg PH, Heller I, Wascher D, Haines-Young R (2009) Assessing landscape functions with broad-scale environmental data: insights gained from a prototype development for Europe. Environ Manag 44:1099–1120

Köppel J, Peters W, Wende W (2004) Eingriffsregelung, Umweltverträglichkeitsprüfung, FFH-Verträglichkeitsprüfung. Ulmer, Stuttgart

Kort J, Collins M, Ditsch D (1998) A review of soil erosion potential associated with biomass crops. Biomass Bioenerg 14:351–359

Koschke L, Fürst C, Frank S, Makeschin F (2012) A multicriteria approach for an integrated land-cover-based assessment of ecosystem services provision to support landscape planning. Ecol Indic 21:54–66

Kroll F, Müller F, Haase D, Fohrer N (2012) Rural-urban gradient analysis of ecosystem services supply and demand dynamics. Land Use Policy 29:521–535

Kuckartz U, Schack K (2002) Umweltkommunikation gestalten. Eine Studie zu Akteuren, Rahmenbedingungen und Einflussfaktoren des Informationsgeschehens. Leske und Budrich, Opladen

Kulozik A (2009) Focus on Landscape–Ein Beitrag zur Stärkung des Landschaftsbewusstseins von Bewohnern der North Isles in Shetland durch ein umweltbildnerisches Programm für Grundschulkinder. Diplomarbeit, FH Osnabrück

Lange H (2005) Lebensstile–der sanfte Weg zu mehr Nachhaltigkeit? In: Michelsen G, Godemann J (eds) Handbuch Nachhaltigkeitskommunikation: Grundlagen und Praxis. Ökom, München, pp 160–172

Lee YH, Bückmann W, Haber W (2008) Bio-Kraftstoff, Nachhaltigkeit, Natur- und Bodenschutz. NuR 30:821–831

Leipert C (1989) Die heimlichen Kosten des Fortschritts–Wie Umweltzerstörung das Wirtschaftswachstum fördert. Fischer, Frankfurt/Main

Liesebach M, Mulsow H (2003) Der Sommervogelbestand einer Kurzumtriebsplantage, der umgebenden Feldflur und des angrenzenden Fichtenwaldes im Vergleich. Die Holzzucht 54:27–31

Lisowski R (2006) Die strategische Planung politischer Kampagnen in Wirtschaft und Politik. Isensee, Oldenburg, Bremen

L.I.S.T. Stadtentwicklungsgesellschaft mbH (ed) (2011) Handbuch zur Partizipation. Berlin

Londo M, Roose M, Dekker J, de Graaf H (2004) Willow short-rotation coppice in multiple land-use systems: evaluation of four combination options in the Dutch context. Biomass Bioenerg 27:205–221

Löwenstein W (1994) Die Reisekostenmethode und die Bedingte Bewertungsmethode als Instrumente zur monetären Bewertung der Erholungsfunktion des Waldes. Ein ökonomischer und ökonometrischer Vergleich. JD Sauerländer's, Frankfurt/Main

Ludwig D (2000) Limitations of economic valuation of ecosystems. Ecosystems 3:31–35

Luhmann N (2008) Ökologische Kommunikation. Kann die moderne Gesellschaft sich auf ökologische Gefährdungen einstellen? 5. Aufl. Verlg für Sozialwissenschaften, Wiesbaden

Lupp G, Albrecht J, Bastian O, Darbi M, Denner M, Gies M, Grunewald K, Kretschmer A, Lüttich K, Matzdorf B, Neitzel H, Starick A, Steinhäußer R, Syrbe R-U, Tröger M, Uckert G, Zander P (2011) Land use management, ecosystem services and biodiversity–developing regulatory measures for sustainable energy crop production (LOBESTEIN). In: Frantál B (ed) Exploring new landscapes of energies. Collection of extended abstracts of papers from the 8th International Geographical Conference & Workshop CONGEO 1.–5. August, Brno, Czech Republic. Institute of Geonics, Academy of Science of the Czech Republic, pp 51–52

Luttmann V, Schröder H (1995) Monetäre Bewertung der Fernerholung im Naturschutzgebiet Lüneburger Heide. JD Sauerländer's, Frankfurt/M

Lütz M, Bastian O, Röder M, Syrbe R-U (2007) Szenarienanalyse zur Veränderung von Agrarlandschaften. Natschutz Landschplan 39:205–211

Marggraf R, Bräuer I, Fischer A, Menzel S, Stratmann U, Suhr A (eds) (2005) Ökonomische Bewertung bei umweltrelevanten Entscheidungen. Einsatzmöglichkeiten von Zahlungsbereitschaftsanalysen in Politik und Verwaltung. Ökologie und Wirtschaftsforschung 55, Metropolis, Marburg, 380S

Matzdorf B, Reutter M, Hübner C (2010) Gutachten-Vorstudie: Bewertung der Ökosystemdienstleistungen von HNV-Grünland (High Nature Value Grassland)–Abschlussbericht im Auftrag des Bundesamtes für Naturschutz, Bonn. ► http://www.z2.zalf.de/oa/46452aad-17c1-45ed-82b6-fd9ba62fc9c2.pdf

McLaughlin SB, Walsh ME (1998) Evaluating environmental consequences of producing herbaceous crops for bioenergy. Biomass Bioenerg 14:317–324

MEA–Millennium Ecosystem Assessment (2005a) Ecosystems and human well-being: policy responses, vol. 3. Island, Washington

MEA–Millennium Ecosystem Assessment (2005b) Our human planet: summary for decision-makers. Island, Washington

Meyerhoff J, Angeli D, Hartje V (2012) Valuing the benefits of implementing a national strategy on biological diversity–The case of Germany. Environmental Science & Policy 23:109–119

Michelsen G (2002) Bildung und Kommunikation für eine nachhaltige Entwicklung: Sozialwissenschaftliche Perspektiven. In: Beyer A (ed) Fit für Nachhaltigkeit? Biologisch-anthropologische Grundlagen einer Bildung für nachhaltige Entwicklung. Leske und Budrich, Opladen, pp 193–216

Michelsen G, Godemann J (eds) (2005) Handbuch Nachhaltigkeitskommunikation: Grundlagen und Praxis. Oekom, München

Mill JS (1848) The principles of political economy with some of their applications to social philosophy. London, 7th edn. 1909 by Longmans, Green and Co.

Müller F (2005) Indicating ecosystem and landscape organisation. Ecol Indic 5:280–294

Müller K, Bork HR, Dosch A, Hagedorn K, Kern J, Peters J, Petersen HG, Nagel UJ, Schatz T, Schmidt R, Toussaint V, Weith T, Wotke A (eds) (2000) Nachhaltige Landnutzung im Konsens. Ansätze für eine dauerhaft-umweltgerechte Nutzung der Agrarlandschaften in Nordostdeutschland. Focus, Gießen, 190 S

Müller F, Baesler C, Schubert H, Klotz S (eds) (2010) Long-term ecological research–between theory and application. Springer, Dordrecht

NABU–Naturschutzbund Deutschland e. V. (2005) Nachwachsende Rohstoffe und Naturschutz: Argumente des NABU an einen naturverträglichen Anbau. NABU Positionspapier 04/2005. ► http://www.nabu.de/imperia/md/content/nabude/energie/biomasse/1.pdf. Accessed 7 June 2010

Nassauer JI, Corry RC (2004) Using normative scenarios in landscape ecology. Lands Ecol 19:343–356

Nassauer JI, Corry RC, Cruse RM (2002) The landscape in 2025: alternative future landscape scenarios: a means to consider agricultural policy. J Soil Water Conserv 57:44A–53A

Natural Capital Project (2012) Startseite. ► http://www.naturalcapitalproject.org/. Accessed 28 March 2012

Nedkov S, Burkhard B (2012) Flood regulating ecosystem services–mapping supply and demand in the Etropole municipality, Bulgaria. Ecol Indic 21:67–79

Nelson E, Polasky S, Lewis DJ, Plantinga AJ, Lonsdorf E, White D, Bael D, Lawler JJ (2008) Efficiency of incentives to jointly increase carbon sequestration and species conservation on a landscape. Proc Natl Acad Sci USA 105:9471–9476

Nelson E, Mendoza G, Regetz J, Polasky S, Tallis H, Cameron DR, Chan KMA, Daily GC, Goldstein J, Kareiva PM, Lonsdorf E, Naidoo R, Ricketts TH, Shaw MR (2009) Modeling multiple ecosystem services, biodiversity conservation, commodity production, and tradeoffs at landscape scales. Front Ecol Environ 7:4–11

Niemeijer D, de Groot R (2008) A conceptual framework for selecting environmental indicator sets. Ecol Indic 8:14–25

Norton B (1988) Commodity, amenity, and morality: the limits of quantification of valuing biodiversity. In: Wilson EO (ed) Biodiversity. National Academy, Washington, DC (Chap. 22)

OECD (2003) Environmental indicators–development, measurement and use. OECD Environment Directorate, Paris

Oppermann B (2008) Zur Kunst der Landschaftsvorhersage. Gedanken anlässlich des FLL-Fachforums zum Thema Zukunftslandschaften. Stadt und Grün 57:35–38

Osborne JM, Turner MA (2007) Cost benefit analysis vs. referenda. University of Toronto, Department of Economics. Working Paper 286. ► http://www.economics.utoronto.ca/index.php/index/research/workingPaperDetails/286

Osterburg B, Rühling I, Runge T, Schmidt TG, Seidel K, Antony F, Gödecke B, Witt-Altfelder P (2007) Kosteneffiziente Maßnahmenkombinationen nach Wasserrahmenrichtlinie zur Nitratreduktion in der Landwirtschaft. In: Osterburg B, Runge T (eds) Maßnahmen zur Reduzierung von Stickstoffeinträgen in die Gewässer–eine wasserschutzorientierte Landwirtschaft zur Umsetzung der Wasserrahmenrichtlinie. Landbauforschung Völkenrode, Sonderheft, p 307

Oudenhoven APE van, Petz K, Alkemade R, de Groot RS, Hein L (2012) Indicators for assessing effects of management on ecosystem services. Ecol Indic 21:110–122

Pachauri RK, Reisinger A (2008) Contribution of working groups I, II and III to the fourth assessment report of the intergovernmental panel on climate change. IPCC, Geneva

Pearce DW, Turner RK (1990) Economics of natural resources and the environment. Harvester Wheatsheaf, New York

Pimentel D, Harvey C, Resoosudarmo P, Sinclair K, Kurz D, McNair M, Crist S, Shpritz L, Fitton L, Saffouri R, Blair R (1995) Environmental and economic costs of soil erosion and conservation benefits. Science 267:1117–1123

Polasky S, Nelson E, Camm J, Csuti B, Fackler P, Lonsdorf E, Montgomery C, White D, Arthur J, Garber-Yonts B, Haight R, Kagan J, Starfield A, Tobalske C (2008) Where to put things? Spatial land management to sustain biodiversity and economic returns. Biol Conserv 141:1505–1524

Pretzsch H (2001) Modellierung des Waldwachstums. Parey, Berlin

Rees WE (1992) Ecological footprints and appropriated carrying capacity: what urban economics leaves out. Environ Urban 4:121–130

Reibnitz U von (1991) Szenariotechnik. Instrumente für die unternehmerische und persönliche Erfolgsplanung. Gabler, Wiesbaden

Riedel R (2000) Strukturen der Komplexität–Eine Morphologie des Erkennens und Erklärens. Springer, Heidelberg

Rieß W (2010) Bildung für nachhaltige Entwicklung. Theoretische Analysen und empirische Studien. Waxmann, Münster

Ringland G (1997) Scenario planning. Managing for the future. Wiley, Chichester

Riitters KH, Wickham JD, Vogelmann JE, Jones KB (2000) National land-cover pattern data. Ecology 81:604

Rode M, Kanning H (2006) Beiträge der räumlichen Planung zur Förderung eines natur- und raumverträglichen Ausbaus des energetischen Biomassepfades. Informationen zur Raumentwicklung 1:103–110

Rode M, Schneider C, Ketelhake G, Reißhauer D (2005) Naturschutzverträgliche Erzeugung und Nutzung von Biomasse zur Wärme- und Stromgewinnung. BfN-Skripten 136, Bonn

Röder N, Grützmacher F (2012) Emissionen aus landwirtschaftlich genutzten Mooren–Vermeidungskosten und Anpassungsbedarf. Natur und Landschaft 87:56–61

Rotmans J, Asselt M van, Anastasi C, Greeuw S, Mellors J, Peters S, Rotman D, Rijkens N (2000) Visions for a Sustainable Europe. Futures 32:809–831

Rowe RL, Street NR, Taylor G (2009) Identifying potential environmental impacts of large-scale deployment of dedicated bioenergy crops in the EU. Renew Sust Energ Rev 13:271–290

Ruhl JB, Kraft SE, Lant CL (2007) The law and policy of ecosystem services. Island, Washington

Ryffel A, Grêt-Regamey A (2010) Bewertung der Ökosystemdienstleistungen von Trockenwiesen und -weiden. Vortrag an der internationalen Naturschutzakademie Insel Vilm. ► http://www.bfn.de/0610_v_oekosystemdienstleistunge.html

Santelmann MV, White D, Freemark K, Nassauer JI, Eilers JM, Vaché KB, Danielson BJ, Corry RC, Clark ME, Polasky S, Cruse RM, Sifneos J, Rustigian H, Coiner C, Wu J, Debinski D (2004) Assessing alternative futures for agriculture in Iowa, U.S.A. Lands Ecol 19:357–374

Schack K (2004) Umweltkommunikation als Theorielandschaft. Eine qualitative Studie über Grundorientierungen, Differenzen und Theoriebezüge der Umweltkommunikation. Ökom, München

Schmidt C, Hage G, Galandi R, Hanke R, Hoppenstedt A, Kolodziej J, Stricker M (2010) Kulturlandschaft gestalten–Grundlagen. Landwirtschaftsvlg Münster, Bonn (In: Bundesamt für Naturschutz (ed))

Schweppe-Kraft B (1996) Bewertung von Biotopen auf der Basis eines Investitionsmodells–Eine Weiterentwicklung der Methode Koch. Wertermittlungsforum 1

Schweppe-Kraft B (1998) Monetäre Bewertung von Biotopen. Bundesamt für Naturschutz, Bonn

Schweppe-Kraft B (2009) Natural capital in Germany–state and valuation; with special reference to Biodiversity. In: Döring R (ed) Sustainability, natural capital and nature conservation. Metropolis, Marburg

Seppelt R, Holzkämper A (2007) Multifunctional use of landscape services. Applications and results of optimization techniques of land use scenario development. Proceedings 7. IALE World Congress. p. Wageningen

Siemer SH (2007) Das Programm der Bildung für nachhaltige Entwicklung. Eine systemische Diagnose mit den Schemata Qualität und Nachhaltigkeit. Dissertation, Lüneburg

Spangenberg JH, Settele J (2010) Precisely incorrect? Monetising the value of ecosystem services. Ecol Complex 7:327–337

SRU–Sachverständigenrat für Umwelt (2007) Umweltschutz im Zeichen des Klimawandels. Umweltgutachten. Berlin

Stallmann HR (2011) Ecosystem services in agriculture: Determining suitability for provision by collective management. Ecol Econ 71:131–139

Stern N (2007) The economics of climate change. The Stern review. Cambridge University, Cambridge

Swetnam RD, Fisher B, Mbilinyi BP, Munishi PKT, Willcock S, Ricketts T, Mwakalila S, Balmford A, Burgess ND, Marshall AR, Lewis SL (2010) Mapping socio-economic scenarios of land cover change: a GIS method to enable ecosystem service modelling. J Environ Manag 92:563–574. doi:10.1016/j.jenvman.2010.09.007

Syrbe R-U, Walz U (2012) Spatial indicators for the assessment of ecosystem services: providing, benefiting and connecting areas and landscape metrics. Ecol Indic 21:80–88

Tallis H, Polasky S (2009) Mapping and valuing ecosystem services as an approach for conservation and natural-resource management. The Year in Ecology and Conservation Biology. Ann N Y Acad Sci 1162:265–283

Tallis H, Ricketts T, Guerry A (2011) InVEST 2.1 beta user's guide–integrated valuation of ecosystem services and tradeoffs. ► http://www.invest.ecoinformatics.org/tool-documentation/InVEST_Documentation_v2.1.pdf/preview_popup/fil. Accessed 12 July 2012

Tappeiner U (2007) Land-use change in the European Alps: effects of historical and future scenarios of landscape development on ecosystem services. Heinz Veit, Universität Bern–Thomas Scheurer, ICAS/ICAR–Günter Köck, Österreichische Akademie der Wissenschaften (ed) Proceedings of the ForumAlpinum 2007, 18.–21. April, ÖAW, Engelberg/Switzerland, pp 21–23

TEEB–The Economics of Ecosystems and Biodiversity (2009) TEEB Climate Issues Update. ► http://www.teebweb.org/Portals/25/Documents/TEEB-ClimateIssuesUpdate-Sep2009.pdf

TEEB–The Economics of Ecosystems and Biodiversity (2010) Ecological and economic foundations. In: Kumar P (ed) Earthscan, London

Theßenvitz S (2009) Wer interessiert sich für Umwelt?–Lebenswelten und Zielgruppen in Deutschland. In: Bayer. Landesamt für Umwelt (ed) Für Natur und Umwelt begeistern. Umweltkommunikation Fachtagung des LfU am 28.04.2009. LfU, pp 5–7

Tötzer T, Köstl M, Steinnocher K (2007) Scenarios of land use change in Europe based on socio-economic and demographic driving factors. In: Schrenk M, Popovich

VV, Benedikt J (eds) Real Corp: to plan is not enough: strategies, plans, concepts, projects and their successful implementation in Urban, Regional and Real Estate Development. Proc. Corporation, Competence Center of Urban and Regional Planning, Wien, 20.–23.5.2007, pp 141–150

Turnhout E, Hisschemöller M, Eijsackers H (2007) Ecological indicators: between the two fires of science and policy. Ecol Indic 7:215–228

UBA–Umweltbundesamt (2000) Weiterentwicklung und Präzisierung des Leitbildes der nachhaltigen Entwicklung in der Regionalplanung und regionalen Entwicklungskonzepten. Texte 59/00. Umweltbundesamt, Berlin

UBA–Umweltbundesamt (2007) Ökonomische Bewertung von Umweltschäden. Methodenkonvention zur Schätzung externer Umweltkosten. Umweltbundesamt, Dessau. ► http://www.umweltdaten.de/publikationen/fpdf-l/3193.pdf

UNEP (2007) GEO-4–Global Environment Outlook. ► www.unep.org/geo/geo4.asp. Accessed: 30. Juni 2012

UNEP (2011) Keeping track of our changing environment: From Rio to Rio + 20 (1992–2012). Division of Early Warning and Assessment (DEWA), United Nations Environment Programme (UNEP), Nairobi

Vihervaara P, Kumpula T, Tanskanen A, Burkhard B (2010) Ecosystem services–a tool for sustainable management of human–environmental systems. Case study Finnish Forest Lapland. Ecol Complexity 7:410–420

Wallace KJ (2007) Classification of ecosystem services: problems and solutions. Biol Conserv 139:235–246

Walz A, Lardelli C, Behrendt H, Grêt-Regamey A, Lundström C, Kytzia S, Bebi P (2007) Participatory scenario analysis for integrated regional modelling. Lands Urban Plan 81:114–131

Wehrspaun M, Schoembs H (2002) Die Kluft zwischen Umweltbewusstsein und Umweltverhalten als Herausforderung für die Umweltkommunikation. In: Beyer A (ed) Fit für Nachhaltigkeit? Biologisch-anthropologische Grundlagen einer Bildung für nachhaltige Entwicklung. Leske und Budrich, Opladen, ppk 141–162

Weitzman M (2000) Economic profitability versus ecological entropy. Q J Econ 115:237–263

Whitehead C, Phaneuf D, Dumas CF, Herstine J, Hill J, Buerger B (2010) Convergent validity of revealed and stated recreation behavior with quality change: a comparison of multiple and single site demands. Environ Resour Econ 45:91–112

Wiggering H, Müller F (eds) (2004) Umweltziele und Indikatoren. Springer, Berlin

Wilson E, Piper J (2010) Spatial planning and climate change. Taylor Francis, New York

Windhorst W, Müller F, Wiggering H (2004) Umweltziele und Indikatoren für den Ökosystemschutz. Geowissen und Umwelt. Springer, Berlin, pp 345–373

Wolf A, Appel-Kummer E (2005) Demografische Entwicklung und Naturschutz. Perspektiven bis 2015. BfN

Wronka TC (2004) Ökonomische Umweltbewertung. Vergleichende Analysen und neuere Entwicklungen der kontingenten Bewertung am Beispiel der Artenvielfalt und Trinkwasserqualität. Vauk, Kiel

Zangemeister C (1971) Nutzwertanalyse in der Systemtechnik. Eine Methodik zur multidimensionalen Bewertung und Auswahl von Projektalternativen, 4th edn. (1976), Dissertation, Technische Universität Berlin, 1970, Wittemann, München, 370 S

Ziemann A (2005) Kommunikation der Nachhaltigkeit. Eine kommunikationstheoretische Fundierung. In: Michelsen G, Godemann J (eds) Handbuch Nachhaltigkeitskommunikation: Grundlagen und Praxis. Ökom, München, pp 121–131

Zimmermann H-J, Gutsche L (1991) Multi-Criteria Analyse. Einführung in die Theorie der Entscheidungen bei Mehrfachzielsetzungen. Springer, Berlin

Governing Biodiversity and Ecosystem Service Provision

K. Grunewald, O. Bastian (eds.), *Ecosystem Services – Concept, Methods and Case Studies*,
DOI 10.1007/978-3-662-44143-5_5, © Springer-Verlag Berlin Heidelberg 2015

"The best way to secure the future is to secure the present (Franz Kafka)."

5.1 Policy Mixes for Biodiversity Conservation and Ecosystem Service Management

I. Ring and C. Schröter-Schlaack

5.1.1 Why Use a Policy Mix?

The ecosystem service concept is closely linked to the conservation and sustainable use of biodiversity (MEA 2005a, b). By focusing on the direct and indirect benefits humans derive from nature the concept may bridge the gap between nature conservation and economic development and help mainstream the sustainable management of ecosystems and their services into public policies and private decision-making. However, it has been emphasised by most authors that biodiversity itself is not an ecosystem service, although there is evidence of a central role of biodiversity in the functioning and resilience of ecosystems (MEA 2005b; Elmqvist et al. 2010).

There may be neutral, but also positive (synergies) or negative (trade-offs) interactions between biodiversity and ecosystem services (ES) as well as between the provision of different ES (Elmqvist et al. 2010; Ring et al. 2010). For example, the MEA (2005b) showed that the emphasis on provisioning services within the past decades, e.g. intensification in agricultural production, has had negative impact on biodiversity as well as on regulating and cultural ES. Fostering the provision of specific ES may thus not always be beneficial for biodiversity conservation. In turn, biodiversity conservation may not equally contribute to ecosystem service provision. This complex relationship has to be considered when analysing policies and governance regimes for biodiversity conservation and ecosystem service provision.

Real-world policies for conservation and sustainable management of biodiversity typically apply multiple instruments at the same time. Justifications for using a policy mix emphasise the distinctive character of biological diversity as inherently complex and dynamic (OECD 1999). The heterogeneity of ecosystems and species involves heterogeneous objectives that naturally call for a range of different instruments capable of addressing the multidimensional aspects of biodiversity loss and ecosystem degradation (Gunningham and Young 1997). Policies for biodiversity conservation and the sustainable provision of ES contrast with the homogeneous characteristics of other environmental solutions that may need to address just a single pollutant. Ignorance, uncertainties and informational failures are central in a way that successful conservation policies need to account for the precautionary principle, the idea of safe minimum standards, and adaptive management to prevent major irreversible losses (OECD 1999).

The focus on policy mix analysis is even more relevant for the sustainable provision of ES. When assessing the institutions influencing ES provision, a wide(r) range of policy sectors has to be considered. As mentioned above, some of these sectorial regulations may have positive impact on biodiversity and other ES, while others may cause negative impact, thereby aggravating trade-offs between different policy goals (Elmqvist et al. 2010; Ring and Schröter-Schlaack 2011a). For example, a reduction or ban on fertilizer use to safeguard drinking water provision will also have positive effects for biodiversity conservation. Agricultural subsidies, on the contrary, may foster provisioning services, i.e. crop yield, but may negatively impact biodiversity and other agricultural ES, e.g. soil fertility or landscape beauty. Promoting the expansion of renewable energies may lead to land-use intensification in agriculture that reduces crop rotation and depletes biodiversity. Moreover, other provisioning services, such as production of food crops, may be crowded out due to subsidies for energy crops (▶ Sect. 4.4.2).

Against the background of these interactions any assessment of policy responses with regard to biodiversity conservation and ecosystem service management has to consider the existing mix of policy instruments (see ▶ Box for a definition of a policy mix). Although most of the existing studies on instrument choice and design focus on single policy instruments, we argue in favour of a three-step policy mix analysis. The first step comprises the identification of the context and the main challenges for a policy response. The second step includes criteria and recommendations regarding the choice of instruments, about the functional role different instruments

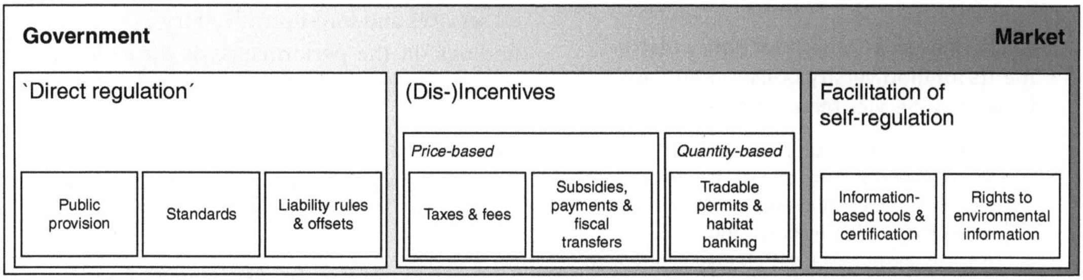

Fig. 5.1 Continuum of policy instruments for biodiversity conservation and ecosystem service management. © Schröter-Schlaack and Ring 2011

might play in addressing the challenges highlighted in step 1 and how interactions between instruments in policy mixes could be considered. Lastly, the third step elaborates specific design issues in order to maximise the value added by single instruments within policy mixes for biodiversity conservation and ecosystem service management (Schröter-Schlaack and Ring 2011). Before presenting these steps in more detail, the following section provides a short overview about the potential policy responses.

What is a Policy Mix?

A policy mix is a combination of policy instruments, which has evolved to influence the quantity and quality of biodiversity conservation and ecosystem service provision in public and private sectors (Ring and Schröter-Schlaack 2011b, p. 15).

5.1.2 A Well-Equipped Toolbox of Policy Instruments

Single instruments that may compose a policy mix can be classified following their distinct functionality. Economic literature on instrument choice and design typically uses the three following categories (e.g. Michaelis 1996; Gunningham and Young 1997; Sterner 2003) (see Fig. 5.1):

- Direct regulation, i.e. command-and-control instruments that directly steer the policy addressee's behaviour by standards, best available technology requirements or spatial planning, including protected area designation.
- Incentive-based approaches, such as environmental taxes, fees or levies that impose a price on environmentally harmful activities thereby internalising external effects of consumption or production patterns. For biodiversity conservation and the sustainable management of ecosystem services, internalising positive externalities is of equal importance (TEEB 2011). Such measures include payments for environmental services and ecological fiscal transfers (Ring 2011, ▶ Sect. 5.2).
- Instruments to support self-regulation of markets by informing and educating people about the environmental impacts of their behaviour or provide motivation towards conservation and sustainable use of ES in consumption or investment decisions.

In practical politics, several instruments from these categories can often be found in combination. Some instruments may have been introduced on purpose to enhance the outcome of another instrument. Informational instruments, for example, are often introduced to provide relevant addressees with the knowledge necessary to enhance the outcome of regulatory or incentive-based instruments. In other cases, incentive-based instruments are introduced as compensation for the costs imposed by regulatory instruments, such as restricted land-use intensity in drinking water catchments or nature protection areas.

In economic literature on instrument choice and design a multitude of criteria has been suggested to analyse and assess policy instruments. In

Challenges and Context for Policy Instruments for Biodiversity Conservation and Ecosystem Service Management

1. What are the important characteristics of biodiversity and ecosystems that will influence appropriateness, applicability and success of certain instruments and their combinations?
2. What are the policy objectives regarding biodiversity conservation and ecosystem service management?
3. What are the drivers of biodiversity loss and ecosystem degradation and how might these be adequately addressed?

the following, these criteria are grouped into four main clusters:

1. Environmental effectiveness, i.e. whether the environmental goal was reached by the use of the instrument.
2. Cost-effectiveness, i.e. whether the environmental goal was reached at the lowest costs. Besides opportunity costs this also comprises implementation and transaction costs associated with the specific instrument.
3. Social and distributional impact, i.e. whether there are positive or negative social impacts associated with the use of the instrument and how the benefits and costs are distributed among actors and social groups.
4. Institutional arrangements, i.e. institutions necessary for the successful implementation and operation of the instrument.

In textbook economics, incentive-based approaches are deemed to be more flexible and cost-effective than command-and-control-type measures (Michaelis 1996; OECD 2007). A comprehensive literature review of policy instruments for biodiversity conservation and a sustainable provision of ecosystem services showed, however, that policy mixes are not only a matter of fact in real-world policy, but combining instruments can also be theoretically justified for efficiency reasons and a range of other motives (Ring and Schröter-Schlaack 2011a). Building on that work, ◘ Table 5.1 presents characteristics of the instruments reviewed (including regulatory instruments, offsets, habitat banking and tradable development rights, easements and tax reliefs, ecological fiscal transfers, payments for environmental services and forest certification) as well as main findings on the performance of the different approaches (Schröter-Schlaack and Ring 2011, p. 178 et seq.). For a detailed discussion of payments for environmental or ecosystem services (PES) and ecological fiscal transfers (EFT) see also ► Sect. 5.2.

5.1.3 Assessing Instruments for Biodiversity Conservation and Ecosystem Service Management in Policy Mixes

In the following sections we develop a stepwise approach to assess instruments for biodiversity conservation and a sustainable management of ES in policy mixes. This will be based on existing frameworks for policy mix assessment in other policy sectors (Ring and Schröter-Schlaack 2011b) and the specific characteristics of biodiversity and ES. The framework's three fundamental steps are built up by the criteria to evaluate the underlying problem, the policy instrument or the relevant policy mix (◘ Table 5.2). These broad assessment categories can be further subdivided into relevant issues to consider in steps 1 and 2, and into fine grain assessment criteria for the detailed evaluation and design of policy instruments in step 3.

Step 1: Identifying Challenges and Context

When it comes to analysing policy mixes, the focus is not on maximising effectiveness or efficiency of individual policy measures but on the complementarity of the instruments involved, their interplay and the ability of the policy mix to address all drivers of the underlying problem (Ring and Schröter-Schlaack 2011b). The appropriate mix of instruments and actors will hence depend upon the nature of the environmental problem, the target groups and wider contextual factors (Gunningham et al. 1998).

Against this backdrop, the first step of the proposed framework consists in gaining a thorough understanding of the policy object, i.e. biodiversity conservation and ES management. Although we believe the questions listed in the ► Box to be neither comprehensive nor exclusive, they may cover the most relevant questions to be answered in a preparatory screening phase of the policy mix analysis.

Table 5.1 Hypotheses on performance of selected single instruments for biodiversity conservation. (Source: Schröter-Schlaack and Ring 2011; based on Schröter-Schlaack and Blumentrath 2011; Santos et al. 2011; Oosterhuis 2011; Ring et al. 2011; Porras et al. 2011; Kaechele et al. 2011)

Instrument type	'Direct regulation', e.g. protected area (PA) designation	Offsets, habitat banking and permit trading	Tax reliefs	Ecological fiscal transfers	Reducing emissions from deforestation and degradation (REDD and REDD+)	Payments for environmental services (PES)	Forest certification
Goal	Safeguard important areas for species and habitat conservation	Account for and mitigate inevitable impacts on biodiversity and ecosystems	Account for positive environmental externalities provided by land users	Compensating decentralised governments for opportunity and/or management costs as well as spillover benefits of protected areas (PA)	Multinational and multilevel policies and measures to reduce deforestation and forest degradation and associated carbon emissions in developing countries (REDD), while considering conservation and co-benefits (REDD+)	Incentivising land users for biodiversity conservation and ecosystem service provision, e.g. by compensating for associated opportunity and management costs	Promote biodiversity and environmentally friendly forest production in accordance with legal codes and certification requirements
Actors adressed	Private and public actors	Private and public actors	Private actors	Public actors	Public and private actors	Mostly private actors/land users	Private actors (consumers)
Baseline and policy context	Protection provided by other primary instruments (e.g. emission/management standards) or existing PA network, very often no protection at all	Impact allowed by (management/emission/performance) standards	Taxpayers behaviour without the tax relief (business as usual might be biodiversity friendly anyhow)	PA coverage when instrument is introduced	Deforestation and degradation rates without REDD (i.e. business as usual defined, e.g. by historical or forecasted deforestation rates or national circumstances)	Land-use practice without incentives by PES schemes (business as usual could be either static, declining or improving)	National forestry regulation, certification process most often progressive and adaptive

Table 5.1 Continued

Instrument type	'Direct regulation', e.g. protected area (PA) designation	Offsets, habitat banking and permit trading	Tax reliefs	Ecological fiscal transfers	Reducing emissions from deforestation and degradation (REDD and REDD+)	Payments for environmental services (PES)	Forest certification
Conservation effectiveness	High–increase in/conservation of biodiversity and ecosystem service provision; however, effectiveness may be at risk due to weak enforcement or may erode in the future due to changing environmental conditions (e.g. climate change)	Medium–although typically designed to allow for a 'no net loss' goal, problems arise in assuring equivalence of mitigation measures and their long-term monitoring	Low–depending on tax burden relieved (existence of tax, actual enforcement of payments, and sufficient tax rate); non targeted approach	Medium to high–increase in quantity and quality of PAs likely (especially when beneficiary of transfers can influence quantity and quality of PAs)	Potentially medium to high–depending on actual design (additionality, avoidance of leakage, permanence and carbon accountability) of scheme once established	Low to high–depending on instrument design regarding baseline, and additionality, leakage, permanence and participation	Medium–impact dependent on rigorousness of standard and framing conditions, such as intensity of investment, difficulties in transport and licensing, land tenure and conflicts with competing land uses
Associated costs and proxies for cost-effectiveness	Medium–though PAs very often show a positive benefit-cost-relationship, local opportunity costs can be substantial	High–in particular the option to trade mitigation measures significantly reduces opportunity costs; however, some ecosystem/habitat types may be (too) costly to restore	Medium–low transaction costs as resting on existing administrative procedure; however, very often incentives provided are insufficient for required change in land-use practice	Medium to low–low transaction costs as it builds on existing mechanism (fiscal transfer schemes and PA designation)	Potentially medium to high–pilot schemes may have underestimated implementation and transaction costs of fully developed REDD architecture	Medium to high–no up-front public investment for buying land, auction-based programmes limit excessive rents; however potentially high transaction costs	Medium–administrative costs of certification scheme may be substantial (in particular in tropical forests)

Table 5.1 Continued

Instrument type	'Direct regulation', e.g. protected area (PA) designation	Offsets, habitat banking and permit trading	Tax reliefs	Ecological fiscal transfers	Reducing emissions from deforestation and degradation (REDD and REDD +)	Payments for environmental services (PES)	Forest certification
Social impacts	Medium–ecosystem services protected by PAs may benefit (local) population; however, substantial opportunity costs and risk to revoke informal rights (e.g. access/abstraction) in area designation	Medium–increase in education/job and income opportunities for rural landowners marketing offsets; compensation of opportunity cost of land conservation (TDR)	Medium–compensation for opportunity costs of environmentally friendly land-use practices; however, only applicable to tax debtors (e.g. landowners)	Medium–depending on entry point of PAs in fiscal transfer systems; fiscal transfers as such address inequalities between jurisdictions	Potentially high–depending on the institutional infrastructure at international and national level to enable broad participation of and within developing countries	Medium–support of rural livelihoods, resource management and social coordination capacities; but enrolment numbers limited by insecure property rights and transaction costs, mixed effect on poverty alleviation	Low to medium–difficult to reach smaller operators except through subsidised schemes; communities are often benefited through workforce participation and engagement in co-benefits
Legal and institutional requirements	Medium to high–easily introducible for a few unique spots; increasingly difficult to implement if demand for land is highly competitive	High–strong public sector involvement necessary in standard setting and monitoring of mitigation measures, high up-front investment for trading architecture	Low–tax deductions are likely to be politically accepted; implementation builds on existing administrative structures	Medium–requires existing fiscal equalisation scheme; introduction of PA indicator often needs constitutional changes and new laws, requiring political majorities	Medium to high–countries need to be able to participate internationally/implement national and subnational-level programmes; this may require broad stakeholder participation, reform of national forest laws and creation of new institutions	Medium to high–definition and enforcement of property rights key for programme success, more effective programmes require high up-front costs for baseline setting, negotiations, fund- and awareness raising	Medium to high–effective forest legislation/laws on property rights; architecture to distribute benefits in case of community involvement

5

Table 5.2 A three-step framework for assessing and designing policy mixes for biodiversity conservation and ecosystem service management. (Soure: Schröter-Schlaack and Ring 2011, p. 184)

	Assessment category	Issues to consider
First step		
Identifying challenges and context Scoping phase	Characteristics of biodiversity and ecosystem services	Potential trade-offs between biodiversity and ecosystem services
		Irreversibility of biodiversity loss
		Tipping points and threshold effects
		Lacking property rights for biodiversity and many ecosystem services
		Defining ecosystem service in question
	Objectives regarding biodiversity conservation and ecosystem service management	Range of ecosystem services utilisation
		Trade-offs between different ecosystem services
	Drivers of biodiversity loss and ecosystem degradation	Direct and indirect drivers from various sectors
		Negative impact of drivers amplified by sectorial policies
	Actors and governance levels	Public and private actors
		Local to global level actors
		Alteration of decision-making processes and inputs across scales–and thus necessary policies
	Cultural and constitutional settings	Local knowledge and traditional practices
		Relative appropriateness of monetary valuation and market-based conservation in cultural context
		Constitutional options and constraints
Second step		
Identifying gaps and choosing instruments for analysis Evaluating the functional role of instruments in the policy mix	Policies in place versus new instruments under consideration	Policy mix across sectors and governmental levels (national/federal versus regional/local)
		Experience with policy instruments
		Persistence of existing instruments
	Context-specific strengths and weaknesses of instruments	Dealing with uncertainty and ignorance
		Lacking property rights
		Spatial targeting of instrument
		Additionality
		Type of ecosystem service
	Instrument interactions	Inherently complementary interaction
		Inherently negative interaction
		Sequencing/path-dependency
		Context-dependent interaction

■ Table 5.2 Continued

	Assessment category	Issues to consider
Third step		
Policy evaluation and design Impact evaluation for existing (ex post) and scenario analysis for new instruments (ex ante)	Conservation effectiveness	E.g. trend in numbers of endangered species and others
	Cost-effectiveness and further efficiency criteria	E.g. increase in transaction costs in relation to higher conservation effectiveness of measures and others
	Distributive impact and legitimacy	E.g. beneficiaries and benefactors of a certain conservation measure and others
	Institutional options and constraints	E.g. constitutional fit and administrative practicability and others

Within this first step, it is necessary to identify relevant actors–both private and public–in the affected political and economic sectors on the relevant governance levels. Moreover, constitutional and legal requirements as well as the cultural perceptions of biodiversity and ecosystem services may open up options or impose constraints on the implementation of potential policy instruments (Brondízio et al. 2010).

Step 2: Identifying Gaps and Choosing Policy Instruments for Analysis

During the second step of the proposed framework gaps in the implemented policy mix have to be identified and potential instrument alternatives or complements have to be chosen, as further assessed in step 3. In this respect, it is necessary to first identify the policies already in place, as most aspects of biodiversity are already covered or at least influenced by existing policies. These policies will not always originate from environmental policies only, but might stem from different sectorial policies, e.g. agri- and silviculture, energy, transport or trade policy as well. Taking stock of existing policies may point to shortcomings, unaccounted trade-offs and blind spots of the currently applied instruments (► Box).

Based on such assessment, policy-makers may have two options or pathways to enhance the overall performance of the policy mix (■ Fig. 5.2): on the one hand, they could aim at improving the existing mix of instruments by explicitly considering the effects of instrument interaction in fine grain design of single components of the mix (ex post analysis). On the other hand, policy-makers may opt for introducing a new instrument into the existing mix in order to account for yet unconsidered aspects of the problem (ex ante analysis). This may include, e.g. actors, activities or sectors so far not explicitly addressed or the acknowledgement of recently evolved ecological knowledge.

Second, the different strengths and weaknesses of instruments are of different importance for different conservation and management goals. 'Direct regulation' will have to play a crucial role in safeguarding a minimum level of biodiversity to avoid crossing critical thresholds of ecosystem functioning. Incentive-based instruments merit particular consideration for managing marketable ES, and sustainably using ES within safe margins that do not endanger ecosystem functioning. Motivational, educational and informational instruments are always an important component of the policy mix as they raise awareness for biodiversity conservation and the consequences of continued loss of biodiversity and ecosystem service degradation, enhance acceptance of policies, and increase participation in voluntary conservation and management measures. In contrast to other fields of environmental

Assessing Existing Policies Against the Challenges for Biodiversity Conservation and Ecosystem Service Management

1. Do the policies in place adequately address the irreversibility of biodiversity loss as well as thresholds of ecosystem resilience that–once crossed–will result in a failure of the ecosystem to deliver its services?
2. Do the instruments in place address the trade-offs between biodiversity conservation and ecosystem service provision on the one hand and between different ecosystem services on the other?
3. Are the drivers of biodiversity loss and ecosystem degradation identified and addressed by existing policies?
4. Are all relevant actors addressed by policy instruments or who is missing?
5. What is the scope of new instruments judged on available experience of policy-makers and policy-addressees and the overall attitude of the society regarding biodiversity conservation, ecosystem service management and public regulation?

regulation (e.g. in controlling air pollution, see OECD 2007) overlap of instruments in biodiversity conservation and ecosystem service management constitutes an insurance against knowledge gaps, policy and implementation failures and should thus not be treated as generally inefficient (Gunningham and Young 1997). The spatial heterogeneity of biodiversity conservation and ecosystem service provision potential often requires a mix of instruments to be applied. Incentive-based instruments may be linked to regulation or planning (eligible areas for PES may be linked to, e.g. protected areas), or provide spatial bonuses in areas targeted for special conservation efforts. The performance of 'direct regulation' can in turn be supported by incentive-based instruments when actors are incentivised to provide conservation and management action beyond regulatory minimum requirements.

Lastly, if instruments are applied simultaneously they will not only work towards the desired policy goal, e.g. biodiversity conservation, but they may also interact and thereby influence the performance of the policy mix. Thus, it is necessary to reconsider the classifications of instrument interactions available, identify the functional role of each approach within a policy mix and choose complementary instruments to the policies already in place (Schröter-Schlaack and Ring 2011).

Step 3: Policy Evaluation and Design

The third step of the proposed framework turns the focus to the evaluation and design of single instruments so that the additional value of the relevant instrument to the existing policies is maximised. To develop policy recommendations we refer to the policy instrument evaluation criteria mentioned above: conservation effectiveness; cost-effectiveness; social impact, fairness and policy legitimacy; and institutional aspects. When dealing with policy mixes, the ultimate goal for instrument design is no longer to develop first-best or second-best single policy solutions, but to optimise design regarding the functional role of the instrument in the policy mix.

Conclusion

Real-world policies and environmental policy in particular are characterised by the existence of policy mixes. This holds especially true for policy responses to the ongoing biodiversity loss and the associated degradation of ecosystems' ability to provide ES. Despite this observation, most of the literature on instrument choice has focused on the analysis of individual instruments rather than policy mixes. Building on the existing literature on policy mixes and a number of reviews on selected individual policy instruments, this chapter has developed a stepwise framework for assessing instruments in policy mixes for biodiversity conservation

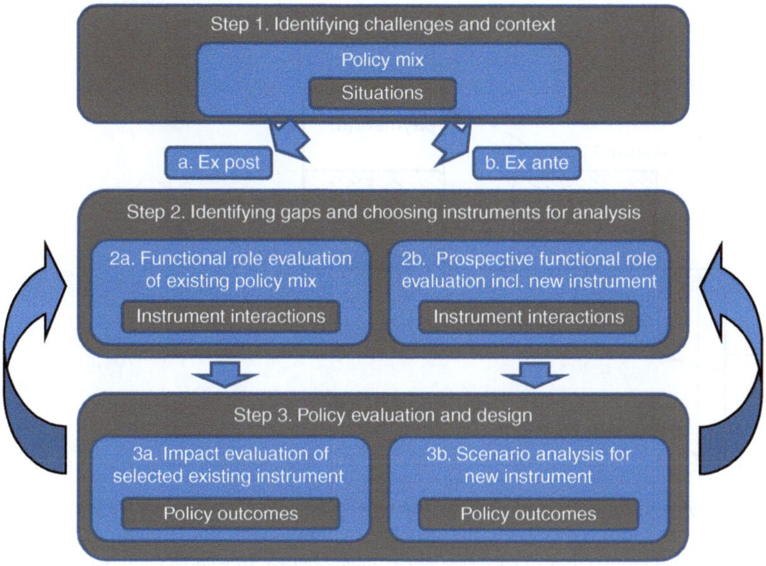

Fig. 5.2 Three-step-framework for ex post and ex ante analysis of policy mixes. © Schröter-Schlaack and Ring 2011

and ecosystem service provision (Ring and Schröter-Schlaack 2011a, Fig. 5.2).

As in any other policy field, there will be no 'blueprint' for optimally designing a policy mix for biodiversity conservation and ecosystem service management as each country is different and relies on biodiversity and ES to a different extent (TEEB 2010a). Moreover, ecosystems may be in different stages of degradation and thus in different proximity to tipping points of critical ecosystem service provision. Finally, each country deals with a different set of policies already in place. Nevertheless, two recommendations on mainstreaming biodiversity conservation and ecosystem service management may apply in almost all cases, irrespective of the specific setting (TEEB 2011):

- The policy mix should not be limited to 'environmental' or 'conservation' policies but should also encompass other sectorial policies, such as agriculture, energy or transport.
- A policy mix can be developed using a stepwise approach that starts with the more easily available opportunities.

5.2 Selected Financial Mechanisms: Payments for Ecosystem Services and Ecological Fiscal Transfers

I. Ring and M. Mewes

The loss of biodiversity and ES is often due to market failures concerning public goods. On the one hand, the destruction and deterioration of habitats as well as pollution (e.g. nitrogen and phosphorus inputs in water bodies) lead to negative impact and represent so-called negative external effects, which are not internalised or not sufficiently internalised. Economic production and consumption, which negatively influences the environment, is still too cheap. This also holds for the intensive production of many provisioning services of ecosystems such as agricultural use at the expense of their regulating services. The social costs of this behaviour are not reflected in the prices of the corresponding goods and services.

On the other hand, services of land users and public actors for biodiversity conservation and the conservation of ecosystems and their services are

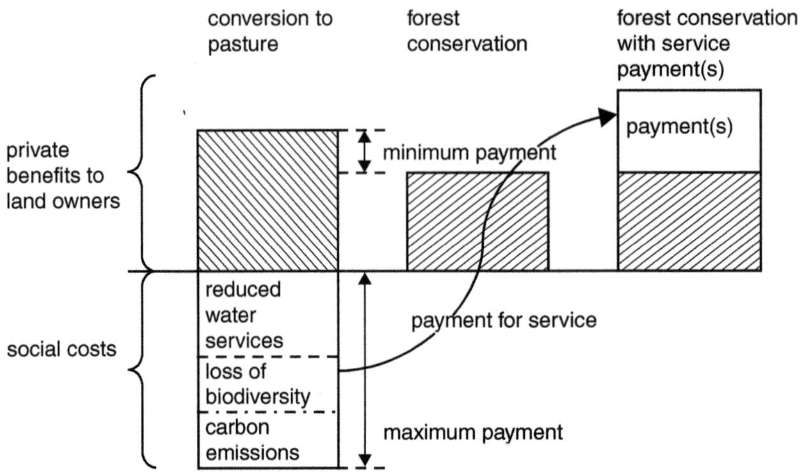

Fig. 5.3 Background of introducing payments for ecosystem services: the conversion of forest to pasture leads both to a higher private benefit to the land owner and social costs due to the loss of ecosystem services. The land owner can be offered a payment to conserve the forest, which has to equal at least his gain in private benefit by a conversion to pasture. (Adapted from Engel et al. 2008)

often linked to positive externalities, representing social benefits. Because of the inadequate internalisation of such positive external effects these social benefits do not often pay off for the suppliers of such services under the current social framework and market conditions. They are not paid adequately for the costs of their implemented measures. Therefore, these socially desirable services are not sufficiently provided, both regarding measures for the conservation of endangered species and the conservation of regulation functions of ecosystems (Ring 2011).

▶ Section 5.1 on policy mixes already introduced potential instruments to solve such problems, including regulatory approaches (law and order), planning law, economic instruments such as taxes, charges or payments for environmental services, as well as informational, motivational and educational instruments. Why economic instruments also make sense in nature conservation and the conservation of ES, has impressively been presented by the results of the global TEEB-study (The Economics of Ecosystems and Biodiversity) (Ring et al. 2010b; TEEB 2010b, 2011).

In Germany, the memorandum 'Economics for Nature Conservation' focused in the same direction (Hampicke and Wätzold 2009). How can economic activities be harmonised with nature conservation goals through greater use of economic instruments? Here, clear incentives, goals, market creation and making use of win-win effects (synergy) play an important role. In this section, two economic instruments are presented in more detail: payments for ecosystem services (PES) and ecological fiscal transfers (EFT).

5.2.1 Payments for Ecosystem Services

For quite some time now, payments for environmental or ecosystem services (PES) have become widely known at a global scale (Wunder 2005; Wunder et al. 2008; Gundimeda and Wätzold 2010; Porras et al. 2011; Ten Brink et al. 2011). The goal of this instrument is the development of economic incentives for the protection of biodiversity and the provision of ES. In general, opportunity and management costs are compensated (**Fig. 5.3**). Scientists have discussed this instrument in special issues of scientific journals (e.g. *Ecological Economics* 2008 (65), 2010 (69)) in a broad manner.

Examples for the implementation of PES exist worldwide on a local, regional or national level. In Europe due to the EU common agricultural policy since 1992 agri-environment schemes are manda-

Case Study 1: KULAP, Brandenburg and Berlin

The goal of the agri-environment scheme of Berlin and Brandenburg is the *'Förderung umweltgerechter landwirtschaftlicher Produktionsverfahren und die Erhaltung der Kulturlandschaft'* (KULAP 2012) (promotion of environmentally sound agricultural production and the conservation of the cultural landscape). Agri-environment schemes of the German federal states are part of the European Agricultural Fund for Rural Development (EAFRD). The scheme comprises amongst others measures to protect the environment as well as to conserve natural resources (more information under: ► www.mil.brandenburg.de/cms/detail.php/bb1.c.213972.de). With a view to the concept of ecosystem services the scheme addresses the environmentally friendly provision of provisioning services as well as the protection of regulating services. EU, national state and federal states finance the scheme. In general farmers are providers.

Case Study 2: Northeim Project, Lower Saxony

In the Northeim project (► www.zlu.agrar.uni-goettingen.de/index.php?option=com_content&view=article&id=47&Itemid=56&lang=de) a tender in form of an auction for a result-oriented remuneration was tested (Bertke 2005; Klimek et al. 2008). The pilot project was initiated by the University of Göttingen, Germany, in cooperation with the responsible public authorities. The goal of the pilot project was the conservation of agro-biodiversity in grassland. Therefore, farmers received a payment for the provision of a defined number of species in the study region.

tory for the EU member states in the framework of plans for rural development (► http://ec.europa.eu/agriculture/envir/measures/index_de.htm; Hartmann et al. 2006; ► Box case study 1). Land users can conclude contracts to implement measures offered in the corresponding agri-environment schemes and get payments for ES (due to simplification reasons we also include biodiversity thereunder). Such agri-environment schemes are directed especially to payments for opportunity costs for the implementation of specific measures in the agricultural sector. As the success and results of such schemes remain mixed at best, for some years result-oriented schemes have been increasingly tested that directly reward the relevant ecological services (Freese et al. 2011, ► Box case study 2).

In the following sections we first provide a general overview of the instrument PES. The introduction of such payments does not per se guarantee goal achievement (e.g. Klejn et al. 2003). Therefore, criteria are necessary to evaluate the schemes. Subsequently, we present some of the important criteria.

Definitions and Design Options

In literature the term PES in general means a market-based instrument and follows the definition of Wunder (2005, p. 3) '(a) a voluntary transaction where (b) a well-defined environmental service (or a land use likely to secure that service) (c) is being 'bought' by a (minimum one) service buyer (d) from a (minimum one) service provider (e) if and only if the service provider secures service provision (conditionality)'.

The first precondition, thus is, that the participation in a scheme with PES is voluntary. Further important preconditions to clarify the following questions resulting out of the definition, are:

1. What is paid for?
   The goal of the instrument is in general the provision of a defined ecosystem service, e.g. clean water. Therefore, one has to determine, how to measure the service and thus the goal achievement. There are services, which are directly measurable, e.g. carbon sequestration, but also services one can determine only by using a proxy, e.g. impacts on biodiversity (Gundimeda and Wätzold 2010). Herewith the question is also narrowly linked if a payment should be action- or result-oriented. If the payment follows the implementation of a measure (action-oriented), it can happen that the ex-

pected and desired result, e.g. more species or a defined, better water quality is not achieved. A generous definition (FAO 2007) concerning the service to be performed guaranties payments for ES, as soon as the provided performance by the management increases the performance provided without payment. Also, the possibility exists to bundle the provision of several ES in one payment.

2. Who pays?
   In general the one, who attains a benefit of the provided ecosystem service(s) pays. There are two different forms: (1) Private user-financed programmes: in these programmes private interests and benefits are represented, e.g. if a water company offers land users payments for a better water quality and availability in its water catchment, usually a company or non governmental organisation pays. (2) Government-financed programmes: such programmes represent public interests. The governmental site as 'buyer' normally does not benefit by the programme but the 'society' as a whole benefits, e.g. concerning carbon dioxide or biodiversity. The government finances the programme (and pays).
3. How is it paid?
   In industrialised countries the mode of payment is usually money, but in principle allowance in kind is possible. The amount of the payment mostly equals the opportunity costs of the provider, which occur due to the provision of the ecosystem service. Additionally, administrative costs can be considered. The payment can be calculated once and be fixed in the contract or be variable. One payment or continuous permanent payments are possible. There are also different design options concerning the time frame such as monthly, yearly, ex ante or ex post payments. The duration of the contract has to be determined (short-, middle- or long-term). Last but not least a spatial differentiation can be done.
4. Who gets paid?
   Addressees of schemes with PES are mainly private actors/land users, who are named (ecosystem service/service) provider or deliverer. Besides the direct negotiation of the scheme and contracts between buyer and service provider, intermediaries between the parties are also possible, e.g. a certification office or research institute.
5. When is it paid?
   It is important to exactly define in the scheme, when the payment is to be made, that means only under the condition that the performance is actually provided as contracted (=*conditionality*).

It becomes apparent that each definition plays an important role in the design for PES and has to be determined carefully. Due to the different design options different types of programmes/schemes are possible. Overviews can be found, e.g. by Wunder et al. (2008), Nill (2011) or Porras et al. (2011). In which case what kind of design should be chosen, depends especially on the fitting of short-, medium- or long-term time horizon as well as the spatial (local, regional or national) level for buyer, provider and ecosystem service provision. Many factors influence the success of PES schemes. In the following different factors are introduced, which can be used to evaluate the scheme.

Ecological Effectiveness

One important criterion to evaluate the success of a PES scheme is its ecological effectiveness. In what way does the scheme contribute to an improvement/increase of the ecosystem service or the halt/stop of deterioration (e.g. stop of biodiversity loss)? The degree of goal achievement is not always easy to determine. It depends especially on the measurability of the ecosystem service (see above: what is paid for?) and the handling of uncertainty. The ecological effectiveness can be low to high depending on the design of the scheme, whereas the following factors play a crucial role:

1. Monitoring: In order to evaluate the goal achievement a monitoring of the ecosystem service(s) is necessary. Different methods are available, e.g. measurements in laboratories, on-the-spot inspections, indicators, etc. It has to be clear which kind of monitoring can and should be implemented, its frequency, who is responsible for the monitoring and who pays

for it. Also, rules have to be defined beforehand in the case that the monitoring reveals that the service provider violates the contract. One can think of no payment at all, only a part of the payment or as the case may be a fine or other sanctions.

2. Additional benefit, that means the determination of the baseline/reference situation and scenarios in conjunction with an additional ecological benefit through the scheme (*additionality*; ◘ Fig. 5.4): It has to be guaranteed that in fact an additional unit of ecosystem service due to the payment arises, which would not exist without the scheme. In order to determine this additional benefit, the reference situation and its expected development without an incentive payment has to be determined. One has to make sure that not beforehand of an expected scheme implementation the reference situation is damaged on purpose, e.g. by an expanded afforestation, in order to gain a higher additional benefit after the start of an 'increase-in-forest' scheme.
3. *Leakage* or *spillover*: It can happen that the introduction of a scheme with PES only leads to a spatial shift of the land use which negatively affects the environment. Therefore, altogether no 'plus' of the desired ecosystem service is provided, because while at one place the ecosystem service increases due to a payment at a different place it decreases due to the new derogation of the land-use shift.
4. *Permanence*: A scheme should come along with a long-term conservation of or increase in ES, especially desirable if the payments run out. This depends crucially on the duration of the negative externality, which the payments should internalise (Wunder et al. 2008).

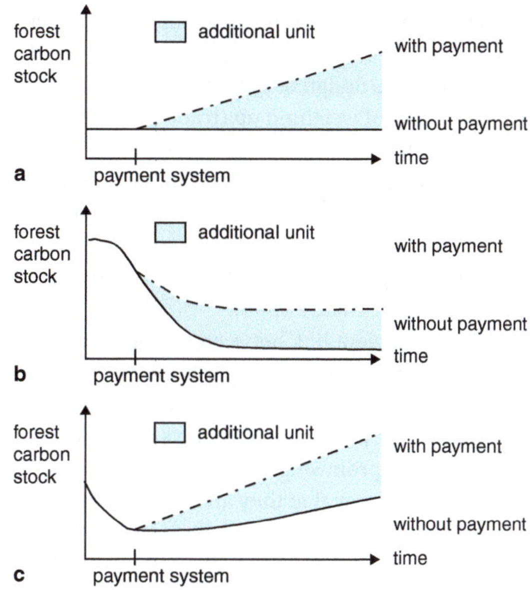

◘ **Fig. 5.4** Three different scenarios for PES systems: **a** static, **b** deteriorating and **c** improving the service-delivery baseline of the reference scenario. *Dash-dot lines* show the service delivery with payment systems, *solid lines* without. The additional unit is the increase in service delivery through the payment system compared to the reference scenario (*baseline*). (Adapted from Wunder 2005, 2007)

Costs and Cost Efficiency

One important requirement for the design of schemes is that they should be cost-efficient. For this purpose goals (e.g. good water quality, high level of biodiversity) should be achieved with a minimal budget, or the other way round the available budget should be spent in a way, that the goal achievement is maximised (Wätzold and Schwerdtner 2005). The cost efficiency of schemes is influenced by different factors:

1. Insufficient knowledge of the costs of the service provider: If a buyer has no or insufficient information about the real opportunity costs of service providers, the payments can be too high and the cost efficiency of the scheme decreases. This problem can be solved, e.g. by the implementation of auctions in which the service providers make an offer for the provision of a service and reveal their costs (▶ Box Example 2; Ferraro 2008). Such differentiated payments can help to better spend the available budget for the scheme, that means, e.g. more contracts can be closed. However, transaction costs (administration and negotiation costs) increase with auctions, leading to conflicts or so-called trade-offs between additional costs of instrument choice and design and the

cost savings within the payments (Wätzold et al. 2010). Possible trade-offs should be considered by the design of schemes.

2. Necessity of weighing up (trade-offs) between different cost categories: a higher participation of potential service providers in the design phase of PES schemes leads in general to higher costs for the buyer beforehand but also to a high acceptance by the later service-providers (e.g. Perrot-Maître 2006).
3. An insufficient link between ecological and economic knowledge: It has been shown that the integration of economic and ecological knowledge in combined models (Wätzold et al. 2006) is a promising approach to design payments in a way that they are ecologically effective and cost-efficient. For example a software-based decision support software can be helpful (e.g. Mewes et al. 2012), which not only can evaluate existing schemes but also can make suggestions for new, cost-efficient schemes with the help of optimisation algorithms.

Further Criteria

Further criteria are social and distributive effects as well as legal and institutional requirements:

1. Social and distributive impacts: Besides the criterion of cost efficiency the question of fairness of the payment systems plays an important role (e.g. concerning acceptance). One should bear in mind critically who can participate in the scheme/participates, if there are obstacles for potential participants and how to deal with them. This also holds for the participation by the design of the schemes. Questions of distribution are especially meaningful, if PES are implemented in structurally weak rural areas or so-called developing countries. If the institutional and legal preconditions can be designed appropriately, payments can support the rural population and contribute to poverty reduction in so called developing countries (Gundimeda and Wätzold 2010).
2. Legal and institutional requirements: For payment systems to function, *property rights* must be defined and be enforceable, which is often difficult mainly in so-called developing countries. In Germany and Europe it is more important, how far the tenure of agricultural areas results in that schemes or individual measures are not implemented.

5.2.2 Ecological Fiscal Transfers

Greening Intergovernmental Fiscal Relations in Germany

By way of fiscal transfers, intergovernmental fiscal relations distribute and allocate public revenues between and across different governmental levels. In the German federal system, the vertical dimension of intergovernmental fiscal relations distributes and allocates public revenues between the national and the state level (the German Länder) as well as between each state and the local level (district-independent cities, districts, municipalities). In addition, fiscal relations fulfil a crucial and constitutionally anchored redistributive function in Germany because an important goal is to reduce fiscal inequalities between the various jurisdictions. In their horizontal dimension, fiscal relations lead to fiscal equalisation between financially strong and financially weak states and local municipalities by redistributing public revenues from the former to the latter. For the calculation of fiscal transfers, the usual procedure is to contrast the fiscal need of a jurisdiction with its fiscal capacity, based on its own revenues. As the number of inhabitants is commonly used as an abstract indicator for fiscal needs reflecting the provision of various public goods and services, it is above all populous jurisdictions that mostly benefit from fiscal transfers today. This makes sense, of course, insofar as numerous public services are provided for the inhabitants of these jurisdictions.

Another purpose of such schemes is to compensate jurisdictions for expenditures incurred in the provision of public goods and services with positive effects on inhabitants in jurisdictions beyond their boundaries. Traditionally, district-independent and large cities provide various educational, health-related and cultural services for adjacent areas and their inhabitants, e.g. universities and secondary schools, hospitals, theatres and operas. The benefits of these services, reaching beyond city borders (so-called spillover benefits), come along with costs that the city itself has to cover. In order to compensate for such spatial externalities relating to costs and benefits and to account for services brought to other areas, many German Länder apply weighting factors for urban compared to rural dwellers in the calculation of the relevant fiscal needs, thereby artificially increasing urban populations.

However, rural and remote as well as suburban areas likewise provide a variety of services for cities. They provide food and drinking water (provisioning services), regulate the climate through their forests, provide areas for flood retention (regulating services) and serve as recreational space for the urban population (cultural services). Besides providing many ES, rural and remote areas also fulfil important ecological public functions, which benefit the entire population (Ring 2004). For this reason the inclusion of ecological indicators into intergovernmental fiscal relations has been demanded for some time in Germany. A prominent advocate was the German Advisory Council on the Environment (SRU) in its 1996 report (SRU 1996). According to studies by Ring (2001, 2002, 2008a), ecological public sector functions are already partly taken into account by way of specific-purpose transfers in some of the state fiscal transfer laws to the local level. These relate foremost to end-of-pipe and infrastructure-related public functions of the local governments such as the provision of drinking water, sewage and waste disposal. Precautionary and intergenerational public functions such as nature and landscape conservation, protection of water bodies and soil conservation only play a minor role (Ring 2002).

Design Options and International Experiences

Apart from the commonly used socio-economic indicators, fiscal transfer schemes should systematically take into account ecological indicators that reflect the provision of ecological public goods and services (Ring 2002). These ecological indicators are the basis for the distribution of ecological fiscal transfers (EFT). Their integration into the fiscal transfer system can have different rationales (▶ Box) (Ring et al. 2011). Ecological fiscal transfers can be specific-purpose transfers, being only allocated for the provision of specific public goods and services. They can also be designed as general purpose or lump-sum transfers without being tied to any conditions. Finally, combinations of general lump-sum and specific-purpose transfers are possible, depending on the relevant costs to be compensated.

Different Possible Rationales for Ecological Fiscal Transfers (Ring et al. 2011)

1. Compensation of management costs for providing ecological public goods and services
2. Compensation of opportunity costs of biodiversity conservation and the conservation of ES
3. Payments for spillover benefits of biodiversity conservation and the conservation of ES beyond the boundaries of a jurisdiction
4. Vertical or horizontal fiscal equalisation between financially strong and financially weak jurisdictions considering ecological indicators (distributive fairness)

Opposite to payments for ES portrayed above, which address mostly private actors, ecological fiscal transfers provide economic incentives for public actors. Depending on their design, the integration of ecological indicators into the German fiscal transfer system could create incentives for the German states and compensate for their above-average contribution for nature conservation or the provision of ES. The integration of ecological indicators in communal fiscal transfer laws of the *Länder* would provide the incentives at the local level, respectively.

International experiences with ecological indicators included in fiscal transfer schemes have been gained since the beginning of the 1990s in some Brazilian states (May et al. 2002; Ring 2008b). So far, 16 out of 26 states introduced ecological indicators in their respective state fiscal transfer laws for the redistribution of state value-added tax from the state to the local level (ICMS Ecológico). 13 states use protected areas for biodiversity conservation as their basic indicator; additionally,

5

Table 5.3 Relevance of ecological fiscal transfers for local budgets of selected municipalities in Portugal (2008). (Source: Santos et al. 2012)

	Municipalities	Share of fiscal transfers as a proportion of total municipal revenue (%)	Share of ecological fiscal transfers (%)	Share of conservation areas to total municipal area (%)
Municipalities with more than 70 % conservation areas	Campo Maior	89	25	100
	Murtosa	78	6	80
	Porto de Mós	75	11	76
	Aljezur	70	16	73
	Barrancos	97	26	100
	Terras de Bouro	94	22	95
	Freixo de Espada à Cinta	93	21	91
	Castro Verde	90	34	76
Municipalities with less than 70 % conservation areas	Lisboa	25	0	0
	Grândola	71	2	9
	Viana do Castelo	60	0.5	24
	Lamego	80	1	33
	Almeirim	62	0	0
	Peso da Régua	87	0.4	12
	Évora	62	1	16
	Vimioso	96	8	38

some states take other ecological indicators into account (Ring et al. 2011). In Europe, Portugal is the first country to introduce protected areas as indicators for fiscal transfers from the national to the local level. According to the new local finances law from 2007, Natura 2000 sites as well as further nationally protected areas are considered (Santos et al. 2012).

Table 5.3 illustrates how especially rural municipalities with a high share of protected areas benefit from ecological fiscal transfers. In the municipality of Castro Verde, for instance, which has 76 % of its municipal area designated as protected areas, 34 % of the municipal budget stems from these new ecological fiscal transfers due to the municipality's share of protected areas. Since fiscal transfers to municipalities depend also on various other indicators (mostly the number of inhabitants, but also municipal area in general or social burden), the share of ecological fiscal transfers in the municipal budget is not proportional to the share of protected areas in relation to the municipal area. Hence, municipalities whose area is completely made up of protected areas can have a lower proportion of ecological fiscal transfers in their budget than Castro Verde.

What Would Be the Effect of Ecological Fiscal Transfers on the Local Level in Germany?

Concrete suggestions for greening the fiscal transfer system in Germany mostly aim at including nature conservation in the communal financial transfer scheme (Perner and Thöne 2005; Ring 2008a). Nature conservation is an important building block in the protection of ES, above all regulating and cultural services. Nature conservation is a public task, which is beneficial at the national and international levels, thus reaching far beyond municipal boundaries. At the same time, costs of nature conservation are unequally distributed in terms of space. This is, amongst others, due to the unequal distribution of protected areas (Ring 2004).

Ring (2008a) suggested two alternatives to include nature conservation as an indicator for fiscal transfers from the state level to the local level within the Saxon communal fiscal transfer system. These alternatives were modelled as scenarios for Saxony. Their impacts were illustrated in a spatially explicit manner using geographical information systems (GIS). The results are based on administrative boundaries and further data for the calculation of fiscal needs (principal approach: number of inhabitants; additional approach: number of schoolchildren) as well as fiscal capacities (municipal revenues) as relevant for the Saxon communal fiscal transfer system as of 2002.

In the variant presented here, the fiscal need is expanded with an approach to nature conservation, which reflects the local ecological services that create spillover benefits beyond municipal boundaries. The approach to nature conservation is based on the designated protected areas measured in standardised conservation units (CUs) while avoiding double counts due to overlapping protected areas of different categories within municipal boundaries. Without any overlaps means the following: If a special area of conservation (SAC) according to the *EU Habitats Directive*, being more valuable from a nature conservation perspective, is located in a nature park, the SAC is counted while the area of the nature park is reduced accordingly in order to avoid double counting. For this purpose, the different protected area categories existent according to the Saxon nature conservation law (national park, Natura 2000 site, nature reserve, biosphere reserve, nature park and landscape reserve) are overlayed with municipal boundaries. The categories are weighted according to their relevance to nature conservation and the associated land-use restrictions (Ring 2008a). For instance, one hectare of a national park is considered with 100 % of its area (1 ha CU) whereas one hectare of a landscape reserve is only considered with 30 % of its area (0.3 ha CU). Comparable to the treatment of general area-related indicators in states such as Brandenburg and Saxony-Anhalt, one hectare CU is set equivalent to a certain number of inhabitants. ◘ Figure 5.5 shows the resulting relative changes of general lump-sum transfers to the municipalities in Saxony if, besides the number of inhabitants and schoolchildren, protected areas as well were considered for calculating fiscal need and one hectare CU corresponded to one inhabitant.

Evaluation of the Environmental Performance of Ecological Fiscal Transfers

Ecological fiscal transfers as a comparably new instrument, which has been introduced only in a few countries so far, is still lacking thorough investigation. Ring et al. (2011) evaluated ecological fiscal transfers for the first time according to the criteria of environmental effectiveness, cost-effectiveness, social impacts, institutional context and legal requirements. In the following text, the first two criteria will be addressed exemplarily. Further, the status quo of implementation in conjuncture with the legal requirements in Germany will be briefly presented.

The evaluation of the effectiveness of ecological fiscal transfers is closely linked to the rationales behind their implementation (▶ Box above). It is appropriate to evaluate the effectiveness by looking at the development of the chosen indicator after its actual implementation. For instance, after the Brazilian state Paraná introduced protected areas as an indicator for fiscal transfers in 1992, the protected area in the state has increased by about 165 % (May et al. 2002; Ring et al. 2011).

As for the cost-effectiveness, usually costs of an environmental policy instrument are contrasted with its environmental effectiveness. Which cost categories (management costs, opportunity costs, transaction costs) apply depends again on the rationale behind the implementation of the instrument (▶ Box above). However, transaction costs such as costs associated with the introduction and implementation of ecological fiscal transfers are compa-

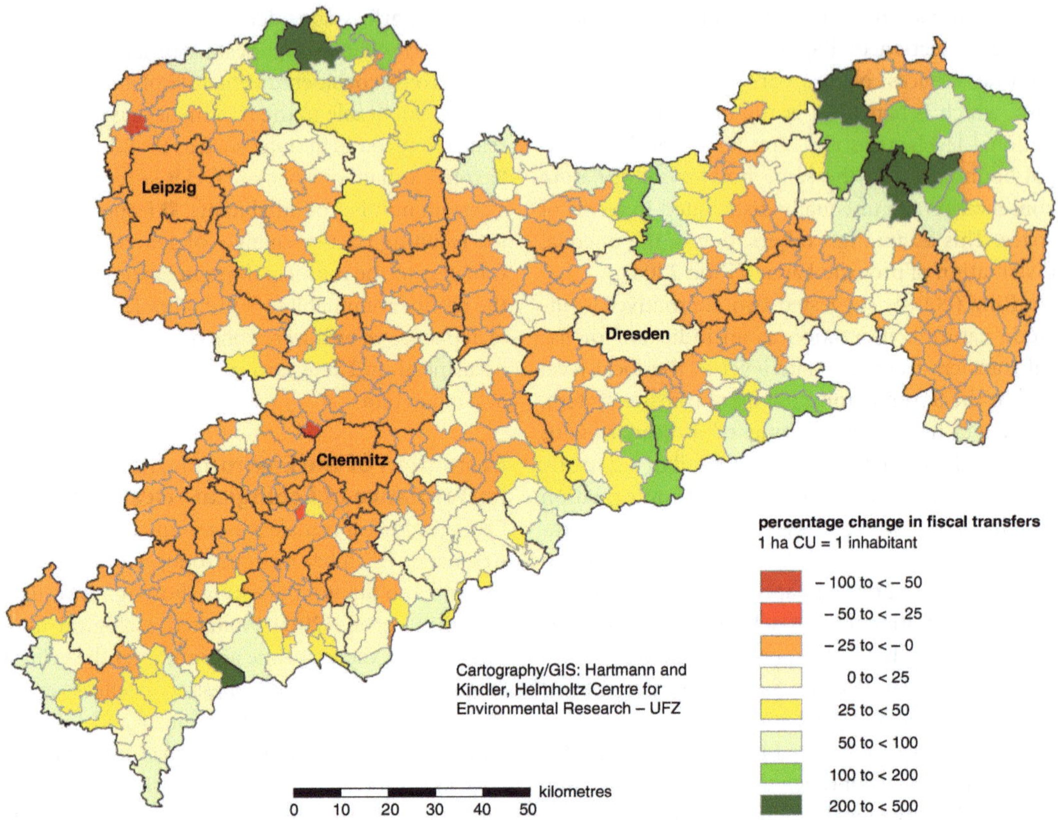

Fig. 5.5 Percentage change in general lump-sum transfers when the Saxon fiscal transfer system 2002 was expanded to include designated protected areas. In this model, conservation units (CUs) are used in addition to inhabitants and schoolchildren to calculate the fiscal need of a municipality, assuming one hectare CU is equal to one inhabitant. (Source: Ring 2008a). Based on protected area data, Saxon State Office for Environment and Geology 2004; Administrative boundaries VG 250, Federal Agency for Cartography and Geodesy 2002. Cartography and GIS: Hartmann and Kindler, Helmholtz Centre for Environmental Research–UFZ)

rably low as this instrument builds up on existing intergovernmental fiscal relations with familiar administrative structures and procedures. This is especially the case if accessible indicators such as protected areas are used (Ring et al. 2011).

The situation in Germany can be summarised as follows: ecological fiscal transfers exist within communal fiscal transfer laws in the form of specific-purpose transfers for mostly end-of-pipe and infrastructure-related ecological tasks (Ring 2001, 2008a). However, there is an increasing number of academic and political voices asking to systematically include ecological services also in the intergovernmental fiscal transfer system from the federal to the state level as well as illustrating the relevant consequences. Czybulka and Luttmann (2005) discuss arguments in favour of considering state-provided services for natural heritage in the federal fiscal transfer system. Schröter-Schlaack et al. (2013) present for the first time the results of including different ecological indicators in the German federal fiscal transfer system from the federal to the state level. Yet without broad political support, the necessary constitutional changes for an ecological fiscal transfer system will not take place. In this sense, it is at least a start that proposals in this regard are seriously examined also in the political sphere. For instance, Till Backhaus, environmental minister of Mecklenburg-Western

Pomerania, repeatedly calls to include services of his state supportive of biological diversity into the fiscal transfer system (e. g. Backhaus 2008). Likewise, the German Green Party (Bundestagsfraktion Bündnis 90/Die Grünen 2012) has formulated in its position paper 'Biodiversity 2020' as one of their core goals, to check 'how a financial equalisation mechanism between the German *Länder* can be designed with respect to their different nature conservation expenditures'.

Conclusions

In ► Sect. 5.2, two economic instruments were presented that aim to better align activities of private and public actors with biodiversity conservation and the provision of ES. PES schemes serve primarily to set incentives for private actors. They are used for a variety of ES now. The comparably new instrument 'ecological fiscal transfers' addresses public actors. It compensates for their costs stemming from nature conservation or the provision of ES or, in other words, rewards these efforts.

5.3 Integrating the Concept of Ecosystem Services into Landscape Planning

A. Grünwald and W. Wende

Looking at international trends and discussions of new ways to assess landscape ecology and protect biodiversity, one necessarily encounters the concept of ES (Jessel 2011, ► Chap. 1). For an enhanced focus on the management of ES, it would clearly appear necessary to improve the methods for planning instrumental aspects (Vasishth 2008; Jedicke 2010; Kienast 2010; von Haaren and Albert 2011). Endlicher (2011) argues that optimizing interactions between humans and the environment should be a priority (Breuste et al. 2011; Richter and Weiland 2012). Hence, the question is whether and how the concept of ES can be transferred to planning practice, and above all how it can be 'spatialised'. Moreover, there are difficulties involved in integrating the ES concept into the legally embedded planning tools and planning practices accepted by various stakeholders, clients and administrative offices.

In Germany, the municipal landscape plan is the valid local landscape policy tool for the cartographic representation of environmental concerns within the borders of municipalities and communities (von Haaren 2004; von Haaren et al. 2008; Heiland 2010). The persuasive and informational effect of this plan with regard to environmental aspects, regardless of whether or not it has been integrated into the preparatory land-use planning process, should not be underestimated (Gruehn and Kenneweg 1998; Wende et al. 2012). Nevertheless, landscape planning can, at least theoretically, be optimised by ES assessment approaches. From the authors' point of view, the anchorage of ES within spatial planning and the decision-making toolkit would appear necessary if the concept of ES is to be successful. Which tools could, in addition to spatial planning instruments (particularly the municipal landscape plan), incorporate these approaches usefully? Hence, the following chapter deals with the practical methodological possibilities for the integration of ES into the German landscape planning structure.

▪ A Methodological Development of Municipal Landscape Planning?

Landscape planning is, apart from several specialist terms, very close to the concept of ecosystem/landscape services[1] (Kienast 2010). Moreover, like the concept of ES, it aims at preserving and developing elements of nature using universally accepted assessment scales.

▪ Term

In their comparison of landscape planning and the concept of ES, von Haaren and Albert (2011) and Albert et al. (2012) show there are not only many

1 Landscape services (► Sect. 3.4): Here, the perspective is broadened beyond ecosystems and emphasises aesthetic, ethical and sociocultural aspects as well as anthropogenic changes. Because of the stronger spatial orientation, it can have a higher relevance for practical spatial planning, particularly landscape planning, and support participative approaches (Kirchhoff et al. 2012).

commonalities, but also different theoretical and methodological emphases. These include primarily the scales, the consideration of public and private goods, and the economic assessment and participation of all stakeholders. Landscape planning can profit by using the strengths of the ecosystem service approach and vice versa. The concept of ES can be positively influenced by the tried-and-tested methods from landscape planning.

Landscape planning can improve the strategies for the communication and implementation of planning measures, for example by embedding economic considerations (Stokman and von Haaren 2012). Conversely, normative legal standards and formal political/administrative decision-making procedures can, in the process of establishing and enacting a landscape plan, provide the basis for transparent and monetary assessment processes which emphasise on 'natural capital'. Therefore, it would seem evident, or at least worth discussion, to combine both approaches.

The advantages and disadvantages of embedding the concept at the local level generally, and at that of local landscape planning in particular, have been addressed, e.g., in TEEB (2010a), Jedicke (2010) or NeFo (2011). There is a broad consensus that the changed mode of communication with both the lay public and with decision-making stakeholders resulting from ES assessments has been positive. One persistent problem is the lack of significance accorded to biodiversity, in comparison with other ES (Anderson et al. 2009; NeFo 2011), and is the risk involved in the assignment of monetary values, both within the ES system, and with respect to commercial market goods. Another is that of communicating of these factors with the stakeholders (von Haaren and Albert 2011).

To date, how ES are to be incorporated in the practice of the drafting of a municipal landscape plan, and how that is then to be implemented in detail has been specified only rudimentarily. Studies at the local and regional levels have addressed only partial aspects, and have not been covered by the formal planning instruments of German planning law. Within the broad framework of spatial and ecological planning, various examples have been investigated, such as the choice of locations for new building areas with the lowest indirect costs (Grêt-Regamey et al. 2008), the cost-related advantages of a flood protection solution, which considers the values of ES (Grossmann et al. 2010), or the effects of various land use options (e.g. Vihervaara et al. 2010; Swetnam et al. 2010). However, to the authors' knowledge, there has as yet been no instance of the direct incorporation of ES into the landscape planning process.

5.3.1 Linking Ecosystem Services with the Landscape Plan

The municipal landscape plan is already a well-developed planning tool. It fulfils highly professional standards and requirements. The landscape plan examines the social values and natural assets for which a public interest exists, and brings together the knowledge of experts. Measures for the protection, maintenance and development of natural assets are derived from the professional demands of the conservation of nature.

In a landscape plan, the central elements of planning are the protected assets and the functions of the ecosystem and the landscape, i.e., primarily assets of public interest. By contrast, the concept of ES also covers public and private goods and services for human existence and human well-being. To date, the monetary quantification of nature and the landscape and their functions, and measures for their protection, maintenance and development, have not been the object of municipal landscape plans.

Landscape planning and ecosystem service assessment initially have different focuses, and can therefore not be directly linked. The municipal landscape plan remains as a planning tool at the local administrative level and should retain the accustomed quality and complexity of ecological information. A comprehensive method approved in practice for the integration of ES into the landscape plan has to date been outlined only rudimentarily, if at all. Given that an established planning tool is to be developed further by means of a concept that is currently being broadly discussed, but is not to be fundamentally changed, the acknowledged development steps will in this case have to constitute the framework for the linkage. Therefore, the assessment of ES should be carried out in parallel with, and according to, the main development steps of the landscape plan (► Box).

The Main Development Steps of the Landscape Plan

At the stage of the 'Basic evaluation/inventory acquisition', primarily the framework conditions of planning, and beyond that, the condition of nature and the landscape and the existing and foreseeable land uses are ascertained. The ascertainment of the services extant in the planning area, of the associated stakeholders, and of the demand for ES can then follow. That is then followed by the 'Analysis of conflicts, the prediction of conflicts, and the assessment'. The capacity of nature and the landscape and the tolerance of present and future utilizations are assessed by using indicator-based methods, according to normative requirements. Concurrently, the estimation of ecosystem services with respect to quantity, quality and spatial distribution, and their present and future users as measured by demand and by the conceptual goals can take place. The quantification of services builds the base for the translation into monetary values, and hence for the spatial mapping of natural capital as one possibility of presentation. The following step is that of designing the development and the conception measure–'target and measure concepts'–sectoral for particular natural assets, and integrated into the overall concept. The emphasis there is on the natural assets.

The connection and feedback between these two concepts can take place at various positions. The ES approach complements the statements of landscape planning, but does not replace its legally required working steps. A better knowledge of the ES in a certain local area and of their condition and their economic value can also be used to prioritise the measures of the landscape plan, to determine cost-benefit ratios of measures, and to give the stakeholders a reason for their decision for the implementation of measures of protection, maintenance and development of nature and landscape. In addition, the results of the local ecosystem service assessments can be used as a basis for communication with the stakeholders and for better understanding landscape planning. Conversely, the data collected on the condition of nature and landscape constitute the basis for the operationalization of ES. Hence, not all data need to be gathered anew, but if necessary, they still have to be monetised.

5.3.2 Implementation in Practice–Testing the Example of the Service 'Erosion Protection'

In a model segment area in the city of Dippoldiswalde, Saxony, the authors tested whether the theoretically composed framework could be used in practice for the integration of the ES approach into the local landscape plan, and examined the difficulties which might arise. For this test, the working steps 'Analysing conflicts' and 'Assessment' were chosen.

Because the examinations of this selected aspect focus on the methodology, the investigation of a representative part of landscape is initially sufficient. By implication, it is possible to transfer the methodology to the entire area of the municipality.

In the area investigated, agricultural use predominates, particularly on sloped surfaces. The borders of the investigation are based on the sheet line system of the topographical map of a scale of 1:10,000, and on the borders of the municipality. The ecosystem service selected for the investigation, 'erosion protection' (from the class regulating services; ► Sect. 3.2), was chosen on the one hand due to the special problems in that area–soil erosion by water–and on the other to the data needed for the operationalization and acquisition effort. The authors of this paper stress that their analysis of erosion protection addressed only one component of ES. The results of this study cannot readily be transferred to other ES. Particularly, the operationalization and monetary assessment of services associated with biodiversity remain difficult if not impossible in the context of landscape planning.

The ultimate goal of the application of the concept of ES cannot consist exclusively of monetary assessment.

Data and Methods

The spatial variations of many landscape functions and services can already be explained with aggregated data of soil type, land use and topography (Willemen et al. 2008). Therefore, for the ecosystem service assessment in the landscape plan, the basic data already gathered was to be used, and not procured anew. For certain questions, an estimate in monetary terms and the connection of these values with economic indicators could then ensue.

De Groot et al. (2010) and Burkhard et al. (2011) propose the 'amount of retained soil' or the 'loss of soil particles by wind or water' as indicators for an operationalization of the service 'Erosion protection'. In the present example, the quantification of the service was carried out by spatial classification into so-called erosion resistance classes, by means of indicators for the description of the mechanical erosion resistance of the soil. Bastian and Schreiber (1999) have proposed these for use even for landscape planning at the municipal level. For this purpose, the natural soil erosion in tonnes per hectare and year (t ha^{-1} a^{-1}) has been ascertained on the basis of erosion resistance by soil type and slope angle, taking precipitation into account. The usage-dependent soil erosion, again in tonnes per hectare and year, can then be calculated by multiplication of the natural soil erosion by a usage-dependent factor. This procedure uses a simplified technique (Bastian and Schreiber 1999, pp. 216 ff.) based on the *universal soil loss equation*. In this case, only soil erosion by water was considered. As described in the valid landscape plan (status: draft 2009), susceptibility to soil erosion by wind is slight to very slight, no model for calculating soil erosion in t ha^{-1} a^{-1} (e.g. by means of the *revised wind erosion equation*), analogous to the soil erosion by water, was carried out here. However, this means that the ecosystem service 'total erosion protection' would be even higher than in the following example.

In landscape plans, erosion resistance is generally rated in Classes I–VI, or in erosion sensitivity levels of 0–5, i.e., in status values, which are based on average soil erosion in t ha^{-1} a^{-1}, which in turn, as the flow value used in this example, constitute the basis for monetization.

For the following monetary assessment, the replacement cost method was used. In this case, the damage caused by soil erosion and the costs for their restoration or replacement, oriented towards current market prices, are considered. The assessment includes both on-site costs, i.e., for the preservation of the current condition, and off-site costs, i.e., those generated outside of the erosion areas, for the elimination of sediments in various forms and at various places.

The removal of organic substance by the erosion of the surface soil reduces the soil fertility and the water-retaining capacity, or water availability. Moreover, the thickness of the topsoil decreases, leaving only soil with rocky content. A study from the USA (Pimentel et al. 1995) on the calculation of on-site costs considered compensation for lost nutrients, water and removed soil, as well as the off-site costs of measures for such infrastructure as roads, tracks, buildings, pipelines or flood and water reservoirs, where sedimentation reduced efficacy; it did not include the cost caused by damage to, or effects upon biodiversity, or the financial losses of fish farms. There is no technologically based way to provide short-term effective alternatives for the characteristic species assemblage lost due to damage to terrestrial or aquatic habitats. Moreover, the effects depend to a great extent on the nature of the biotope affected and on other conditions. Therefore, it is very difficult to calculate these costs by using the replacement cost method. Moreover, no commercial fish farms or fishing grounds are known in the examination area, so this calculation was not relevant.

The on-site costs include all the positions named above. For the calculation of the off-site costs for the damages which typically occur in the planning area, an average value was generated, since it would be too complex at this level of scale to carry out a calculation for each area segment. The calculation was based on current prices in the gardening/landscaping industries (application/removal of soil) and in agriculture (irrigation, fertilizing).

The calculation of on-site replacement costs per tonne of lost soil comes to € 58.84 per year, and off-site replacement costs are € 15.83 per year; the total is thus € 74.67 t^{-1} (Grünwald 2011).

The service was quantified by the two-dimensionally and spatially precise mapping of usage-dependent soil erosion. The link to monetary values was accomplished on the basis of the average value of soil erosion in t ha^{-1} a^{-1}, in accordance with the erosion resistance classes. The ascertained replacement cost of approximately € 75 t^{-1} of soil erosion was transferred inverted to the erosion resistance classes, so as to illustrate benefits lost. The decrease of harvest on-site and the ascertainment of the damage off-site can be compensated by technological means (replacement). Hence, the calculated costs equal the value of the ES, had they been preserved and generated by nature. That means that the lowest erosion resistance class (VI– 'very low resistance') is assumed to have the benefit 'zero', so that the monetary value of natural capital in such areas equals €0. On the basis of this linkage, the value of the service–natural capital could be spatially precisely mapped (◘ Fig. 5.6).

In order to avoid excessive generalization due to overlaying the relief, the soil type and the land use at the outset, the operationalization used a basic grid size of 15 × 15 m^2. With respect to the aggregation of the values, additional services and greater generalization should be carried out, as a result of which the relatively detailed information for the several services would be lost again.

The grid in ◘ Fig. 5.6 is therefore still 15 × 15 m, so that changes of the service within the same land use type, e.g. farm fields, are still visible, due to a difference in the slope. The calculated monetary values are per hectare and year.

▪ Discussion

In this example erosion protection, quantification and monetary valuation of ES at the communal level is generally possible. Various additional studies (Egoh et al. 2008; de Groot et al. 2010; Willemen et al. 2010) show how the spatial precise delineation, quantification, and linkage to monetary values can also be realised for other services (▶ Chap. 4).

The map in ◘ Fig. 5.6 shows the spatial arrangement of the existing natural capital of the ecosystem service 'erosion protection'. Provided there is such a spatial arrangement for all services in the given examination area, an assessment of the entire natural capital of the municipality can be shown. Generally, such maps indicate that nature and the landscape have significant social value, which is often not noticed. But they are not an assessment of the service itself, nor do they provide any information about the multi-functionality of a landscape. They may prove helpful as guides for decisions on certain issues, and they can, e.g. in case of the development of a strategy, constitute the basis for building scenarios or for comparing alternatives with regard to the benefits of ecosystems for humans.

The service 'erosion protection' is quite easy to ascertain, since its assessment builds on data provided in the landscape plan, which may already have been quantified. The calculation of monetary values can then flow directly from that. Nevertheless, the additional effort needed to ascertain the amount of soil erosion and economic valuation is high. If this procedure is to be consistently applied to all services, a high input of effort and documentation is necessary, and will require a detailed knowledge of the concept of ES by the planner. Moreover, it is not possible for all services to be identified in terms of their spatial characteristics with accurately defined boundaries (Willemen et al. 2008). Sometimes models must be used which require special knowledge of other scientific fields.

The concept of ES is complex, and makes high demands on the planner, particularly with regard to operationalization, the calculation and transfer of the values of ES into economic values, and the interpretation and analysis of the results. At present, decision-makers at the municipal level and also planners who develop local landscape plans are probably not adequately familiar with the concept. No legal requirement for implementation exists.

Regarding the assessment of the benefits of ecosystems for humans, the concept of ES is determined, amongst others, by objectively verifiable, preferably numerical or monetizable factors, whilst the landscape plan is to some extent also affected by normative-qualitative standards, and by standards of evaluation. For the integration of the ecosystem service approach into the landscape plan, the latter should focus more strongly on quantitative and monetizable standards of evaluation.

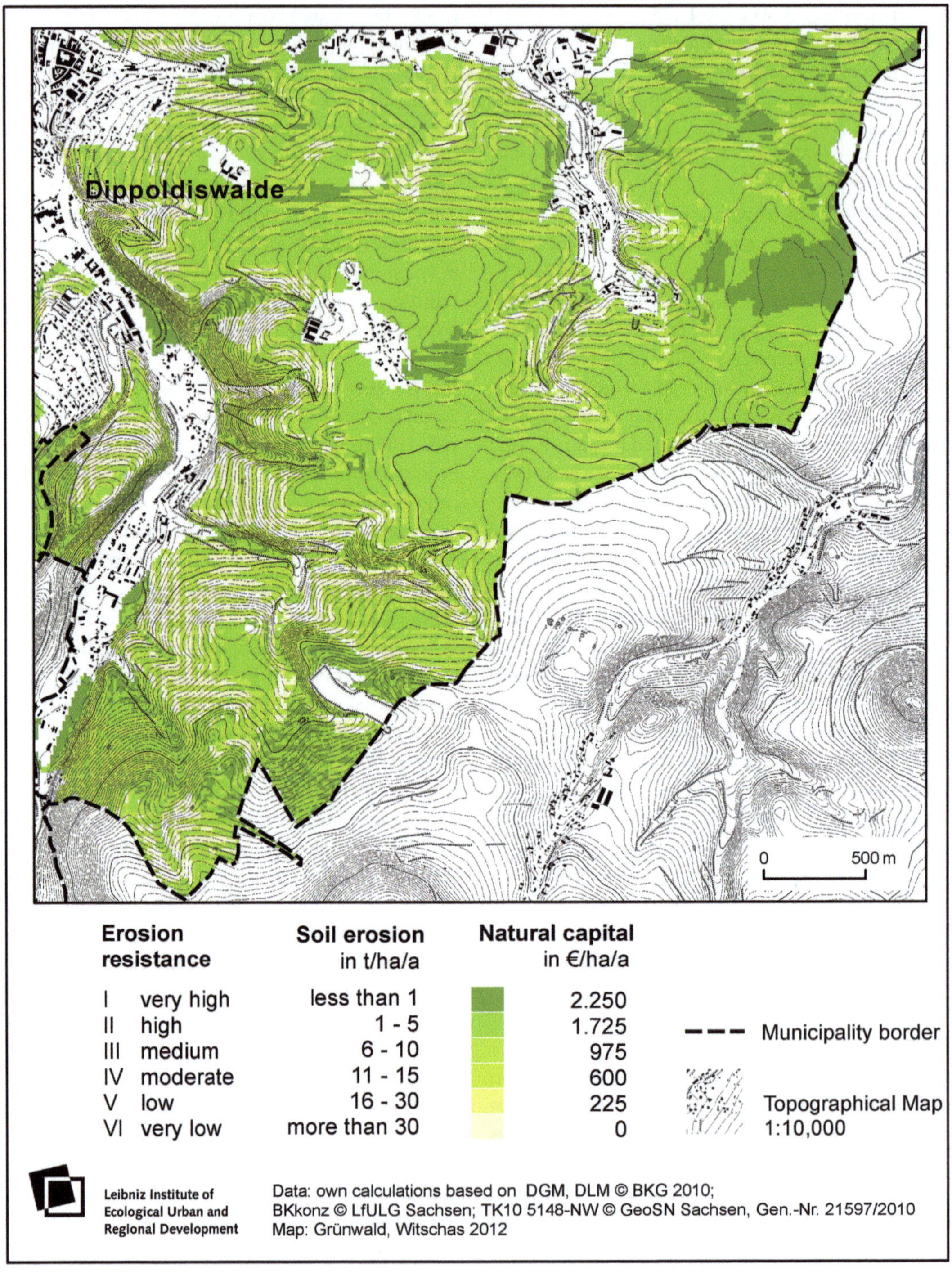

Fig. 5.6 Capital of ES 'erosion protection'–spatial arrangement of monetary values in tonnes per hectare and year compared to soil erosion caused by water © IÖR/Grünwald, Witschas

Therefore, the ES approach in its present form is not fully suitable to the concerns and framework requirements for broadly based practical use in the legally established landscape planning process, including use by nature conservation authorities. The following preconditions will have to be fulfilled to enable use of the concept in practice in future, in order to guarantee high planning quality and to make the contents comparable:

- Standardization/rules for a clear delineation and traceability of concepts and methods (quantification as well as monetization)
- Standardization of the availability and quality of data, and/or the examination of data
- Formulation and anchorage of the standards of evaluation of ES
- Professional training of the planners and, as appropriate, also the policy-makers who develop the plans
- Establishment of a payment base for the additional expenses resulting from this kind of planning (e.g. in the fee structure for architects and engineers).

The analyses (Grünwald 2011) have shown that no overall ascertainment of all cost factors is possible. For example, the damage to terrestrial habitats resulting from the input of soil and nutrients could not be calculated, since the effects of nutrient input depend on a variety of factors, and cannot be resolved by a standardised technical solution. Therefore, this component was not calculated, so that real replacement costs would be higher than the costs ascertained.

The attempt carried out for that purpose to transfer values via 'benefit transfers' from an existing study should also be assessed critically; however, it appears to be the only practicable way, considering the budget of German landscape planning. Independent investigations of costs in the planning area are not provided in the present scope of services of the fee structure for architects and engineers. It is also clear that the stated costs, particularly if they need to get interpolated on larger areas, suggest a degree of accuracy, which is not given in this case. The overall monetary assessment of natural capital ultimately depends on the size of the spatial relations considered. Values calculated for ES containing decimal places are nonetheless only approximate, so that these fractions should be passed on only in the totals. That raises the question as to whether a monetary assessment is generally necessary, or makes sense, and if so, in what detail.

Quantification and monetization ensure that especially regulation services and cultural and recreation prevention services are assigned a monetary value. This information provides a weighty argument for political decision-makers for the implementation of nature conservation measures. To this extent, the ES approach definitely delivers additional and possibly far-reaching important arguments in favour of the implementation of measures for the protection, maintenance and development of nature and the landscape.

Conclusion

Landscape planning as a planning instrument is generally appropriate for the integration of at least certain ES. The standard practical application of the concept is still in its initial stages. The testing of only one service for practicability has already raised numerous questions.

For an application of the ES approach to the landscape plan, additional methods for the operationalization of single services and for economic assessment must be adopted. Syrbe and Walz (2012) mention landscape structure indices as an example of a favourable possibility for a location-based estimation of certain ES by GIS. With such landscape-based measures of habitats, surfaces and landscape structure as a cost-effective indicator, the assessment can be realised more easily and accurately. If the concept is to be implemented in planning practice over the long term, planning tools need to be developed, incentives for implementation provided, or legal requirements established. In Switzerland, the Federal Office for the Environment has already published such a recommendation for implementation (Staub et al. 2011). In addition to the general classification of the ES relevant to Switzerland, it also includes the criteria for operationalization and the data sources for those criteria.

Another special challenge, which, like comprehensive quantification, can be realised only with great effort, is the monetization of services. Moreover, monetary values can be calculated only approximately, both for single services, and in sum for all services. On the one hand, it makes sense to use the specific values for the planning area; on the other, the high cost and effort in relation to the result to should be viewed critically. For reasons of labour economy, the effort and complexity used to calculate monetary values should be kept low. On the other hand, if the effort is reduced too greatly, e.g. by an overly great approximation in the estimation of values, or by benefit transfer from other contexts, the validity of the results will be limited.

Fundamentally, the concept of ES aims at emphasizing the beneficial effects of ecosystems for humans. The conversion of natural capital into monetary values is only one aspect, however. As one addresses ES, the stakeholder moves ever more into focus, and participation becomes more important. Emphasizing the services per se, and, the more intensive contact with the stakeholders can give the intended incentives for incorporation of the ecosystem service approach into landscape planning. The total abandonment of monetization would be possible, but would then fail to exhaust the potentials which this approach offers. Therefore, the conclusions should not depend exclusively on the results of monetization, but should rather in certain cases see these results as an important basis for the guidance of decision-making.

Moreover, the problem remains that ES such as biodiversity, which are difficult or impossible to quantify or monetise, are systematically underrepresented in the overall assessment of ES in the spatial contexts selected, or are even completely 'blanked out'. Here, a certain degree of overall imbalance of assessment between abiotic and biotic services is apparent, due to inevitable methodological deficits. Jessel (2011) points to the fundamental differences between the two concepts of 'ES' and 'biodiversity', and stresses primarily the fact that the anthropocentric perspective of ES is incompatible with the concept of 'biodiversity'.

Landscape planning depends on a greater appreciation for nature and the landscape as a basis for the cultural, social and economic development of a society. With the integration of the concept of ES into municipal landscape planning, it is possible to enhance this appreciation. The local landscape plan is already an important tool for the development of nature and the landscape at the municipal level in Germany, and it can, in combination with a further establishment of the ES concept, gain in importance. Despite the existing difficulties, it is worthwhile to continue this debate over whether and how the concept of ES might be integrated into spatial and ecological planning.

5.4 Governance in Nature Conservation

O. Bastian

5.4.1 Governance and Protection of Biodiversity

In view of the ongoing loss of biodiversity, the destruction of natural ecosystems, and the reduced supply of ES, it is becoming increasingly urgent to identify suitable strategies to conteract these unfavourable trends. It is necessary to apply a wide range of policy instruments: state rules (laws, regulations, etc.), the engagement and the participation of (groups of) persons and organizations, as well as property and market-based approaches (Kenward et al. 2011; Southern et al. 2011; ► Sect. 5.1).

People are acknowledging at a progressive rate that regional and local actors cannot rely only on solutions provided by state authorities. State activities have to be complemented by voluntary local and regional efforts to improve urban and regional development and to foster innovations and creativity. Consequently, practitioners and researchers alike pay due attention to the concept of regional governance in its manifold manifestations (e.g. Danson et al. 2000; Diller 2002; Knieling 2003; Wirth et al. 2010).

Governance in its broadest sense can be understood as 'a process of coordination of actors, social groups and institutions in order to attain appropriate goals that have been discussed and collectively defined in fragmented, uncertain environments' (Le Galès 1998; see also the definition in ► Sect. 2.1). It is important to distinguish between different types or mechanisms of coordination (e.g., markets, hierarchies, networks, hybrid forms of coordination), institutional levels (e.g., local, regional, state), and between actors from different spheres of society (e.g., economy, education, politics, and so forth). By providing a framework the concept of governance helps to analyse the complicated constellation of actors in a social-ecological system (Wirth et al. 2010).

While management describes the actions in a certain area or ecosystem governance tackles the questions of responsibility: who is responsible, who is making the decisions, and how this is done (Kenward et al. 2011). Graham et al. (2003) define governance as 'the interactions among structures, processes, and traditions that determine how power is exercised, how decisions are taken on issues of public concern, and how citizens or other stakeholders have their say.' *Collaborative governance* is the integration of values (economic, social, as well as environmental ones) through "*...a collaborative, multipartner decision-making process*" (Lamont 2006).

Government and governance have similar roots, but government refers only to bodies and processes that are largely separate from citizens, the private sector and civil society. Governments are key players in governance processes, but are only one among the many possible players. In other words:

'Governance includes the state, but transcends it by taking in the private sector and civil society. All three are critical for sustaining human development. The state creates a conducive political and legal environment. The private sector generates jobs and income. And civil society facilitates political and social interaction–mobiliziling groups to participate in economic, social and political activities. Because each has weaknesses and strengths, *good governance* is to promote constructive interaction among all three' (Kenward et al. 2011).

Governance settings depend in large part on formal mandates, institutions, processes, and relevant legal and customary rights. But they are more complex and nuanced phenomena than one may imagine. Regardless of formal authority, decisions (concerning biodiversity and ES) may be influenced by history and culture, access to information, basic economic outlook, among other things (UNDP 1999).

◘ Table 5.4 gives an overview of the various governance types, which differ greatly in parts depending on the role of actors, the decision level, and other factors (Hahn et al. 2008; Kenward et al. 2011).

5.4.2 The Project GEM-CON-BIO

The CBD recommends to using not only legal but also economic and social instruments for an effective protection of biodiversity and ES. However, the pros and cons of the instruments are still being discussed, e.g. protection regulations versus social and economic incentives (James et al. 1999; Ferraro and Kiss 2002; Adams et al. 2004). Many of these regulative instruments (like access or restrictions), social and economic instruments (like moratoria, taxes, and subventions) are applied, but their effectiveness has not been fully investigated and underpinned by enough studies, yet. To decrease this deficit the EU-project GEM-CON-BIO (*Governance and Ecosystems Management for the Conservation of Biodiversity*–Manos and Papathanasiou 2008; Simoncini et al. 2008; Kenward et al. 2011), in which the author was involved, was launched. The project addressed the question, which governance types and institutions are most suited to contribute to sustainable development and the maintenance of biological diversity.

Objectives

In GEM-CON-BIO 34 case studies were analysed on two scales to identify governance strategies that may benefit three outcomes, namely: (1) enhancing delivery of ES; (2) ensuring sustainable use of natural resources; and (3) maintaining biodiversity. This so-called *biodiversity governance* was defined as the way, how society on all levels aligns and regulates its political, economic and social concerns with regard to the use and the protection of biodiversity and ES.

Table 5.4 Governance types identified within the GEM-CON-BIO project, and their defining characteristics. (from Hahn et al. 2008)

	State controlled				Community based	Policy Network based	Market based
	National/federal	**Decentralised**	**Delegated**	**Corporatist**			
Description	Strong centralised control over management through State agencies	Management delegated to most appropriate administrative level	Management delegated to a non governmental body, e.g. academic, NGO, private sector. Remains within govt. policy objectives	Management employed through negotiated agreement between state agents and organisations interested	Objectives and main processes of ESM are defined by self-organised communities depending on ecosystems for their livelihood	Objectives and policies are negotiated and implemented among local stakeholders, government agencies and NGOs	Objectives oriented towards economic returns
Main management objectives	Regulatory compliance and economic development				Ecosystem resilience, economic development	Regulatory compliance, ecosystem resilience, economic development	Economic development
Key policy instruments	Legislation and policy guidance				Decentralised formal and informal institutions	Mixture	Economic incentives
Main Ownership structure	State	State/mix			Community/mix	State/mix	Private
Level of vertical integration	High (with state agencies)			Medium	Variable		Low
Level of horizontal integration	Low	Medium		Low	High		Variable

Table 5.4 Continued

	State controlled				Community based	Policy Network based	Market based
	National/federal	Decentralised	Delegated	Corporatist			
Generation of knowledge	Low	Medium		Low	Variable		Low
Local community participation	Low	Variable		Low	High	Variable	
Adaptive management	Low	Medium		Low	Variable	High	Low
Multilevel governance	Limited	Possible			Important		Unimportant
Leadership	Limited	Possible			Important	Unimportant	
Market/financial tools	Medium			High	Medium		High
Regulatory tools	High	Medium	Low	Medium	Low		High
Societal tools	Low		Low	Medium	Low	High	Low

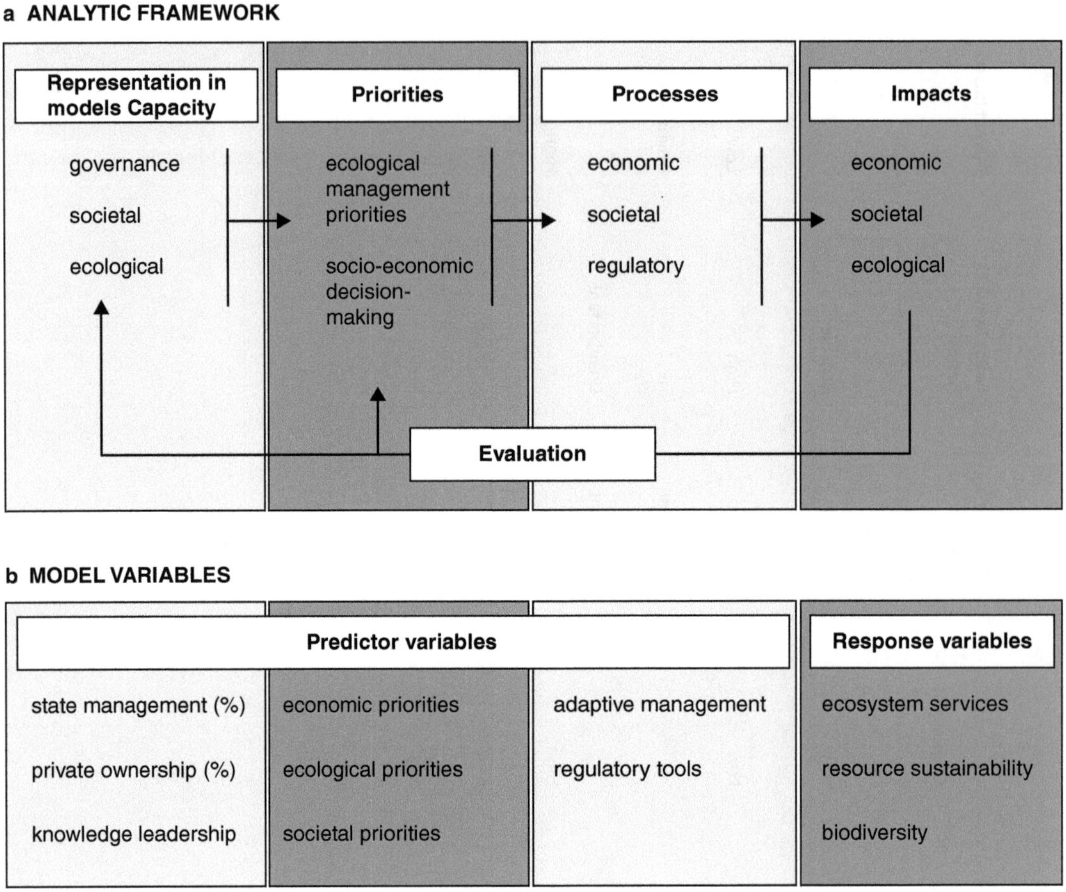

Fig. 5.7 **a** Conceptual framework for analysing the performance of different governance strategies; **b** model variables. (According Kenward et al. 2011)

- **Methods**

An analytic framework (Fig. 5.7) was developed to assess the relative importance of a suite of governance strategies for effective biodiversity conservation, based on measuring indicator variables in four main categories:

1. initial capacity;
2. management priorities;
3. main processes and tools aimed at those priorities; and
4. environmental response variables that potentially depend on (1)–(3).

For statistical analyses, we selected those variables that represent the logical structure of the analytical framework. We used information modelling techniques (IT) to examine which factors, whether on their own or in combination, best explain the various circumstances with regard to ES, sustainable resource management, and maintenance of biodiversity.

Using standardised questionnaires and expert judgment, we collected continuous and categorical data for 34 case studies on two different scales. A total of 26 cases examined the management of study areas from the local to subnational scale: 15 of these studies came from eight European countries, 2 from the USA, and 9 from different developing countries. Eight additional case studies involved the use of specific ES at an international scale, including organic agriculture around the Baltic Sea, North Sea fisheries, and a 27-country European

Union-wide survey of six recreational activities dependent on wild resources (hunting, bird watching, collecting berries, etc.).

For each of the case studies, 70 research questions split in five clusters were answered. Natural, social, economic, institutional resources, external driving forces, and pressures in the study areas were considered as key factors of the *governance initial capacity* and as a basis for management goals and decision-making. *Initial capacity* has a decisive influence on the applied governance types, which, in turn, retroact on the economic, financial, social, and ecological situation in the areas.

▪ Results

In the analysed case studies from the EU and the USA, the following governance types were important (◘ Table 5.4):

1. State controlled (a) national/federal, (b) decentralised, (c) delegated, (d) corporatist
2. Community-based
3. Politicy network-based
4. Market-based

The case studies showed that it is useful to

- consider and coordinate natural, social, cultural, economic and institutional resources and capacities as comprehensively as possible, to reach a high level of governance with respect to the protection of biodiversity
- apply a mixture of different governance types for ecosystem management according to the specific ecological, social, and economic demands, as it is also useful for the protection of biodiversity. Mixed governance types are more efficient than individual types.

An appropriate mixture includes regulative, participatory and economic/financial, social/cultural instruments, and it involves public administration, citizen participation and market-based approaches. Market instruments and/or quasi-market measures (e.g. agri-environmental measures) are especially useful if conservation measures cause opportunity costs and compete with economic activities. Thus, markets can influence biodiversity and ES both positively and negatively. For example, the implementation of biotope connections between Natura 2000 sites can be promoted by mixed strategies being able to develop long-term strategies and to prepare and implement management plans.

In the future, adaptive management may become increasingly important. Adaptive management can be defined as the structuring of policy and management options as a set of empirically testable hypotheses, which help to learn from the implementation of decisions and to achieve a higher adaptation capacity regarding unavoidable system changes (Lamont 2006 in Manos and Papathanasiou 2008). Adaptive management, incorporating monitoring and feedback, have long been proposed as powerful tools to ensure successful conservation outcomes (Holling 1978; Walters 1986). The case studies confirm this high efficiency. Various recent international agreements take the fact into consideration that implementing adaptive management and concomitant devolution of governance are needed to ensure the sustainable use of biodiversity (Convention on Biological Diversity 2004; Bern Convention 2007).

Most of the studies showed that it is relatively easy to reach positive results for biodiversity and ES in areas where state property and forest cover predominate. Governance structures can be improved when capable and engaged persons and/or organizations take strong leadership. Sustainable use and maintenance of biodiversity and ES benefit from the existence of effective institutions, especially if there is a high level of vertical and horizontal integration between them.

The protection and development goals must be formulated clearly, combined with social and economic goals, and fixed in management and sector plans. Suitability and appropriateness of the goals are an important guarantee for their success. All relevant ES should be included. If only provisioning services are included, which have the character of private goods, without considering regulation services based on public goods, and being fundamental for human well-being, serious risks for biodiversity and ES will rise. For example, in agriculture the production of goods but not of general interest services are regulated by markets. For those services compensation payments are offered, but these are often too low and economically not attractive, whereas the one-sided orientation towards provisioning services is stimulated by incentives, recently additionally for energy cropping, which further distorts the missing balance between the various ES (▶ Sect. 6.2).

The case studies have also underpinned that there is a strong demand for regulations and environmental standards (e.g., Water Framework Directive, Natura 2000), if negative impacts on ecosystems and ES as well as serious risks and dangers exist. Where markets for biodiversity and ES can be used (e.g., organic farming, tourism) or quasi-markets for the exchange of a public good between businessmen and states can be created (e.g., specific agri-environmental measures), market mechanisms may be effective. Vice-versa, higher pressures on biodiversity can be observed where market instruments are not used sufficiently and where the actors have insufficient knowledge.

Even though the protection of biological diversity benefits from the setting of ecological priorities and suitable regulations, it became obvious that the supply of ES from the local to the international level is strictly bound to economic priorities. These results confirm the necessity of a dual approach, which includes both protection and use.

Targeted monitoring is useful to successfully apply adaptive management strategies, to exclude or at least diminish negative influences and to support positive ones. The question is: What is the effect of biodiversity and ES on management measures? Effective monitoring of biodiversity needs the development and application of new governance indicators (e.g., type- and quality-related variables to evaluate participation). To stop the loss of biodiversity, not only pressures and their driving forces but also the speed and efficiency of policy reactions must be considered (Manos and Papathanasiou 2008; Kenward et al. 2011).

No blanket solutions can be prescribed, but governance and management of ecosystems have to take the diversity of ecological, social, economic, cultural, historical, and institutional aspects within and between countries into account. It is also important to enhance the communication on decisions on governance and ecosystem management taken on national and international levels in order to improve the cooperation of stakeholders on the different levels (horizontal and vertical). Moreover, there is an urgent need to enhance public awareness of the biodiversity values as a precondition of human life quality and economic activities. In 2007, a survey showed that, only 35 % of European citizens knew what the term biodiversity means (Manos and Papathanasiou 2008).

▪ Case Study Moritzburg Hilly Landscape

One of the case studies of the GEM-CON-BIO project was located in the Moritzburg hilly landscape north of the Saxon capital Dresden (characteristics of the study area ▶ Sect. 6.2).

The vicinity of Dresden has land-use interferences between agriculture, settlement, traffic, tourism/recreation, and the demands of nature conservation.

The most important initial conditions and available resources, which are influencing the results of governance in terms of biodiversity conservation for this case study are as follows: the main ES is the production of food (crops or livestock) by private farmers and a huge agricultural enterprise. External drivers, especially economic ones, which are mainly affected by the EU-Common Agricultural Policy (CAP) (market prices, subsidies), are influencing the management to a large extent. Due to the rights of landowners in terms of use and management of natural resources and the low enforcement of available regulations concerning nature conservation issues, the external drivers push back most of the state or private endeavours for governance for biodiversity conservation. The overriding interests of economic development also have effects on the major threats on biodiversity, such as agro-industrial farming, e.g. large field plots, monocultures, mechanization, increased use of chemicals, maize and rape for energy production, and infrastructure development/traffic routes.

Regarding the ecosystem management objectives and decision-making, a huge number of partially overlapping sector plans governing the use and management of natural resources we have to mention also here. In reality, most of these nature- or biodiversity-related plans do not have a significant influence on the management. The management of the area is predominantly governed by individual economic decisions of landowners or leaseholders according to the legal framework. Only some of the small areas influenced by contracts about the environmentally friendly management of natural resources between land users and state agencies are managed directly for biodiversity conservation. Several monitoring activities, especially on birds, are carried out. So far, the results, e.g. the rapid decline of field-birds, do not influence the main agricultural activities.

Generally, all sector plans with a focus on biodiversity are developed by the state to counter the economics-based individual decisions about management that mainly cause the loss of biodiversity. Some plans are oriented strongly towards the ecological dimension (contracts regarding the management of natural resources, the Natura 2000 management plan, ordinances for protected areas, etc.), others are looking for a well-balanced trade-off between the economic, ecological and social dimension (land-use planning, regional plan, etc.). The plans are developed by state agencies under voluntary involvement of local stakeholders. However, despite the contracts about the management of natural resources, the plans are weakly implemented in real management activities.

Among the governance processes related to biodiversity and ecosystem management in the area, the following factors belonging to three complexes are the most widely implemented:

1. Economic/financial instruments (market tools and incentives): funds as subsidies or financial incentives are the most relevant instruments for the stimulation of nature-related management measures in this area. As it is usual in Germany, the financial means for landscape management and biodiversity conservation are provided by state authorities and benefit private land users (compensation for income losses).
2. Legislative tools, regulations: the management of ecosystems is also driven by regulations for protected areas according to the Saxon Nature Conservation Act and the Federal Soil Conservation Act and other derived regulations as well as by agricultural regulations related to CAP: Cross Compliance.
3. Social processes collaboration among local stakeholders; leadership role in management processes): comprehensive cooperation in terms of biodiversity management results from the continuous work of the very active 'NGO of ornithologists in Großdittmannsdorf' (bottom-up approach). This organization successfully keeps contact with and cooperates with all stakeholders involved, and, therefore, holds the leadership role in local biodiversity management processes. Unfortunately, its influence on the dominant resource management activities of the farmers or agriculture enterprises is limited. But the participatory processes exercised for more than 30 years have resulted in positive social effects in the area. It is mainly caused by their constant, collaborative and successful work in nature conservation issues. Since the NGO has established contacts to the responsible state agencies, also the level of vertical trust within the managed area has increased. This continuous growth of horizontal as well as vertical trust is generating a permanent background for further endeavors in nature conservation. The described social effects generated by this local NGO distinguish the 'Moritzburg hilly landscape' from other areas, where no or only few stakeholders engaged in nature conservation and claim compliance with the law.

Nevertheless, also in the Moritzburg hilly landscape negative changes concerning the state of biodiversity and the supply of ES occurred. Monitoring data show losses or population declines of the most significant bird species for which the area is especially worthy of protection, and for which a European bird conservation site (Special Protection Area–SPA, in the framework of Natura 2000) had been designated. A major cause is the ongoing intensification of agriculture, such as the recently increasing cultivation of energy crops, particularly maize and rape (Bastian and Schrack 2007; Schrack 2008; Lupp et al. 2011).

The impacts described above are mainly caused by the influence of external drivers on the local ecosystem management, which is dominated by intensive agricultural practices. Particular governance activities do not lead to essential success in biodiversity conservation. Despite the high biotic value of the 'Moritzburg hilly landscape', the extensive network of different protected sites and the number of sector plans regarding or respecting nature conservation issues, in fact the described efforts in nature conservation, are too weak to meet the huge challenge of biodiversity loss. The effectiveness of the governance processes in relation to the official–mostly well balanced–management objectives, named in the sector plans, is low. Besides the formal planning, there are no adequate instruments for implementing their contents. Economic interests are mostly dominating practical management decisions. The endeavours in ecosystem management for biodiversity by state agencies and by farmers do not meet the challenges of land use, which preserves biodiversity and ES. Based on voluntary work, the leading nature protection association is not able to fill this gap.

Conclusion

Maintenance and development of biodiversity and ES may be, in principle, taken forward by suitable governance structures and processes. This is particularly the case for regulation and sociocultural ES, while provision ES follow market-based mechanisms. Altogether the case studies from many countries show that the application of a wide range of interlinked instruments and governance types, from economic via legal to social and participatory approaches, are the most fruitful ones.

References

Adams WM, Aveling R, Brockington D, Dickson B, Elliott J, Hutton J, Roe D, Vira B, Wolmer W et al (2004) Biodiversity conservation and the eradication of poverty. Science 306:1146–1149

Albert C, von Haaren C, Galler C (2012) Ökosystemdienstleistungen. Alter Wein in neuen Schläuchen oder ein Impuls für die Landschaftsplanung? Nat Schutz Landsch Plan 44:142–148

Anderson BJ, Armsworth PR, Eigenbrod F, Thomas CD, Gillings S, Heinemeyer A, Roy DB, Gaston KJ (2009) Spatial covariance between biodiversity and other ecosystem service priorities. J Appl Ecol 46:888–896

Backhaus T (2008) Backhaus begrüßt Erklärung des Bundes und der Länder zur biologischen Vielfalt. Natur & Umwelt vom 7. Mai 2008, ► www.mv-schlagzeilen.de/

Bastian O, Schrack M (2007) Energie vom Acker – Traum oder Albtraum? – Mitt. Landesverein Sächs. Heimatschutz 3:57–66

Bastian O, Schreiber K-F (1999) Analyse und ökologische Bewertung der Landschaft (2. edn.). Spektrum Akademischer Verlag, Heidelberg

Bern Convention (2007) European charter on hunting and biodiversity (Council of Europe, Strasbourg, France) ►http://www.facenatura2000.net/conference2009/2.4.Lasen-Dias.pdf. Accessed 19 July 2012

Bertke E (2005) Ökologische Güter in einem ergebnisorientierten Honorierungssystem für ökologische Leistungen der Landwirtschaft. Herleitung – Definition – Kontrolle. PhD-thesis, University Göttingen, ibidem Verlag, Stuttgart

Breuste J, Haase D, Elmqvist T (2011) Urban landscapes and ecosystem services. In: Sandhu H, Wratten S, Cullen R, Costanza R (eds) ES2: ecosystem services in engineered systems. Wiley-Blackwell, Oxford

Brondízio ES, Gatzweiler FG, Zografos C, Kumar M (2010) Socio-cultural context of ecosystem and biodiversity evaluation: biodiversity, ecosystems and ecosystem services. In: Kumar P (eds) The economics of ecosystems and biodiversity: ecological and economic foundations. Earthscan, London, pp 149–181

Bundestagsfraktion Bündnis 90/Die Grünen (2012) Biodiversität 2020. Das grüne Handlungskonzept zum Schutz der biologischen Vielfalt. ► www.gruene-bundestag.de/themen/biologische-vielfalt/der-gruene-aktionsplan-fuer-den-schutz-der-biologischen-vielfalt.html

Burkhard B, Kroll F, Nedkov S, Müller F (2011) Mapping ecosystem service supply, demand and budgets. Ecol Indic 21:17–29

Convention on Biological Diversity (2004) Addis Ababa Principles and Guidelines for the Sustainable Use of Biodiversity. Secretariat Conv Biol Divers, Montreal, ► www.cbd.int/doc/publications/addis-gdl-en.pdf. Accessed 19 July 2012

Czybulka D, Luttmann M (2005) Die Berücksichtigung von Leistungen der Länder für das Naturerbe im Finanzausgleichssystem des Bundes. Natur und Recht 2:79–86

Danson M, Halkier H, Cameron G (eds) (2000) Governance, institutional change and regional development. Ashgate, Aldershot

de Groot RS, Alkemade R, Braat L, Hein L, Willemen L (2010) Challenges in integration the concept of ecosystem services and values in landscape planning, management and decision making. Ecol Complex 7:260–272

Diller C (2002) Zwischen Netzwerk und Institution. Eine Bilanz regionaler Kooperationen in Deutschland. VS-Verlag, Opladen

Egoh B, Reyers B, Rouget M, Richardson DM, Le Maitre DC, van Jaarsveld AS (2008) Mapping ecosystem services for planning and management. Agric Ecosyst Environ 127:135–140

Elmqvist T, Maltby E, Barker T, Mortimer M, Perrings C, Aronson J, de Groot R, Fitter A, Mace G, Norberg J, Sousa Pinto I, Ring I (2010) Biodiversity, ecosystems and ecosystem services. In: Kumar P (ed) The economics of ecosystems and biodiversity: ecological and economic foundations. Earthscan, London, Washington, pp 41–111

Endlicher W (2011) Introduction: from urban nature studies to ecosystem services. In: Endlicher W et al (eds) Perspectives in urban ecology. Springer, Heidelberg
Engel S, Pagiola S, Wunder S (2008) Designing payments for environmental services in theory and practice: an overview of the issues. Ecol Econ 65:663–674
FAO–Food and Agriculture Organization of the United Nations (2007) The state of food and agriculture. Paying farmers for environmental services. FAO Agriculture Series No. 38. Rome, Italy
Ferraro PJ (2008) Asymmetric information and contract design for payments for environmental services. Ecol Econ 65:810–821
Ferraro PJ, Kiss A (2002) Direct payments to conserve biodiversity. Science 298:1718–1719
Freese J, Klimek S, Marggraf R (2011) Aktionen- und ergebnisorientierte Honorierung bei Agrarumweltmaßnahmen. Nat Landsch 4:156–159
Graham J, Amos B, Plumptre T (2003) Governance principles for protected areas in the 21st century, a discussion. ► www.earthlore.ca/clients/WPC/English/grfx/sessions/PDFs/session_1/Amos_plenary.pdf. Accessed 8 June 2012
Grêt-Regamey A, Walz A, Bebi P (2008) Valuing ecosystem services for sustainable landscape planning in Alpine regions. Mt Res Dev 28:156–165
Grossmann M, Hartje V, Meyerhoff J (2010) Ökonomische Bewertung naturverträglicher Hochwasservorsorge an der Elbe, Abschlussbericht des F + E-Vorhabens "Naturverträgliche Hochwasservorsorge an der Elbe und Nebenflüssen und ihr volkswirtschaftlicher Nutzen. Teil: Ökonomische Bewertung naturverträglicher Hochwasservorsorge an der Elbe und ihren Nebenflüssen". BfN-Heft 89
Gruehn D, Kenneweg H (1998) Berücksichtigung der Belange von Naturschutz und Landschaftspflege in der Flächennutzungsplanung. Ergebnisse aus dem F + E-Vorhaben 80806011 des Bundesamtes für Naturschutz. Bundesamt für Naturschutz, BfN-Schriftenvertrieb im Landwirtschaftsverlag
Grünwald A (2011) Zukunft Landschaftsplan – Perspektiven einer methodischen Weiterentwicklung unter Anwendung des Konzeptes der Ökosystemdienstleistungen. Masterarbeit, TU Dresden
Gundimeda H, Wätzold F (2010) Payments for ecosystem services and conservation banking. In: Wittmer H, Gundimeda H (coordinators), TEEB (eds) The economics of ecosystems and biodiversity for local and regional policy makers (Chap. 8, pp 141–160). (► www.teebweb.org)
Gunningham N, Young MD (1997) Toward optimal environmental policy: the case of biodiversity conservation. Ecol Law Quart 24:243–296
Gunningham N, Sinclair D, Grabosky P (1998) Smart regulation: designing environmental policy. Clarendon, Oxford
Haaren C von (ed) (2004) Landschaftsplanung. Eugen Ulmer, Stuttgart
Haaren C von, Albert C (2011) Integrating ecosystem services and environmental planning: limitations and synergies. Int J Biodivers Sci Ecosys Services Manage 7. doi:10.1080/21513732.2011.616534
Haaren C von, Galler C, Ott S (2008) Landscape planning. The basis of sustainable landscape development. Bundesamt für Naturschutz, Bonn
Hahn T, Galaz V, Terry A (2008) The GEM-CON-BIO analysis framework. In: Manos B, Papathanasiou J (eds) GEM-CON-BIO Governance and ecosystem management for the conservation of biodiversity. Aristoteles-Universität Thessaloniki, Griechenland, pp 39–52
Hampicke U, Wätzold F (Spokeperson for the initiative) (2009) Memorandum economics for nature conservation–harmonising economic activities with protecting and conserving biodiversity, Greifswald, Leipzig, Bonn. ► www.bfn.de/fileadmin/MDB/documents/themen/oekonomie/MemoOekNaturschutz.pdf
Hartmann E, Thomas F, Luick R (2006) Agrarumweltprogramme in Deutschland. Anreiz für umweltfreundliches Wirtschaften in der Landwirtschaft und Kooperationen mit dem Naturschutz. Nat.schutz Landsch.plan 38:205–213
Heiland S (2010) Landschaftsplanung. In: Henckel D, Kuczkowski K v, Lau P, Pahl-Weber E, Stellmacher F (eds) Planen – Bauen – Umwelt. Ein Handbuch. VS-Verlag, Wiesbaden, pp 294–300
Holling CS (1978) Adaptive environment assessment and management. Wiley, London
James AN, Gaston KJ, Balmford A (1999) Balancing the earth's accounts. Nature 401:323–324
Jedicke E (2010) Kann die TEEB-Studie den Landschaftsplan reanimieren? Nat Schutz Landsch Plan 42:289
Jessel B (2011) Ökosystemdienstleistungen – Potenziale und Grenzen eines aktuellen umweltpolitischen Konzepts. In: BBN (ed) Frischer Wind und weite Horizonte – Jahrbuch für Naturschutz und Landschaftspflege, Bd 3, Naturschutz und Gesellschaft, Bonn, pp 72–87
Kaechele K, May P, Primmer E, Ludwig G (2011) Forest certification: a voluntary instrument for environmental governance. In: Ring I, Schröter-Schlaack C (eds) Instrument mixes for biodiversity policies. POLICY MIX Report 2/2011. Helmholtz Centre for Environmental Research – UFZ, Leipzig, pp 162–174. ► http://policymix.nina.no
Kenward R, Whittingham M, Arampatzis S, Manos B, Hahn T, Terry A, Simoncini R, Alcorn J, Bastian O, Donlan M, Elowe K, Franzén F, Karacsyonyi Z, Larsson M, Manou D, Navodaru I, Papadopoulou O, Papathanasiou J, von Raggamby A, Sharp R, Söderqvist T, Soutukorva A, Vavrova L, Aebischer N, Leader-Williams N, Rutz C (2011) Identifying governance strategies that effectively support ecosystem services, resource sustainability, and biodiversity. Proc Natl Acad Sci U S A 108:5308–5312

Kienast F (2010) Landschaftsdienstleistungen: Ein taugliches Konzept für Forschung und Praxis? Forum für Wissen:7–12

Kirchhoff T, Trepl L, Vicenzotti V (2012) What is landscape ecology? An analysis and evaluation of six different conceptions. Landscape Res (38:33–51). doi:10.1080/01426397.2011.640751

Kleijn D, Sutherland WJ (2003) How effective are European agri-environment schemes in conserving and promoting biodiversity? J Appl Ecol 40:947–969

Klimek S, Richter-Kemmermann A, Steinmann H-H, Fresse J, Isselstein J (2008) Rewarding farmers for delivering vascular plant diversity in managed grasslands: a transdisciplinary case-study approach. Biol Conserv 131:2888–2897

Knieling J (2003) Kooperative Regionalplanung und Regional Governance: Praxisbeispiele, Theoriebezüge und Perspektiven. Informationen zur Raumentwicklung 8/9:463–478

Krönert R, Steinhardt U, Volk M (eds) (2001) Landscape balance and landscape assessment. Springer, Berlin

KULAP–Kulturlandschaftsprogramm (2012) Richtlinie des Ministeriums für Infrastruktur und Landwirtschaft des Landes Brandenburg zur Förderung umweltgerechter landwirtschaftlicher Produktionsverfahren und zur Erhaltung der Kulturlandschaft der Länder Brandenburg und Berlin (KULAP 2007) vom 27. August 2010, geändert mit Erlass vom 29. Juli 2010 und vom Erlass 30. Januar 2012

Lamont A (2006) Policy characterization of ecosystem management. Environ Monit Assess 113:5–18

Le Galès P (1998) Regulations and Governance in European Cities. Int J Urban Regional 22:482–506

Lupp G, Albrecht J, Darbi M, Bastian O (2011) Ecosystem services in energy crop production–a concept for regulatory measures in spatial planning? J Landscape Ecol 4:49–66

Manos B, Papathanasiou J (2008) GEM-CON-BIO Governance and Ecosystem Management for the Conservation of Biodiversity. Aristotle University Thessaloniki, Greece

May PH, Veiga Neto F, Denardin V, Loureiro W (2002) Using fiscal instruments to encourage conservation: Municipal responses to the "ecological" value-added tax in Paraná and Minas Gerais, Brazil. In: Pagiola S, Bishop J, Landell-Mills N (eds) Selling forest environmental services: Market-based mechanisms for conservation and development. Earthscan, London, pp 173–199

MEA–Millennium Ecosystem Assessment (2005a) Ecosystems and human well-being: biodiversity synthesis. World Resources Institute, Washington DC

MEA–Millennium Ecosystem Assessment (2005b) Ecosystems and human well-being: synthesis. Island Press, Washington DC

Mewes M, Sturm A, Johst K, Drechsler M, Wätzold F (2012) Handbuch der Software Ecopay zur Bestimmung kosteneffizienter Ausgleichszahlungen für Maßnahmen zum Schutz gefährdeter Arten und Lebensraumtypen im Grünland. UFZ-Bericht 01/2012. Helmholtz-Zentrum für Umweltforschung – UFZ, Leipzig

Michaelis P (1996) Ökonomische Instrumente der Umweltpolitik. Physica-Verlag, Heidelberg

NeFo–Netzwerkforum zur Biodiversitätsforschung Deutschland (2011) Endbericht zum F & E-Vorhaben. Workshop: Ökosystemdienstleistungen – warum ein sperriges Konzept Karriere macht. Georg-August-Universität Göttingen, ► www.biodiversity.de/index.php/de/fuer-presse-medien/experteninterviews/1657-interview-barkmann-2. Accessed 10 Feb 2012

Nill D (2011) Bezahlung von Ökosystemdienstleistungen für den Erhalt der landwirtschaftlichen genetischen Vielfalt. Konzepte, Erfahrungen und Relevanz für die Entwicklungszusammenarbeit. GIZ, Bonn

OECD (1999) Handbook of incentive measures for biodiversity: design and implementation. OECD, Paris

OECD (2007) Instrument mixes for environmental policy. OECD, Paris

Oosterhuis F (2011) Tax reliefs for biodiversity conservation. In: Ring I, Schröter-Schlaack C (eds) Instrument Mixes for Biodiversity Policies. POLICY MIX Report 2/2011. Helmholtz Centre for Environmental Research – UFZ, Leipzig, pp 89–97. ► http://policymix.nina.no

Perner A, Thöne M (2005) Naturschutz im Finanzausgleich. Erweiterung des naturschutzpolitischen Instrumentariums um finanzielle Anreize für Gebietskörperschaften. FiFo-Berichte Nr. 3, Mai 2005, Finanzwissenschaftliches Forschungsinstitut an der Universität zu Köln

Perrot-Maître D (2006) The Vittel payments for ecosystem services: a "perfect" PES case? International Institute for Environment and Development, London

Pimentel D, Harvey C, Resoosudarmo P, Sinclair K, Kurz D, McNair M, Crist S, Shpritz L, Fitton L, Saffouri R, Blair R (1995) Environmental and economic costs of soil erosion and conservation benefits. Science 267:1117–1123

Porras I, Chacón-Cascante A, Robalino J, Oosterhuis F (2011) PES and other economic beasts: assessing PES within a policy mix in conservation. In: Ring I, Schröter-Schlaack C (eds) Instrument mixes for biodiversity policies. POLICY MIX Report 2/2011. Helmholtz-Zentrum für Umweltforschung – UFZ, Leipzig, pp 119–144. ► http://policymix.nina.no

Richter M, Weiland U (eds) (2012) Applied urban ecology – a global framework. Wiley-Blackwell, West-Sussex

Ring I (2001) Ökologische Aufgaben und ihre Berücksichtigung im kommunalen Finanzausgleich. Z Angew Umweltforschung 13:236–249

Ring I (2002) Ecological public functions and fiscal equalisation at the local level in Germany. Ecol Econ 42:415–427

Ring I (2004) Naturschutz in der föderalen Aufgabenteilung: Zur Notwendigkeit einer Bundeskompetenz aus ökonomischer Perspektive. Nat Landsch 79:494–500

Ring I (2008a) Compensating municipalities for protected areas. Fiscal transfers for biodiversity conservation in Saxony, Germany. GAIA 17(S1):143–151

Ring I (2008b) Integrating local ecological services into intergovernmental fiscal transfers: the case of the ecological ICMS in Brazil. Land Use Policy 25:485–497

Ring I (2011) Economic Instruments for Conservation Policies in Federal Systems. UFZ Habilitation Nr. 1/2011; zugleich Habilitationsschrift, Wirtschaftswissenschaftliche Fakultät der Universität Leipzig, Leipzig, 219 p. ► www.ufz.de/index.php?de†=13920

Ring I, Schröter-Schlaack C (eds) (2011a) Instrument Mixes for Biodiversity Policies. POLICY MIX Report 2/2011. Helmholtz Centre for Environmental Research – UFZ, Leipzig. ► http://policymix.nina.no

Ring I, Schröter-Schlaack C (2011b) Justifying and Assessing Policy Mixes for Biodiversity and Ecosystem Governance. In: Ring I, Schröter-Schlaack C (eds) Instrument Mixes for Biodiversity Policies. POLICY MIX Report 2/2011. Helmholtz Centre for Environmental Research – UFZ, Leipzig, pp 14–35. ► http://policymix.nina.no

Ring I, Drechsler M, van Teeffelen AJ, Irawan S, Venter O (2010a) Biodiversity conservation and climate mitigation: what role can economic instruments play? Curr Opin Environ Sustain 2:50–58

Ring I, Hansjürgens B, Elmqvist T, Wittmer H, Sukhdev P (2010b) Challenges in framing the economics of ecosystems and biodiversity: the TEEB initiative. Curr Opin Environ Sustain 2:15–26

Ring I, May P, Loureiro W, Santos R, Antunes P, Clemente P (2011) Ecological fiscal transfers. In: Ring I, Schröter-Schlaack C (eds) Instrument Mixes for Biodiversity Policies. POLICY MIX Report 2/2011. Helmholtz Centre for Environmental Research – UFZ, Leipzig, pp 98–118. ► http://policymix.nina.no

Santos R, Clemente P, Antunes P, Schröter-Schlaack C, Ring I (2011) Offsets, habitat banking and tradable permits for biodiversity conservation. In: Ring I, Schröter-Schlaack C (eds) Instrument mixes for biodiversity policies. POLICY MIX Report 2/2011. Helmholtz Centre for Environmental Research – UFZ, Leipzig, pp 59–88. ► http://policymix.nina.no

Santos R, Ring I, Antunes P, Clemente P (2012) Fiscal transfers for biodiversity conservation: the Portuguese local finances law. Land Use Policy 29:261–273

Schrack M (eds) (2008) Der Natur verpflichtet. Projekte, Ergebnisse und Erfahrungen der ehrenamtlichen Naturschutzarbeit in Großdittmannsdorf. Veröff. Mus. Westlaus. Kamenz, Sonderheft, 180 p

Schröter-Schlaack C, Blumentrath S (2011) Direct regulation for biodiversity conservation. In: Ring I, Schröter-Schlaack C (eds) Instrument mixes for biodiversity policies. POLICY MIX Report 2/2011. Helmholtz Centre for Environmental Research – UFZ, Leipzig, pp 36–58. ► http://policymix.nina.no

Schröter-Schlaack C, Ring I (2011) Towards a framework for assessing instruments in policy mixes for biodiversity and ecosystem governance. In: Ring I, Schröter-Schlaack C (eds) Instrument Mixes for Biodiversity Policies. POLICY MIX Report 2/2011. Helmholtz Centre for Environmental Research – UFZ, Leipzig, pp 175–208. ► http://policymix.nina.no

Schröter-Schlaack C, Ring I, Möckel S, Schulz-Zunkel C, Lienhoop N, Klenke R, Lenk T (2013) Assessment of existing and proposed policy instruments for biodiversity conservation in Germany: The role of ecological fiscal transfers. POLICY MIX Report No. 1/2013. Helmholtz Centre for Environmental Research – UFZ, Leipzig, 95 p. ► http://policymix.nina.no

Simoncini R, Borrini-Feyerabend G, Lassen B (2008) Policy guidelines on governance and ecosystem management for biodiversity conservation. GEM-CON-BIO project report, IUCN Regional Office for Europe, Brüssel

Southern A, Lovetta A, O'Riordana T, Watkinson A (2011) Sustainable landscape governance: lessons from a catchment based study in whole landscape design. Landsc Urban Plan 101:179–189

SRU–Der Rat von Sachverständigen für Umweltfragen (1996) Konzepte einer dauerhaft-umweltgerechten Nutzung ländlicher Räume, Sondergutachten. Stuttgart

Staub C, Ott W, Heusi F, Klingler G, Jenny A, Häcki M, Hauser A (2011) Indikatoren für Ökosystemdienstleistungen: Systematik, Methodik und Umsetzungsempfehlung für eine wohlfahrtsbezogene Umweltberichterstattung. Bundesamt für Umwelt, Bern. Umwelt-Wissen Nr. 1102

Sterner T (2003) Policy instruments for environmental and natural resource management. RFF Press, Washington DC

Stokman A, von Haaren C (2012) Integrating science and creativity for landscape planning and design of urban areas. In: Richter M, Weiland U (eds) Applied urban ecology–a global framework. Wiley-Blackwell, West-Sussex, pp 170–185

Swetnam RD, Fisher B, Mbilinyi BP, Munishi PKT, Willcock S, Ricketts T, Mwakalila S, Balmford A, Burgess ND, Marshall AR, Lewis SL (2010) Mapping socio-economic scenarios of land cover change: a GIS method to enable ecosystem service modelling. J Environ Manage 92:1–12

Syrbe R-U, Walz U (2012) Spatial indicators for the assessment of ecosystem services: providing, benefiting and connecting areas and landscape metrics. Ecol Indic 21:80–88

TEEB–The economics of ecosystems and biodiversity (2010a) Mainstreaming the Economics of Nature. A synthesis of the approach, conclusions and recommendations of TEEB. ► www.teebweb.org

TEEB–The economics of ecosystems and biodiversity (2010b) The economics of ecosystems and biodiversity for local and regional policy makers. ► www.teebweb.org

TEEB–The economics of ecosystems and biodiversity (2011) The economics of ecosystems and biodiversity in national and international policy making. Ten Brink P (ed), Earthscan, London

ten Brink P, Bassi S, Bishop J, Harvey CA, Ruhweza A, Verma M, Wertz-Kanounnikoff S (2011) Rewarding benefits through payments and markets. Section 1: payments for ecosystem services. In: TEEB–the economics of ecosystems and biodiversity in national and international policy making. Ten Brink P (ed), Earthscan, London, pp 181–199

Termorshuizen J, Opdam P (2009) Landscape services as a bridge between landscape ecology and sustainable development. Landsc Ecol 24:1037–1052

Troy A, Wilson MA (2006) Mapping ecosystem services: practical challenges and opportunities in linking GIS and value transfer. Ecol Econ 60:435–449

UNDP–United Nation Development Programme (1999) Human development report 1999–globalisation with a human face. united nations development programme, New York

Vasishth A (2008) A scale-hierarchic ecosystem approach to integrative ecological planning. Progress in Planning 70:99–132

Vihervaara P, Kumpula T, Tanskanen A, Burkhard B (2010) Ecosystem services–a tool for sustainable management of human–environment systems. Case study in Finish Forest Lapland. Ecol Complex 7:410–420

Walters K (1986) Adaptive management of renewable resources. Macmillan, New York

Wätzold F, Schwerdtner K (2005) Why be wasteful when preserving a valuable resource? A review article on the cost-effectiveness of European conservation policy. Biol Conserv 123:327–338

Wätzold F, Drechsler M, Armstrong CW, Baumgärtner S, Grimm V, Huth A, Perrings C, Possingham HP, Shogren JF, Skonhoft A, Verboom-Vasiljev J, Wissel C (2006) Ecological-economic modeling for biodiversity management: potential, pitfalls, prospects. Conserv Biol 20:1034–1041

Wätzold F, Mewes M, van Apeldoorn R, Varjopouro R, Chmielewski TJ, Veeneklaas F, Kosola MJ (2010) Cost-effectiveness of managing Natura 2000 sites: an exploratory study for Finland, Germany, the Netherlands and Poland. Biodivers Conserv 19:2053–2069

Wende W, Reinsch N, Jülg D, Funke J (2005) Kommunale Landschaftspläne – Rahmenbedingungen der praktischen Umsetzung von Erfordernissen und Maßnahmen. Landschaftsentwicklung und Umweltforschung – Schriftenreihe der Fakultät Architektur Umwelt Gesellschaft TU Berlin, Nr. 126

Wende W, Marschall I, Heiland S, Lipp T, Reinke M, Schaal P, Schmidt C (2009) Umsetzung von Maßnahmenvorschlägen örtlicher Landschaftspläne – Ergebnisse eines hochschulübergreifenden Projektes in acht Bundesländern. Nat Schutz Landsch Plan 41:145–149

Wende W, Wojtkiewicz W, Marschall I, Heiland S, Lipp T, Reinke M, Schaal P, Schmidt C (2012) Putting the plan into practice: implementation of proposals for measures of local landscape plans. Landscape Res. 37:483-500

Willemen L, Hein L, Verburg PH (2010) Evaluating the impact of regional development policies on future landscape services. Ecol Econ 69:2244–2254

Willemen L, Verburg PH, Hein L, van Mensvoort MEF (2008) Spatial characterization of landscape functions. Landsc Urban Plan 88:34–43

Wirth P, Hutter G, Schanze J (2010) Flood risk management and regional governance–the case of Weisseritz Regio (Germany). In: Kluvánková-Oravská T et al (eds) From government to governance? New governance for water and biodiversity in an enlarged Europe. Alfa Nakladatelství, Prague, pp 128–141

Wunder S (2005) Payments for environmental services: some nuts and bolts. Occasional Paper No. 42, Center for International Forestry Research, Nairobi, Kenya. ► www.cifor.org/publications/pdf_files/occpapers/op-42.pdf

Wunder S (2007) The efficiency of payments for environmental services in tropical conservation. Conserv Biol 21:48–58

Wunder S, Engel S, Pagiola S (2008) Taking stock: a comparative analysis of payments for environmental services programs in developed and developing countries. Ecol Econ 65:834–852

Land Use, Maintenance and Protection to Ensure ES

K. Grunewald, O. Bastian (eds.), *Ecosystem Services – Concept, Methods and Case Studies*,
DOI 10.1007/978-3-662-44143-5_6, © Springer-Verlag Berlin Heidelberg 2015

6.1 Concept for the Selection of Case Studies

K. Grunewald and O. Bastian

"Nature does not rate."

The ES concept is not designed to end in itself, but also helps to develop and implement better activity for utilization and protection of nature. Links and differences between ES and biodiversity were already pointed out in ▶ Chap. 1. ◘ Figure 6.1 illustrates that ES, biodiversity, sustainability and land-use management including natural resources and the protection of soil, water and climate are presenting different accesses in the area of tension and "ensure growing needs of human and ecosystem/natural capacity" which are overlapping to a greater or lesser extent. One should indicate the complementary in its issue not simply differentiating concept whereby considering the indicators and methods are different in its main focuses and perspectives.

Science and politics need to ensure that the practical part will be supported: What is actually meant, which interactions are relevant, what is important? For example, the National Committee for Global Change Research (NKGCF 2011) demands good practise definition of regionally specific indicators and monitoring strategies, including the comparison of measurable variables of biodiversity, functions and services of ecosystems, land-use and socio-economic trends, or "development and verification of models and concepts in sustainability research with a view to biodiversity and ecosystem services" (ES). As a result high research policy expectations are aroused in a complicated, integrated area. It is possible that different target systems, e.g. of agriculture and protection of species, might show different behaviour. Therefore, fundamentals need to be developed to negotiate considerations (◘ Fig. 6.1).

Sustainable land management and landscape conservation are crucial factors of our basis of life. That' is why we focused on ES case studies concerning the subject land use/land-use changes as well as protection and maintenance of landscapes.

The land surface represents the primary human habit at which mankind has influenced and actively created for centuries. Recent developments of global changes like demographic change, climate change plus globalization of economic systems constitute enormous challenges with this limited resource to treat. Besides the area of nutrition this also concerns the supply of energy and habitat or preservation of ecosystems.

Predominantly land use and driving forces of land-use changes constitute a socio-economic category. Humans as acting entity of interventions in nature are set to land utilization (Ott 2010). This implies it is not about whether but how the intervention into ecosystems is made and how certain interventions can be evaluated. The ES concept shall and can help to understand contexts. Again references to biodiversity are relevant, especially within the German National Strategy on Biodiversity (BMU 2007) and within the Federal Biological Diversity Programme (BMU 2011).

Drawbacks in the natural environment like loss of animal and plant species, penetration of Neobiota, enhanced pollutant concentrations in soils, water and air, soil erosion, sealing and fragmentation of natural habitats, loss and damage of landscape elements or increasing noise because of traffic routes is mostly due to cause of land uses of humans. For decades these ecosystem changes are subject of environmental research discussions, but to what extent are they influencing utilization, user and the human well-being? Even if objective facts of land use are analysed, systematised and represented according to scientific categories (biophysical methods, ▶ Sect. 4.1), troubleshootings and decisions are taking place in public discourse. In this specific aspect the integrative ES concept is said to help and enable innovative realizations (▶ Sect. 4.5).

In democratic countries like Germany, land use or cause of potential abuses (see above) are controlled on the basis of rules of what they should and should not do. The legislation can be considered extremely great in the EU and Germany. Tedious social negotiation processes are normally preceded to legal regulations. These assume ecological analysis as well as perception and estimation of the risks or the hazards for humans and environment (=constructivistical social process). Condition for adequate social trade is the timely recognition of land-use problems but it is not a guarantee for 'right' responses.

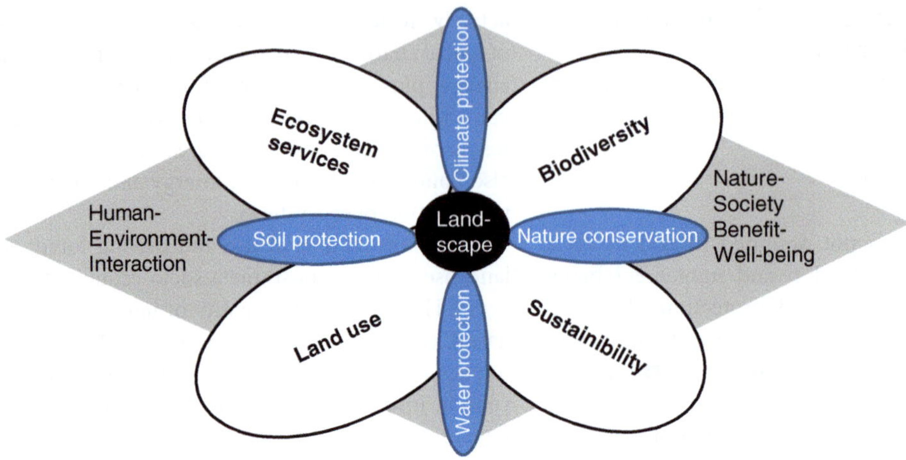

■ **Fig. 6.1** Scheme of overlying terms and concepts in the human-environment sector. © Grunewald

Integrative management which in addition is made to contribute a balance between objectives of protection, sustainable use and equitable sharing of gains derived by use, is the matter of the interface ES–land use (Jessel 2011). Mankind is considered an explicit part of ecosystems (landscape approach, ▶ Sect. 3.4). This corresponds to the principles of the ecosystem approach of the Convention on Biological Diversity (CBD 2010), the so-called Malawi-Principles (Häusler and Scherer-Lorenzen 2002).

The objective target, which is linked to case studies presented hereafter, is mainly consistent of:

1. Demonstration of multifaceted applications of the ES concept: terms, categories, approaches of analysis and assessment, cost-benefits considerations, mechanisms for controlling and finance (aspects of methodology)
2. Presentation of possibilities, how ES approaches could contribute to sustainable land use (new way of looking at concepts, design options, possibilities and limits of the concept)
3. Discussion of the current status of ES capturing in Germany (regional and ecosystem-/land-use type specific aspects).

The different professional backgrounds of the case study authors caused various perspectives and emphases. Articles had to be kept as short as possible so that individual problems could not be presented explicitly and in a detailed manner. In accordance to settings of priorities on land use and ecosystem types the case studies have initially been classified to areas of the main land cover categories in Germany (agrarian, forestry and urban ecosystems). Marine, coastal and high mountain ecosystems were left aside. All ecosystems are representing a production, living and regeneration space although with different emphases so, in principle, all three categories are relevant in ES.

Wherever humans need to intervene in nature to protect their own existence, a target-orientated landscape management is necessary for the preservation of values and services of ecosystems. The necessary expenses are describing a minimum-indicator for the valuation of ecosystems, because their existence is not secured without these accomplishments. Such analyses are focused in the landscape management (*Landschaftspflege*) accounting evaluations (▶ Sect. 6.5). Completing specific aspects of nature conservation, soil-, water-, and flood protection as well as climate- and peat protection will be discussed. Thereby, aspects of hemeroby, structural characteristics but also processes and matter balances are figured.

Furthermore, the case studies have been selected according to the following criteria:

- ES are assessed in projects and the results have been discussed in public ('wealth of existing data').
- Representation and transferability: ES were processed on a regional basis, they are typical, verified and validated ('exemplary representation').

■ Figure 6.2 illustrates the location of the case studies, which are mainly situated in Central and East Germany (states Saxony and Mecklenburg-Western Pomerania/Brandenburg, urban area of Leipzig, county Goerlitz, Ore Mountains, Mulde-Loesshuegelland, floodplain and catchment area Elbe). In ► Chap. 3, 4 and 5 individual cases of methods and techniques with regional examples have already been visualised (■ Fig. 6.2).

6.2 Assessment of Selected Services of Agro-Ecosystems

6.2.1 Introduction

O. Bastian

Utilised agricultural areas are currently taking up about half of the territory of the EU. Over many centuries, due to ongoing development created by humans, these agro-ecosystems have been and are partly still treasured for their biological variety and as producers of diverse ES. However, the continuous and ever-increasing intensification of agricultural production in favourable areas (land consolidation, large-scale application economy, mechanization, chemical-based approach) plus the mission and reforestation in disadvantaged regions has resulted in serious decline of biodiversity and many regulating and (socio)cultural ES over decades–a process which is expected to continue.

To give the farmers and to spread public understanding of the biological diversity, as well as the positive effects and achievements coming from numerous gentle cultivated agricultural lands, the implementation of the ES concept is suggestive for a number of reasons (Plieninger and Schleyer 2010):

- No other ecosystem has been as well researched in how management measures may influence ES (e.g. reduction of the input of fertilisers and pesticides into surface waters through growing of woods in the agro-culture).
- Many ES are placed as paddock jointly with agro-products; in few cases entire waiver on agricultural production is required to support ES.
- In the European agriculture significant experiences with economic incentive instruments, which can specifically be targeted as supply for ES, are already available.
- Many agro-ecosystems are disposed of high potential to strengthen ES. Agriculture relies highly on ES (regulation capacities), otherwise it provides efforts in significantly extents (provision performances). The community might impose external costs according to cultivation management in terms of habitat losses, nutrient translocations or greenhouse gas emissions.

Still, it should be noted that the ES term is not very common in the European agricultural policy so far.

In the following sections three case studies on ES or comparable issues in the area of agriculture will be presented:

1. The development of local agri-environmental programmes and measures (► Sect. 6.2.2)
2. The agro-economic evaluation of implementing a landscape plan (► Sect. 6.2.3) and
3. The identification of ES in extensively used grassland rich in species (so-called High Nature Value-Farmland, HNV; ► Sect. 6.2.4)

6.2.2 Agri-Environmental Measures: The AEMBAC Methodology

O. Bastian

To maintain or enhance biodiversity, ES and sustainability of agro-ecosystems, the European Union provides incentives for environmentally friendly

Fig. 6.2 Spatial location of the case studies. The number in brackets refers to the section where they are discussed. © IÖR/Grunewald and Witschas

farming. The Common Agricultural Policy of the EU (CAP) consists of two main pillars: Pillar 1 includes direct payments to the farmers (to support their income under the condition that they respect minimum requirements of environmental compatibility, so-called Cross-Compliance rules). Pillar 2 aims to improve the competitiveness of agriculture and forestry, biodiversity, environment and landscape, and the living conditions in rural areas.

Payment mechanisms are set into force, so-called Payments for Ecosystem Services (PES), for which well-defined services are performed (directly or indirectly) on a voluntary basis against paying a defined monetary amount.

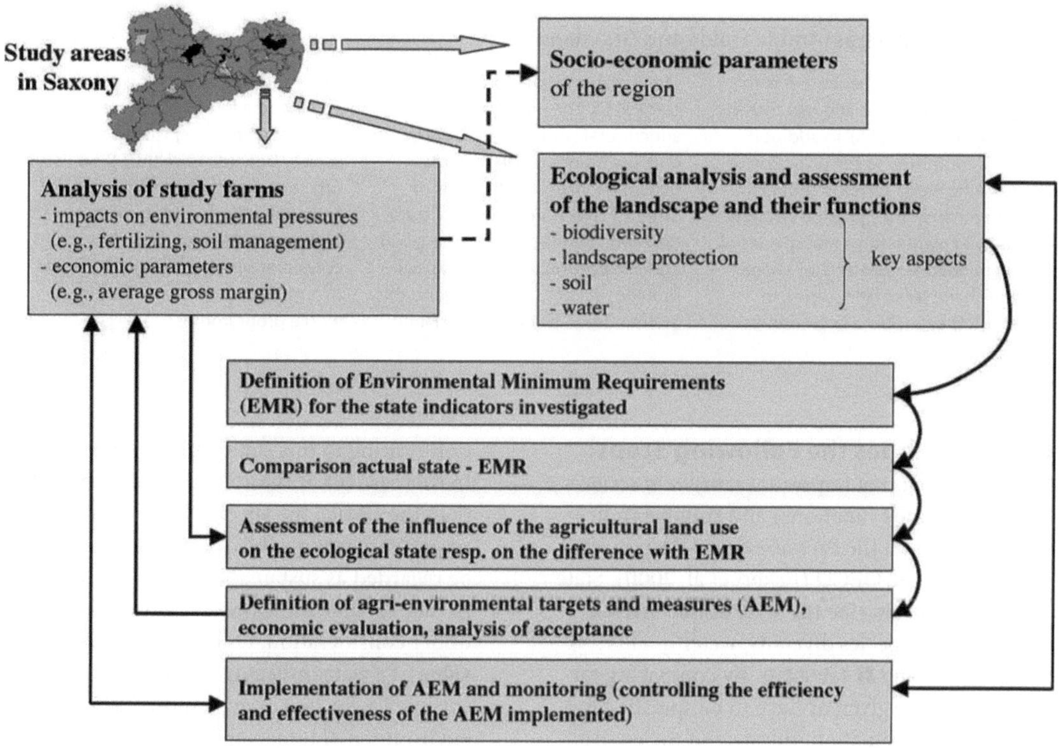

Fig. 6.3 Working steps of the AEMBAC methodology on the example of study areas in Saxony

Agri-environmental programmes include a wide range of measures to improve the ecological situation in agricultural areas and, finally, the maintenance and enhancement of biodiversity and ES. For example, the conservation soil tillage shall reduce soil erosion, and the use of meadows according to nature conservation viewpoints shall maintain and increase the diversity of species in grassland ecosystems. As Plieninger and Schleyer (2010) argue, however, the specific ES to be delivered are mostly not defined clearly.

Apart from the fact that agri-environmental programmes hardly refer to ES, their reference to landscape units is also poor, i.e. the regional peculiarities and requirements are not taken into consideration enough. To overcome such deficits, the AEMBAC methodology provides a promising approach that was developed in the framework of an EU-project and tested in several European countries, including Germany (AEMBAC=*Definition of a common European framework for the development of local agri-environmental programmes for biodiversity and landscape conservation*) (Bastian et al. 2003, 2005, 2007; Lütz et al. 2006).

The consideration of regional/local peculiarities or the character of an area and the consideration of existing ecosystem properties, potentials and functions (or ES) belongs to the key points of the AEMBAC methodology. It can be divided into three phases (Fig. 6.3):

- Phase I: Assessment of the ecological capacity of the agricultural landscape based on various landscape functions (or ES, with the main areas of focus 'biodiversity', 'scenery', 'soils', 'waters'), analysis of positive and negative environmental impacts and assessment of the ecological sustainability of the current agricultural production in the study areas
- Phase II: Identification of local agri-environmental measures
- Phase III: Agreement of the suggestions with farmers and authorities

Agricultural Areas (Arable Fields and Grassland)

Agricultural areas' main task is the production of food and raw materials (provisioning ES). In addition, they contribute to the provision of drinking water through groundwater recharge, they provide habitats for wild plant and animal species of unwooded areas, and they shape the character of landscapes. In short, they supply a large number of provision, regulation, and socio cultural ES. The management of agricultural lands has to avoid the one-sided orientation towards maximum yields on the costs of other ES. The 'normal' level of these demands is prescribed by the 'good agricultural practice'. If the farmer provides services going beyond this 'normal' level, he can make claims to compensation for reduced yields and income losses. These claims against society are justified as society benefits from the ES. Society can stimulate farmers by financial means that they voluntarily meet higher requirements of nature conservation. This is done by means of agri-environmental measures or whole programmes.

Phase I Includes the Following Steps:

1. Identification of important *ecosystem services* (or landscape functions) and suitable indicators based on the Pressure-State-Response-Model of the OECD (Eckert et al. 2000). State indicators describe the state of the environment (e.g. species diversity, water quality; ◘ Fig. 6.4 and ◘ Table 6.1). By comparing targets that are given or have to be specified for the specific situation, the state of the environment can be assessed (◘ Table 6.2). Pressure indicators address risks for the environment and the reasons for them (e.g. N-balance, application of fertilisers and biocides, nitrate loads of groundwater, disposition for erosion, crop diversity, size of field-plots, crop rotation, methods of livestock breeding). Response indicators address the consequences society and politics are ready to bear to improve the given situation.
2. Definition of *Environmental Minimum Requirements (EMR)* for selected indicators with respect to the maintenance of ecosystem functioning (including agro-ecosystems) to facilitate the definition of agri-environmental targets and measures. EMR are referring to the carrying capacity of a landscape.
   An EMR is a single value (a threshold), a range, or a set of values of a state indicator that is assumed to be sufficient for the satisfactory performance of the landscape function analysed. If the actual value of a state indicator achieves its EMR, no impacts (either positive or negative) on the particular landscape function relating to this state indicator are detected. In this case (or, if the actual state is even better than the EMR), the specific land-use practice or measure responsible for this situation can be regarded as sustainable in relation to the particular state indicator or landscape function under consideration. One and the same EMR value of a state indicator can be applied for a specific ES for two or even more ES. EMR or environmental targets for agricultural landscapes are listed in literature (e.g. Breitschuh et al. 2000; Knickel et al. 2001). Several targets are also written into laws (e.g. nature conservation acts) and the different instruments of spatial planning (e.g. regional plans, land-use plans), mostly following a process of political consideration. The definition of local EMR should be oriented on specific–first ecologically justified–EMR tailored for the particular areas. The so-called good agricultural practice is not identical with EMR. The current good agricultural practice may well cause ecological damages and violate the principles of sustainability.
3. Analysis and evaluation of (negative but also positive) *environmental impacts* caused by agriculture. It is interesting whether an agricultural system or a measure (in form of a pressure indicator) impairs one or several ES. Based on this analysis, priorities can be set towards negative/positive influences (from agricultural practices), which have to be curbed or encouraged urgently.

Landscape function and state indicators / Pressure indicators	**Farm structure**	Median field size	Proportion plant / animal production	**Plant production**	Crop species diversity	Cereals in crop rotation	Rape in crop rotation	Root crops in crop rotation	Maize (corn) for silage	Legumes and grass mixture	Plant protection	Share and type of fallow fields	**Livestock**	Cattle (units / ha)	Forms of housing	**Energy and nutrient management**	Humus balance	Nitrogen balance
Habitat function																		
Biotope value			-									+		-				-
Vegetation and Flora		-	0		+	+	-	+	-	+	-	+					0	-
Fauna		-	0		+	+	-	+	-	+	0	+			+		0	0
Sustainability of land use	**NS**	**-2**	**0**	**NS**	**+1**	**+1**	**-1**	**+1**	**-2**	**+1**	**-1**	**+1**	**S**	**0**	**+1**	**NS**	**0**	**-2**
Landscape conservation																		
Linear biotopes		-																
Landscape diversity		0	+															
Cultural landscape elements		?																
Recreation value		0	+		0		0								0			
Sustainability of land use	**S**	**-1**	**+1**	**S**	**0**		**0**						**S**		**0**	**A**		
Soil function																		
Biotic yield potential (productivity of the site)		-	0		0	+	+	0	-	-		0		+	+		0	0
Erosion		-	0		0	+	+	0	-	-		0		0	0		0	
Transfer of nutrients and biocides		-			0	+	+	0	-	-	-	-		0	0		0	0
Sustainability of land use	**NS**	**-1**	**0**	**S**	**0**	**+1**	**+1**	**0**	**-1**	**-1**	**-1**	**0**	**S**	**0**	**+1**	**S**	**0**	**0**
Water function																		
Water quality		-			0	+	+	0	-	-	-	-		0	0		0	-
Water body morphology			-															
Sustainability of land use	**NS**	**-1**	**-1**	**S**	**0**	**+1**	**+1**	**0**	**-1**	**-1**	**-1**	**0**	**S**	**0**	**0**	**S**	**0**	**-1**

Sustainability with regard to the realization of EMR

- S sustainable
- NS not sustainable
- A no or very low influence ("absent")

Impact of the pressure indicator on the state indicator

- no impact
- low impact
- medium to high impact
- very high impact
- ? unknown

Effect (related to the realization of the EMR)

- \- negative effect
- 0 realization of the EMR
- \+ positive effect

Fig. 6.4 Ecological impact matrix: comparison of pressure and state indicators (or landscape functions/ES) and assessment of sustainability

4. Assessment of the *carrying* capacity of the agro-ecosystems/the landscape against agricultural activities and evaluation of their sustainability, based on the comparison (deficit analysis) between the current state (state and pressure indicators) and the EMR (Fig. 6.4). Statements about sustainability aim to assess whether the current land-use practices are long-term compatible for nature and society or whether alterations are necessary, e.g. by the implementation of agri-environmental measures. Sustainability in this context means the long-term maintenance of agricultural production (provisioning services) without hampering other ES of the agro-ecosystem or of adjacent ecosystems, i.e. production and the maintenance of the environment are not considered as contradictions. In agro-ecosystems, assessment standards that only consider the natural environment are equally questionable as such standards which only count the economic success. The comparison between the current state and Environmental Minimum Requirements reveals deficits, which should be remedied by agri-environmental measures. If the current state is in line with the EMR value, no measure is necessary.

Table 6.1 Blending single action recommendations to a multifunctional package of actions in the study area "Jahna" (Central Saxonian Loess Area; Agreement of measures for several state indicators, avoiding reduncancies)

Measures			Quantity (ha)
Bufferung through grassland stripes: – Near-natural, valuable floodplains – Near-natural, wooded wet biotope complexes – Existing wood structures including orchards			265
Establishment and management of orchards on arable fields			580
Establishment of linear or areal woods in accordance with the potential natural vegetation			1000
Conversion of arable land to extensively used permanent grassland in floodplains and on arable land (hollows) endangered by erosion			3500
Establishment of permanent field margins (without plant protection, N-fertilisers, and with wider plant spacing)			250
Introduction of organic farming			2000
Conservation soil cultivation (including 1000 ha organic farming)			6500
Reduction of fertiliser impacts per area (further investigation on current and desired state necessary)			–
Comparison of current and desired state			
	Area (ha)	Area (%)	
Arable land (status quo)	19,770	100	
Arable land (target)	14,425	73	including 250 ha field margins, 2000 ha organic farming, 5500 ha conservation soil cultivation (assumption that in organic farming applies conservation soil cultivation)
Grassland (status quo)	1751	100	
Grassland (target)	5516	315	
Conversion	1570	8 % (of arable land)	Withdrawal of arable land from agricultural prodcution for the plantation of coppices and orchards

Phase II Includes the Following Steps:

1. *Definition of local agri-environmental targets* basing on minimum requirements (EMR from phase I) and socio-economic conditions in the study areas.
2. Definition of the *most suitable EMR* to achieve these targets.
   One should be aware, however, that it is possible to identify the need for action and the content of agri-environmental measures (AEM) leading to more sustainability in the agricultural landscape. A scientifically valid quantification of the extent of an AEM for a specific area, however, seems to be almost impossible. Therefore, it seems appropriate to define grey areas and safety margins, which have to operate without secured knowledge, if–from a nature conservation point of view–with sufficient probability inacceptable impairments may be expected after the limit zone is reached or surpassed (precautionary principle) (Dröschmeister 1998). Priority should be given

Table 6.2 Ecological demands of the landscape plan for the rural municipality Promnitztal and their possible implementation by agricultural measures

	Requirements of landscape plan	Modelled measures
Arable fields	General decrease of production intensity	Reduction of nitrogen fertilisers and biocides by 20 %
Field margins	Establishment on all plots of land	No nitrogen fertilisers and biocides
Buffer zones	Establishment around valuable biotopes (running waters, woods, wet areas)	Natural succession or planted woods on 50 % of all buffer areas, reduction of nitrogen fertilisers and biocides by 40 % on the remaining 50 % of buffer areas
Protection of reptiles	Reduction of land-use intensity at some places	Arable fields: reduction of nitrogen and biocide application by 40 % grassland: development of rough grassland, renunciation of nitrogen and biocide application
Grassland	General decrease of production intensity	Reduction of nitrogen and biocide application by 20 %, mowing pasture (two cuts + pasturing)
Planting of woods	Development of hedges typical for the area as well as rich-structured forest edges	Calculation of income-losses due to land abandonment, consideration of positive influences of hedges on yields

to such measures, which counteract especially serious environmental loads, while improving several ES (e.g. increasing biological diversity + the aesthetic value of the landscape), and remaining financially feasible. An intelligent selection of AEM (e.g. the plantation of hedgerows) may influence several indicators or ES positively at the same time. The definition of specific AEM (disclosed separately according to single indicators or ES) may also lead to redundancies, partly even to an oversized claiming of agricultural area for nature conservation and landscape management. Therefore, it is important to blend the single proposals to a package of action purged from redundancies. These measures have to be located within the study areas (Table 6.1). An action plan results from this, which, first, incorporates only environmental considerations. It is an intermediate step, which has to be evaluated in monetary terms during a following working step (phase III) and which is provided to the farmers and other stakeholders to check their acceptance.

3. **Proposition of legal and/or economic stimuli:** Assessing the applicability of economic (market-oriented), legal and controlling tools as incentives for farmers to maintain environmental goods and to give up unsustainable agricultural practices.

Phase III Includes the Following Tasks:

1. Analysis of the *acceptance and feasibility* of the proposed agri-environmental measures (AEM) by farmers and authorities
2. Calculation of the *costs for the authorities* (implementation of agri-environmental programmes) and farmers (economic aspects, gross margins)
3. Overall evaluation of all *economic or financial aspects of the implementation of AEM.*

Through interviews with farmers and authorities measures are identified, which may contribute especially effectively to improving the state of the environment. Obstacles are revealed, and–if necessary–corrections or alternatives for certain measures are identified (participatory approach).

Checklists serve to assess the economic efficiency and applicability of the suggested measures.

Conclusion

The AEMBAC methodology is in general suitable to analyse agri-environmental problems comprehensively, to link ecological and socio-economic aspects by referring to environmental functions or ES, and to show and strengthen the relation between the natural environment and human well-being. The approach may contribute to maintain the diversity of rural regions and to support the competitiveness of agriculture against the background of the increasing liberalization and globalization of agricultural markets. Although AMEBAC was developed for agricultural areas, in principle it can also be applied to other economic branches, e.g. to forestry and fishery.

6.2.3 Agro-economic Evaluation of Landscape Plans

O. Bastian

For the implementation of proposed agri-environmental or nature conservation measures it is necessary to determine the costs and to reach a consensus between the involved stakeholders.

For the integration of nature conservation objectives into land use, the instrument of landscape planning can be used. As the guiding planning of space-related environmental protection, landscape planning pools the various specific activities and specialist contributions of environmental protection and nature conservation. The present practical implementation of landscape plans in rural regions, however, cannot satisfy (► Sect. 5.3). One point of criticism is that proposed measures are often reasoned only from a nature conservation point of view without taking the interests and economic capacities of land users into account (Geisler 1995; Marschall 1998).

Thus, it may be useful to underpin landscape plans (and protected areas concepts) by economic evaluations. In the following section, an investigation on the example of the (former) rural community Promnitztal (State of Saxony, Germany) will be presented, to which extent the nature conservation objectives and stipulations of a landscape plan are realistic, i.e. if and under which preconditions the agricultural enterprises would be able to meet the demands.

For this purpose, an agro-economic evaluation of the measures laid down in the landscape plan was performed. Taking the economic frame conditions into consideration, and basing on the landscape plan, concepts for a nature-compatible and economically sustainable land use were elaborated. It was shown that the integration of nature conservation objectives into the agricultural production may be reasonable also from an economic point of view (Lütz and Bastian 2000, 2002).

The *project area*, the former rural community Promnitztal–independent until 31 December 1998, now part of the small town Radeburg–is immediately north and adjacent to the Saxon state capital Dresden and covers c. 2085 ha. The territory of Promnitztal is almost totally protected (landscape protection area 'Moritzburg small-hill landscape') and belongs–from a physical-geographical point of view–to the Western Lusatian Hills and Low Mountain Range. The 'Moritzburg small-hill landscape' is characterised by a small-scale pattern of small hills and low ridges with exposed rocks and flat hollows. The bedrock is dominated by monzonits, but granodiorite, sandy and holocene substrates also occur. The basic geomorphological pattern causes a high diversity of soil, water and climatic conditions which is responsible for the present vegetation cover and land use. Effective agricultural production is hampered by complicated natural site conditions. Forests and woods are concentrated on the crests of the rocky and stony hills, arable fields on slopes and grassland in moist hollows. Land improvements (especially drainage) have tried to diminish this natural heterogeneity but with little success. Drainage facilities fell into disrepair after a few years, and the thin soil cover on the hills is an insuperable obstacle for ploughing. The result is a rich-structured rural landscape with a notably high biodiversity and interesting scenery. The area is particularly rich in species which are adapted to less intensive agriculture, e.g. rare arable weeds, plants of field margins, edges and small coppices, birds breeding in hedges, woods, grassland and arable fields; amphibians, reptiles and many insect species (Neef 1962; Mannsfeld 1972; Bastian and Schrack 1997; Schrack 2008).

During the investigation period (1999) six full-time and four part-time farmers had fields within the study area. Mainly market crops and fodder plants (for milk production) were cultivated. The proportion of crops between 1996 and 1998 was as follows: 55 % cereals (winter rye, winter wheat, winter barley, oats), 20 % oilseeds (sunflowers, rape, flax) and 25 % field-fodder plants (maize). The proportion and pattern of crops were determined by natural conditions, variations in crop rotations, the need for animal food, and environmental restrictions through the programme "Agriculture harmless to the environment" of the Federal State of Saxony. The latter compensates for environmental measures in the cultural landscape, financial compensations for economically deprived areas, subsidies and actual market prices. The predominating crop rotations in the 1990s were as follows: winter wheat–winter barley–rape–winter rye or maize on soils better provided with water and nutrients; winter rape or maize–cereals (no wheat) on hills and near cowsheds.

▪ Methods

To evaluate the effects of the nature conservation measures proposed or demanded by the landscape plan, a business management analysis in selected enterprises was carried out (◘ Fig. 6.5). The standard variable margin (=gross margin) per hectare was used as the basis for the evaluation of arable crops. This was done by comparing the inputs and the outputs of each production method. Thus, the variable margin is: Agricultural yield (sum of market prices, subsidies from the EU including agri-environmental incentives, and compensations for deprived areas), less costs of production (seeds, fertilisers, biocides, costs for machines and human labour). The variable margin was established for each crop for three consecutive years (between 1995 and 1998) and generalised to the average variable margin per hectare over the 3 years. The evaluation for grassland and maize was not carried out in monetary terms as for the arable crops, but with the method of fodder supply of metabolizable energy ('net-energy lactation').

The demands and restrictions of the existing landscape plan (◘ Table 6.2) were incorporated into the variable margins and fodder evaluations. The consequences of the landscape plan implementation were expressed by the difference from the initial situation. The losses in yields were calculated according to Zeddies et al. (1997; the production function), and by comparisons with data from literature (e.g. Diercks and Heitefuß 1990; Mährlein 1993).

A constraint assumption in the study was that the sum of energy supply from the basic fodder areas (maize, grassland) should stay constant, i.e. the stock of cattle and the milk production should not be affected by these measures. To compensate the reduced yields, the increase of maize fields was assumed (3x higher fodder production than on grassland). Agricultural subsidies were considered as well (◘ Table 6.3).

▪ Economic Evaluation of the Landscape Plan

The application of all the proposed measures led to an increase of the variable margin on arable fields. The reason for this was the high level of compensation payments provided for arable field margins (strips at the margins of arable fields which are not treated with chemicals to favour the development of a rich community of arable weeds). The average variable margin of arable fields increased in this model from 104 €/ha (20 %) up to 629 €/ha (◘ Table 6.3).

These positive influences on the variable margin, however, were balanced by other demands of the landscape plan: creation of hedges, forests and forest edges, areas for natural successions, buffer zones, grassland and revitalization of waters. To compensate for the measures of extensification (especially, the reduction of utilization intensity) 69 ha of arable fields were needed for additional field-fodder cultivation (maize). The effect of splitting these area losses on the variable margin was an annual monetary loss of 59,465 € totally or 131 €/ha of remaining arable fields. The average variable margin, therefore, decreased by 26 €/ha (5 %) to 498 €/ha.

The calculations showed that an almost income neutral realization of the landscape plan was possi-

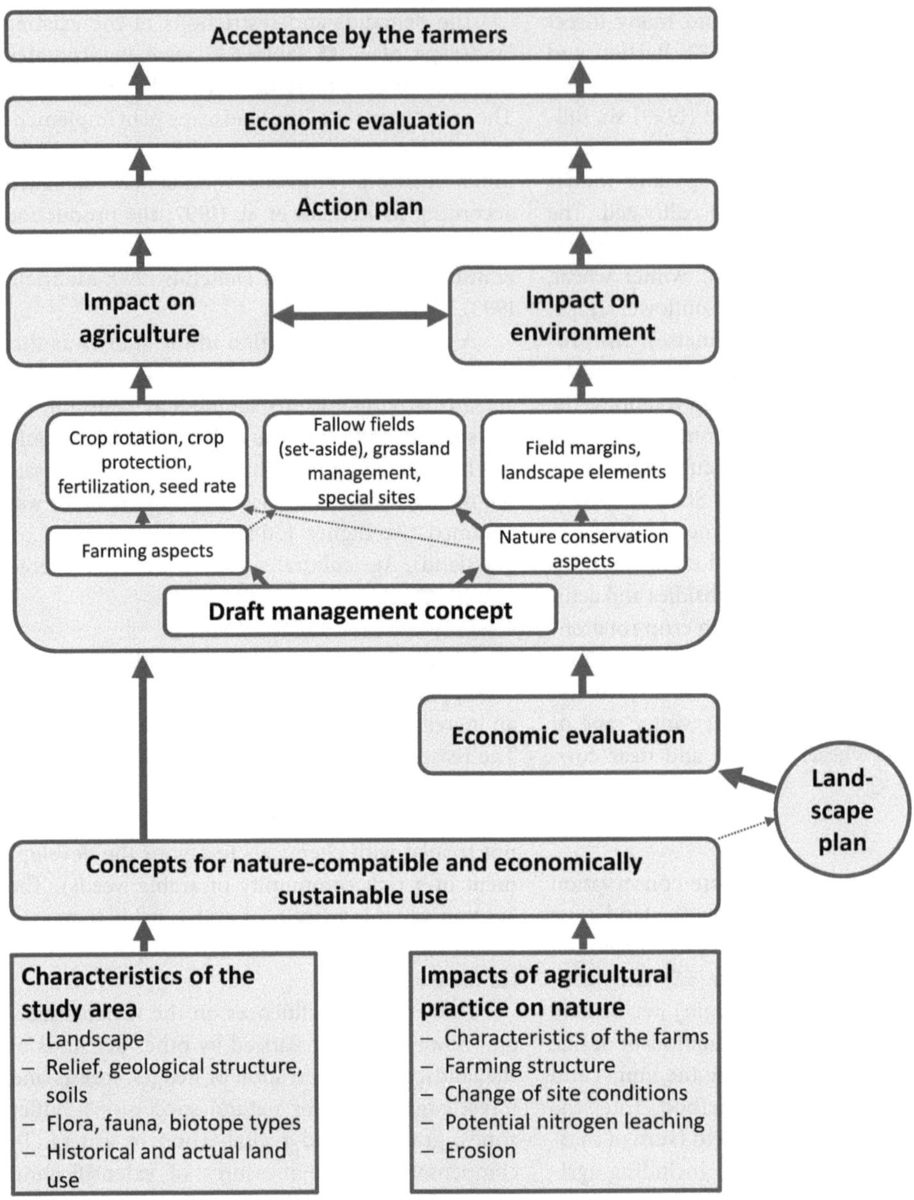

Fig. 6.5 Steps of economic evaluation of the landscape plan

ble (Table 6.4). In order to compensate for financial losses (which are not acceptable from the farmers' point of view), it was necessary to reduce the 49.7 ha of land removed from agricultural production in favour of hedges and other ecological measures as proposed in the landscape plan to only 45.1 ha. This means that by this calculation 5.8 % of the agricultural area could be withdrawn from cultivation without negative influences on the income of the farmers. This fact contradicted the usually poor perception of landscape plans (not only) among farmers.

▪ **The Acceptance of the Farmers**

The resulting new management concept was presented as a whole to farm-managers for examination, and the acceptability of the proposed mea-

Table 6.3 Economic effects of the demands set up in the landscape plan Promnitztal

	Measures area (ha)	Benefit/loss (€ ha^{-1})	Share of the variable margin (%)
Arable fields (without maize) (N-fertiliser and biocide reduction by 20 %)	373.9	−3.9	−0.2
Buffer zones and 'reptile protection area' (1.4 ha) (N-fertiliser and biocide reduction by 40 %)	17,1 (21 km × 18 m)	−18.3	−3.5
Field margin (without maize plots) (total renunciation of N-fertilisers and biocides)	81,7 (45 km × 18 m)	+389.7	+74.2
Average variable margin of arable fields	–	+104.3	+20.0
Losses of area Direct losses: 49.7 ha Reduction of intensity (indirect losses): 69.1 ha	–	−130.9	–
Average total variable margin	–	−27.1	−5.0

Table 6.4 Economic evaluation of the measures (selection)

Measure	Evaluation parameters	Gain/loss (€)
Intercropping and undersown crops	Subsidies and nitrogen saving–Costs of the measures – autumn catch crops – winter catch crops – undersown crops	 +61 bis +82 −31 bis +37 +5 bis +26
Less application of N-fertilisers in farming	Subsidies and nitrogen saving–Changes in yields	+8
Field margins	Subsidies and nitrogen saving–Changes in yields – normal sowing density – reduced sowing density	 +383 +547
Fallow-land	Subsidies–gross margin – temporal set-aside – permanent set-aside	 −97 Gain on poor sites
Grassland	Subsidies and cost saving–Loss of arable land for fodder cultivation – no use of synthetic N-fertilisers – extensive pasture – extensive meadow	 −26 +41 +56
Plantation of hedge-rows	Losses in arable land–higher yields by wind-breaks – yearly effect per 100 m hedge (width: 10 m)	 −46

sures was ascertained. The attitude of farmers to the proposed concepts was not based only on economic aspects. Of course, their receptiveness was greater for such measures when compensating programmes supported them. If agricultural land was demanded irreversibly (e.g. for woods), the farmers' attitude was less favourable.

Conclusion

The existing incentives of the agri-environmental programmes can be regarded as the reason for the positive economic balance. As in the period of investigation the farmers did not use all incentives consequently despite potential income gains; there was a favourable constellation for the assessment of the landscape plan in terms of its economic feasibility. Thus, agri-environmental programmes may be effective mechanisms to integrate objectives of nature conservation/environmental protection into the agricultural land management, because farmers receive compensation payments from society for income losses caused by an increased supply of (regulation and sociocultural) ES.

Supported by stated subsidies, deprived agricultural regions may maintain nature-compatible agriculture. This is all the more important as such areas are often very valuable for nature conservation. Farmers, however, are less willing to participate in agri-environmental programmes and to manage their land in the sense of multifunctionality and ES, the higher the chances of income generation from market crops (fertile soils) and/or higher demands, e.g. through the boom of energy crops, if–at the same time–the financial allocation of the environmental programmes stagnate or decrease.

Only the monetary evaluation of the measures and the following discussion with the farmers enabled the implementation of selected measures by better participation in agri-environmental programmes. Although subjective reasons played a role in the decisions of farmers, they assessed the proposed measures mainly from an economic perspective.

The growing interest of society in multifunctional and ecologically intact (agricultural) landscapes should extend the spectrum of tasks and the responsibility of land users in future significantly (Vos and Meekes 1999). In view of the partially low ecological efficiency, the high administration expense and the low allocation of financial means in the existing subsidy programmes, the financial means should be moved to defined benefit plans for environmental protection in agriculture (Bronner et al. 1997). Conversely, environmental damages caused by land-use practices, which are not in line with the professional standards, should lead more to monetary consequences for those who cause the damages (Simoncini 1998). Finally, land-use forms, in which the maintenance of ecosystems and their manifold services and sustainable, resource-economical production are an inherent part should be identified.

6.2.4 Species-Rich Grassland Services

M. Reutter and B. Matzdorf

Introduction

Internationally, the stated objective is to halt the loss in biodiversity (European Commission 2011). One of the consequences for agricultural landscapes is that species linked to agricultural use that are typical for European cultivated landscapes are to be maintained (BMU 2007). Species-rich grassland plays a crucial role in this context (EEA 2004; BfN 2008; BMU 2010a). According to the latest status review of *high nature value* (HNV) farmland, species-rich grassland accounts for some 16.8 % of the total grassland area in Germany (BfN 2010), corresponding to an area of approximately 1 million hectares (with regard to ATKIS, data basis for the status review) or 0.8 million hectares (as regards the statistical data) (◘ Fig. 6.6 and 6.7). Based on the definition in the context of HNV mapping, species-rich grasslands are extensively used grassland expressions that are particularly species-rich by regional standards (BfN 2008). The abundance of species is identified via indicator species in terrestrial mapping. This type of identification is already used in a number of federal states (Briemle and Oppermann 2003; Keienburg et al. 2006; Matzdorf et al. 2008). The method was primarily devised for mesophilic, moist and moderately dry grassland; in this way, all species-rich grassland areas and those that are valuable from a nature conservation perspective are captured together with FFH habitat types (BfN 2009).

◘ Figure 6.7 shows the distribution of species-rich grassland in Germany as a result of HNV mapping. The spatial basis of the survey is site-specific space structuring (BfN 2004). The percentages determined range from 5 to 30 % within the spatial units. In this connection, regions rich in grassland are not always those with a high proportion of species-rich grassland. The relatively small proportions in grassland-rich regions in the northern part of Lower Saxony and in the foothills of the Alps are striking in this respect. The uplands of southern and central Germany exhibit particularly high proportions.

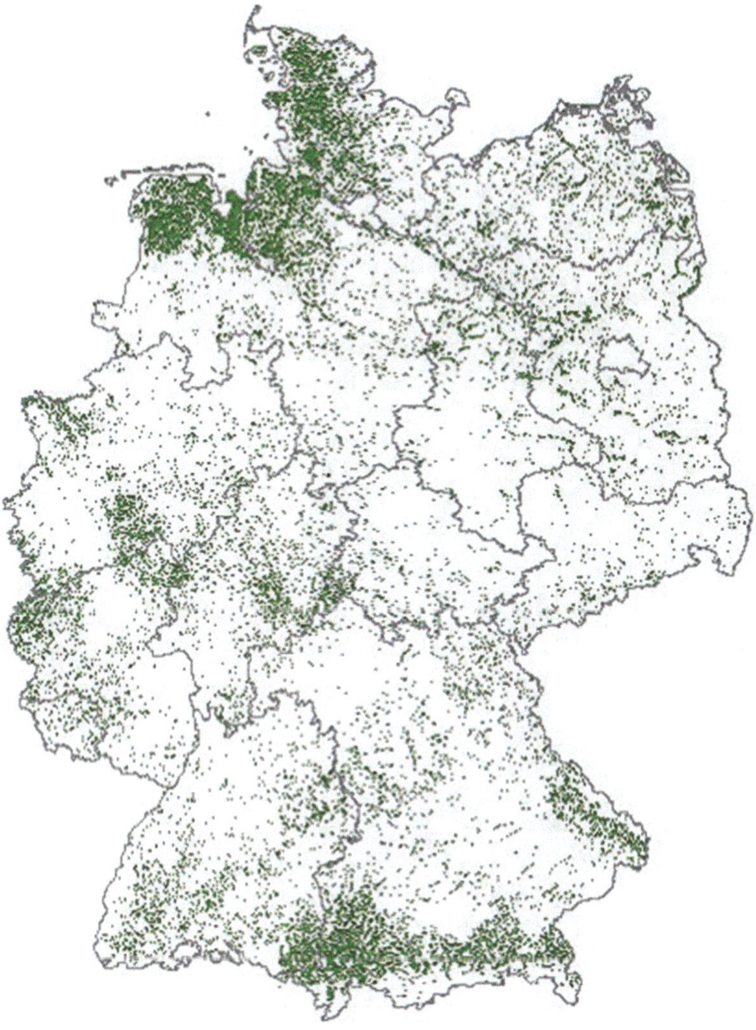

Fig. 6.6 Grassland (green) in Germany (authors' design, data basis CORINE 2000)

Irrespective of the current proportion of valuable grassland, the aim in all regions is at least to maintain it. Ultimately, however, the goal is not only to preserve species diversity of the agricultural landscape, but to achieve a trend reversal. One of the measures that seeks to achieve this is to increase the total percentage of HNV farmland from the current level of 13 to 19 % (BMU 2010a). Against this backdrop, it is also interesting to address potential synergies with regard to other environmental objectives, not least to provide arguments in favour of maintaining biodiversity in connection with the financial resources required to this end.

The aim of the following article is to highlight the services provided by HNV grassland areas to human beings. In particular, the monetary assessment of individual services is explored. If the concept of ES is applied in the case of cultivated landscape, account must be taken of the fact that many such services are not merely ES. It goes without saying that ecosystem processes are necessary for the species diversity of grassland, in addition to human services. In Germany, grassland is predominantly not a pure 'natural product'–it only arises and exists on account of the human impact on natural, ecosystem processes.

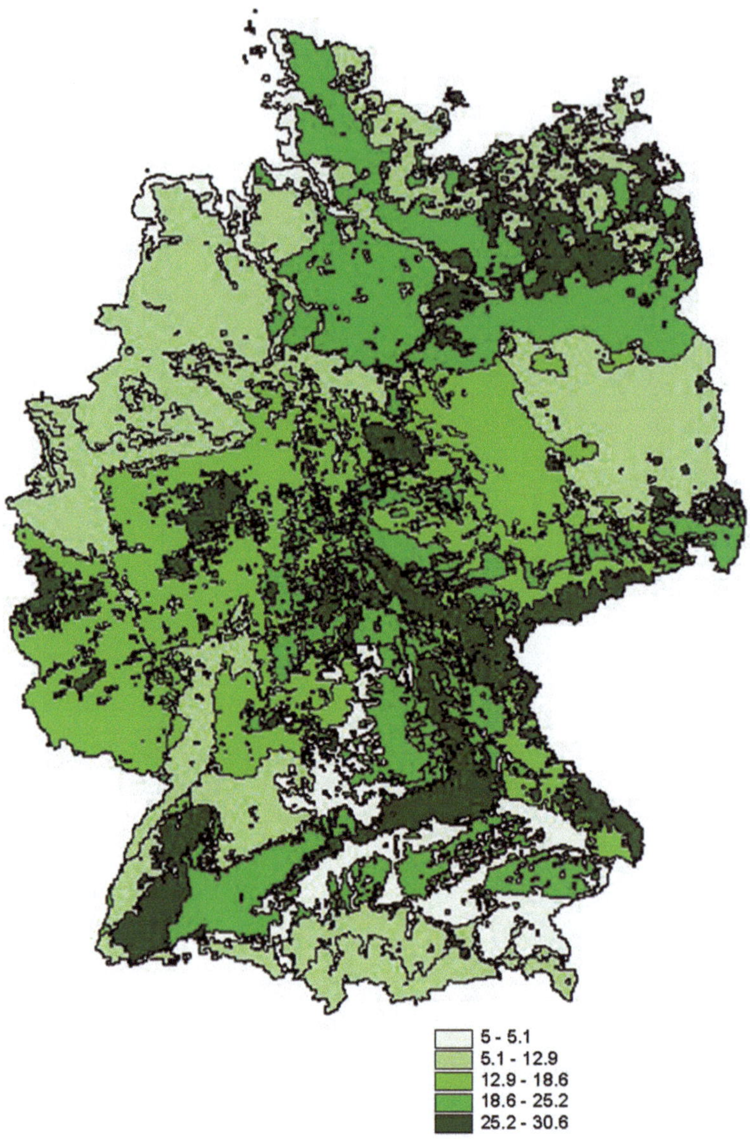

Fig. 6.7 Species-rich grassland as a percentage of all grassland (authors' design on the basis of site-specific space structuring, BfN 2004, and data from initial mapping of HNV farmland, BfN 2010)

Selection of Services and Assessment Approach

The following two criteria were taken into consideration in the selection of services: which services are produced by species-rich grassland, and for which services there is a specific demand.

In accordance with the Methodological Convention of the Federal Environment Agency, agreed targets are an expression of a socio-political preference (UBA 2007), and hence an expression of such a demand. We regard the implementation of the Water Framework Directive and the implementation of the Kyoto Protocol to be two key socially and economically relevant target agreements. Agricultural use plays a role in both cases (BMU 2010b; UBA 2011). Social relevance is highlighted by the fact that "improving the status of groundwater and surface waters" and 'climate change' are considered important challenges besides the objective of maintaining and developing biodiversity when it comes to the allocation of financial resources from the second pillar of agricultural policy (The Council of the European Union 2009). These objectives are ideal for highlighting synergies generated by the preservation of species-rich grassland.

If one assumes, given the current pressure on agricultural land (Lind et al. 2008; Nitsch et al. 2012), that species-rich areas would possibly be used as intensive grassland or arable land if no additional protective measures were taken, then this would be likely to result in a potential increase in water- and climate-relevant emissions, based on the current state of knowledge (Kühbauch 1995; Kersebaum et al. 2006; Osterburg et al. 2007; UBA 2010). This can have a negative impact on achieving the objectives mentioned above; conversely, the preservation of species-rich areas can have a positive effect. Owing to the current high level of importance of these potential services, these two ES ('groundwater protection' and 'climate protection') were selected with the aim of determining their value below.

Note that these services can only be recorded if a more intensive land use is assumed for reference purposes. The assumption that all species-rich grassland would be intensively used or converted into arable land does not appear to be realistic under the present conditions. In the context of this analysis, therefore, a more conservative scenario was taken as an example in which it was assumed that around 50 % of species-rich grassland would be used as more intensive grassland and some 5 % would be converted into arable land. On the basis of information available about the spatial distribution of HNV grassland, a total of approximately 1 million hectares was specified.

The economic value of the climate-relevant emissions that could potentially be avoided is balanced on the basis of damage costs and set as opportunity costs and market values in the comparison with avoidance costs. An avoidance cost approach was chosen as the method used to quantify the services for groundwater. Whilst climate-relevant emissions involve the avoidance of any emissions regardless of the geography, the demand for avoiding emissions under the Water Framework Directive is dependent on the specific location of the areas. With regard to the implementation of the Water Framework Directive, social preference only exists, strictly speaking, in areas in which the emissions generated would endanger the good status of water bodies.

The physical supply (potential or capacity) determined in the first step serves as the basis for the economic assessment. The materials and methods used are explained in further detail in the respective sections below.

Quantifying Emissions Reduced due to Species-Rich Grassland

▪ Contribution to the Protection of Groundwater Bodies

▪▪ Capacity of Areas

Within the context of the Water Framework Directive, measures are to be taken in virtually all coordination areas in Germany to reduce diffuse inputs of substances for groundwater bodies (BMU 2010b). The results generated by Osterburg et al. (2007) are built upon in order to assess the service provided by species-rich grassland. These authors assessed different agriculturally relevant measures in terms of their potential for reducing nitrate in groundwater against the backdrop of the Water Framework Directive. Variabilities for the impact of extensive use, typical for species-rich grassland, were assessed at the level of Germany compared to intensive use or the impact of converting arable land to grassland. We use these values below to demonstrate the capacity of species-rich grassland for groundwater protection.

According to Osterburg et al. (2007), extensive use reduces the nitrogen load in the leachate by 0 to 20 kg ha^{-1} compared to intensive use. Abstaining from sward renovation with ploughing up and resowing, which would be necessary to maintain species-rich grassland, increases the value by 40 to 80 kg ha^{-1}. Compared to use as arable land, it is assessed that extensive grassland use decreases this value by 30 to 70 kg ha^{-1}.

If one assumes, based on these figures, that some 50 % of the estimated extent of species-rich grassland is intensively used and a further 5 % would be converted into arable land, then up to 13,500 t of

Table 6.5 The impact and costs of an alternative land use of species-rich grassland as well as an alternative measure to reduce the nitrogen load in the leachate. Data source: Osterburg et al. 2007

Impact and costs compared to extensive grassland use				
	Additional kg N ha^{-1}		Costs in € ha^{-1}	
Alternative land use	min.	max.	min.	max.
Intensive grassland use	0	20	80	150
Sward renovation	40	80	20	50
Use as arable land	30	70	370	600
Alternative for reducing the nitrogen load from utilised agriculture areas				
	Reduction in kg N ha^{-1}		Costs in € ha^{-1}	
	min.	max.	min.	max.
Catch crops as winter greening as opposed to winter fallow	25	50	40	120

additional nitrogen load could occur annually in the leachate or 40,000 additional tonnes for sward renovation associated with intensive grassland use.

Economic Value

Costs would be incurred if the additional emissions of 13,500 t of nitrogen in the leachate would have to be saved elsewhere. If, in turn, one assumes the range of measures described in Osterburg et al. (2007) is taken, it would make sense to grow catch crops with relatively good cost-effectiveness. Depending on the cost-effectiveness relation, costs of € 10.8 million (cost-effectiveness relation: 50 kg N per € 40) or costs of € 64.8 million (cost-effectiveness relation: 25 kg N per € 120) are yielded for the avoidance of 13,500 t of nitrogen. Table 6.5 shows that the preservation of extensive grassland use can by all means constitute a cost-effective avoidance of emissions, depending on the concrete situation.

The avoidance costs are particularly important in regions where contaminated groundwater bodies already exist. Figure 6.8 designates the chemical status of groundwater bodies in Germany, and shows where excessive levels of contamination already exist. It must be said, however, that the illustration is an overall assessment that does not explicitly reveal the necessity of a reduced nitrogen load in the leachate. No further correction of the values calculated above is undertaken.

If, however, nitrogen emissions are too high at present and mitigation measures are already necessary, then the prevention measures calculated above under the assumption of the lowest costs may under certain circumstances be considerably more expensive. Maintaining species-rich areas therefore becomes increasingly more profitable.

In *summary*, the value of species-rich grassland for water quality must be determined with full knowledge of the location and characteristics of areas, the load situation and other development objectives. However, the values show how important species-rich grassland is in implementing the Water Framework Directive (Fig. 6.8).

Contribution to Climate Protection

Capacity of Areas

Below we use a simplified method of climate reporting (area of land-use change; UBA 2010), taking into account the current distribution of species-rich grassland, to show how much CO_2 would be released if 5 % of such grassland was converted into arable land. The 5 % scenario was chosen because such ploughing up of grassland is permitted at the federal state level in the current CAP funding period. Federal states must only take action to prevent a further net loss if the 5 % level is exceeded.

Scientists are currently of the opinion that converting grassland into arable land leads to the release of CO_2, and hence to the loss of organic carbon (UBA 2010). The quantity released is dependent not only on use, but also on site factors such as the soil type, hydromorphy, plants and climate.

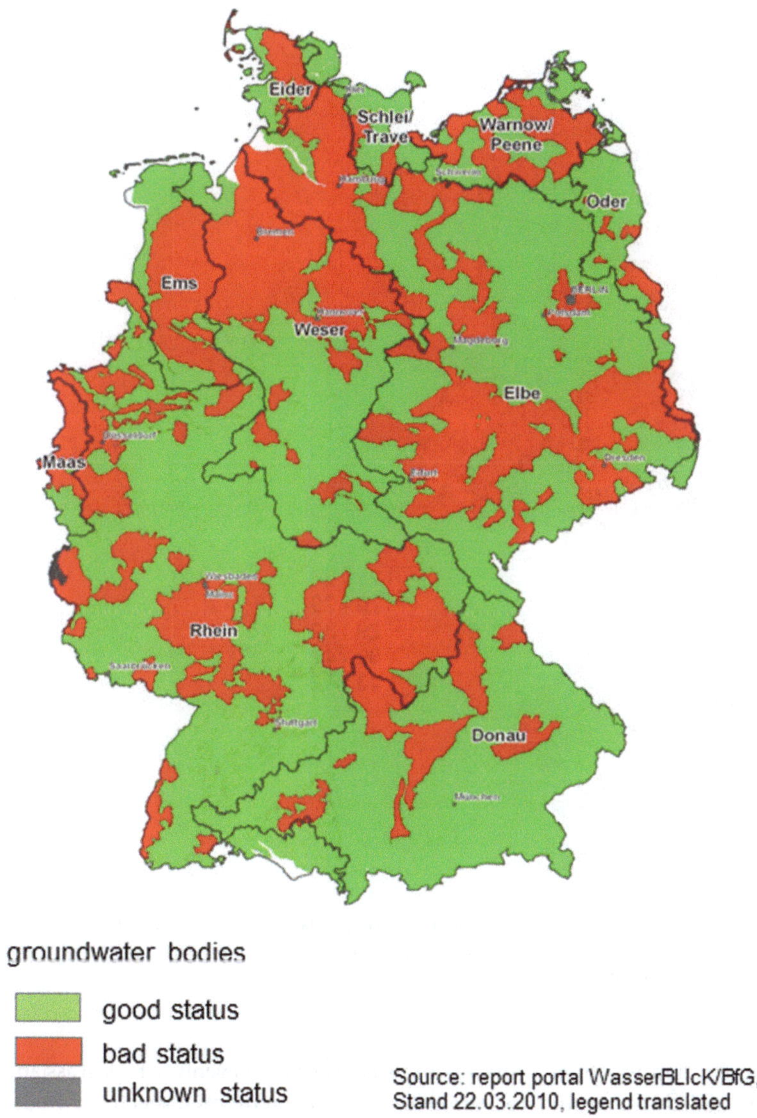

Fig. 6.8 Chemical status of groundwater bodies in Germany (the red areas indicate poor status). Source: BMU 2010b

In this connection, moor soils have the highest carbon stocks. If management is changed from grassland to arable land, it is assumed, so far with great uncertainty, that the annual release increases from an estimated value of approximately 5 to 11 t ha^{-1} (UBA 2010); for mineral soils, a loss of 30.43 % of carbon stocks is assumed following the conversion of grassland into arable land.

The key data basis for the calculation was the general soil map for Germany (BÜK 1000), which combines different types of location into soil associations. Each soil association is described in a guiding profile. On the assumption that these guiding profiles are considered to be representative, the data given was used to calculate site-differentiated carbon stocks. The C_{org} contents were multiplied by the respective crude densities and horizon thick-

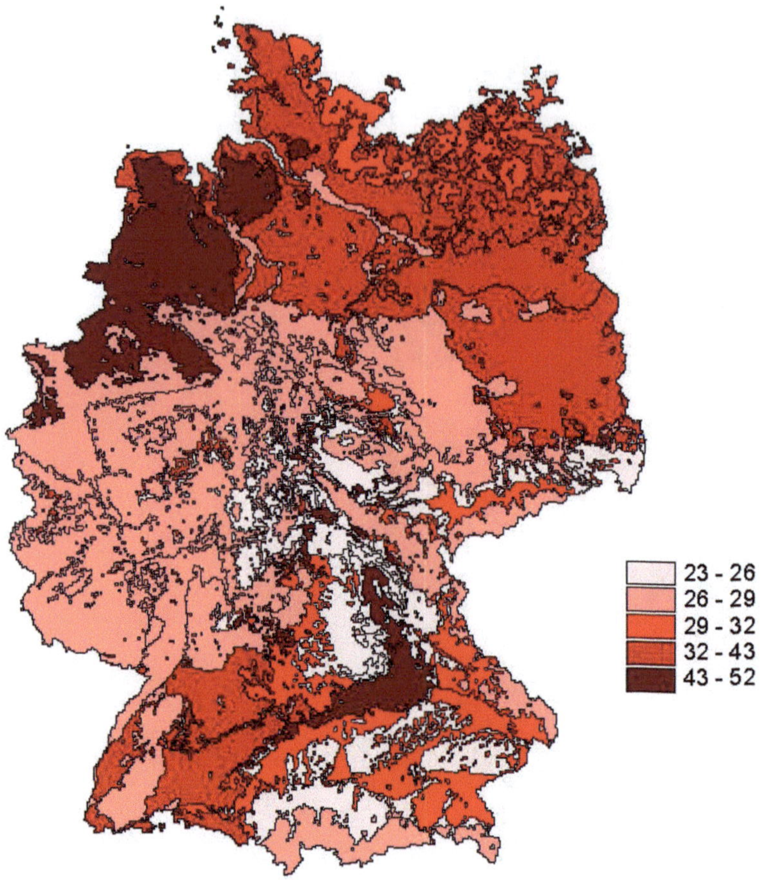

Fig. 6.9 Average loss of organic carbon stocks C_{org} (t ha^{-1}) in the conversion of grassland to arable land (authors' calculation on the basis of site-specific space structuring by the Federal Agency for Nature Conservation (BfN); area-weighted mean calculated from blending BÜK 1000 with CORINE 2000)

nesses, from which the skeleton fraction was then subtracted. Horizon stocks were added to a depth of up to 30 cm. In order to integrate the estimated value for moor soils mentioned above into the calculation, we used a period of 10 years. Carbon stocks were determined specifically for grassland areas by blending the general soil map with land-use data. Due to the high computational effort involved with ATKIS, the relation to land use was achieved using CORINE 2000.

Figure 6.9 shows the area-weighted, average loss of organic carbon stocks resulting from the calculation in the event of a conversion from grassland to arable land within site-specific spatial units. If one assumes that species-rich grassland within site-specific units is distributed evenly across grassland occurrence, a conversion of 5 % respectively would release approximately 6 million tonnes of CO_2.

It should be borne in mind that this is a simplified calculation method compared to the method involved in climate reporting. The result should be considered as an approximate value that nevertheless attempts to integrate the resulting highly decisive locational capacities. It is particularly difficult–but crucial–to identify moor locations. The German-wide general soil map is highly generalised. Emission levels and a better understanding of carbon stocks are currently being investigated scientifically in various projects. In particular the data basis and assumptions concerning release rates are to be improved further in connection with climate reporting (UBA 2010). On the other hand, there is a barrier because little specific location information exists for species-rich areas. The uniform distribution used within site-specific spatial units has not yet been verified. Consequently, the calculation was made based on the latest available information, but ought to be improved further (◘ Fig. 6.9).

▪▪ Economic Value

If damage costs of € 70 per tonne of CO_2 equivalent (CO_2 eq.) published in the Methodological Convention of the Federal Environment Agency (UBA 2007) to estimate external environmental costs are assumed, then the calculated conversion of 5 % of regional species-rich grassland would entail costs of € 420 million. Owing to the major uncertainty about damage costs, the Federal Environment Agency recommends considering the potential range from € 20 to 280 per tonne of CO_2 eq. The damage costs would then have a margin of uncertainty from € 120 to 1680 million.

Avoidance costs are set at € 20 per tonne of CO_2 for hydro electric power plants, for example, and € 40 per tonne of CO_2 for biomass power plants or wind turbines (Herminghaus 2012). If, however, the average prices of auctioning emission permits in Germany are assumed, then this is just under € 7 (DEHSt 2012). Hence the range of monetary value is very wide. For this reason, we renounce from the subsequent method of discounting. It would be more sensible to specify orders of magnitude.

Based on the initially specified average damage costs of € 70 per tonne of CO_2 eq., an area-related, regional mean value between around € 6000 and 13,000 per hectare is yielded, without taking into account discounting (a sum that, in view of those of current premiums within the framework of agri-environment measures, for example, such as in the range mentioned above in ◘ Table 6.5, no longer appears to be particularly high). If the figure of € 7 specified above is assumed to be the "market value", then area-related, regional mean values of around (only) € 650 and 1300 per hectare are yielded, without taking into account discounting.

In *summary*, the value of species-rich grassland for the climate is dependent to a great extent on its current level of stored organic carbon. In addition, it is also influenced in this case by the alternative possibilities and costs involved in compensating for the emission of climate-relevant gases or to ensure they are not increased by an alternative land use. In any case, the maintenance of species-rich grassland exhibits an undoubtedly interesting importance for the implementation of climate objectives, particularly since it arises as an add-on effect for water protection and biodiversity objectives.

Discussion and Outlook

Maintaining species-rich grassland is a declared objective of the national biodiversity strategy (BMU 2007). The social significance of this objective is also illustrated by the fact that of HNV-Farmland is an indicator in the context of allocating financial resources to promote rural development. The success of rural development programmes is therefore measured in terms of the maintenance of these areas, amongst other things (The Council of the European Union 2005).

The analyses demonstrate that economically relevant, environmental services in the area of climate and water-body protection are associated with the maintenance of species-rich grassland. The amount depends on the concrete local situation. Thereby the avoidance of additional emissions may be of particular economic importance in regions that already have an unfavourable status of water bodies. The avoidance of climate-relevant carbon emissions on these areas is in principle also economically relevant. It is essential that emissions are reduced further, which means avoiding additional emissions.

It must be borne in mind that the services considered cannot be offset equally for all species-rich grassland areas and the references assumed for balancing are crucial. In the context of this study, a scenario was applied that assumes that intensive grassland is used on around 50 % of the land and approximately 5 % of the land is converted. Particularly for 'extreme locations' that are very valuable from a nature conservation perspective, such as wet meadows, arid grassland or hillside locations, forestation and the formation of scrub constitute an alternative option, which may even be more realistic. If the current payment for species-rich grassland were abandoned, some areas would be taken out of use completely under the current framework conditions under certain circumstances, and scrub encroachment would occur leading to the successive development of woodlands. In case of this reference (succession), there are no synergies between the maintenance of species-rich grassland areas and environmental services in the area of climate and groundwater protection, because forest development is not assessed as negative for these environmental services, and may even be considered positive. Thus locational differences would have to be taken into greater account to achieve more precise balancing. It must also be borne in mind that the reference depends crucially on political and economic framework conditions. For example, sustained pressure on grassland areas due to the production of biomass can also lead to changed reference scenarios. Then greater proportions of more intensive use or the ploughing up of grassland than applied here would have to be assumed.

In everyday decision-making, for example involving funds to promote ecologically sound grassland use, it is therefore crucial to factor the original value of diversity into the equation. In this connection, a differentiation must be made between the direct value of biodiversity and the value that diverse habitats provide for tourism and recreation, for example. However, the direct value can only be estimated methodologically in economic terms as the so-called *nonuse value* via contingent valuation or *stated preference* methods (▶ Sect. 4.2). Great reservations are often harboured against these methods in Germany. Against this backdrop, it will continue to be necessary, particularly in the area of traditional nature protection, to include other forms of assessment as the foundation for decision-making. This could be possible in a monetary and verbally argumentative manner on the basis of *nonuse values*. Databases are available as guidance for Germany in the form of contingent valuation by Hampicke et al. (1991) and recent studies by the Federal Agency for Nature Conservation (BfN 2012c, Meyerhoff et al. 2012). With regard to methods, it must be noted with these studies that the calculated values by all means reproduce the *nonuse value*, as well as utility values for local recreation, for example. If one follows the argumentation that objectives enshrined in law are assessed as politically framed social preferences (UBA 2007), then reinstatement costs such as for the habitat types that constitute important target areas in the Habitats Directive, also requested by Schweppe-Kraft (2009), would also be legitimate from an economic perspective.

Conclusion

Species-rich grassland is not only important regarding biodiversity objectives. Avoiding the intensive use of grassland or conversions into arable land can support the objectives of implementing the Water Framework Directive and climate change mitigation targets. The potential involved depends on the specific local situation. Based on the calculation of avoidance costs, potential damage costs and willingness to pay, we demonstrated that species-rich grassland generates a range of considerable benefits. These monetary values can be included in the line of argument to support the desired maintenance of particularly species-rich grassland. In this connection, the ES concept encourages a holistic view of these areas that are valuable from a nature conservation perspective. With pro active use via a systematic, locationally differentiated assessment, it could be very useful as decision-making support for nature protection, particularly as part of targeted financial support, e.g. via agri-environment measures.

6.3 Economic Benefit Valuation of the Influence of a Forest Conversion Programme on Ecosystem Services in the Northeastern Lowlands of Germany

P. Elsasser and H. Englert

6.3.1 Introduction

Increasing the naturalness and the resilience of forests through 'forest conversion' is an important topic in German forestry for a number of reasons (Knoke et al. 2008). Many of today's forests consist of rather uniformly structured conifer stands which may be easier to manage, but which feature low biodiversity rates and are often particularly damaged in disasters like storms, fire, and insects. Moreover, droughts are of growing concern, especially in the eastern part of the North German Plain which already suffers from low annual precipitation–a situation which might be even further aggravated due to climate change. Public as well as private forest enterprises aim to stabilise homogenously structured forests by investing in large forest conversion programmes which convert purely coniferous stands into mixed and broadleaved forests. These efforts are financially supported by subsidisation programmes at federal and at state level (BMELV 2011b). Arguments in favour of such forest conversion programmes are that they not only reduce risks through the diversification process, but also enhance the supply of ecological services like watershed and climate protection, biodiversity and recreation opportunities for the population (Fritz 2006).

This case study analyses the economic value of ES which is modified through a forest conversion programme in the northeastern lowlands of Germany. The background information used in this case study comes from the interdisciplinary research project 'Newal-Net', funded by the Federal Research Ministry between 2005 and 2009. The Newal-Net project involved silviculturists, climatologists, ecologists, cultural and education scientists, and economists as well as practitioners. Project partners and stakeholders developed an overall concept ('*leitbild*') of future landscape development for a region in northeastern Germany which is currently dominated by purely pine forests.

The region abuts Berlin to the north and extends 17,500 km^2, encompassing about a third of the federal states of Brandenburg and Mecklenburg-West Pomerania. 30 % of the region is covered by forests. The '*leitbild*' (called "climate adaptive deciduous mixed forest") was discussed and further developed together with regional stakeholders. Following these discussions two scenarios for regional forest development, up until 2100, were modelled. Specifically the scenarios were

1. A stepwise realisation of the '*leitbild*' whenever a forest stand is harvested
2. The continuation of the current forest management plans, i.e. 'business as usual' ('bau')

The '*leitbild*' scenario was envisaged as a continuous reduction of the conifer area in the region, from an original 76 % in 2006 down to only 13 % in 2100 and correspondingly enhancing the area of mixed and deciduous forests up to 87 % in 2100. In the 'bau' scenario a reduction in conifer area was also planned, however at a much more conservative rate (from 76 % in 2006 to 67 % in 2100). In this scenario, the final share of mixed and deciduous forests is 23 % in 2100.

The goal of the project partners was to use this generated data from the scenario modelling to quantify and analyse, according their expertise, the impact of applying the '*leitbild*' concept in practice and compare this to the 'bau' situation. The central question for the economists in the project is: Are there substantial changes in the range of services provided by the forest and their subsequent economic values and does this speak for or against implementing the concept of "climate adaptive deciduous mixed forest"? Thus, the focus was on the 'benefits' side of the problem, specifically on the impacts on regional timber and biomass supply, landscape values, recreational values, and carbon sequestration, rather than on the 'cost' side of forest conversion. The remainder of this chapter summarises the main results of the study (the description is partly based on Elsasser et al. (2010a); further details on methods, analyses and results can be found in Elsasser et al. (2010b).

6.3.2 Raw Wood Production

The production of raw wood as a "provisioning ecosystem service" (MEA 2003) is the basic source of income for most forest enterprises. Methodologically, the valuation of raw wood production was based on forest development and forest utilisation models in combination with price data derived from market observations. Note, a similar approach was used for valuing the carbon sequestration service in the next chapter. To simulate forest development until the year 2100 we used forest growth models based on yield tables, which include assumptions about forest management planning. The results from the scenarios revealed different development outcomes. These results are primarily dependent on the spe-

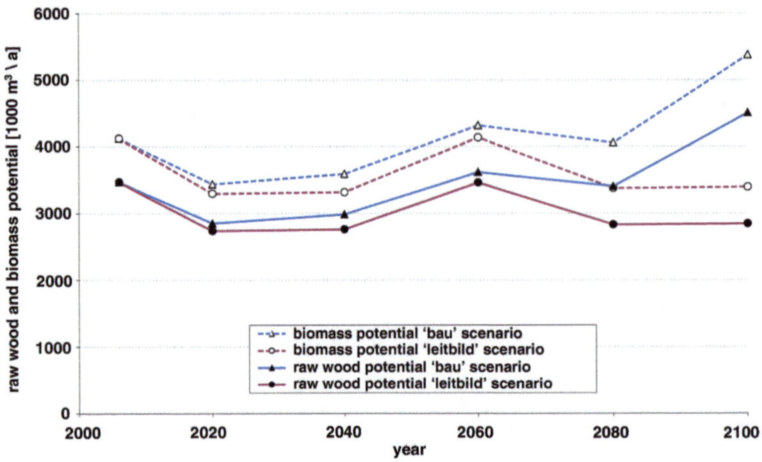

Fig. 6.10 Development of raw wood (in 1000 m^3 of merchantable timber) and biomass potential (in 1000 m^3)

cific tree species compositions in the respective scenarios. Our estimates of the expected raw wood revenues were based on forest condition descriptions, which are derived from the forest growth models in the reference years 2006, 2020, 2040, 2060, 2080 and 2100. Starting from each reference year, we simulated potential harvests in each forest stand for 5-year periods using the same yield tables and rotation cycles which were applied in the forest development models. The potential yield of raw wood and biomass (called 'raw wood potential' and 'biomass potential', respectively) was calculated by aggregating the estimated yield volumes.

The potential yield of raw wood and further aboveground biomass was then allocated to different assortments according to common assortment tables. These assortments are the actual products of a forest enterprise and therefore comprehensive price data is available. Total biomass volume was extrapolated from coarse wood volume using tree species and age specific expansion functions (Dieter and Elsasser 2002). Finally, the value of the raw wood and biomass potential was calculated by combining the price data with the assortment specific volumes.

The different silvicultural concepts, which both scenarios are based on, particularly affect the development of the forest area stocked with the tree species group 'pine and larch'. In the '*leitbild*' scenario, the area stocked with 'oak and other deciduous trees' increases considerably at the expense of the 'pine and larch' area. In contrast, the tree species groups 'beech' and 'spruce, fir, douglas fir' evolve similarly until 2080 under both scenarios, differing only in the year 2100.

Figure 6.10 compares the development of the raw wood and biomass potential in both scenarios. The potential raw wood volume is approximately 3.5 million m^3 in the starting year, 2006. Until the year 2060, the raw wood potential of the scenarios do not differ substantially. It becomes apparent that even a significant change in the management concept (as is simulated in the '*leitbild*' scenario) only affects the potential raw wood volume appreciably after a time lag of 50 years. The potential raw wood volume in 2060 is more or less similar to that in the starting year. Only from 2080 onwards the raw wood potential develops differently under both scenarios: it increases under '*bau*', but remains at the same level under the '*leitbild*' scenario. The biomass potential develops in a similar manner (Fig. 6.10) but constantly exceeds the raw wood potential by about 18 to 20 %.

The long-term impact of altered silvicultural concepts also becomes apparent when comparing the development of the assortment structure over time. Only from 2080 onwards, the smallwood supply from broadleaves increases considerably. A corresponding increase in the supply of large dimension hardwood is not (yet) visible within the analysed period.

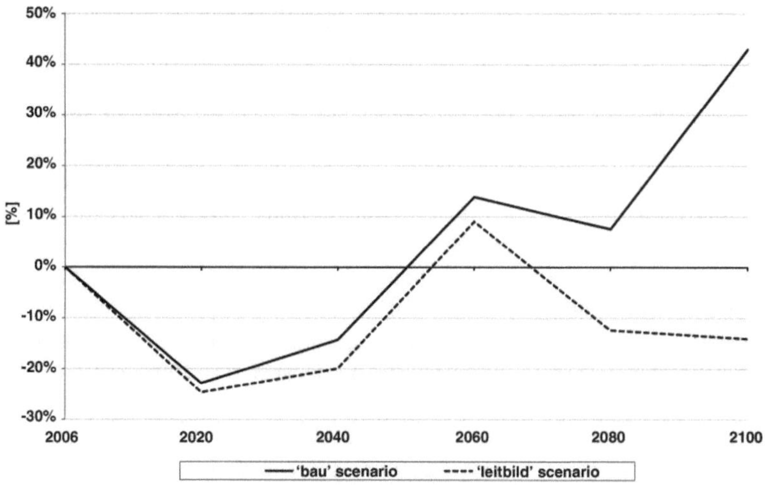

Fig. 6.11 Value development of the raw wood potential compared to the year 2006

Figure 6.11 shows the development of the raw wood potential's economic value. Here the total volume of the raw wood potential has been evaluated at 2005 prices (ZMP 2009; due to the lack of price information for wood residues, we assumed a respective price of 15 €/m³ (exempt from harvest costs). The curve therefore reveals only those value changes which are caused by changes in the quantity or the structure of the raw wood potential due to the assumption of constant timber prices during the period under consideration.

The value development is different between both scenarios. Under '*bau*', values increase from € 120 m in 2006 to € 171 m in 2100, an increase of 43 %. In contrast, values slightly sink by 14 % under the '*leitbild*'-scenario during this time, amounting to only € 103 m in 2100. The average revenues from a cubic metre of raw timber remain broadly constant in both scenarios during the given period (this can be concluded from the fact that quantity and value developments are almost proportional; Fig. 6.10 and 6.11). At least until 2100, the reduced raw timber supply in the '*leitbild*' scenario is not compensated by higher values of the produced hardwood; rather it results in a loss, amounting to € 70 m in 2100.

6.3.3 Carbon Sequestration

The sequestration of carbon in trees is a 'regulating ecosystem service' according to MEA (2003) which was brought into wide public awareness by the UN framework convention on climate change. As has been discussed in the previous section, the change in the tree species composition clearly affects timber and biomass stocks and therefore the amount of carbon stored. In order to describe the carbon stock development, the volume of the aboveground tree biomass (merchantable wood, lop and needles) as well as the tree root biomass was extrapolated from the standing stock of merchantable wood. Volumes were converted into masses according to their respective wood density (Kollmann 1982). The carbon content was then transferred according to Lamlom and Savidge (2003).

The development of carbon stocks is very similar to that of the raw wood stocks. Carbon continues to accumulate in both scenarios until 2040 due to the age structure in the project area. However, net carbon sequestration decreases continually over this period. Starting in circa 2060, the harvest of mature forest stands will cause net emissions of carbon in both scenarios. The reduction is more pronounced in the '*bau*' than in the '*leitbild*' scenario. Accordingly, the '*leitbild*' scenario causes less sequestration in the beginning (by 200 Gg C/a in the period 2006–2020), but fewer net emissions in the end (the difference to '*bau*' amounting to just under 150 Gg C/a in the period 2080–2100).

In regards to the monetary value of the sequestration service, some important caveats have to be kept in mind, especially given the length of the period under consideration. First, the influence of altering silvicultural concepts in northeastern Germany on global climate is only marginal, given the global dimension of the problem. Thus the impact of mitigation on the local population will be almost negligible. Second, prices for certificates at the carbon markets emerging under the European Trading System and the framework convention on climate change may be used as value indicators; however forest sink certificates are not (yet) being accepted. Their acceptance depends on complex international negotiations and it is therefore uncertain if this will change in the future. Even if it does, the agreements reached in these negotiations will heavily influence the scarcity, and thus the price, of carbon emission permits. As a consequence of both considerations, assigning a positive monetary value to regional forest sequestration services is based on optimistic assumptions.

We addressed the associated uncertainty by allowing for a broad value range for the sequestration service. These are oriented at avoided damage costs, on the one hand, and at prices at the various markets for emission permits, on the other. However, even when we assumed a value of 100 €/t CO_2 for the whole time span until 2100 (a value which appears rather extreme from today's perspective), the effect on the total value balance was quite small. In comparison to the '*bau*' scenario, a forest conversion following the '*leitbild*' scenario would cause a sequestration value loss of about 5.5 m €/year in the period 2006–2020. By the end of the considered period (i.e. in 2080–2100), it would lead to a gain of about 4 m €/year. In comparison to the losses in raw timber values described above, this effect is negligible.

6.3.4 Scenic Beauty and Recreation Values

The MEA counts a landscape's scenic beauty and its recreation service as 'cultural ecosystem service'. In order to value these, we conducted a regional population survey. The survey asked several basic questions about the respondents' attitudes towards the landscape where they live and its management. It also contained a choice experiment to determine a monetary valuation for changes in the scenic beauty and recreation in the landscape.

Choice experiments are a method of preference determination where respondents choose between alternative bundles of goods (▶ Sect. 4.2.3). Each of these goods is characterised by different attributes which assume varying levels. Multinomial logit models are then used to estimate how much a change in the level of each attribute has influenced the choice probability of the respective goods. If one of the attributes is the price or the cost of a good, then a marginal willingness-to-pay (WTP) for the other attributes can be calculated from the obtained statistical measures (see e.g. Hensher et al. 2005 for details).

In this study, 999 inhabitants were selected for the survey. We displayed to each respondent choice cards describing three alternative residential environments. In each alternative, three attributes were varied, namely the view of the landscape (as visualised by computer generated images), the possibility of entering meadows and forests for recreation purposes, and the cost of living there (as a price indicator). The computer images showed various landscape views typical of the region in summer or in winter. Alternatively, a pine forest, a deciduous forest or a mixed forest, each with high or low structural diversity or a meadow without trees were shown. Interviewees were asked to choose their most preferred landscape among the three presented choice options.

The analysis showed that the ecosystem service 'recreation' (here defined as the possibility to enter meadows and/or forests for recreation purposes) has a substantial monetary value. It amounts between 55 and 90 € per household and year. This is in line with the result of comparable studies. The valuation of the landscape views revealed that all alternatives which contained forests were clearly preferred over the situation without trees. This result is independent of tree species and the level of structural diversity in the forest images. A closer comparison between the various forest images show that households have a significant WTP for deciduous and mixed forests rather than coniferous ones–but only if the images show forests in their summer aspect. Summer aspect WTP ranged between more than 40 and more than 85 €/household/year. If the images additionally showed high structural diversity, this resulted in an extra 20 €/household/year. In contrast to this finding, no general preference for deciduous and mixed forests in their winter aspect could be confirmed. However, structural diversity of forest stands played an even more important role in winter than in summertime: The additional WTP for high structural diversity under winter conditions was between 90 and 160 €/household/year.

Based on these results, it was possible to examine the long-term impacts of forest conversion on the landscape views over time and then to extrapolate WTP of the regional population. For this purpose we weighted the valuation results for the summer and winter aspect, respectively, at a rate of 7:5 (according to the length of the vegetation period). Two variants were calculated in order to account for the influence of structural diversity on results. For a 'lower' variant we supposed that future forest conversion will always produce forests of low diversity. In contrast, a 'diversified' variant was calculated using the forest views which showed high structural diversity. For simplicity we assumed the pace of the forest conversion will be driven by silvicultural considerations only (i.e. no concentration along settlement centres, for example), and that population numbers as well as their preferences remain constant over time.

According to this extrapolation, landscape values will increase over time in both scenarios because both envisage a forest conversion of purely conifer stands into stands with higher shares of deciduous trees (even though the extent of the conversion is considerably lower in the '*bau*' scenario). Because of the stepwise realisation of the conversion programmes, landscape values are highest at the end of the period. In a short to medium term view the consumer surplus difference between both scenarios is comparatively small; for 2020 it amounts to 3.0 m €/year in the 'lower' variant (or 6.2 m €/year respectively, if high diversity of the converted stands is assumed for both scenarios). Until 2100 the difference increases to 16.0 m €/year (or 34.1 m €/year respectively, in the 'diversified' variant). When interpreting these last-mentioned numbers it should however be kept in mind, that such projections into a distant future inevitably imply substantial uncertainties.

6.3.5 Synopsis and Discussion

The balance of the monetary values of all forest ES considered here changes over time. A look at the respective temporal development reveals that no appreciable monetary losses occur until about 2060, if the '*leitbild*' scenario is put into practice; the balance is even slightly positive in the 'diversified' variant of landscape development which assumes a conversion into highly diverse forest stands. After 2060 however, neither the positive influences on landscape values nor the (weakly) positive influences on carbon sequestration are able to compensate the significant losses due to decreased wood production. This is true even if unrealistically high sequestration values are being assumed: in 2100, the realisation of the '*leitbild*' scenario would cause a utility loss of 30–50 m €/year in comparison to '*bau*'-oriented silviculture, depending on calculation variant. ◘ Figure 6.12 illustrates this finding (a: without carbon sequestration values; b: assuming a carbon sequestration value of 100 €/t CO_2. The very small differences between both parts of the figure demonstrate how weakly the carbon sequestration service impacts the overall result) (◘ Fig. 6.12).

Although the losses uncovered here will be relevant in a more distant future only, such a result violates norms of sustainability as well as intergenerational justice–even if substitutability between different forest services is permitted, as is implied by 'weak' sustainability approaches. When interpreting this result it should be kept in mind that the value balance presented here, and its temporal development, are based on a number of simplifying assumptions:

1. In the absence of more reliable information, the long-term considerations above assume continuousness in the overall economic development, in the sense of constant value relations between timber market values (which dominate the overall result) and other goods and services, including the forest ES valued here.
2. Constancy in the long run is assumed with regard to growth conditions for trees, too. If tree growth will decrease in the future due to changing environmental conditions (e.g. because of increasing drought stress), then the influence of wood production on the overall result will diminish in relation to other forest ES–at least under the condition that the value of these other services will be less affected

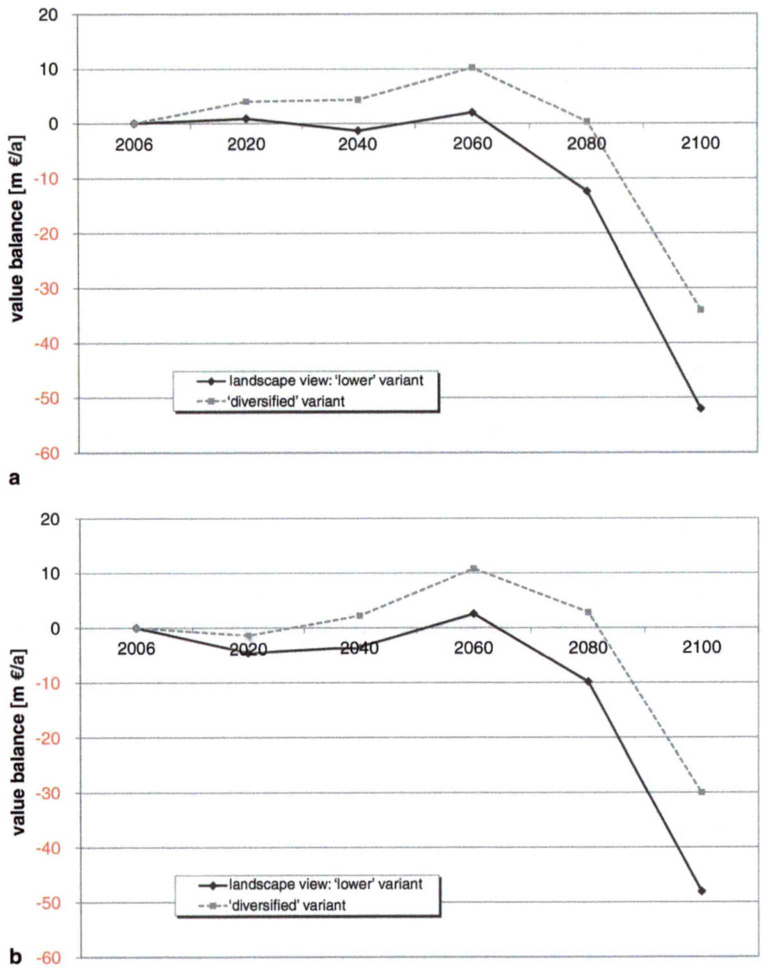

Fig. 6.12 Development of the value balance of wood production, landscape view and carbon sequestration service under forest conversion according to the *'leitbild'* as compared to the *'bau'* scenario (a: CO_2 base price = 0 €/t; b: 100 €/t)

by the growth decrease, or not affected at all. Under enhanced growth conditions for trees (e.g. due to extended annual growth periods) this will be the other way round.

3. Production risks as well as their possible shifts due to climate change are not taken into account in the present investigation–neither natural risks (e.g. droughts, fire, storms, insect calamities) nor financial ones (including changing price relations between tree species). Generally, forest conversion in the sense of the *'leitbild'* can lead to a better distribution of unknown risks and may be economically rational in this sense (see e.g. Knoke et al. 2008), thus qualifying the long-term utility loss associated with a realisation of the '*leitbild*'. However, the present silvicultural strategies underlying the '*bau*' scenario provide for some forest conversion, too, albeit to a lesser extent. It is not possible to determine objectively which degree of forest conversion is most advantageous for distributing production risks, since this depends also on the risk aversion of today's decision-makers.

4. In addition to timber markets the present investigation captures landscape views, recreation and carbon sequestration as important ES which would be affected if the *'leitbild'* was realised. Nevertheless important impacts on other potentially affected ES remain unconsidered, like e.g. water supply and consumptive use of forest biodiversity. However, a well comparable study on the influence of a forest conversion programme on regional biodiversity values is available from two regions in Lower Saxony (Liebe et al. 2006 in Meyerhoff et al. 2006). According to the mentioned study, a third to half of the respondents in the regions of Lüneburger Heide and Solling were generally prepared to pay for increasing the amount of deciduous trees and related biodiversity improvements (the programme provided for increasing the share of deciduous trees from 30 and 40 %, respectively, to 60 %, corresponding to the state forest administration's conversion programme LÖWE (Niedersächsische Landesregierung 1991). With regard to the motives for the decision to pay, the attributes 'biotope protection' and 'species protection' ranked highest, even higher than 'increasing landscape diversity'. This finding emphasises the relevance of biotope protection services. Nonetheless, mean WTP for the conversion programme (about 6 to 15 €/person/year depending on valuation method) was in the same order of magnitude as the WTP identified in the present study for landscape views, even slightly lower. This implies that even if biodiversity values were additionally included here, the abovementioned sustainability violation would likely not be dispelled: It persists even if it is assumed that the *'leitbild'* scenario's positive influences on biodiversity values might be much higher than those on landscape values.[1]

1 The overall balance becomes positive over the whole period only if the most optimistic of each calculation variants is chosen (i. e. the 'diversified' variant for landscape view valuation, and a carbon sequestration value of 100 €/t CO_2) and if a biodiversity value of at least 108 €/household/year is assumed (which is more than three times the value determined by Liebe et al. 2006). All three assumptions are quite unrealistic.

5. Finally it must be stressed that the present investigation did not aim at a comprehensive benefit-cost analysis. Specifically, investment costs of forest conversion have not been an issue here; therefore the presented value balances do not contain such costs. The higher these costs are in reality (e.g. for planting and maintaining deciduous trees, protecting them against damage from game animals, etc.), the more adverse will the balance become with regard to a realisation of the *'leitbild'* of 'climate adaptive deciduous mixed forest'. Likewise it will become increasingly questionable whether the positive balance indicated here until 2060 under inclusion of public goods (and under partly optimistic assumptions about their values) will persist at all. Even when allowing for positive impacts on ES it can be inferred, that a forest conversion according to the *'leitbild'* will most probably only be compatible with economic and sustainability goals if conversion costs can be restricted to a minimum.

Conclusion

What are the conclusions which can be drawn from this case study, for MEA's ecosystem approach (MEA 2003) and for the economic valuation of ES in general? The MEA approach focuses at the entirety of services which may be affected by management and utilisation (here: of forests), and thus it helps keeping essential environmental aspects in view, which go beyond mere production goals. Thereby the MEA approach takes up the important concern of environmental economics that external effects be identified and quantified, in order to be incorporated in decisions of environmental relevance. Even though including *every* ecosystem service possibly affected by any intervention might turn out impossible due to data limitations, the present case study demonstrates that quantifying and valuing important provisioning, regulating, and cultural ecosystem services (CES) is indeed possible. Generally this is also true for 'supporting services'. However, most supporting services do not have direct consumption benefits, but rather serve as inputs for the production of other goods. Thus their value should not be added to the values of the produced goods, in order to avoid double counting (► Sect. 3.2.5).

Table 6.6 Urban ecosystem services and indicators for the quality of life in sustainability dimensions

Sustainability dimension	Urban ecosystem service	Indicator for urban quality of ilfe
Ecology	Air filtering Climate regulation Noise reduction Rainwater drainage Water supply Wastewater purification Food production	Health (clean air, protection against respiratory illnesses, death from heat exposure and hypothermia) Safety Drinking water Food
Social	Landscape Recreation Cultural values Environmental education	Aesthetics of the environment Recreation, stress reduction Intellectual enrichment Communication Living area
Economy	Allocation of land for economic activities and transport	Accessibility Income

Own compilation from Haase (2011) after the MEA (2005), TEEB (2010), Fisher et al. (2009) as well as Santos and Martins (2007)

6

6.4 Urban Ecosystem Services: Leipzig as a Case Study

D. Haase

Urban ecosystem services (Bolund and Hunhammar 1999; TEEB 2010; Elmqvist et al. 2013; Haase et al. 2014) describe ecosystem functions (processes, structures), which are provided by nature or the urban ecosystem and used by the inhabitants of a city and/or a metropolitan area. Examples of urban ES are the provision of freshwater and drinking water from precipitation and the natural filtration of soils, the regulation of peak run-off during periods of extreme precipitation and the resultant reduction of high water in urban areas, food production (fruit, vegetables) in urban gardens or garden allotments, the pollination of fruit trees by urban bees or the provision of fresh (cool) air from open spaces and recreational areas.

Without ES, our life as we know it today, would not be possible in the city (Guo et al. 2010; Haase 2011). Some studies have attempted to express urban ES in terms of monetary value to emphasise the economic importance of and dependency on nature–particularly in cities (Gómez-Baggethun et al. 2010; Bastian et al. 2012). In addition, non-monetary models and assessment approaches (as presented in the following) are very well suited to emphasise the importance of urban ES.

According to the Millennium Ecosystem Assessment (MEA 2005), the TEEB study Manual for Cities (2010) as well as Fisher et al. (2009), urban ecosystem services can be divided into four categories (Table 6.6): provisioning services (food, water, raw materials, genetic resources, medicinal resources, etc.), regulating services (regulation of climatic extremes such as heavy precipitation or extreme summer heat, floods, diseases, water and air quality, waste management, etc.), cultural services (recreation, aesthetics, spiritual fulfilment, environmental education, etc.), and supporting services (in the broader sense, ecosystem processes and functions such as soil formation, biodiversity, pollination, nutrient cycles, etc.). Urban ES are very closely to the urban quality of life (Schetke et al. 2012; Santos and Martins 2007), which is based on the three dimensions of sustainability and rather expresses city dwellers 'needs' for services, which are partially covered by ecosystems and/or ecosystem functions. The complementary role of both concepts is presented in Table 6.6.

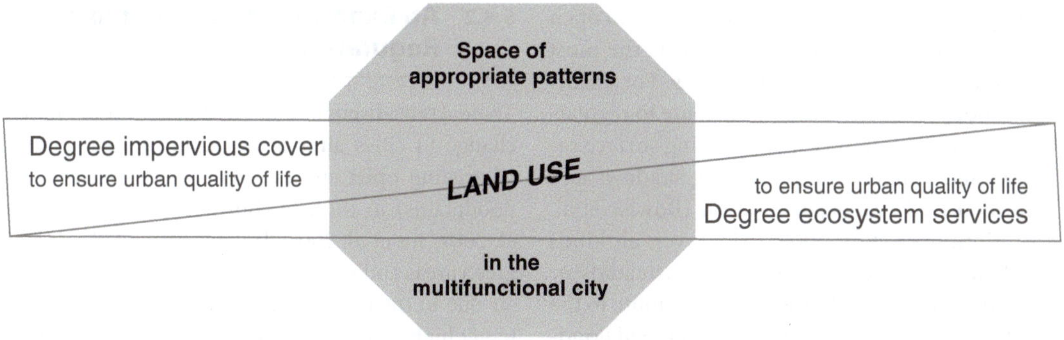

Fig. 6.13 The nexus between urban land use and urban ecosystem services–the options for defining urban land use to secure urban ecosystem services will always be a compromise to suit both socio-economics and the ecosystem. This 'search' for compromises has to be (re-)negotiated. Multi-criteria assessment, trade-offs as well as partial optimisation can all be useful methods in defining compromise

Fig. 6.14 Restriction and potential of urban ecosystem services from urban development and land sealing (Munich, *left*) as well as land-use perforation from shrinkage (Rabet in Leipzig, *right*). © Dagmar Haase 2010

6.4.1 Urban Ecosystem Services and Urban Land Use: A Complex Nexus

Urban ES are very closely linked to urban land use (Fig. 6.13). On a global scale, urban land use constitutes at the most 4 % of the Earth's surface, and yet more than half of the world's population now live in cities; with trend still on the rise (Seto et al. 2011). From an economics point of view, 95 % of today's cumulative global gross domestic product is generated in cities. The sealing and conversion of semi-natural areas and agricultural land into settlements and roads belong to the most significant effects (mostly of a negative ecological nature) all over the world, with such effects often being irreversible. The rural–urban gradient that is emerging is characterised by dispersed land-use development with an increasing amount of land sealing in city centres (Haase and Nuissl 2010). Sealed/partly sealed soils can no longer (or at least only to a restricted degree) fulfil the urban ES described above (Haase and Nuissl 2007; Fig. 6.13 and 6.14). Therefore, possible courses of action in the field of urban land use to secure ES are always part of a multi-criteria compromise, which has to be (re-) negotiated time and again (► Sect. 5.1, 5.4).

In particular, the land-use types of urban open spaces, urban green and forests provide the most diverse urban ES for urban residents. For example, wooded areas and parks contribute to regulating extreme temperatures, by reducing surface radiation and temperature by casting shadows and from increased evapotranspiration (Bowler et al. 2010, Kottmeier et al. 2007). Moreover, all kinds of urban green areas, even urban wastelands or brownfields and waterbodies can contribute to the well-being of urban residents. Undeveloped floodplains primarily regulate flooding (Haase 2003). Unsealed land surfaces are suitable for rainwater retention and percolation, and are therefore able to regulate rapid aboveground discharge runoff from heavy rain events. Rainwater seepage installations especially designed for this purpose in residential areas can additionally serve as in situ rainwater utilisation (Haase 2009). With respect to the increasingly more important debate on anthropogenically induced climate change, urban green areas (above all trees and forests) can contribute to local carbon sequestration. Admittedly, current studies only refer to 1–2 % of the emissions resulting from cities that can be neutralised by urban vegetation (Strohbach and Haase 2012 for Leipzig, Nowak and Crane 2002 for US cities). In spite of this small contribution, however, land uses with tree cover still contribute towards reducing the 'ecological footprint' of a city.

Recently, another process (that is moving in the opposite direction from urban growth) has been given increasingly more attention: urban shrinkage. Cities labelled with economic problems and population losses are characterised worldwide by a decrease in the intensity of urban land use, vacant premises as well as urban wastelands (◘ Fig. 6.14; Haase et al. 2007). This land-use perforation (Lütke-Daldrup 2001) provides the unique opportunity for the revitalization of (inner)-city areas (Lorance Rall and Haase 2011) and an associated 'revitalization' of urban ES (Haase 2008).

A prime example of simultaneous urban growth and shrinkage is the city of Leipzig situated in one of the new German federal states of the Federal Republic of Germany (Lütke-Daldrup 2001). The following examples of the analysis and evaluation of urban ES refer to Leipzig and are internationally published.

6.4.2 An Example of Local Climate Regulation

The current discussion about adaptation to climate change in cities aims at reducing the temperature of existing open spaces (green areas, waterbodies, floodplains) in the city (Bowler et al. 2010; Gill et al. 2007; Jin et al. 2005; Tratalos et al. 2007). Urban green areas and waterbodies are able to produce cool air due to their specific heat evaporation and counteract high summer temperatures (Gill et al. 2007).

In addition, shaded areas play an important role: here the temperature reduction compared to a sunny location can be up to a maximum of 5 K per 10-minute interval during diurnal variation. As shown in ◘ Fig. 6.15 for the City of Leipzig, an increase in the amount of shade (i.e. the percentage of park areas with tree cover and/or the number of trees in parks) of urban green spaces benefits the temperature regulation effect (Bowler et al. 2010). The cooling effect of 3 K (on average) in the diurnal variation from shading is thereby empirically determined: on the one hand it is measured and on the other hand it is extrapolated to the urban area by using data from the aerial photography of shaded areas in parks.

Using indicators such as *shade*, but also *surface emissivity* or *f-evapotranspiration*, the effects of local climate regulation of urban areas and land use patterns can be estimated. Schwarz et al. (2010) quantified the effects of town-planning measures for the local government District of Leipzig such as economic development, the designation of industrial land, land reclamation from open-cast mining as well as the effect of regional green belts (◘ Fig. 6.16).

6.4.3 An Example of Flood Regulation

The urban ecosystem service 'flood regulation' that is particularly important in cities, can be simplified by the following eq. 6.1, in which the precipitation N is the sum of discharge A (base runoff [A_b], intermediate [interflow A_i] and surface runoff [A_o]), evaporation ETP and intermediate storage S:

$$N = (A_b + A_i + A_o) + ETP + S \qquad (6.1)$$

Unlike river catchment areas, the amount of sealing and the size of the sewer systems play an important

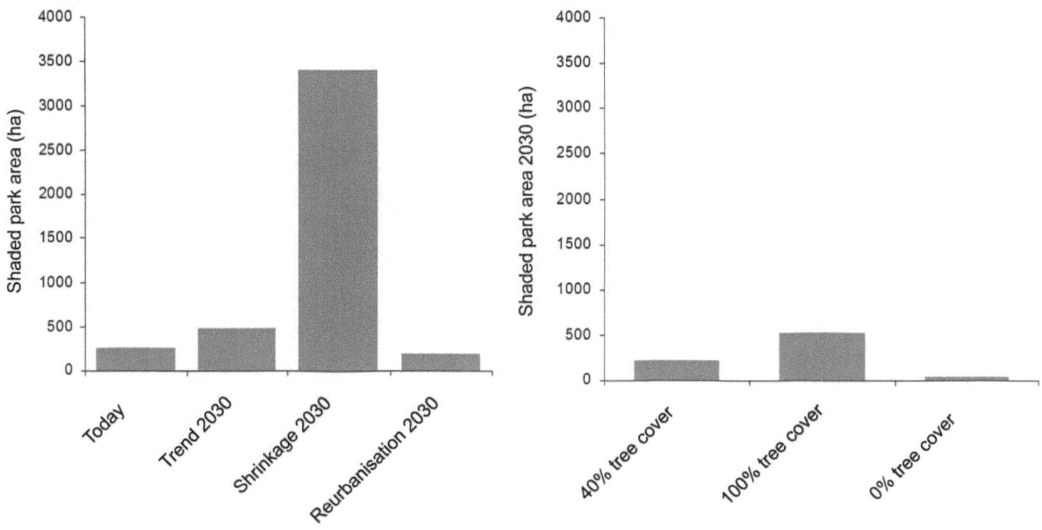

Fig. 6.15 The temperature difference between areas exposed to the sun and shaded areas of urban parks in late sum mer in Leipzig (own data from data collection in August 2009). Air temperature was measured with temperature sensors and recorded by means of a data logger. An average temperature difference of 3 K was empirically calculated between shaded and nonshaded areas for different parks and the 40 % shade determined by remote-sensing data was transferred to all parks in the city. Hence, it was possible to simulate the effect of increased and/or decreased shade on the regulation of the local climate (cooling) in public parks for scenarios of urban shrinkage, urban restructuring and re-urbanisation. Apart from which, the urban ecosystem services effect "climate regulation from shading" could be calculated for various reforestation measures in urban parks

role, as they determine how much precipitation water is available to the ecosystem and/or how much directly gets into the receiving water from direct runoff (Haase 2009). Here, it applies: the higher the degree of soil sealing, the lower the base and/or intermediate runoff, the higher the surface runoff and the lower the flood water regulation. Urban ecosystems are very vulnerable to discharge points and local flooding and/or ground water flooding.

An efficient method for calculating water regulation within an urban area is the empirically determined Bagrov-relationship (eq. 6.2), with which the total runoff (Q) is calculated from the real evapotranspiration (ETa) of an area (Glugla and Fürtig 1997). The total runoff (Q) is calculated by the difference between the real evapotranspiration (ETa) and the long-term precipitant (N). With increasing N, ETa is closer to the potential evaporation (ETp), while with decreasing N, ETa is closer to this. The intensity, with which these boundary conditions are reached, is altered by the storage properties of the evaporating surface (effectiveness parameter n), which are determined by the land use, soil sealing and soil type:

$$\frac{dETa}{dN} = 1 - \left(\frac{ETa}{ETp}\right)^n \tag{6.2}$$

From the base runoff it is then possible to calculate the direct runoff by determining the proportion p of surplus water (the difference between precipitation and evaporation), where p is derived from the initial parameters, slope inclination, soil type, depth to the water table and the land use and/or the degree of sealing (eq. 6.3; Haase 2009):

$$A_0 = (N - ETa)\,p / 100 \tag{6.3}$$

One of the results from calculating flood regulation is shown in **Fig. 6.17** using the example of the City of Leipzig over the period 1870–2006.

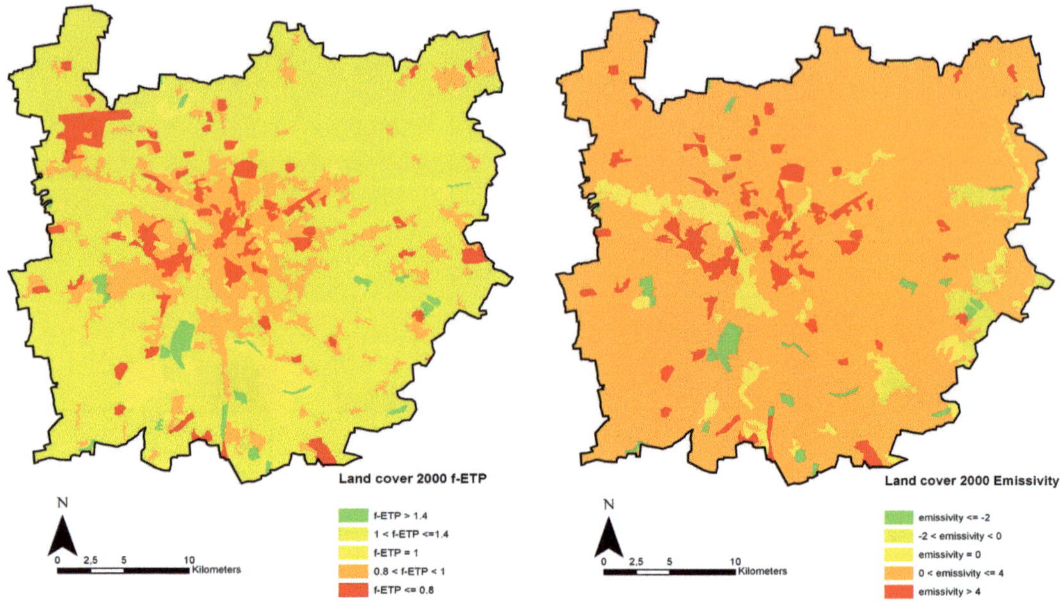

Fig. 6.16 The effects of different planning instruments on the urban ecosystem service 'local climate regulation' for the local government District of Leipzig (compared with Corine Land Cover Data in the base year 2000 (= CLC 2000)). On the right the indicator *surface emissivity* (*), in the center the indicator *f-evapotranspiration* (**) and on the right the aggregated indicator values for CLC2000 and the planning measures considered. Land-use changes resulting from planning measures were incorporated into the GIS, by editing the areas with measures as polygons and accepting an appropriate new/altered land-use form. © Schwarz et al. 2011

6.4.4 An Example of Carbon Sequestration in the Urban Area–Reducing the Ecological Backpack of the City?

On the one hand, cities belong to the main emitters of carbon dioxide (Churkina 2008). On the other, they are also able to fix some of these carbon dioxide emissions–primarily in soils and trees. Carbon sequestration belongs to the global most important urban ES that are provided in cities, even if this contribution is quite small in terms of figures. How can one assess the contribution (balance) of a city to the global carbon balance?

For the city of Leipzig carbon storage from urban vegetation (trees) was empirically determined and modelled with the goal of

- Obtaining a spatially explicit representation of carbon storage and
- Establishing comparative values of carbon sequestration from trees for typical urban land-use types (Strohbach und Haase 2012).

For this a stratified *random sample* of 190 locations was laid over 19 land-cover classes, for which the percentage of trees, the age of the trees, the species as well as the DBH were all determined. Moreover, the tree crown cover was derived from colour-infrared images (CIR) by means of object-oriented *random forest* algorithms (for details on the method see Strohbach and Haase 2012). The tree crown cover of the entire city area was quantified with 19 %. Using the data on the tree trunk diameter, it was possible to determine the aboveground biomass of the trees as well as stored carbon by means of allometric equations.

By applying the above methodology, a total carbon storage of 316,000 Mg C or 11 Mg C per hectare could be determined in the aboveground tree biomass of Leipzig. The highest values were determined for residential areas and floodplain areas (Fig. 6.18). Compared to values from US American and Chinese cities (Hutyra et al. 2011) and compared with the annual CO_2 emissions of the city, the carbon storage values for Leipzig are

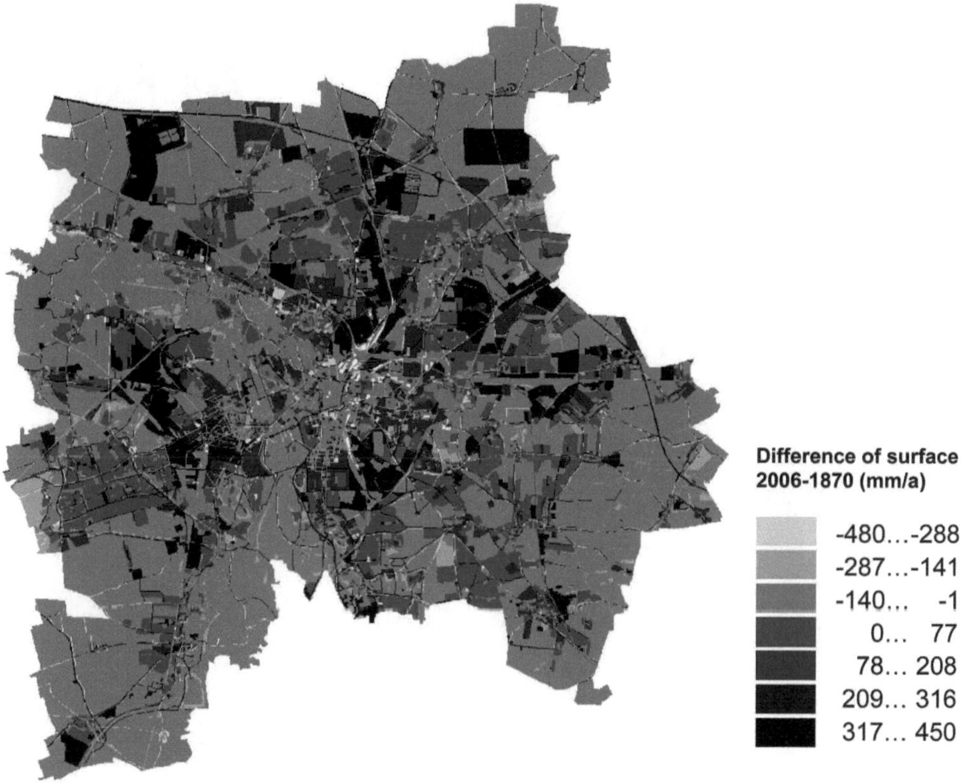

Fig. 6.17 Change in the surface runoff A_0 in the City of Leipzig between 1870 and 2006, calculated after the procedure explained in ► Sect. 6.4.3. The urban ecosystem services in this case are the difference between the values from 2006 to 1870, displayed in the legend below. One notices a considerable loss of flood regulation services due to sealing in Leipzig, shown as positive differences (growth)

rather small–and yet they should not be underestimated. This quantification method is particularly suitable for assessing the carbon mitigation potential of urban restructuring projects for cities in decline, as Strohbach et al. (2012) found for the demolishing and subsequent use projects 'dark forest' and 'floodlit grove' in the east of Leipzig.

6.4.5 An Example of the Recreational and Nature Experience

Urban green areas (including their habitats and biotopes) provide a multitude of ecological functions and thus also provide many urban ES (Haase 2011). Public green spaces include urban forests, parks with tree cover, cemeteries, sports fields with little or no tree cover and successional wastelands. Even the green spaces along the sides of roads act as urban biotopes (Breuste et al. 2007). In addition, there are also private urban green spaces such as back gardens, allotment gardens or golf courses. Their significance in terms of their function not only as habitats but also as regulators is by far underestimated. Generally speaking, the urban ES of urban green spaces range from the biotope formation function in the true sense to the nature conservation function (species diversity), the bio-indicator and information function (clean air), the climate regulation function (provision and storage of cool air), the soil protection function (filtering, buffering) as well as some particularly important functions for urban inhabitants including recreation, noise buffering, the townscape, as well as

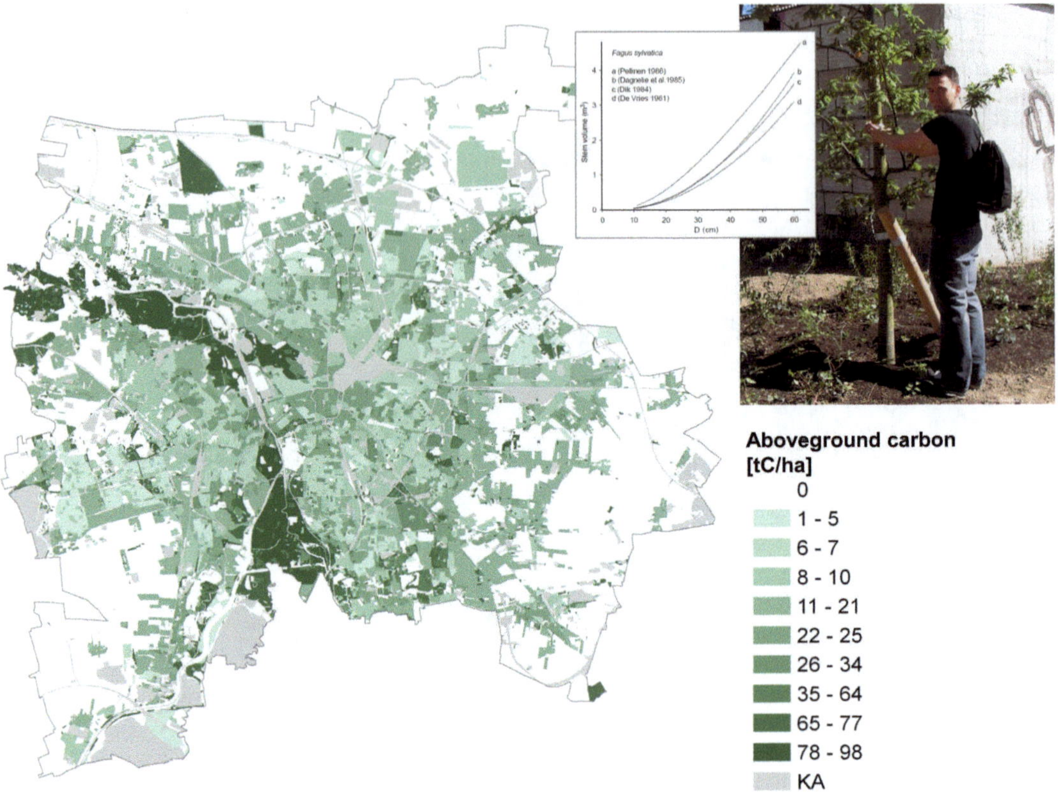

Fig. 6.18 Spatial distribution of the carbon sequestration capacity of the urban vegetation in Leipzig after urban structure and/or land-use types (Strohbach and Haase 2012). The lowest values came from urban agricultural areas, the highest came from the floodplain forest that runs through the city. Similarly, areas with Wilhelminian architecture and villas showed above average values for carbon sequestration. The locations were determined by *random sampling* and the carbon data were recorded in the field by measuring the DBH of trees as well as the respective allometric curves of the relationship between DBH and aboveground biomass established

educational and historical functions (Burgess 1988; Bolund and Hunhammar 1999). Parks, playgrounds, meadows, cemeteries, green wastelands and urban forests secure the recreational function of humans in the city. They help to maintain mental and physical health as well as making it possible to experience nature (Yli-Pelkonen and Nielemä 2005; Chiesura 2004). Consequently, they need to be classed as considerable determinants of the urban quality of life (Santos and Martins 2007). Besides which, they provide aesthetic pleasure and inspiration in the city (Breuste 1999).

One can determine the supply and demand of urban green spaces for recreation very efficiently using geographical information systems (e.g. ArcGIS) on the basis of land-use data and local government statistics (Comber et al. 2008). Indicators, which were determined for the City of Leipzig on the other hand (**Fig. 6.19**) are *area* of, *percentage* of and *per capita area* of urban green as well as its accessibility (buffer or network analysis). Many cities have specified threshold values for these indicators (Kabisch and Haase 2011). Furthermore, gradient analyses of green area requirements and potential demand as well as dissimilarity indices such as *Gini* or *Theil* are suitable for describing the concentration (fairness) of the recreation function.

Gruehn et al. (2006) analysed the effect of urban green spaces on the value of properties and real estate (► Sect. 4.2). The results show that there are numerous positive effects from open spaces and green spaces on the standard land value of up to

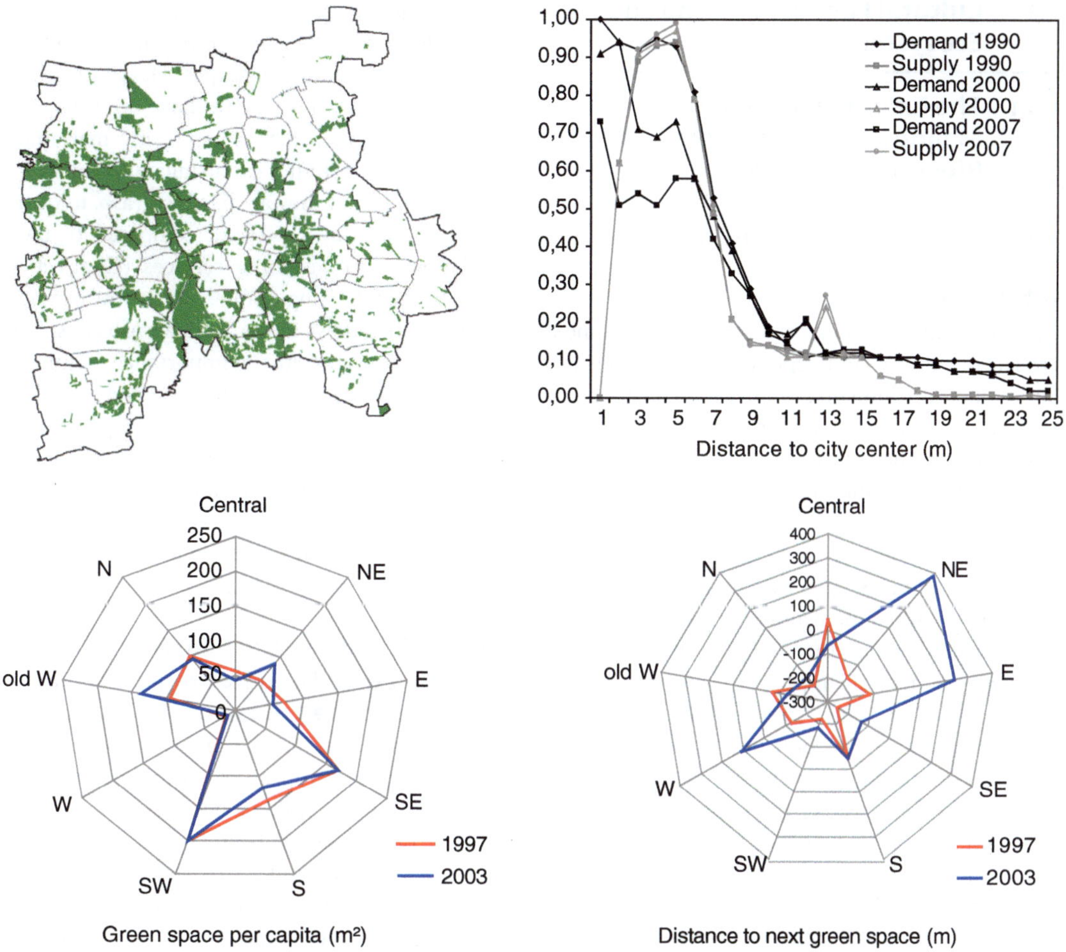

Fig. 6.19 Recreation function in Leipzig: portrayed as a green area map for 2003 (*upper left*), as a gradient of supply and demand (*top right*) as well as equity on the municipality level for two land-use time steps (*below*)

20 %, depending on the function, configuration, accessibility, residence quality and spatial configuration of the area. It should be noted however that these values depend on the city in question and are therefore to be taken as approximate values.

The Status of Urban Ecosystem Services Implementation in Germany

Despite the heated international discussion on ES and their enhancement (see MEA 2005; TEEB 2010), so far administrative institutions responsible for the planning of German cities have been hesitant to take up the issue. There is a German participation and numerous activities on the *IPBES*-platform (*Intergovernmental Science-Policy Platform on Biodiversity and Ecosystem Services*) but contrary to the USA, the UK, the Netherlands (Termorshuizen and Opdam 2009) or Switzerland (Staub et al. 2011) for example there are still no generally applicable guidelines for the implementation of approaches using urban ecosystem services for (urban) landscape planning. Nevertheless, governance structures of town and city development and landscape/nature protection often apply the principles of the urban ES concept like for example the temporary use projects in Berlin or Leipzig (here: licence agreements on interim land use, Lorance Rall and Haase 2011).

6.5 Cultural Landscapes and their Ecosystem Services

6.5.1 The Example of Orchard Meadows in the Swabian Alb Biosphere Reserve

B. Ohnesorge, C. Bieling, C. Schleyer and T. Plieninger

Introduction

Europe's land cover has been shaped to become a mosaic of cultural landscapes by wide-ranging and comprehensive human use for centuries (Schaich et al. 2010). The term 'cultural landscape', on the one hand, emphasises the aspect of anthropogenic cultivation and visible changes to the natural landscape due to continuous use. On the other hand, it includes a cognitive dimension pointing to the cultural meaning (► Sect. 3.4) humans assign to the space surrounding them and to the natural and anthropogenic elements therein (Jones 2003). Human influence on natural landscapes is often associated solely with disturbance and negative influences. But mankind can also contribute in many ways to the diversity and uniqueness of a landscape, be it through socio-economic, emotional or intellectual input (Moreira et al. 2006).

While a large part of scientific literature deals with capturing ES on the ecosystem scale, this chapter points to ES delivered on the landscape scale. Cultural landscapes usually embrace a multitude of different habitats. Forest patches, meadows, hedgerows and streams, for example, all deliver specific ES locally. Their composition on the landscape scale delivers additional services such as the regulation of the landscape water regime, groundwater renewal or pollination of cultural crops by wild insects. On this scale, the CES play an important role. Due to the long history of human use and the intense interaction between man and nature, cultural landscapes are closely linked to a sense of place and cultural heritage, but also to values, biographies and identities (Schaich et al. 2010). Moreover, recreation, inspiration and spiritual edification by enjoying cultural landscapes become ever more important, as most Europeans spend an increasing amount of their time in cities and inside buildings.

The cultural services should, thus, be taken especially into account whenever the ES concept is applied to cultural landscapes. Despite the central influence cultural services have on well-being and quality of life, they are still only marginally accounted for in studies capturing and valuing ES. One reason is that they are hard to grasp, to quantify or to spatially represent, compared to most regulating or provisioning services (Chan et al. 2012). In the following chapter, we use the example of orchard meadows in the Swabian Alb Biosphere Reserve to demonstrate the importance of cultural services and introduce a method for the spatial recording of their manifestations. We explain some examples of instruments and initiatives for the conservation of orchard meadows (*Payments for Ecosystem Services*) and discuss the implementation of the ecosystem service concept in landscape management.

From Efficient Food Production to Threatened Cultural Heritage: Orchard Meadows in the Course of the Centuries

Orchard meadows are common landscape elements in eleven European countries and comprise a total of about 1 million hectares. Their distributional range extends from Northern France to Southern Germany and Switzerland to Poland. In the German state of Baden-Württemberg, orchard meadows cover approximately 116,000 ha and 7.1 % of the agricultural land. The largest contiguous area of orchard meadows occurs at the northwestern border of the Swabian Alb Biosphere Reserve, approximately 30 km southeast of Stuttgart, situated in the foothills of the Swabian Alb mountain range. This region exhibits a mild climate and a diverse landscape scenery, which is characterised by extensive, richly structured orchard meadows around settlements, small-scale agriculture, and deciduous hillside forests. Due to its location within the Stuttgart metropolitan area, the area has a high population density of 719 inhabitants per km^2. Therefore, the orchard meadow landscapes are not only of importance for biodiversity conservation but also for landscape-related recreation (LUBW 2009). The ecoregion harbours around 600,000 scattered fruit trees and 6,000 ha of orchard meadows.

Orchard Meadows

Scattered fruit tree or orchard meadows are an excellent study area as they combine many attributes of traditional cultural landscapes: a mixture of low-intensity orchards and agricultural use, a small-scale landscape mosaic and structural richness favour a high biological diversity. This type of landscape also brings forward considerable cultural services such as recreation and a 'sense of place' (◻ Fig. 6.20).

Lucke et al. (1992) define scattered fruit trees as "generally tall growing trees of different fruit species, varieties and age brackets, standing 'scattered' in irregular intervals on fields, meadows and pastures. The term also includes singular trees along paths, roads and slopes, small groups of trees, tree rows and extensive sites with trees standing in regular, but large intervals". Apple, pear, prune plum and sweet cherry are the most common species.

Orchard meadow landscapes are appreciated for the diverse ES that they provide to society. Among the provisioning services fruit production is important–despite a widespread conversion to high-performance fruit plantations (Weller 2006). Appreciated products are dessert fruit, fruit juices, cider and liquors. Also, orchard meadows provide cultural services as scattered fruit trees are outstanding components of an aesthetically pleasing landscape. Due to their location close to villages and towns, they contribute to recreation and often attract day tourists. Substantial regulation services are supplied by positively influencing local climates (e.g. through wind protection or shading), by halting soil erosion on hillsides, and by reducing nutrient inputs into water bodies. The large diversity of fruit varieties (more than 3000 fruit varieties have been recorded in Germany; MLR 2009) represents an important genetic resource. Finally, orchard meadows offer habitats to many plant and animal species (Herzog 2000).

Orchard meadows were introduced in many German regions as innovative land-use systems in the eighteenth and nineteenth centuries, often supported by the respective government. The principal aim was to improve profitability of agriculture and the provision of food to local people. In a first stage, fruit trees were planted in home gardens and on crop fields; at a later stage, these were often converted from croplands to meadows. Important land uses on the ground were livestock grazing as well as cultivation of fodder, cereals, root crops, horticultural crops, and berries. The orchard meadow economy culminated in the middle of the twentieth century. The last large-scale plantations of orchard meadows took place during and shortly after World War II, in order to supply people with fresh fruit (Müller 2005).

Orchard meadows have been strongly declining since the 1950s. While agricultural statistics for the federal state of Baden-Württemberg recorded 18 million scattered fruit trees in 1965, only 11.4 million trees were counted in 1990 and 9.3 million trees in 2009 (MLR 2009). Many orchard meadows were replaced systematically by modern and intensive fruit plantations, which allow a rational management based on machinery. In the 1960s and 1970s the federal state of Baden-Württemberg and the European Economic Community subsidised the clearing of orchard meadows in order to promote the 'modernisation' of orcharding. In consequence, approximately 15,700 ha of orchard meadows were removed between 1957 and 1974 (Weller 2006). Many of the orchard meadows close to urban areas vanished through establishment of settlements and industrial areas. The construction and upgrading of roads implied a loss of many scattered fruit tree rows.

The conservation of orchard meadows depends on active management, in particular regular pruning and renewal of trees. With the rising competition of intensive plantations and changing consumption patterns (e.g. increasing quality demands towards dessert fruit), traditional orcharding practices often lost profitability. Due to land abandonment and lacking regeneration of trees, many orchard meadows were encroached by bushes and shrubs while many fruit tree stands are overaged

Fig. 6.20 Orchard meadows are characteristic of many Central European cultural landscapes. © Tobias Plieninger

and deceased through lack of management. The remaining scattered fruit tree stands are usually not managed by professional fruit growers, but by hobby and self-sustaining farmers (Weller 2006). The fruit trees are overwhelmingly found on meadows and pastures, while scattered fruit trees have largely vanished from croplands.

Manifestations of Cultural Ecosystem Services of Orchard Meadows

Despite, or maybe because of their decline, the significance of orchard meadows as an extremely valuable landscape feature is increasingly stressed by the local population, but also by particular stakeholder groups like conservationists or representatives of tourist associations. Their values are most commonly described in terms of CES like aesthetics, sense of place and cultural heritage. In the following section, we present an approach that delivers concrete insights into the character and prevalence of CES related to an area covered with orchard meadows in the foothills of the Swabian Alb in southwestern Germany.

Manifestations of CES in the Foothills of the Swabian Alb

The method applied begins with the consideration that, like in the case of agricultural land use, the use of cultural services also leaves traces in the landscape. Campfire places indicate recreational use, wayside crosses reflect religious values, and well-tended historical features show that cultural heritage is relevant. These are visible manifestations of CES, and the basic idea of the proposed approach is to record such visible signs of people actually using ecosystem features in an intangible manner. The approach corresponds with the notion highlighted, for example, by Eiter (2010) and Stephenson (2008) that the experiencing of landscapes, i.e. their use, is a crucial constituent for landscape values.

We explored the potentials and constraints of this approach using the example of an area covered with orchard meadows. In a systematic field-walking procedure, all visible signs of use, which are predominantly related to intangible services, were recorded and documented in a map, which was later transferred to a GIS (for a detailed de-

■ **Fig. 6.21** Bench as manifestation of aesthetic and recreational services. © Claudia Bieling

scription of methods and results see Bieling and Plieninger 2013). In the investigated area, manifold and numerous manifestations of CES were identified and were grouped into seven categories of similar uses: benches (■ Fig. 6.21), subsistence gardens, hunting facilities, recreational facilities (e.g. small private huts, barbecue places), manifestations indicative of cultural ties to the past (e.g. commemorative plaques), trails and signs for hikers and cyclists as well as 'other' (e.g. a small Christmas tree plantation likely to indicate religious values, but as a source of subsistence also connected to identity) (■ Table 6.7). Subsistence gardens and hunting facilities were interpreted as predominantly indicating intangible services, because gardening and hunting entail strong connections to questions of identity, but also to social relations and recreation, which typically outweighs the significance of the products obtained (e.g. vegetables, game meat).

In a second step, the categories of recorded manifestations were connected with the types of CES as described in the MEA 2005 (■ Table 6.7). Manifestations of recreational values and identity were particularly frequent, whereas inspiration and spiritual/religious values seemed to be of little relevance in the investigated area.

Two aspects have to be considered when drawing conclusions from these findings: first, this exploratory study shows that visible manifestations are good indicators for some, but not for all types of CES. For instance, it was not possible to find any indicator for inspirational services, even though it is highly probable that people do actually derive inspiration from the investigated orchard meadows. In this regard, the landscape is probably not the right place to look for manifestations, and indicators might be better found in places like local art exhibitions, regional literature and poetry or children's kindergarten drawings. Second, manifestations of CES are highly context-dependent, and it would not be appropriate to compare results between different natural and cultural contexts. For instance, we hardly found any manifestation for spiritual/

Table 6.7 Types of cultural ecosystem services (CES), associated visible manifestations and their relevance in the investigated area

CES types	Associated visible manifestations	Relevance (recorded manifestations/ total number of manifestations) (%)
Identity	Subsistence gardens, hunting facilities, pond, Christmas tree plantation	28
Cultural heritage	Memorials, commemorations, historical sites	5
Spiritual/religious services	Christmas tree plantation	1
Inspiration	–	0
Aesthetic services	Benches	11
Recreation	Hiking trails and signs; recreational facilities; benches; subsistence gardens; hunting facilities	55

religious values in our case study area, which belongs to a predominantly Protestant municipality. In Protestant culture it is not common to express religious values through manifestations such as wayside crosses, which can often be observed in Roman Catholic-dominated areas. However, this does not necessarily mean that people in this Protestant municipality attach less spiritual values to their surroundings than their Catholic neighbours a few kilometres away, where one would certainly find more indicators of spiritual services.

Visible manifestations can be analysed in a spatially explicit way, which offers several benefits: On the one hand, hot spots of CES provision are revealed, on the other hand, places are indicated which have relatively little relevance in this regard. Moreover, as can be seen from visualising the results in a GIS and applying statistical analyses, the approach helps to identify common overlaps of different ES. In our example, places that provided recreational services typically also were important in terms of identity and cultural heritage. Likewise, the linkages between CES and variables like topography or exposition may be investigated. However, the complexity of assigning CES to a specific place has to be considered. For example, aesthetics like a beautiful view depend both on the place of 'consumption' (the place where the view is enjoyed) and the place of provision (i.e. the landscape being viewed).

▪ Cultural Ecosystem Services as a Facilitator of Sustainable Land-Use Practices

The example of the orchard meadows shows that cultural landscapes are tightly linked to CES. This applies to tourists, but even more to the local population and particularly land owners, which are most relevant for the implementation of sustainable land-use practices. Nonmaterial values play an outstanding role in land owners' land-use decisions. This is, for instance, revealed in a study on how nonindustrial private forest owners view so-called close-to-nature management practices. It concludes that the implementation of such approaches is not so much driven by financial incentives like subsidies, but depends on the compliance of management practices with nonmaterial values and cultural background (e.g. family traditions) held by the land owner (Bieling 2004). Therefore, CES may be seen as a starting point and most relevant argument in incentives that strive for management approaches which sustain a broad range of ES. Another important issue in this context is that CES, in contrast to most provisioning and regulating services, are typically linked to a specific place and cannot be replaced by technical or socio-economic solutions (MEA 2005). In Central Europe, the high economic standard allows to import agrarian goods from all regions of the world. Services like local recreation, sense of place or cultural heritage, however, are inextricably linked to the cultural landscapes people are attached to in their daily life–if they are not sustained in these places, they are lost (compare Guo et al. 2010 for the increased dependence on CES in economically well-developed countries).

Policy Instruments and Civil Society Initiatives for Preserving the Manifold Services Provided by Orchard Meadows

Given the manifold cultural and other ES provided by orchard meadows, political and economic instruments are critical to counter the degradation and drastic decline of these important elements of the regional cultural landscape. Unlike in other German federal states like Saxony, orchard meadows in Baden-Württemberg are not protected by specific regulatory state policies. However, orchard meadows are often either part of Natura 2000 areas, situated within nature protection or landscape protection areas, or can be found in the so-called buffer or transition zones of biosphere reserves. Outside nature protection areas, orchard meadows can be used for agricultural purposes, yet only in a nonintensive manner and in compliance with further specific nature protection regulations. Such land-use restrictions, however, are often perceived by land users as unduly intervention by state authorities. Even more importantly, these legal restrictions do not contain any obligations to carry out the necessary maintenance of the orchard meadows through nonintensive land management in a regular fashion.

To counteract the decline and increasing degradation of orchard meadows, a number of public and private funding programmes and other incentive-based instruments have been launched in recent years. These reward mechanisms–in the international scientific literature usually referred to as 'Payments for Ecosystem Services'–provide fixed payments for land users to voluntarily carry out concrete measures, such as a specific bird-friendly pruning. In Baden-Württemberg, the state government is spending about 10 million Euros per year for direct and indirect measures for the preservation of orchard meadows (MLR 2009). While in many cases CES are positively affected by these preservation measures, they are hardly targeted directly. Promoting cultural services more directly often fails because they are much harder to grasp compared to provisioning and regulatory services as well as more difficult to link to specific forms of land management and to quantify.

Preservation of orchard meadows is mainly supported by the agri-environmental programme of Baden-Württemberg (MEKA III). Land users receive 2.50 € per tree for the mandatory management and maintenance of grassland under and between the fruit trees and for the regeneration of standard fruit trees. Other grants for the conservation and maintenance of meadow orchards, but also for planting new fruit trees are available in certain funding and project areas in the context of either land consolidation measures or the Directive on Countryside Conservation (Landschaftspflegerichtlinie) of the federal state of Baden-Württemberg. In the future, establishing and maintaining orchard meadows will be considered as compensation measures for interventions in nature and landscapes and will entitle to obtain appropriate compensation payments (so-called eco-accounts or habitat banking).

Other state-funded assistance programmes support the processing and marketing of fruits from orchard meadows in the biosphere area and beyond: This includes promoting investments in diversification measures of agricultural enterprises, for example for establishing small distilleries, but also for creating efficient networks of (juice) wineries and for the purchase of mobile juice presses and bottling machines (MLR 2008, 2009).

Other incentive-based forms of promoting orchard meadows in the biosphere reserve are initiatives where, for example, apple juice from scattered fruit trees is marketed at a premium to compensate for maintenance efforts. Here, the redemption price for the producer is higher than the current or normal price if the fruits come from orchard meadows that are managed nonintensively or even ecologically and/or that are situated in specific regional production areas. For example, in the context of the PLENUM project (which started in 2001) of the federal state of Baden-Württemberg which aims to preserve and develop nature and the environment the apple juice and liqueur brand 'ebbes Guad's' was developed in collaboration with local (juice) wineries and fruit-growing associations. Here, the fruits come from about 200 ha of orchard meadows where only specific production practices are used and land users receive a premium of 3 €/dt beyond the current market price. PLENUM funds are also used to support the apple juice initiative 'Feines von Reutlinger Streuobstwiesen' ('Fine fruits from orchard meadows in Reutlingen') where land-users receive an even higher premium to adhere to additional and more stringent ecological criteria. In addition to these premium price initiatives, in the PLENUM area Swabian Alb also infrastructure measures for cider and wineries are cofinanced as well as projects promoting the integration of actors along the entire value chain (producers, processors, retailers, restaurants, and consumers) and improving public relations and fostering the education and training of orchard farmers (PLENUM 2008).

The wide range of payment and reward programmes to promote orchard meadows illustrates the increasing importance of incentive-based instruments for the conservation of cultural landscapes. This seems all the more 'trend-setting' since the majority of ES provided by traditional orchard meadows are public goods which are not traded at 'free markets'. Without any doubt, the funding schemes and initiatives mentioned above have certainly provided a number of impulses for the preservation and maintenance of orchard meadows. Due to a number of reasons, however, it appears to be doubtful that the existing incentive-based instruments–both, content-wise and with respect to the financial resources available–will be able to stop or even reverse the decline and degradation of orchard meadows. For example, only 1.67 million of altogether 9.3 million scattered fruit trees in Baden-Württemberg were promoted in the framework of the agri-environmental programme MEKA III (MLR 2010).

The poor uptake of this funding instrument is caused, first, by the fact that a large part of the orchard meadows is not maintained by farmers, but by amateur cultivators who are not eligible to receive payments from EU cofinanced agri-environmental programmes. Second, the plantation and maintenance of orchard meadows are associated with high investment and production costs, which are only partly covered by the respective premiums of state-financed schemes. Third, the farming requirements that have to be met to receive payments through these schemes as well as the funding periods are often very inflexible and the measures are hardly tailored to the specifics of the regional ecosystems and farming systems.

Further, competing policies, such as schemes promoting the cultivation of energy crops, also increase the opportunity costs of managing orchard meadows and, thus, increase the land-use pressure in areas with orchard meadows (cf. Schleyer and Plieninger 2011). Finally, before the background of an increasingly rigid government spending it seems useful to make more use of private financial sources for such programmes. Here, price premium initiatives, private investments in marketing and processing infrastructures as well as compensatory payments for interventions in nature and landscapes have a high potential.

Conclusion

Orchard meadows combine many features of traditional cultural landscapes and deliver a multitude of ES. Although they were originally developed for the sake of efficient food production, today especially their cultural services are acknowledged, such as a sense of place, recreation and inspiration. But these landscapes are also subject to the comprehensive land-use trends we could observe during the last decades: intensification of production, settlement and infrastructure development on the one hand, urbanisation and land abandonment on the other (Plieninger et al. 2006) pose a threat to the survival of these meadows.

The ES approach can offer valuable perspectives in the struggle for conserving traditional cultural landscapes and in balancing different demands on land use. In order to understand the societal value of a landscape comprehensively, its cultural services must be taken into account adequately.

We introduced mapping of landscape elements which represent the cultural meaning of the landscape as a possible way of operationalising cultural services. Deliberately linking values such as recreation, identity, cultural heritage and a sense of place to tangible landscape elements may result in a stronger motivation to conserve the landscape than what could be achieved by directives and financial incentives alone. The ES concept in landscape management bears great potential for visualising the links between cultural landscapes and human well-being. Awareness-raising for the interrelation between land use and the immaterial parameters of well-being must be just as strong as it is for the material aspects in order to conserve cultural landscapes' ES and to foster them by means of targeted policies or initiatives.

6.5.2 Calculation of Landscape Management Measures and Costs

K. Grunewald, R.-U. Syrbe and O. Bastian

Objectives and Methodology

A broad social consensus exists for the permanent preservation and development of our cultural landscapes and habitats, which expresses itself in a growing demand for biodiversity and intact cultural landscapes, as well as in the willingness to provide the financial means to do this (e.g. Hampicke 2006; Spangenberg and Settele 2010). However, in order to secure their own existence, human beings have to intervene in nature, altering it by various forms of land use. In addition to provisioning services (e.g. generation of foodstuffs and raw materials through agriculture and forestry), cultural landscapes and their ecosystems also furnish many regulation and sociocultural services. To make the broad range of ES permanently available, i.e. to ensure that biodiversity and productivity of the ecosystems are preserved, targeted landscape management is necessary, something which entails financial expenditure for society (Grunewald et al. 2014).

If we take the politically specified need for the preservation of species and habitats as contained in treaties, guidelines, laws and regulations into account (e.g. Convention on Biological Diversity, EU-Natura 2000 Guidelines, EU Biodiversity Strategy, Measures Program for Biological Diversity, etc.), then suitable measures have to be taken in accordance with the technically derived requirements. The social expenditures and costs for landscape management therefore represent an indicator of the economic valuation of ecosystems, since their existence is not secure without these performances. Apart from ethical, aesthetic and informational values, which are very difficult to determine in monetary terms, landscape management accounting as a tool for indicating the need of action to maintain ecosystems can help to negotiate the level of socially agreed demand for nature as well as the willingness to pay for nature protection. The ES approach addresses calls for incorporation of such economic valuations in ecological management decisions (Carpenter and Turner 2000; Farber et al. 2006).

Landscape management is defined as the totality of all measures for the safeguarding, maintenance and development of natural habitats for indigenous species of plants and animals, and for the maintenance and renaturalization of ecosystems and landscapes in the event of damage (Jedicke 1996). In this context, one important task is the preservation and development of ecological and landscape diversity. Landscape management is particularly concerned with the safeguarding and provision of general interest services for society (particularly regulation and sociocultural ES).

Even if appropriate databases have gradually been improved, comprehensible calculation models relating to the valuation of landscape maintenance services are not very common. However, knowledge of the magnitude of required financial resources are necessary to plan and to ensure the expenses for an extensive land management, at least in part, and/ or additional use, maintenance and development of habitats. Payments for ecosystems and benefits from ES must be considered in context (allocation options). For example, if farmers get payments for landscape conservation measures, such as the extensive use of grassland, they should have entitled only to the compensation of additional expenses or income compensation. As they are participants in the market, distortion must be avoided in favour of payments, which are justified by adequate performances and not by appropriate consideration covered subsidies.

Any investigation of the requirements for landscape management includes a comprehensive recording of landscape management objects or tasks (habitats, structures, species, as well as any existing

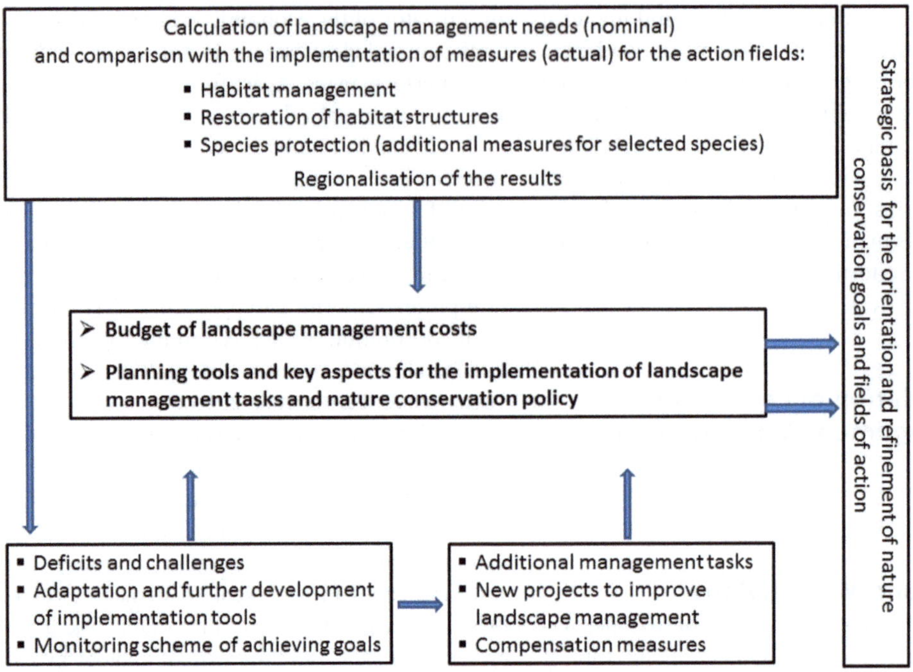

Fig. 6.22 Scheme of the key aspects of a landscape management accounting system. (According to Grunewald and Syrbe 2013; Grunewald et al. 2014)

deficits) and an estimation of the management costs (related to year, type and object). The total financial requirements consist of the costs for the management, developmental and investment measures for each type of habitat, as well as of special expenditures for the protection of particular species.

Based on approaches of LfUG (1999) and Döring (2005), we developed a methodology for a regional (state-wide) calculation of landscape management measures and costs. It was compiled in an exemplary manner for the Federal State of Saxony (ca. 18,420 km^2) in Germany. Figure 6.22 illustrates the contents and the focal points of such a regional landscape management accounting system with the possibility of regional differentiation on the administrative level and on the level of the physical region (Fig. 6.22).

The objective fields of practical action of a landscape management strategy include, above all, open-land areas, water structures, woods, as well as measures for species protection relating to object and habitat. In this context, reference must be made to the funding practices of individual countries and of the EU with regard to nature conservation and landscape management.

The data sources which could be used for the creation of a landscape management accounting system are of varying quality (survey status, regionally different mappers, etc.). This applies to the whole analytic section of the landscape management accounting, which is based on facts and assignments. Consequently, error margins and uncertainties have to be taken into account in the overall context. However, the results were verified using both data obtained from literature and alternative approaches (in particular expert knowledge), and may be regarded as generally reliable, as well as providing a sufficient level of accuracy for the purposes of the landscape management accounting system (Grunewald and Syrbe 2013).

The first step in the frame of the methodology is the mapping of management-relevant habitat types of a region or of a country as "regional scenario of habitat types, which are relevant for the landscape management" (groups of habitat types, see Grunewald et al. 2014):

- Determination of habitat areas on the basis of existing habitat mappings: selective mapping of the habitats (biotopes) in Saxony (SBK; *German: Selektive Biotopkartierung*), which are important for nature conservation
- Expansion in accordance with the habitat types mapped within the Natura 2000 sites of the EU (Habitat Directive 1992), as far as these surpass the SBK-habitat mappings
- Integration of the so far almost neglected arable land, which (still) does not comply with the criteria of the habitat type "extensively used fields rich in wild herbs", but which could contribute to the preservation of important species of our cultural landscapes by means of careful agricultural management: estimation of a realistic area using the HNV (High Nature Value) farmland indicator of the German Federal Agency for Nature Conservation (► http://www.bfn.de/0315_hnv.html).

Subsequently, the maintenance status of habitats is to be estimated and subdivided into regular and periodic measures. 'Regular management' refers to management of 100 % of the habitat areas at least twice or more often per decade. 'Periodic management' (investment costs) refers to either so-called once-off measures, e.g. land clearance or treetop pruning in the case of old fruit trees, or to cases where the management interval is so large that the relevant habitat type is 'affected' by a measure once at most within the usual period of 10 years, e.g. in the case of hedge/copse maintenance.

Depending on the evaluations in the framework of the Fauna-Flora-Habitat (FFH) monitoring activities within the framework of Natura 2000, the additional devolvement requirements were estimated for each habitat or habitat type.

In the case of habitat areas in a condition of preservation which can be rated as average to bad condition (C), regular management measures are not sufficient to restore an excellent (A) or a good (B) condition of preservation of the habitat type. In these cases, some once-off management measures with a greater expenditure of time and cost are necessary, such as land clearance and repeated mowing. Consequently, once-off management measures (=investment) are designed for open land habitats such as grassland, heathland and neglected grassland, if these were rated only in preservation condition 'C' (bad). In such cases the cost expenditure of approximately 1000 €/ha (plus rate of increase) is distributed over an implementation period of 10 years.

Not all habitat types are covered by measures and costs in FFH management planning. In such cases, plausible cost estimations were undertaken by means of comparisons. Among other factors, this concerns the habitat types YS (clearance cairns) and UR (vineyard extensively), whose management costs are taken as analogous to BH/BA (field hedge/field trees) or YM (natural stone walls).

The cost rates specific to regular management measures were determined on the basis of the currently valid funding rates for habitat management (EU funds, combined with cofinancing by the country). In addition, the 5242 individual funding measures for the Saxon guidelines NE ("Natural Heritage" 2007) and AuW ("Agricultural Environmental Measures and Forestation" 2007) were evaluated in the year of reference 2009.

The records were filtered according to habitat types requiring maintenance. Subsequently, the claimed area per measure and the corresponding funding area for each habitat type were determined. The possibility of adaptations of premiums on the basis of cost increases occurs regularly in the specified planning period of 10 years (at least once upon expiry of the funding period) as well as irregularly, according to special adaptation requirements. A total increase of the premium rates of 10 % was assumed for the period under consideration (10 years), which gives a mark-up of 5 % on the current calculation sum.

Furthermore, an estimation of 5 % transaction costs was stipulated. Transaction costs are described as "expenses which arise in order to obtain information, to make decisions or conduct negotiations, to monitor agreements and rules, as well as implementing them or adapting them to modified framework conditions" (Masten 2000). Therefore, within the framework of landscape management, expenses which arise in connection with information, consultation, conclusion and accompanying bureaucracy within the environmental programmes should be remunerated at a general rate per hectare.

In the case of premium rates in the field of habitat use and management, we are dealing with fixed cost rates. Neither planning and management costs nor incentive components are contained therein. Calculation of an incentive component (actually held to be necessary) was not carried out as this has been excluded by the EU for the foreseeable future (period under consideration).

Results Exemplified for Saxony

▪ Measures and Cost Calculation for Habitat Management

The habitat area in Saxony relevant in terms of landscape management was identified at about 162,000 ha. This is equivalent to 9.3 % of the total area of Saxony. More than half of the relevant habitat area is to be balanced with regular management measures (grassland habitats, heathland and neglected grassland) and with use and management in keeping with conservation objectives (arable land, grassland). The management measures described as episodic are relevant for approximately half of the identified habitat area in Saxony, including all forest habitat types. The habitat types of near natural lakes/ponds, orchard meadows, dwarf shrub heathland, fen and marsh require both regular and episodic management measures.

A landscape management assessment for the Federal State of Saxony was first put forward in 1999 (LfUG 1999). At that time, an area was identified whose habitat management requirements were lower by 4.1 % in comparison to today (2011). The difference is caused principally by the present inclusion of the category "arable land which is to be used less intensively", since the intensity of use and, therefore, the necessity of action in the area of arable land have increased significantly. This caused, among other factors, the need for improvement of the population of ground-breeding species or the protection of rare segetal weeds. The arable land which should be used less intensively alone represents 33,000 ha, that is, approximately 3 % of Saxony's agricultural area. But the proportions of forest habitats, trees, hedges and bushes requiring maintenance have also increased, and the habitat types of the FFH Directive which were formerly only partially recorded by means of selective habitat (biotope) mapping have now also been integrated.

It has been calculated that about 49 million euros are needed annually to cover the costs of management and development of habitats in Saxony (Grunewald and Syrbe 2013). Of this, more than 17 million euros per year would be needed as compensation payments for "arable land which is to be used less intensively". Focal points of action were identified at the district, physical region, and natural region level, indicating a need for resources and accountability. Accordingly, in rural districts with a large surface area and a high proportion of habitats, which require maintenance, a sum of up to 7.7 million euros per year must be spent, while in the urban municipalities of Leipzig, Dresden, and Chemnitz, a considerably lower level of financial need exists in this regard. That would be a starting point for the reorganization of ecological fiscal equalization discussed in Sect. 5.2.2.

▪ Restructuring Requirements for Landscape Elements and Habitat Structures

The agricultural landscape offers various points of approach with regard to restructuring (Syrbe and Grunewald 2013). First the watercourses including their accompanying structures, should be in a good ecological condition, so as to fulfil habitat functions and to contribute to a steady water and mass balance. Second apart from the bodies of water, agricultural areas with low amounts of forest and water also need field trees, lines of trees, hedges and various sorts of linear or field margin structures (Ringler et al. 1997), in order to provide cover, food and nesting opportunities to the organisms of agricultural ecosystems, to upgrade the landscape, and to avert the dangers of erosion. The objectives of landscape management demonstrate that, among other tasks, the edges of forests are to be upgraded by means of staged boundary structures, a minimum proportion of wetland habitats are to be preserved in meadows, and terraces and clearance cairns/stone walls are to be conserved or reconstructed in the mountains.

In the case of running waters and their accompanying structural elements, there was a need for the opening of piped flowing water sections at

300 km, the restructuring of trees along approximately 680 km of watercourses, and the discontinuance or modification of uses on 21.3 thousand ha in the environment of the water body. In addition, a quantitative deficit analysis suggests the reestablishment of 2500 km of linear structural elements in the agricultural landscape, which is a wholly realistic dimension when compared with activities of reestablishment previously carried out, e.g. in connection with impact mitigation regulation. In this way, a structural density characterizing the landscape before agricultural collectivization would be achieved, at least in regard to the wooded areas.

The suggested measures are qualitatively supplemented by a determination of focal regions within Saxony for extensive restructuring measures. Because of acute lack, it is recommended that additional field trees and extensively used areas should be established there (Syrbe and Grunewald 2013). Some landscape regions display very high restructuring requirements with above-average costs. In total, annual restructuring costs of 12 million euros were assessed for Saxony, which are to be implemented over an area of 25,000 ha. This includes both changes of use as well as linear measures, which have been applied generally to all areas in order to avoid double payments.

With regard to restructuring, the areas in question have increased in comparison with the assessment made in 1999 (LfUG 1999). The cost estimation, however, turns out to be moderately lower, since the focus is now on the transformation of the use of arable land, instead of the more comprehensive planting of trees as planned in 1999, which has so far been partially implemented. With regard to the restructuring of water bodies, it is important to note that significant individual tasks within the framework of the measure plans according to the EU Water Framework Directive (WFD) are meanwhile being executed and are no longer being assigned to landscape management in the cost assessment, although they still have to be performed (Grunewald et al. 2014).

▪ Specific Measures for Species Protection

Special conservation measures for endangered animal and plant species are necessary if the management as part of 'Good Practice' or the care and protection of habitats is not sufficient to the preservation of this species in the long term. At present conservation programmes exist for freshwater pearl mussel (*Margaritifera margaritifera*), white stork (*Ciconia ciconia*) and otter (*Lutra lutra*) in Saxony. Necessary costs to address specific issues of species protection in Saxony were estimated at 2.43 million euros per year (approximately 1.7 million euros for regional and 0.75 million euros for state-wide significant species). Substantial additional funds for these types of protective measures can be generated via the engagement and compensation scheme or project funding and endowments.

▪ Aspects of Implementation and Financing

Undeniably, many successes have been made through the recent efforts of the landscape maintenance stakeholders. Thus, the preservation of semi-natural habitats, such as dry and medium-dry grassland, dwarf shrub heath and Nardus-grasslands, mountain and marsh meadows or stone and vineyard walls was a success. Also, populations of previously highly threatened species, such as sea eagles, have recovered and once extinct species such as wolves have returned to Saxony, or they could be resettled, as in the case of salmon.

On the other hand, the statement must be made that a fundamentally nature-friendly orientation of farming has not yet been reached, at all. The inventories in the FFH habitat sites underline this fact. For example, six of 15 FFH habitat types of agricultural ecosystems, occurring in Saxony, have an unfavourable state (Hettwer et al. 2009).

In the context of agricultural management a satisfactory distinction between necessary use processes and ecologically disadvantageous engagement is insufficiently resolved till this day, especially as the management intensity and, thus, the risks to nature have steadily increased in recent years. The reference level of 'good practice' is not yet defined in a satisfactory manner in terms of the requirements of nature conservation and landscape management both in the agricultural and the forestry sector. For that reason alone the recognition of ecological services in agriculture and forestry is difficult, e.g. as compensation and restoration measure.

Thus, land users can be obtained as reliable partners of nature conservation. In order to

Table 6.8 Expenditure (state budget titles) for nature conservation and landscape management of selected German states. (Source: Statistics of the federal states; SMUL 2010; Gottschall 2011; StMUG 2011)

	Baden-Wuerttemberg	Bavaria	Saxony
State budget (2011)	35.1 billion €	42.5 billion €	15.5 billion €
Thereof nature conservation and landscape management	ca. 30 million €	ca. 39 million €	ca. 10 million €
%	0.09	0.09	0.06
€ per inhabitant	2.78	3.11	2.44
€ per km^2	840	553	544

reach conservation goals together, a corresponding range of flexible, goal-oriented and actionable conservation funding measures is needed, because these measures are the central elements for the implementation of objectives of nature conservation and landscape management on substantial proportions of land. Through financial means a high commitment and importance is achieved. In this way, the protection of natural elements and cultural landscapes can be ensured through a co-operative approach.

For the implementation of objectives of nature conservation and landscape management, a range of financing instruments is available, especially with programmes in the fields of environment and nature conservation, agriculture, forestry, fishing and rural development. The impact of international legal standards to funding offers by the federal states has risen in recent years. The 2007–2013 period was largely financed by European sources. The main basis for the use of EU funds for nature conservation and landscape management is the European Agricultural Fund for Rural Development (EAFRD). Tools that contribute to tasks of biodiversity and nature conservation in cultural landscapes are available in the rural development programmes with its variety of areas, measures and project-related subsidies and funding.

The tasks of landscape management are co financed by German federal states. ◘ Table 6.8 gives an orientation to the (relatively small) expenses for nature conservation and landscape management in state budgets by comparing the German federal states of Baden-Wuerttemberg, Bavaria and Saxony. Although the main part of funding is now being paid from EU funds, in the last decade Saxony has increased expenditure by about 2 million euros (from approximately 8 million euros in 1999 to ca. 10 million euros in 2011). The state funding also contributes to the preservation of jobs and regional development, especially in rural areas. Note the so-called leverage for the financing of nature conservation and landscape management. In the next EU funding period starting in 2014, Saxony will only receive 50 % co financing share instead of the previously 75 %. This is already the case in Baden-Wuerttemberg and Bavaria in the current EU funding period.

The measures of landscape management are not arranged (principle of voluntariness). However, the administrative procedures for promotion are usually very expensive simply due to the requirements of the integrated administration and control system (IACS) of the EU (application, examination, control). Efforts to maximise simplification of measures and related management and monitoring procedures cannot always agree with the demand for customised solutions. The limited availability of funds is another reason for the discrepancy between actual funding and calculated needs (target).

The fact that the EU has failed its set goal to halt the loss of biodiversity by 2010 requires new or modified approaches to be successful in the current decade. Starting from the social responsibility of ownership it seems urgently necessary to specify the regulatory framework of land use. In economic competition, the guiding principle of voluntariness can work only if the limits of use are fixed. To prevent nature damaging procedures, a new definition of 'good practice' is necessary as well as the setting of obligatory treatment guidelines, do's and don'ts in protected areas including Natura 2000 sites. Since a strong political will and a high social demand for biodiversity and ES exist, it should not be left solely to the discretion of the relatively small group of land owners and land managers, whether they correspond to this concern.

▪ Target-Actual Comparison on the Example of Saxony

It has been calculated that about 67.4 million euros per year are currently needed for landscape management in Saxony (estimated minimum cost of landscape maintenance; Grunewald and Syrbe 2013). Of this total, the management and development of habitats was calculated with 49 million euros, restructuring measures with 16 million euros and specific measures for endangered species with 2.4 million euros.

Overall, the required target sum is about 23.4 million euros higher than the one calculated a decade ago. This is due to a larger scope of landscape management tasks, the rate of inflation, but also due to methodological differences. Among the strategically defined areas of action, the ranking of needs is as follows: first habitat management, second restructuring and third protection of species. The focus has continued to change in the direction of maintenance (habitat management: four fifths of the cost requirements) compared to development (restructuring). The maintenance of grassland continues to play an important role, but has significantly exceeded by the financial resources needed for "arable land which is to be used less intensively" (see above) in the current assessment.

The utilization of funding in the State of Saxony was exemplarily analysed for the reference year 2009, for which the data of the agricultural report (SMUL 2009b) as well as the funding guidelines "Agricultural environmental measures and forestation" (RL AuW/2007), "Timber and forestry" (RL WuF/2007) and "Natural heritage" (RL NE/2007) have been evaluated. Overall, the actual flow was approximately 12.5 million euros in the field of nature conservation and landscape management for the reference year 2009 from these programmes. They are predominantly attributable to careful land use and habitat management. Nature conservation measures have been implemented in about 36,000 hectares of land in Saxony during the reference period (Grunewald and Syrbe 2013).

Koch et al. (2011) have analysed and evaluated the status of current participation in nature conservation funding in the State of Saxony. Key messages are:

- The currently funded habitat management area is approximately 1900 ha and therefore around 1000 ha below the number funded in the years 2003 to 2006. The delayed entry into force and early exposure of the funding opportunity, the costs of biomass disposal and pre financing of management measures are to be seen as problematic in this context.
- In about 14 % of permanent grassland the conservation-oriented grassland management is currently being funded, a slight increase compared to the previous programme (despite the criticism of the inflexibility of the funding scheme compared to the former programme).
- The trend of participation in measures of nature conservation-oriented agricultural use is positive, but the area of nature conservation fallow fields is much too small to address the loss of biodiversity on arable land. Less than 0.5 % of arable land in Saxony is supported by measures of nature conservation-oriented management and arrangement.
- For years, the funding area of nature conservation-oriented management of ponds has been at a high level, covering 84 % of Saxon pond surface currently.
- The implementation of investments in nature conservation measures in open spaces, in waters, in forests and species protection measures is far below the determined nature conservation needs (targets).

With regard to ES, these facts can be interpreted as follows: Participation in funding programmes is currently not very attractive; most likely it is associated with forms of use, which in this way can help to cover costs and income security (grassland, ponds), while in agriculture relatively high profits can be made through the sale of food crops and biomass (leaving aside funding resources and without dealing with the complicated application).

Challenges and Future Tasks of Landscape Management

The targeted goal of landscape management consists of the preservation and stimulation of wild species of plants and animals, as well as their habitats and biocoenoses, and the identification, preservation and stimulation of ecosystem (complexes) worthy of protection, landscape elements, landscape sections or even complete landscapes. Measures and funding to achieve these goals must be readjusted again in accordance with the available resources.

The quantitative, normative fixing of values is helpful to policy makers and administration in terms of implementation and supervision. Within the framework of the landscape management strategy, situation analyses and justification contexts in particular were drawn up and specific nature conservation goals were proposed. However, the question how far certain conditions or effects can be classified as desirable requires the formation of public opinion and cannot be ascertained on the basis of scientific reasons alone (Valsangiacomo 1998). It should also be remembered that the goals of nature conservation repeatedly come up against the limits of the possible in a narrow system framework, in the form of so-called practical constraints. This can result in individual cases that environmental authorities abandon ambitious and comprehensive targets.

Landscapes are changing rapidly. Turnaround in energy policy, climate change, economic globalisation and demographic changes will be the main drivers in this respect in Germany in the coming decade. A predictive location-based planning and control should help to make this development in an environmentally friendly way. New challenges from the perspective of nature conservation and landscape management relate mainly to the balance between protection and managing the cultural landscape with increasing pressure on land surface.

Landscape management as a societal task requires the cooperation of many actors, especially on a voluntary basis. For this purpose, the financial support of nature conservation measures is the main instrument of government action in the sense of compensation for costs and benefits, as well as opportunity payment for loss of earnings to those which implement landscape management.

But it must not be overlooked that in addition to funding a variety of other fields of action must be used to achieve the conservation objectives. Such alternative instruments include (Grunewald and Syrbe 2013):

- The contractual maintenance of state-owned land, land acquisition and possibly also exchange of areas to fine tune the location and distribution of these focus areas to the management objectives (applies, inter alia, in the forest); an additional effort free of charge to more ecological services is demanding for state and local governments because of the exemplary role of the public sector on state-owned land compared to private lands.
- The targeted arrangement of offerings in the context of eco-accounts, where funding and measures of eco-accounts exclude each other.
- The award (or support) of regional or eco-labels and certification of nature-friendly producing companies and their products; this can help to cover the abovementioned costs via a market with growing ethical demands.
- Regulatory options for action of any kind with prohibitions and requirements up to the application of the protection of objects and territories, such as the legal protection status of care-depending habitat types.
- Projects for compensation for ES, in which the beneficiaries of such services (for example, tourism providers, water companies, associations) reimburse certain expenses or loss of profits by the land users on the basis of private law agreements.
- Nature conservation law governing of land management as legal requirements of 'good practice'.
- Providing support and acquisition of alternative funding: sponsorship, professional volunteer management.

The integration of nature conservation objectives in the area of management of farmers (as all land users–a requirement of the Federal Program on

Biological Diversity; BMU 2011) involving the utilisation of funding is to estimate as not sufficient. One reason for this is the agro-economic conditions and the high requirements when participating in EU-funded support programmes. Number and scope of funding programmes are appropriate to the diversified nature conservation goals and the related landscape maintenance tasks. The application for the promotion of voluntary services is connected with large efforts for the actors (partly due to EU requirements). In contrast to large agricultural cooperatives with specialised employees, private managers are often overstrained and may pass on requesting important nature conservation measures. The effort in the context of conservation funding is estimated to be very high for control authorities as well (Koch et al. 2011). Therefore, as part of the programme planning 2014–2020, appropriate attention should be paid to aspects of simplification of programme participation. Koch et al. (2011) demand the following additional requirements from the State of Saxony for the next funding period:

- For measures of nature-friendly land use and habitat management, the targeted steering of nature conservation funding should be maintained at suitable sites. In particular on arable land a major expansion of funding should be achieved.
- There is greater flexibility in funding required, such as the general possibility of change in a more appropriate nature conservation measure. It should be easy and uncomplicated to have a certain proportion of fallow-fields in order to create suitable habitats for wild animals.

The goal of nature conservation funding is, inter alia, to stimulate the interest of stakeholders in the success of working long-term nature friendly. Thus, a system of incentives for such services in the sense of adequate remuneration of ES and not only the pure reimbursement of expenses is required to cover at least the transaction costs. The priority should be a nature conservation consulting instead of too high control expenses (mandatory by EU laws), which can be perceived as patronizing. It should not be expected that farms responsible for their staff carry out measures for long periods while financing only compensates for the effort. After all, companies that generate low profit might go bankrupt and, thus, will not be available as partner for landscape management.

As long as incentive components cannot be financed by EU funding, other sources of funding need to be developed. Hence, two successfully tested systems can be considered: privately funded bonus payments and (regional) eco-labels. From the perspective of agricultural economic additional revenues through environmentally certified (quality upgraded, higher-priced) agricultural products of food, beverage and spa market could cover at least half of the cost for services of public welfare provided by agriculture. Recent surveys of the Leibniz Institute of Ecological Urban and Regional Development (IOER) show that a large number of consumers would be willing to contribute to a compensation of ES to such services by purchasing certified products (Grunewald et al. 2012).

Conclusion

Landscape management contributes, in particular, to secure and deliver public goods and services to society. This includes the right to know how much nature conservation costs. Monetary aspects play an important role in the landscape management accounting system, since money is a measure of the achievement of structural and functional goals. However, in addition to the sum, the institutional side is of decisive importance for a better implementation.

In the meantime a large amount of ES-related studies exist which includes the costs and benefits of measures for the protection of nature and biodiversity. The usefulness of such measures often significantly exceeds the associated costs. More nature protection and protection of ES, therefore, lead to human welfare. Unfortunately, it does not work 'automatically' because in the context of ES and biodiversity, especially due to the problem of 'public goods', the economic principles can only have an inadequate impact here.

In detail, different degrees of states of maintenance and operating conditions affect the costs and lead to a great number of possible variations. Basically, however, the developed accounting approach appears to be suitable for estimating the costs that are necessary for the preservation and development of target species and habitats at regional, state-wide level (Grunewald et al. 2014). The basic procedure should be transferable to other regions, albeit priorities, habitat types and cost rates may vary.

The management of landscape development is actor-, action- and goal-oriented and is marked by political, economic and institutional conditions (norms, rules, patterns of behavioral, concepts, *leitbilder*, etc.). Nature conservation policy is the result of many negotiations between numerous state, public and private actors (governance of ES–Sect. 5.4). The interconnections are often very complex and difficult to understand. Therefore, the landscape management strategy needs to be readjusted continually. A decadal rhythm for reorientation of landscape management accounting appears to be adequate in this regard.

The global approach to improving the services of the landscape and their contribution to the biodiversity programme of a state leads to the conclusion that, overall, a greater commitment of society for nature and biodiversity is urgently needed at all levels. Furthermore, it must be possible to create a partnership manner of cooperation between the state and private conservationists and the land users, which is not achievable solely through the instrument of a reasonable fee for services rendered.

Finally, a landscape management strategy can develop a high effectiveness, only if it is possible to further integrate the goals of nature conservation and the maintenance of the ES in land use. Nature conservation and landscape management must find consideration in all areas of policy. It is, therefore, a priority task to upgrade and develop the instruments of nature conservation as a whole, which means–if necessary–designation and management of protected areas, investments concerning land purchase, improving compensation and replacement instruments, expanding the contractual nature conservation, tendering of nature conservation achievements, monitoring and promoting result-oriented successes (Jedicke 2010).

6.6 Specific Nature Protection and Development Strategies

6.6.1 Nature Conservation and Ecosystem Services

O. Bastian

Introduction

Nature conservation is often regarded simply as a negative cost factor. Such an opinion ignores the realities of the situation. Rather, nature conservation is, in many respects, extremely useful to human society, also from an economic point of view (cp. Jessel et al. 2009). Nature conservation measures may contribute to value creation or to create jobs, for example, in the areas of ecological landscape management, care of protected areas, or in the increasing tourism in national parks.

The TEEB study (TEEB 2009) analysed that USD 10–12 billion are spent annually on the protection of the approximately 100,000 protected areas. For effective nature conservation, however, US$ 40 billion would be needed. This amount is low in comparison to the US$ 5 trillion, the ecosystems of these protected areas are worth by suppling natural goods and other services. This is more than automotive, steel and IT industries of the world produce together.

Protected areas, such as national parks, biosphere reserves and others can be a framework for successful sustainable rural development, and can contribute to safeguarding jobs, especially in economically lagging areas. As many studies have shown (e.g., Getzner et al. 2002; Popp and Hage 2003; Job and Metzler 2005; Neidlein and Walser 2005; Kettunen et al. 2009; Gantioler et al. 2010), they can provide the basis for various forms of economic activity in a region, such as in agriculture and forestry, nature-based tourism and environmental education. It has been estimated that around 125,000 jobs in the EU were supported through conservation-related activities in 1999 and that this trend was increasing; around 100,000 of these were direct jobs and 25,000 indirect, with around two thirds of the direct jobs related to operational expenditures and one third related to investments (IEEP 2002).

Thus, eco-tourism shows annual growth rates of 20–30 %, compared to 9 % of tourism in general (EU Kommission 2008). Natura 2000 has become a label for an attractive landscape–on a European level! Some future-oriented tourism managers already recognised this and advertise with the label.

In the following, the application of the ES concept will be shown on the example of case studies in Natura 2000 sites of the Ore Mountains, namely

1. On analyses of potentials of FFH sites
2. Referring to FFH habitat types and species
3. With regard to several economic aspects

The investigations were performed mainly in the framework of the Ore Mountains Green Network project funded by the European Union (EFRE Objective 3/INTERREG IV A). The goal of this project was to apply the concept of ES to Natura 2000 sites in the Ore Mountains on both sides of the German-Czech border. The study aimed to reveal the various services and benefits such protected areas provide and to identify and strengthen synergies between nature conservation (Natura 2000) and rural development, especially in the spheres of conservation-friendly agriculture and forestry, eco-tourism and environmental education. The project area consisted of the ridge area of the Ore Mountains in the three districts Sächsische Schweiz-Osterzgebirge, Mittelsachsen and Erzgebirgskreis in the German Federal State of Saxony and the Bohemian districts of Ústecký kraj and Karlovarský kraj in the Czech Republic.

Starting from the present situation of selected Natura 2000 sites, a SWOT analysis revealed the strengths, weaknesses, opportunities and threats with regard to interdependencies between nature conservation and rural development. Strategies and concepts were prepared in close cooperation with local stakeholders to enhance the status of Natura 2000 in rural development. These concepts were designed to show how Natura 2000 sites can be maintained in a favourable state by permanently integrating economic and educational aspects.

Study Area

The Ore Mountains (*Erzgebirge/Krušné hory*) have the shape of a slanted writing desk some 150 km in length, formed by tectonic forces. On the southern side, the mountains slope steeply down towards the Ohře river valley. On the northern side, they drop away gradually over a distance of 30–45 km to the foothills. The ridge of the Ore Mountains, averaging between 800 and 1000 m above sea level, has long constituted the border between Saxony and Bohemia, and today between the Federal Republic of Germany and the Czech Republic. Acid rocks such as gneiss, phyllite and granite are typical as are the raw climate, many raised bogs, mountain meadows and spruce forests. It is a traditional cultural landscape of European significance, especially shaped by ore mining.

The Ore Mountains are rich in beautiful landscapes and natural assets with characteristic ecosystems, such as raised bogs and bog forests that give the impression of pristine nature, but also 'man-made' mountain meadows with their blooming and smelling herbs, matgrass meadows, tall subalpine herbaceous vegetation, agricultural clearance cairns mixed mountain forests and near-natural watercourses (■ Fig. 6.23).

Several rare and threatened species are among the remarkable flora, such as arnica (*Arnica montana*), ragged pink (*Dianthus seguieri*) and several orchid species. The local fauna includes the black grouse *(Tetrao tetrix)* and the corncrake (*Crex crex*). The black grouse (■ Fig. 6.24), which is threatened by extinction, is very important on the European scale. The biggest Central European black grouse population outside the Alps lives in the Ore Mountains, especially on the Czech side of the border. The birds prefer large undisturbed landscapes covered by sparse woods with berry bushes (bilberries/*Vaccinium myrtillus*) and pioneer shrubs (rowan/*Sorbus aucuparia*, birch/*Betula pendula*). The major reasons for the decline of black grouse populations include the afforestation of clearings and forest meadows with spruce monocultures, the increase in predator populations (e.g. red fox, wild boar) and disturbance, e.g. by tourists and wind turbines.

■ Analysis of Selected Potentials of Natura 2000

Regarding the manifold services of nature (not only in protected areas), it is necessary to distinguish between the supply, which is not or

■ **Fig. 6.23** Landscape with agricultural clearance cairns in the Eastern Ore Mountains with mount Geisingberg. © Olaf Bastian

■ **Fig. 6.24** Great efforts are necessary to maintain the Ore Mountains' black grouse population, which is the biggest in Central Europe outside the Alps. © Jan Gläßer

not yet used (potentials, 2nd pillar of the EPPS framework, ▶ Sect. 3.2.2), and the actually used or demanded services (3rd pillar of EPPS). The goal is to display the service capacities of an area as a field of options available to society for use, and also to take into account such categories as risks, carrying capacity and the capacity to handle stress (increasingly summarised today in the term 'resilience'), which limit or may even exclude certain intended uses (■ Table 6.9; Bastian et al. 2010). In the classification of ES and potentials, we follow a trinomial scheme. This breakdown into provision, regulation, and sociocultural services has the advantage that it can be linked to the concept of sustainability using the established ecological, economic and social development categories (▶ Sect. 3.2).

The information used stems from the management plans for the Natura 2000 sites (SCI)(only for the German part) and nature reserves (elaborated–as a rule–by consultants by order of the state environmental authorities) and from governmental agencies. The results confirm that the range of potentials and services delivered by the Natura 2000 sites in the Ore Mountains is very wide and diverse, which goes far beyond the original purpose of Natura 2000 to protect endangered species and habitats.

Natura 2000

Natura 2000 is an EU-wide ecological network of conservation areas with the goal of maintaining and restoring endangered habitats and species of Community importance, i.e. Europe's most important species and habitats. Since its creation, nearly 20 % of Europe's territory has been included in the network–about 25,000 sites in all 27 member states. Natura 2000 represents one of the world's most ambitious approaches for halting the loss of biodiversity.

The main focus of Natura 2000 is to select targets of conservation, such as certain natural and semi-natural habitat types and also species, which are endangered Europe-wide. These tasks are covered by the European Habitats Directive 92/43/EEC (listed in Annexes I and II) and the Birds Directive 79/409/EEC (listed in Annex I).

Consideration must also be given to flora and fauna of Community importance in need of strict protection (Annex IV of the Habitats Directive), and to flora and fauna of Community importance whose taking in the wild and exploitation may be subject to management measures (Annex V of the Habitats Directive). The annexes to the Habitats Directive List 231, natural Habitat Types (Annex I) include more than 1000 animal and plant species (Annexes II, IV and V). The lists contained in Annexes I and II of the Habitats Directive differentiate according to priority and nonpriority species and habitats. Classification is subject to extremely stringent protection requirements to deal with potential impacts (Article 6, Habitats Directive).

The goal of the Birds Directive is to preserve all bird species that occur naturally in EU member state territories and to secure adequate stocks to allow their survival and reproduction over time. Annex I of the Birds Directive lists species that are at special risk and subject to special conservation measures, and currently includes 190 species and subspecies. Almost 100 of these occur in Germany (BfN 2012b).

Germany has proposed 4619 FFH sites of three bio geographical regions (Alpine, Atlantic and Continental) to be registered at the EU in Brussels. This corresponds to a share of 9.3 % of the whole German terrestrial territory. 740 bird sanctuaries (11.2 % of the terrestrial area) must be added, as well as 21,222 km^2 of large waters (Lake Constance, open Sea, Bodden and Wadden Sea) (state: 30.09.2011). Among the protected marine areas, 943,984 ha belong to the exclusive economic zone of Germany. FFH sites and bird sanctuaries may overlap spatially. Together they cover 15.4 % of the terrestrial and 45 % of the marine territory of Germany (BfN 2012b). Natura 2000 sites in Saxony cover approximately 2930 km^2. This corresponds to 15.9 % of the territory of the country.

The 25,000 Natura 2000 sites of the EU cover c. 18 % of the member states' territory (state 2009). The situation varies from country to country: the lowest is 7.2 % FFH sites in Great Britain and the highest 35.5 % in Slovenia (with regard to the whole terrestrial area); the European average is 13.6 %. The marine areas (131,459 km^2) are not included in the percentages (BfN 2012a). Germany bears a special responsibility for the protection of 91 habitat types and 133 plant and animal species (without birds).

In principle, the selection of FFH sites should follow only nature conservation aspects. The FFH habitat types of Annex I of the directive were selected according to the following criteria:

- Representativeness of the natural habitat type in the region
- Relative size of the habitat type with respect to the total stock in the EU member state
- Conservation status of structure and functions of the habitat type and its recoverability
- Overall evaluation of the site with regard to the maintenance of the habitat type concerned

The selection of FFH species according to Annex II of the Habitats Directive were based on the following criteria:

- Population size and density of the species in relation to the entire population in the EU member state
- Conservation status and recoverability of the habitat elements important for this species
- Degree of isolation of the population occurring in the area in relation to the total natural range of this species
- Overall evaluation of the site for the protection of the species concerned

For the Natura 2000 sites, conservation objectives were identified, which are important for the management and to assess the compatibility of projects. The conservation status of habitat types and species is monitored regularly. Regular monitoring and submitting the results to the EU shall ensure the practical implementation of Natura 2000.

Table 6.9 Selected Fauna-Flora-Habitat sites (FFH) along the Ore Mountains ridge, and their role for supplying ES (potentials, services, risks)

Declaration number	Name	L	G	T	M	B	S	W	A	F	Ec	SF	Ae	R	E
FFH sites at the German side															
004E	Buchenwälder und Moorwald bei Neuhausen und Olbernhau	s	Sp	Sr	s			S	Sr	S	S		Sr	Sr	s
007E	Mothäuser Heide		Sp	sr				S	s	S			sp	Sr	Sp
010E	Erzgebirgskamm am Kleinen Kranichsee	sp	S	sr				S	S	S	s		Sr	Sr	s
012	Zweibach		Sp	sr	r		s		S	sp	Sp		Sp	s	p
016E	Erzgebirgskamm am Großen Kranichsee	sp	Sp	Sr	s			S	S	S	S		S	Sr	s
039E	Geisingberg und Geisingberg-wiesen	Sp	s	sr		pr	p		s	sp	sp	s	Sp	Sr	sp
040	Hemmschuh	sp	Sp	Sr			s		S	Sp	Sp	sp	Sp	Sr	p
042E	Mittelgebirgslandschaft um Oelsen	Sp	Sp	sr		pr	sp	S	S	Sp	Sp	Sr	Sp	Sr	Sp
044E	Fürstenauer Heide und Grenz-wiesen Fürstenau	Sp	s				p	s	s	Sp	s	S	Sp	Sr	sp
070E	Wiesen um Halbmeil und Brei-tenbrunn	Sr						S	s	sp	s	Sp	Sp	S	sp
071E	Fichtelbergwiesen	Sr				pr	p	Sp	Sr	s	S	S	Sr	Sr	s
083E	Gimmlitztal	Sr	s	sr				S	S	Sp	Sp	S	Sp	Sr	s
084E	Kahleberg bei Altenberg			sr				s	sr	sp	sp		s	Sr	sp
174	Georgenfelder Hochmoor	s						S	sr	Sp	sp		sp	Sr	Sp
176	Bergwiesen um Schellerhau und Altenberg	Sr				pr	p	s	sr	s	sp		Sp	Sr	sp
177	Bergwiesen um Dönschten	Sr							s	s	s		sp	S	s

Table 6.9 Continued

Declaration number	Name	L	G	T	M	B	S	W	A	F	Ec	SF	Ae	R	E
252	Oberes Freiberger Muldetal	sr	Sp	Sr				S	S	sp	sp	sr	Sp	S	s
262	Bergwiesen um Rübenau, Kühnhaide und Satzung	Sr						s	s	s	s		Sp	Sr	sp
263	Moore und Moorwälder bei Satzung		sp	r	s			S	S	Sp	sp		Sr	S	sp
264	Kriegwaldmoore		Sp	sr				S	s	S	s		S	S	
265	Preßnitz- und Rauschenbachtal	Sr	Sp	Sr			p	S	S	Sp	S	Sp	Sp	S	s
266	Pöhlbachtal	Sr		Sr				S	S	Sp	Sp	S	Sp	S	Sp
271	Kalkbruch Hammerunterwiesenthal							s					s	sr	s
283	Mittelgebirgslandschaft bei Johanngeorgenstadt	Sr	s	Sr			p	S	S	Sp	Sp		Sp	S	S p
FFH sites at the Czech side															
CZ0410040	Pernink	sr		r				S	s	F	s		S	p	p
CZ0410046	Šibeniční vrch									S			s	sr	p
CZ0410155	Rudné	Sr		r				s	s		s	s	sp	Sp	p
CZ0410168	Vysoká Pec	sp		sr		p	p	s	s	s	s		sp	sp	p
CZ0414110	Krušnohorské plató	sp	sp	Sr		p	p	S	Sp	s	sp	s	Sr	Sr	sp
CZ0420021	Kokrháč–Hasištejn		p	sr		p		S	s	Sp	S	s	S	s	p
CZ0420035	Na loučkách	sp	p	sr			p	s	sp	S	sp	s	S	s	p
CZ0420053	Rašeliniště U jezera–Cínovecké rašeliniště			r		p	sp		s	sp		Ss	s	sr	sp
CZ0420074	Grünwaldské vřesoviště		sp	r	sr			s	sp	sp	sp	sp	Sp	sr	p

Table 6.9 Continued

Declaration number	Name	L	G	T	M	B	S	W	A	F	Ec	SF	Ae	R	E
CZ0420144	Novodomské a polské rašeliniště		sp	sr		p		s	sp	sp	sp	sp	Sp	sr	sp
CZ0420160	Podmilesy	s		sr		p		s	sp	sp	s	s	s	s	p
CZ0420171	Údolí Hačky	s		sr		p		s	sp	S	s	S	S	s	p
CZ0420528	Klínovecké Krušnohoří	sp	sp	Sr		p		S	S	Sp	sp		Sp	Sr	p
CZ0424030	Bezručovo údolí			r			p	s	s	s	s	s	s	s	p
CZ0424127	Východní Krušnohoří	sp	sp	Sr		p	p	S	sp	Sp	sp	sp	Sr	sp	sp

Categories:
S, s–services (high/medium significance)
p–potentials (for development)
r–essential risks, threats
Provisioning (economic) services:
L–livestock (husbandry), G–game, T–timber, M–wild fruits (berries, mushrooms), B–biochemical/medicinal resources, S–genetic resources (seeds), W–drinking water
Regulation (ecological) services:
A–air quality/local climate regulation, F–flood protection, Ec–erosion control, SF–self-purification of waters
Sociocultural services:
Es–aesthetic values (scenery), R–recreation, eco-tourism, E–environmental education

Among the *provisioning services* are the categories "Provision (or production) of animal and plant biomass" and the "provision of drinking water".

Regarding *animal products* like milk, meat and wool of domestic animals it is relevant that numerous Natura 2000 sites in the upper Ore Mountains contain grassland biotopes, especially mountain meadows, which depend on careful use and management: regular removal of biomass by mowing, including hay-making or pasturing cattle and sheep. At least some parts of mountain meadows in several FFH sites are managed (cut), but deficits (no mowing; abandonment) exist and potentials and opportunities for careful economic exploitation are unused. Even if the mowing of mountain meadows can be organised, in some cases no customers for the harvest can be found. The results of insufficient grassland management are not only a decline in sensitive meadow species, but also unused economic potentials. Production and marketing of high-quality hay from mountain meadows may be extended, as well as the utilisation of plant biomass from landscape management for energetic purposes (cp. Peters 2009).

Fishery is not well developed in the Ore Mountains Natura 2000 sites. However, anglers could cause vegetation damage along the river banks. There is no potential for more intensive forms of fishery.

The situation of *hunting* is quite different. Although hunting takes place in all forests, even in protected areas, the stock of game, especially red deer, row deer and wild boar, is too high, and exceeds the carrying capacity of the forest ecosystems. Vegetation damage due to peeling and reduced natural regeneration of forest trees are the result. Red deer like to wallow in sensitive raised bog waters. Feeding game in higher altitudes of the mountains during winter can impair valuable biotopes (eutrophication, dissemination of invasive plant species) and sensitive animals (e.g. black grouse). Therefore, hunting should be intensified to reduce the stock of game, which would also develop economic potentials not sufficiently exploited to date.

Most of the Natura 2000 sites are covered by *forests*, at least partially. The forests are used more or less intensively, with the exception of very small reserves. Both the state and the private forest enterprises are geared towards economic benefits. Conflicts with nature conservation result from timber harvesting, including in valuable forest habitats, heavy machines, construction of excessively sized forest roads, the lack of lumbermen (who would be necessary for manual work), drainage, tillage of forest soils, afforestation with foreign tree species. Other threats (also in SCI 263) are caused by large-scale liming of acidic forest soils from aircraft, which damages the pH-balance and the vegetation of raised bogs and bog forests, and afforestation of open areas, including the black grouse habitats. The Natura 2000 sites provide essential potentials for sustainable forestry, but almost no reserves for intensification. Conflicts result from the efforts of nature conservation to recover the original hydro-regime in wet biotopes, and to close ditches in raised bogs.

Wild berries and mushrooms are collected in almost all forest areas, with sporadic threats to sensitive biotopes and animals. The exploitation of biochemical and medicinal resources has hardly been developed at all. Some mountain meadow reserves have potentials for harvesting spignel and other *medicinal plants*. However, dangers of overexploitation and risks for the biotopes and the populations should not be ignored.

With regard to the high biodiversity of the Natura 2000 sites, the use of *genetic resources* should be considered. At present, the seeds of such forest trees as bog pine, spruce, beech and fir are harvested at several sites.

Numerous Natura 2000 sites contribute to the provision of *drinking water* since they are also water protection and headwater areas. Increased water recovery would conflict with the goals of nature conservation.

Among the category of *regulation services, air purification and local climate regulation* (especially by forests) should be mentioned. The vitality and filtering function of forests are threatened by diffuse inputs of nutrient and other matters, and supra-regional climate changes. Already today, spruces show essentially reduced growth in periods of extreme weather conditions (especially heat and drought) (SMUL 2009b).

Due to the steep slopes, the capacity for *run-off regulation* in mountain areas is very important. Natural forests, meadows, swamps and especially bogs balance water runoff, store water during dry periods and prevent flooding. To increase the capacity for water balancing, various measures have been proposed and in fact carried out: changing the tree composition in forests, closing ditches in raised bogs (see above for the conflicts with intensive forestry interests and with the drinking water concerns mentioned), but also providing necessary financial resources.

Forests have a high potential to prevent *soil erosion*. This capacity can be increased by restructuring the forests towards a more natural state. Sometimes there are conflicts with demands for more intensive forestry. Grassland can also reduce erosion risks but harmful damages caused by trampling (overgrazing) may occur.

As natural, richly structured streams are better suited for *self-purification* than canalised waters, their potential can be improved by establishing hydrological buffer zones, supporting water dynamics, and reducing nutrient inputs from adjoining farmlands and settlements. Presently, such objectives are conflicting, e.g. with technical flood prevention measures.

In the category of *sociocultural services, aesthetic values* are very important, especially for enjoying the scenery and for eco-tourism. The large forests of the Ore Mountains at higher altitudes support such goals. Beech forests, to some extent spruce forests, and bog forests are of great interest. The landscape's attractiveness, especially for tourism, stems from a small-scale pattern of various biotopes, e.g. raised bogs, bog forests, headwater areas, mountain meadows and pastures and stone walls, which are also habitats for rare and beautiful species. There are also Natura 2000 sites with historically valuable cultural landscape elements, such as monuments dedicated to the transportation and mining history, and also ancient forests which are cultural monuments. The development of tourism can suffer from the construction of wind turbines on the mountain ridge.

There is a wide variety of touristic activities in the Natura 2000 sites of the Ore Mountains: walking, cycling, mountain-biking, swimming, climbing and collecting wild berries, mushrooms and minerals. Additional opportunities for *nature-based tourism* arise especially from the networking between the German and the Czech side. However, it has to be noted that overexploitation problems (e.g. trampling valuable vegetation cover, disturbance of black grouse and other animal species, waste disposal) already exist. Conflicts result from the increasing utilization of the landscape for sports, especially for skiing, mountain-biking and even quad-biking. Especially Natura 2000 sites with black grouse populations are not suitable for touristic developments.

In the area of *environmental education*, the experience of rare and valuable species and ecosystems is the main aspect. There are nature trails (e.g. across raised bogs) and presentation signboards. Guided tours are offered, and school education includes Natura 2000 sites. 'Scientific' tourism has also developed but in some cases this poses serious threats to the fragile ecosystems (mountain meadows–SCI 039E, 071E) and sensitive species (black grouse).

Conclusion: The Natura 2000 sites of the upper Ore Mountains do not represent pristine nature but they have been shaped and changed by humans for centuries. Also today, they are used for multiple purposes, and they provide a wide range of provisioning, regulation and sociocultural services. Partly there are still development opportunities and unused potentials, but the objectives of nature conservation under the conditions of NATURA 2000 must be respected.

▪ The Role of FFH Habitat Sites and Natura 2000 Relevant Species

The FFH habitat types represent the core of the FFH sites, They are embedded like a mosaic in the 'matrix' of the entire area. In addition to the assessment of ES supply (analysis of potentials) of the respective site, it is interesting to which extent the specific habitat types and species contribute to this. The supply of ES by habitat types of the upper Ore Mountains was estimated semiquantitatively according to three scores. ◘ Table 6.10 shows the results on the example of some forest types.

It is obvious that many ES, especially several provisioning and regulation services, depend on 'rough' vegetation structures or land-use forms, while other

Table 6.10 Ecosystem services provided by Habitats of Community Interest of Natura 2000 sites (FFH) in the upper Ore Mountains (Saxony, Germany). Valuation: 2–high and very high; 1–medium; without score–low or insignificant; * considerable conflicts (with other services, esp. with habitat function/biodiversity)

Habitat types	Ecosystem services																				
	Provisioning										Regulation							Sociocultural			
	L	Fi	G	CF	T	M	B	S	W	EW	C	A	F	Ec	SF	Po	H	Ae	R	E	I	LH
9			2		2*	2	1		2		2	2	2	2	2	1	1	1	2	1	1	
91			2*		2*	2*	1		2		2	2	2	2	2	1	1	2	2*	2	1	
9110			2*		2*	2*		2	2		2	2	2	2	2	1	2	2	2*	2*	2	
9130			1*		2*	1*	1		1		2	2	2	2	2	1	2	2	2*	2*	2	
9160			1*		2*	1*			1		2	2	2	2	2	1	2	2	2*	2*	2	
9180			1*		2*	1*	1		1		2	2	2	2	1	1	2	2	1*	2*	2	
91D			1*		1*	1*			1		2	2	2	1	1	1	2	2	1*	1*	2	1
91E0			1*		2*		1		2		2	2	2	2	2	1	2	2	2*	2*	2	
94			2*		2*	2*			2		2	2	2	2	2	1	1	2	2	1	1	
9410			2*		2*	2*			2		2	2	2	2	2	1	2	2	2*	2*	2	

Habitat types
9 FORESTS
91 Forests of Boreal Europe
9110 Luzulo-Fagetum beech forests
9130 Asperulo-Fagetum beech forests
9160 Sub-Atlantic oak-hornbeam forests (Stellario-Carpinetum)
9180 Tilio-Acerion forests of slopes, screes and ravines
91D0 Bog woodland
91E0 Alluvial forests with *Alnus glutinosa* and *Fraxinus excelsior* (Alno-Padion, *Alnion incanae, Salicion albae*)
94 Temperate mountainous coniferous forests

6

Table 6.10 Continued

Habitat types	Ecosystem services																					
	Provisioning										Regulation							Sociocultural				
	L	Fi	G	CF	T	M	B	S	W	EW	C	A	F	Ec	SF	Po	H	Ae	R	E	I	LH

9410 Acidophilous Picea forests of the montane to alpine levels (Vaccinio-Piceetea)

Ecosystem services

Provisioning services

Supply of animal products

L Livestock (products: milk, meat, wool)

Fi Fish

G Game

Supply of plant products

CF Crops, fodder

T Timber

M Wild fruits (berries, mushrooms)

B Biochemical/ medicinal resources (e.g. Spignel–*Meum athamanticum*–and other herbs)

S Provision of genetic resources: Seeds of forest trees, herbs, grasses (e.g. for plant breeding, hay mulching)

W Drinking water (water protection areas/ headwaters)

EW Energy from water power

Regulating (ecological) services

C Carbon sequestration

A Air quality regulation/local climate regulation

F Water balance regulation (flood mitigation)

Ec Erosion control

SF Self-purification of waters

Po Pollination

H Habitat function

Sociocultural services

Ae Aesthetic values (e.g. scenery)

R Services for recreation and eco-tourism

E Services for environmental education (e.g. cultural-historical aspects)

Information services

I Bioindication

LH Landscape history (e.g. pollen analysis)

ES, particularly several sociocultural ones, also depend on specific habitat or vegetation types.

A remarkable number of particularly animal species listed in the Annexes of the Habitats Directive live in the Ore Mountains. They all have quite different demands upon their habitats and environmental conditions. Only a few species can be unambiguously assigned to particular habitat types. Either they live equally in several habitat types–in some cases using them for different parts of their life cycles, e.g. habitats for breeding and for feeding–or, as in the case of bats, they are bound to certain vegetation structures, vegetation classes or land-use forms. For example, the otter *(Lutra lutra)* needs intact standing and running waters (3150, 3260). Bats need old beech forests (9110, 9130), as do the black stork *(Ciconia nigra)* and the black woodpecker *(Dryocopus martius)*. The corncrake *(Crex crex)* prefers semi-natural tall-herb humid meadows (6410, 6430), the great crested newt *(Triturus cristatus)* lives in standing waters (3150) and the European bullhead *(Cottus gobio)* and the brook lamprey *(Lampetra planeri)* in favourable running waters (3260). The dusky large blue butterfly *(Glaucopsyche nausithous)* needs lowland meadows (6510) with flowering great burnets (*Sanguisorba officinalis)*. Thus, only some of these species can be regarded as indicators of Natura 2000 habitat types and related ES (◘ Table 6.11).

Both the literature survey and the case study of Natura 2000 habitat types in the Ore Mountains has enabled us to show many strong relationships between habitat types and ES regarding some regulating and sociocultural services, but not for most of the provisioning and regulating services. For these ES 'rough' vegetation structures, vegetation classes and land cover are generally more important. It could also be shown that only a small portion of the particular Natura 2000 species (listed in the Annexes of the Habitats directive) is bound to particular Natura 2000 habitat types. Most of these species cannot be regarded as indicators for (changes in) particular habitat types and the ES provided by them (Bastian 2012).

The fact that some services are not dependent on particular species is in agreement with Schwartz et al. (2000), who found among a range of empirical and modelling studies that few studies supported the hypothesis that there was a simple, direct linear relationship between species richness and some aspects of ecosystem functioning, such as productivity, biomass, nutrient cycling, carbon flux or nitrogen use. They concluded that biodiversity as represented by measures of species richness may be important for ecosystem functioning, but other aspects of ecosystem structure might be equally significant. The results also seem to support the criticism of the ES concept by Ridder (2008), who argues that this approach as a justification for conservation is flawed, as only a relatively small number of species provide services. Mostly, these are not rare species, but rather particular groups of species with specific functional characteristics. He argues that land management practices have a much greater effect on most ES. For example, he notes, many ES supplied by forests, such as carbon sequestration, supply of drinking water, flood mitigation, and erosion control, depend principally on the presence of trees and undergrowth. While these processes have a great impact on biodiversity, diminished biodiversity does not in itself constitute a threat to these services.

That should, however, not suggest that diversity of habitat types and species are dispensable since we must be aware that all species and habitat types constitute the assets of several regulating and sociocultural services. Service-providing species are embedded in an ecosystem and to separate out some species from the rest is generally biologically unrealistic as many unobserved interactions take place (e.g. food chains). There is also the necessity for large numbers of species to fulfil the inherent multifunctionality of ecosystems (Haslett et al. 2010). Therefore, the entire ecosystem or habitat should be considered in terms of its supporting role for the SPs (Rounsevell et al. 2010).

Greater species diversity within an ecosystem obviously implies a potentially even greater supply of ecosystem goods which might be used, e.g. medicinal plants (Mertz et al. 2007). Inasmuch as the provision of genetic diversity can be viewed as a service in itself, biodiversity is fundamental to that. Due to remaining knowledge gaps, it may still not be possible to determine conclusively whether a

Table 6.11 Habitat requirements (Natura 2000 habitat types) of several plants and animals of Community importance (Habitats Directive 92/43/EEC and Birds Directive 79/409/EEC) in the upper Ore Mountains. (Data from Steffens et al. 1998; Hauer et al. 2009)

Species	Preferred Habitat Types				
	Waters	Heaths	Grassland	Swamps, bogs	Forests
Annex II Habitats Directive					
Mammals					
Otter *(Lutra lutra)*	3150, 3260				
Greater Mouse-ear Bat *(Myotis myotis)*					9110, 9130
Barbastelle Bat *(Barbastellus barbastellus)*					
Bechstein's Bat *(Myotis bechsteinii)*					9110, 9130
Amphibians					
Great Crested Newt *(Triturus cristatus)*	3150				
Fishes					
European Bullhead *(Cottus gobio)*	3260				
Brook Lamprey *(Lampetra planeri)*	3260				
Insects					
Green Club-tailed Dragonfly *(Ophiogomphus cecilia)*	3260				
Jersey Tiger *(Euplagia quadripunctaria)*			6430		
Dusky Large Blue *(Glaucopsyche nausithous)*			6510		
Annex IV Habitats Directive + Annex I Birds Directive + other species to be protected (selection)					
Mammals					
Dormouse *(Muscardinus avellanarius)*					91
Birds					
Corncrake *(Crex crex)*			6410, 6430		
Meadow Pipit *(Anthus pratensis)*			6230, 6410	7110, 7140	
White-throated Dipper *(Cinclus cinclus)*	3260				
Boreal Owl *(Aegolius funereus)*					9410, (9110)
Eurasian Pygmy Owl *(Glaucidium passerinum)*					9410
Black Grouse *(Tetrao tetrix)*		4030	64, 65	71	
Black Stork *(Ciconia nigra)*					9110
Black Woodpecker *(Dryocopus martius)*					9110
Grey-headed Woodpecker *(Picus canus)*					9110, (91)
Red-breasted Flycatcher*(Ficedula parva)*					9110, 9130 (91E0, 9180)
Insects					
Large White-faced Darter *(Leucorrhinia pectoralis)*	3160				

Legend of habitat types ◘ Table 6.12
3150–Eutrophic standing waters; 3160–Dystrophic standing waters
64–Tall herb communities; 65–Mesophilic grassland; 71–Acidic Sphagnum bogs; 91–Boreal European forests

high level of biodiversity is in fact necessary for the supply of a certain ecosystem service.

Taking the complexity of ecosystems and the variety of interactions with the natural human socio-economic environment into account, to date it is impossible to identify all benefits that might arise from biodiversity. The same is true for predicting the effects that the loss of a single species or a population and/or ecosystem service has. Hence, a precautionary approach is required, and ecosystems should be maintained intact as far as possible to ensure continued service provision in the face of changing environmental conditions and biotic interactions, even if there is presently insufficient supporting scientific evidence (Cooney and Dickson 2005).

▪ Monetary Aspects

It is known that the goal to assess nature and biodiversity economically is limited, because we are dealing mostly with nonuse values, which are not traded on markets (▶ Sect. 4.2). These values arise from the ethical or religious will to maintain nature for its own sake (*existence value*) or for posterity (*bequest value*). Empirical studies have shown that nonuse values represent the greater part of human appreciation of threatened ecosystems. Therefore, alternative evaluation methods must be applied (Macke and Schweppe-Kraft 2011).

The calculation of costs for maintaining and restoring habitat types through the willingness-to-pay approach is common, as the readiness of society is reflected to provide financial means for Nature conservation for ethical or aesthetical reasons (Spangenberg and Settele 2010).

If evaluations on the base of individual preferences are not possible or appropriate, evaluations by society (e.g. sustainability goals) or expert judgments should be used as orientation. Reasons for it are e.g. effects affecting several generations, high uncertainty, or damages hidden to individual persons. There may also exist overarching goals, e.g. for emission reductions (climate protection) or biodiversity targets (CBD, National Biodiversity Strategy, Natura 2000). The costs to reach these goals may be regarded as a measure for the society's willingness-to-pay for the degree of the existing environmental damages (UBA 2007).

The value of Natura 2000 sites and species is also expressed in the consensus of the society to maintain them *(revealed public preferences)*. This is underpinned both by EU and national legislations. It is clear that from the obligation to maintain a 'favourable conservation state' follows the need of costly management measures (Schweppe-Kraft 2009), among them labour costs, e.g. incentive payments for farmers or wages for persons performing these measures. This is also an expression for job and business opportunities in rural regions. The costs calculated for management measures in selected FFH sites of the upper Saxon Ore Mountains (14 FFH sites, total area 6054 ha) can be seen in ◘ Table 6.13. The size of these sites is between 21 ha (Georgenfeld raised bog) and 999 ha (Großer Kranichsee). Not all costs were clearly defined, partly rather wide ranges were presented.

The total financial need (just under € 2 million for only 14 FFH sites in Saxony) is very high compared to the actually available financial means. Güthler and Orlich (2009) identified annual financial requirements for Natura 2000 in Germany for the amount of c. € 620 million (52.7 €/ha), i.e. € 4.34 billion for the EU funding period 2007–2013. Within this funding, which represents the most important financial source of nature conservation, only € 1.86 billion are dedicated to nature conservation measures in Germany. This is only 3–4 % of the total agricultural budget in Germany.

Kettunen et al. (2009) estimated the financial need for the management of the Natura 2000 sites in the EU at € 5.8 billion annually, which is fourfold of the current budget. The European Commission calculates the costs for Natura 2000 sites in Europe with an average of 63 €/ha; this amount is by far exceeded by the benefits from Natura 2000 (e.g. CO_2-fixation, tourism), even though not all ES are included in the calculation of the total benefit.

There are interesting WTP analyses (Hampicke et al. 1991), which found that the German citizens would be ready to spend between € 99 and 123 per

Table 6.12 Calculated costs for management measures in FFH habitat types (HT) in 14 selected FFH sites of the upper Saxon Ore Mountains. (Source: FFH management plans, Saxon State Office for Environment, Agriculture and Geology, data research: M.-L. Plappert)

HT-No.	HT-Name	Occurrence of the HT in the 14 FFH sites	Number of FFH sites with data on the costs for HT management	Total costs per HT (€)
3260	Water courses of plain to montane levels with the Ranunculion fluitantis and Callitricho-Batrachion vegetation	1	1	8000–16,000
4030	European dry heaths	5	5	27,585–30,360
6010	Basophilic pioneer meadows	1	1	1806
6230	Species-rich *Nardus* grasslands, on silicious substrates in mountain areas	10	10	83,875–98,722
6410	*Molinia* meadows on calcareous, peaty or clayey-silt-laden soils (*Molinion caeruleae*)	2	2	646–1964
6430	Hydrophilous tall herb fringe communities	4	4	5407
6510	Lowland hay meadows	3	3	16,885
6520	Mountain hay meadows	10	10	222,796–223,023
7110	Active raised bogs	2	1	39,100
7120	Degraded raised bogs still capable of natural regeneration	4	3	284,304
7140	Transition mires and quaking bogs	7	6	4247
8150	Medio-European upland siliceous screes	2	2	4441
8160	Medio-European calcareous scree of hill and montane levels	1	1	126
8210	Calcareous rocky slopes with chasmophytic vegetation	1	1	1164
9110	Luzulo-Fagetum beech forests	6	4	120,592–178,910
9130	Asperulo-Fagetum beech forests	1	0	–
9180	Tilio-Acerion forests of slopes, screes and ravines	2	1	20,460–22,705
91D1	Bog woodland with downy birches	5	3	91,559–283,003
91D3	Bog woodland with mountain pines	3	2	88,861
91D4	Bog woodland with spruces	5	4	132,969–377,098
91E0	Alluvial forests with *Alnus glutinosa* and *Fraxinus excelsior* (*Alno-Padion, Alnion incanae, Salicion albae*)	2	1	239,41–31,833
9410	Acidophilous *Picea* forests of the montane to alpine levels (Vaccinio-Piceetea)	9	6	203,325
			Total sum	**1,380,500–1,979,900**

Table 6.13 Selected monetary aspects of the evaluation and the maintenance of FFH habitat types on the example of the FFH site 'Geisingberg und Geisingbergwiesen' (Eastern Ore Mountains, Saxony)

Code	Habitat type	Area (ha)	Share (of the FFH site) (%)	L (€ ha^{-1})	L (€)	Sc (€ m^{-2})	Sc (€)	Se (P m^{-2})	Se (P) 10,000 scores	Se (€) x 10,000 €
3150	Natural eutrophic lakes	0.01	<0.01	320	3.2	48.9	0.49	47	0.47	0.0884
3260	Water with the *Ranunculion fluitantis* and *Callitricho-Batrachion* vegetation	0.04	0.01	570	22.80	k. A.	k. A.	41	1.64	0.656
6230	Species-rich Nardus grasslands	1.21	0.37	570	689.7	41.8	50.6	53	64.13	25.652
6510	Lowland hay meadows	0.31	0.09	360	111.6	6.1	1.9	33	10.23	4.092
6520	Mountain hay meadows	67.75	20.85	380	25,745	6.1	416.0	50	3387.5	1355
7140	Transition mires and quaking bogs	0.04	0.01	390	113.1	127.4	5.1	56	2.24	0.896
7230	Alkaline fens	0.25	0.08	390	113.1	127.4	31.85	56	14.0	5.6
8150	Medio-European upland siliceous screes	0.59	0.18	k. A.	k. A.	k. A.	k. A.	43	25.37	10.148
9110	Luzulo-Fagetum beech forests	5.14	1.58	200	1028	18.4	94.8	38	195.32	78.128
9130	Asperulo-Fagetum beech forests	5.18	1.59	200	1036	18.4	95.9	62	321.16	128.464
9180	Tilio-Acerion forests of slopes, screes and ravines	10.58	3.25	200	2116	18.4	195.1	42	444.36	177.74
91E0	Alluvial forests with Alnus glutinosa and Fraxinus excelsior	0.31	0.09	200	62	18.4	5.7	42	13.02	5.208
Total		91,41	28.13		30,927.40		897.0554		4479.44	1791.776[a]

L – annual costs of managing the habitat type per hectare (€/ha) and per total area of the habitat type (€); data from the "Landscape Management Strategy Saxony 2020" (Grunewald and Syrbe 2013)
Sc – biotope values after Schweppe-Kraft (2009) (Habitat-Equivalency-Investment- Model) (in €/m and in € 10,000 for the total area of the habitat type)
Se – biotope values (scores–P) after Sejak et al. (2010) per m^2, per total area of the habitat type (10,000 P) and monetary value per total area of the habitat type (€ 10,000 €, the value of 1 score amounts € 0.4), [a]total value of € 17,917,760 corresponds with 5 % annual discount rate 895,888 €/a (discounted over 20 years)
n. i. – no information

household and year (total € 3.9–4.9 billion) for a conservation programme to maintain biodiversity in Germany (only ethical and aesthetic aspects are reflected). This amount is double of the estimated costs of all measures necessary for the maintenance of biodiversity (€ 1.7–2.3 billion). More recent investigations (Meyerhoff et al. 2010; Schweppe-Kraft 2009) showed a higher WTP (€ 192 per household and year). Each hectare protected area would benefit from c. € 1000 annually.

Why are such nature conservation programmes not implemented into practice? The general mistrust of contingent valuations and the lack of 'hard' facts in the area of biodiversity–unlike with the problem area flood protection, etc.–may be one of the reasons (Schweppe-Kraft 2009).

Closely related with management costs are, for instance, monetary values used for habitat types in the Impact Mitigation Regulation. If values and functions of nature are impaired by an avoidable impact (e.g. by construction measures), the damage can be calculated as the equivalent of the costs (substitution and compensation costs) of compensation and replacement measures necessary for the recovering of the functions of nature (replacement costs). The starting point for the derivation of monetary values are nature conservation requirements on the dimension of compensation and replacement measures (nature conservation right of compensation obligation, German Nature Conservation Act–§ 8). This obligation can be seen as the consensus of society to maintain the functions of nature (or the ES). The costs of the compensation measures are adequate to society's willingness-to-pay. With regard to the evaluation of impacts on the one side and compensation measures on the other side, such numbers set quite general benchmarks that let enough freedom of action for type- and single case-related planning solutions (UBA 2007). The value of natural and semi-natural ecosystems in Germany calculated with the Habitat Equivalency-Investment-Model (Schweppe-Kraft 2009, ► Sect. 4.2) (9.5 % of the country's surface area) amounts to € 740 billion (restoration costs while taking the necessary time or duration of development into account).

Based on a biotope assessment method elaborated in the German federal state Hesse (the so-called 'Hessian Model'), which was recommended by the EU to be applied in Natura 2000 sites, Sejak et al. (2010) created a scoring system for biotope types basing on data from 136 restoration projects in the Czech Republic. They awarded scores between 1 and 6 each for the criteria age (maturity), naturalness, diversity, species diversity, rarity of the biotope type, rarity of species, vulnerability, impacts on the biotope type. The authors assigned monetary values to the scores (1 score=€ 0.40). The financial value of a score corresponds to the arithmetic means of the costs of all analysed restoration projects in the Czech Republic spent to improve the ecological state by one score. They combined the scores representing the 'ecological benefit' with the biotope-specific restoration costs (◘ Table 6.14).

Case Study Mount Geisingberg

On the example of the FFH site 'Geisingberg und Geisingwiesen' (325 ha, identification number DE 5248-303) different value and cost calculations will be shown and compared (◘ Table 6.13, 6.14, ◘ Fig. 6.23). This protected area is situated north of the small towns Altenberg and Geising in the rural district Sächsische Schweiz–Osterzgebirge, in the physical region Eastern Ore Mountains at the altitude between 545 and 824 m a. s. l. (on average 700 m). This FFH site overlaps with other categories of protected areas: the landscape protection area 'Oberes Osterzgebirge' (Upper Ore Mountains), the bird sanctuary (SPA) 'Geisingberg und Geisingbergwiesen' and the nature reserve 'Geisingberg'.

The specific conservation targets are listed in the FFH management plan (Böhnert et al. 2005): e.g. "maintaining the supra-regional important 'basalt cap' of the Geising mount with different succession phases of species-rich montane spruce-(fir-)beech forests, surrounded by a large complex of species-rich montane grassland communities of different levels of trophy and moisture, with agricultural clearance cairns, mountain meadows, moist and wet meadows, fen areas and mat grassland, including the flora and fauna typical for this region. The site is part of the characteristic clearance cairns landscape of the Eastern Ore Mountains with traditional, small-scale landscape structures and extensive land use".

The maintenance and development of selected biotopes is particularly important:

- Protection, maintenance, management and partly restoration of an important complex of species-rich mountain grassland communities, particularly mountain meadows, mat grassland, moist meadows, wet meadows rich in sedges and rushes, and fens of high importance for flora, vegetation and fauna, especially considering the extremely rare chalk fens, and
- Maintenance and targeted development of near-natural tree species combinations, age and spatial structure of the forest, and with special promotion of the richness in old and dead wood and White Fir.

The FFH site 'Geisingberg und Geisingbergwiesen' (total area 325 ha, among them 91.41 ha FFH habitat types) includes 67.75 ha mountain meadows (=20.85 % of the total area). If 380 €/ha are estimated for the annual management, the need of financial means amounts to 25,745 € per year. To manage all FFH habitat types of the site, the financial need would amount at 30,927 € annually. This number is not necessarily in line with the costs mentioned in the management plan as the Saxon management strategy lists average costs for Saxony while the management plan refers to the specific situation.

The Habitat-Equivalency-Investment-Model after Schweppe-Kraft (2009) assigns values of 6.14 €/m^2 of extensively used grassland; this corresponds to € 4,160,000 for the total area of the habitat type 'Mountain meadow' in this FFH site. The Czech model after Sejak et al. (2010) assigns 50 scores per m^2 for mountain meadows. Thus, the habitat type 'mountain meadow' at the Geisingberg reaches $3.387.5 \times 10^4$ scores or € 13.55 billion. The reason for the difference to Schweppe-Kraft (2009) is the much lower amount calculated for meadows.

After Schweppe-Kraft (2009) all FFH habitat sites at the Geisingberg mount combined would reach a value of € 8,970,554, after Sejak et al. (2010) € 17,917,760 with an annual discount rate (5 %) 895,888 €/a (x20 years).

Discussion

Protected areas, among them the FFH and SPA sites of the European Natura 2000 network, supply a wide range of ES. The reference to the manifold services being very important for humans, may support the arguments of nature conservation.

A key challenge for the evaluation of the benefits of a Natura 2000 site is to be able to present and interpret the full range of its benefits. The general underlying idea is that the total (long-term) benefits provided by an ecosystem increase with conservation and sustainable use. ◘ Figure 6.1 presents an overview of how the different value derived from a given system change with increased conservation efforts. In general, despite of the costs of conservation and reduced extraction of biodiversity resources it is foreseen that the net socio-economic benefits provided by the ecosystem remain positive (Kettunen et al. 2009).

If the focus is only on benefits that can be estimated in monetary terms the overall socio-economic picture might not appear favourable to the sites conservation. It is also important to understand, for example, how the identified benefits relate to the conservation goals of the site (e.g. do they conflict with site management plans) and how different stakeholders are affected by these benefits.

It was already mentioned that several dimensions of nature cannot or should not be measured in monetary terms (► Sect. 4.2, TEEB 2009). As many ES cannot be traded on markets this would result in them being under evaluated from an economic perspective (Mertz et al. 2007; Bayon and Jenkins 2010; de Groot et al. 2010). A monetary valuation is generally accepted as nonfeasible for environmental goods and services with a religious or spiritual value or amenity (Spangenberg and Settele 2010). There are societal reasons for not defining nature conservation through market mechanisms only (Ring et al. 2010).

The idea of exact monetary values may appear fascinating but the examples of different results for mountain meadows (see above) indicate fundamental and methodical weaknesses. Moreover, the restoration of destroyed biotopes is not always possible especially if the site conditions are changed irreversibly and if some species became extinct. The calculated value of natural and semi-natural ecosystems in Germany of € 740 billion (Schweppe-Kraft 2009) may appear very high. But: is the value of these ecosystems really below the value of the fixed capital and the production facilities in Germany? And: is it even justified to compare this?

Nevertheless, economic valuation can help policy-makers by shedding light on the contribution made by various ES, whether directly or indirectly, and, thus, serve an informational function. Economic valuation, however, is not the adequate method for determining the goals or priorities of conservation policies. Nonetheless, economic instruments can be applied as effective incentives for biodiversity conservation and ecosystem management to maintain ES (Spangenberg and Settele 2010). In nature conservation, economic valuation of biodiversity and ES is an essential element for making conservation efforts financially sustainable over longer periods of time as it stimulates the perceived need for investing in conservation, be it through the establishment and management of protected areas, through traditional economic instruments such as taxation, licence fees, etc., or through the development of markets and agreements on payments (incentives) for environmental services (Mertz et al. 2007).

6.6.2 Soil and Water Protection

K. Grunewald

Functions and ES of Soils and Waters

Soil and water are components in the landscape system, characterised by high complexity and numerous interfaces with the other geo-components. The resulting significance in ecosystems is expressed on the one hand in the environmental goal of soil and water conservation in the Federal Nature Conservation Act (§ 1 BNatSchG 2009). On the other hand, soil and water are essential prerequisites for the generation of diverse use functions (ES) and, thus, gain an existential importance for human society.

Hydrological ES (water regulation and water purification) as well as pedological ES (erosion and maintenance of soil fertility) form separated subgroups within the group of regulatory ES (► Sect. 3.2). Moreover, freshwater presents one of the essential supply ES (direct market good). Less attention is mostly paid to indirect, usage-independent ES, for example, natural soil profile as an education ES. Soil and Water conservation form an inseparable unit (◘ Fig. 6.25), e.g. the soil type is an indicator for both the ES water regulation and water purification (► Sect. 3.2).

◘ **Fig. 6.25** Land use induced soil erosion on arable land and matter input into the waters of the Kleine Jahna rivulet near the town Riesa. It is hard to define the on-site- und off-site damages and costs. © Karsten Grunewald

Soils are natural components of nature. The processes of soil formation and regeneration of soils occur extremely slowly. Therefore, soils are rated as one of the nonrenewable resources. Currently, land use (increase in the amount of land used for human settlements and the transport infrastructure and dissection of the landscape), erosion and compaction, soil contamination and impoverishment of soil biodiversity must be considered as the main problems of soil protection in Central Europe. The deterioration of soil quality has a direct impact on the quality of water and air, biodiversity and climate change. Furthermore, this can affect the health of the population and the safety of the production of food and feed. Particularly affected by this are soils of very different capacity and functionality (Grunewald 1997). Thus, in addition to site-specific use preventive *soil protection* needs to include above all, the protection of the natural soil functions. This is important especially from the point of view of the functional importance of soils in the landscape balance (Kramer et al. 1999).

Land use-related influences have led to more homogeneous soil physical and chemical conditions of topsoils over natural, place-based soil patterns, for example, in large-area farming. Land use associated with constant supply of substances and energy (soil threatening, tillage, fertilization) and the removal of forms of disadvantage (land improvement, e.g. hydro-melioration) cause relatively homogeneous crop stocks under normal weather processes. Much effort is needed for that. Therefore, for example, modern techniques of modern precision agriculture differentiate small plots so that negative consequences are reducible. However, they do not lead to the abolition of patterns of properties among others in sub-soils so that risks of destructive processes increase compared to soils with natural vegetation cover. As a result of agricultural land use, regarded processes have led to a large heterogeneity of soil-forming substrates and, thus, soil properties to this effect, e.g. through erosion (extensive capped profiles and completely eroded areas on upper slopes or slope shoulders). Certainly it is spatially differentiated whether heterogenizing or homogenizing processes dominate. Statements in this regard are generally related to individual aspects of soil characterization and assessment. However, the heterogenizing, unwanted side effects increase the management effort and, thereby, burden the economics of land use. Therefore, business management calculations weigh up the costs and benefits. In that regard a short-term calculation without follow-cost analysis or the ignoring of external costs arising from the land-use system is dishonest and unsustainable. These ratings cannot be uniform and unambigious, since they are crop-dependent, in the production system differentiated and market-determined (▶ Sect. 6.2.3).

In respect to soil functions another question is of concern: is enhanced pedodiversity (analogous to biodiversity) valuable, beneficial and preferable? This question, however, cannot be answered consistently with respect to partial functions and for different soil landscapes.

With the German Federal Soil Protection Act the implementation of soil conservation has been placed on a uniform legal basis at national level. Core of the law is a strictly functional definition of soil with the distinction between natural functions (as foundation for life and habitat, part of the ecosystem, and decomposition, compensation and generation medium for material impacts, thanks to its filtering, buffering and transforming properties), functions as an archive of natural and cultural history as well as utility functions (source of raw materials, location for various uses as well as areal for human settlement and recreation).

In this context a counteractive effect between the natural soil functions and use functions in terms of the environmental impact should be noted. The utilisation of specific use functions of soils, e.g. the function as source of raw materials, causes the limitation in terms of natural functions. The conflict between soil integrity and specific forms of land use is also expressed in problems such as soil erosion, which overloads natural soil functions and initiates damage processes in ecosystems (Grunewald and Mannsfeld 1999).

The fact of harmful soil changes is of particular importance in this context (§ 2(3) BBodSchG 1998), which are related to impairments of soil functions. Consequently, this leads to the obligation to avoid damages (§ 4 (3) (4) BBodSchG 1998) and area-based soil protection. For the implementation of particular importance as subordinate regulation is the Federal Soil Protection and Contaminated Sites Ordinance (Bundesbodenschutz- und Altlasten-verordnung; BBodSchV 1999).

Another point to note is that soil is a limited resource and their functionality is affected by various forms of exploitation. Its regeneration, is extremely difficult, if not impossible, and sometimes very costly. Basically, in countries such as Germany or Austria the question is arising how the previously strictly functional managed soil protection can be transformed and updated to the ES concept.

A European Soil Framework Directive with the aim of protecting soils is rejected among others by the government of Germany, in particular for reasons of subsidiarity (subordination of rules to others, e.g. WFD). In addition, it is also feared that the implementation of a Soil Framework Directive causes a disproportionate administrative burden and high consequential costs (Kluge et al. 2010).

Water protection is defined as the totality of efforts, conserving waters (coastal waters, surface waters and groundwater) for the purpose of purification of water for drinking or industrial water and the protection of aquatic ecosystems as a subtask of nature protection. Water protection is operated partly usage-oriented, partly detached by user interests and between the commercial and conservation interests numerous areas of conflict exists.

However, the benefits of ES for water protection have become a high priority in policy making:

- The Water Framework Directive (WFD, Directive 2000/60/EC) follows the principle of cost recovery of water services, including environmental and resource costs, which requires a balancing of the costs and benefits (explained in ► Sect. 3.3.2).
- The EU's strategy for protection of the marine environment (2005) requires the implementation of cost-benefit-analyses.
- The Marine Strategy Framework Directive (RL 2008/56/EG) requires Member States to produce an economic and social analysis of the use of their marine waters and of the cost of degradation of the marine environment.

6

ES Valuation on the Example of the Jahna River Catchment in Saxony

The catchment area of the Jahna River in Saxony (Germany), mainly used for agriculture, was selected as the study area for assessing of ES in order to achieve environmental objectives of the WFD, since it has already been studied extensively. The goal of the case study was to analyse the cost-effective combination of measures of agriculture to reduce water erosion and diffuse nutrient inputs in water and to assess selected, mainly not market-based ES. The fundamental approach is based on the EPPS conceptual framework (► Sect. 3.1.2). The core is, that through the (natural) scientific bases the professional requirements of use/conservation of natural resources can be expressed (interdisciplinary, multiple political consideration). These characteristics and pressures on ecosystems are analysed using ecological indicators and reduction potentials are simulated by models. Furthermore, changes in agricultural land use and management form are also evaluated monetarily in regard to reduction in nutrient input into waters (details in Grunewald and Naumann 2012).

▪ Ecosystem Properties and Selected Pollutions in the Catchment Area

The catchment area of the river Jahna is 244 km^2 in size, located between the towns of Döbeln and Riesa (◘ Fig. 6.2), is part of the natural region Lößgefilde (loess-region) in Saxony and has a rural typical, relatively low population density. Due to the very fertile soils the river basin is primarily used for agricultural purposes since time immemorial. About 90 % of the land is occupied by agricultural land. The share of only 6 % grassland suggests that pure arable farms are predominantly located in the study area. Main crops grown are wheat, corn, rapeseed and root crops (sugar beet). About 14 % of the Jahna catchment is designated as protected area which partially overlap (7 % drinking water protection areas, 6.1 % areas of protected landscape, 3.8 % bird protection areas (SPA), 2.4 % habitat protection areas (FFH), 0.2 % Nature Reserves).

Luvisols occur as soil types over a large area, Cambisols, Albeluvisols and Luvisol-Planosols/Stagnosols are also found in smaller plots. In the valleys and depressions partly mighty colluvisols are found. That indicate a high erosion deposition in the area. Accordingly, at the upper slope or slope shoulder locations extensively capped profiles and completely eroded areas are distributed. Water erosion is a problem in the study area since the beginning of the intensive land use of the landscapes. This is documented in the colluviums and high-flood loams respectively in the altered sediment load of waters.

The river Jahna has a length of about 35 km and flows into the Elbe River in Riesa. Numerous interventions in the water system were undertaken in the catchment area, such as longitudinal and transverse profile barriers, run-straightening and relocations, melioration, etc. About 40 dams/reservoirs respectively ponds currently characterise the surface water system; the reservoir Baderitz with 15.8 ha is the largest among them.

According to the WFD, the biological components fish, macroinvertebrates and macrophytes/phytobenthos are relevant for assessing the surface water bodies. Without exception, all of these were assessed as deficient in the catchment area of the Jahna River in the period 2005–2007. The nutrient pollution reflects the poor biological evaluation in the catchment. With the exception of one river water body, the guidance values for total-P and orthophosphate-P were exceeded two to threefold in all years. Currently there are no reference values for the WFD-relevant elements total-N and nitrate-N available. Compared to the nitrate-quality standard for the chemical status it is clear, however, that all surface water bodies are significantly affected by nitrogen. With a mean of 97 mg L^{-1} during the period 2007–2009, the quality standard for nitrate (50 mg L^{-1}) is significantly exceeded in the groundwater body Jahna. The monitoring results of the inventory lead to the conclusion that the objectives of the WFD cannot be achieved in the groundwater body Jahna and all eight river water bodies in the catchment area Jahna by 2015 (Grunewald and Naumann 2012).

Cause analysis for the diffuse nutrient sources were calculated using the model STOFFBILANZ. The Web-GIS-based model STOFFBILANZ (▶ www.stoffbilanz.de) is a method for quantification of sources and path-related nonpoint source pollution (nitrogen, phosphorus and sediment) from the surface (emission) in catchments of mesoscale size. In addition, the quantification of the immission resulting from matter inputs to surface waters is possible using simple estimation methods.

Modelling results for phosphorus (P): in average 52 % of the total P-emissions of 14.5 t yr^{-1} of the Jahna catchment originate from agricultural land. The majority of the agricultural P-discharge is caused by particulate phosphorus (PP) input via water erosion (Haygarth et al. 1998, ◘ Fig. 6.25). Almost 80 % of PP-inputs into surface waters are from the critical source areas. This means the majority of loss comes from a small part of the catchment where areas of high potential for supply (source) and transport (e.g. surface runoff) overlap. These areas were termed critical source areas (CSAs, cf. Heathwaite et al. 2005; Halbfaß and Grunewald 2008; Qui 2009). The estimation of P-concentrations from the total P-loads (including upstream) resulted in a span from 0.33 to 0.73 mg L^{-1} for emissions. By an average P-retention of about 70 % an immission load of 3.9 t y^{-1}, and P concentrations from 0.09 to 0.23 mg L^{-1} were determined for the surface waters.

Modelling results for nitrogen (N): According to the assessment by the STOFFBILANZ model in average 95 % of total N-emissions in the catchment area Jahna (574 t yr^{-1}) originate from agricultural land. In contrast to phosphorus, nitrogen is discharged almost all dissolved on the different flow types. The underground drainage component baseflow dominates followed by intermediate and drain discharge. If a catch crop of 4 % (as in 2005/2006 normal) of arable land in the catchment area Jahna is considered, STOFFBILANZ calculated a reduction of the total diffuse N-emissions (including from settlements) on watershed level of about 8 to 11 % (N-removal by the intercrops of 80 or 100 kg yr^{-1}). Based on the N-loads (including upstream), there were emissions-based N-concentrations from 11 to 23 mg L^{-1}. Taking an average retention of 62 % into account, a total load of 208 t yr^{-1} (immission) and total N-concentrations between 4 and 8 mg L^{-1} were calculated. With respect to the groundwater flow in the Jahna aquifer an N-input of 349 t yr^{-1} in the surface waters was determined. This corresponds to an average load of 7.9 kg ha^{-1} yr^{-1}. An average N-concentration of approximately 69 mg t L^{-1} in groundwater discharge results from a modelled average base flow of 51 mm yr^{-1}.

Different agricultural measures and sets of activities were simulated to estimate the nutrient-reduction potential (◘ Table 6.14). The reduction of P-input is shown on the basis of PP-emission (particulate phosphorus) since a large proportion of the agricultural P-discharge takes place via this path (Pimentel et al. 1995; Halbfaß and Grunewald 2003). For nitrogen, however, the total diffuse N-emissions are shown.

The individual measures with the greatest reduction potential of P-input (40–70 %) are the conversion of arable land into grassland on the CSAs (Halbfaß and Grunewald 2008), conservation tillage and buffer zones (water protection stripes). If measures are combined, however, a reduction of PP-emission of 90 % is possible.

The measures intercropping and reduced fertilization have the largest reduction potential for N-input with 30–50 %. The combination of measures can reduce N-input up to 77 %. Looking at

Table 6.14 Selected results of scenarios for the river catchment Jahna modelled with STOFFBILANZ for the Jahna river catchment. (Grunewald and Naumann 2012)

No.	Scenario/variant	Particulate P-input			Diffuse N-input		
		[t a^{-1}]	[kg ha^{-1} a^{-1}]	[%]*	[t a^{-1}]	[kg ha^{-1} a^{-1}]	[%]*
1	Increase of conservation tillage to 100 % on CSAs	3.5	0.14	−31	537	21.9	−3
2	Increase of conservation tillage to 100 % on the whole arable land	2.8	0.11	−45	451	18.4	−19
3	No maize and root crop cultivation on CSAs	3.7	0.15	−27	555	22.7	0
4	Greened runoff pathways on CSAs	4.6	0.19	−9	553	22.6	0
5	Water protection stripes (buffer zones)	2.7	0.11	−47	543	22.2	−2
6	Land-use change on CSAs	1.6	0.07	−68	538	22.0	−3
7	Land-use change on areas with the highest N-leaching	5.0	0.20	−2	489	20.0	−12
8	Catch crop cultivation–actual state +5 % in the catchment	–	–	–	449	18.3	−19
9	Catch crop cultivation–actual state +16 % in the catchment (max.)	4.9	0.20	−3	324	13.2	−42

* decrease of P- or N-input compared to the actual state; CSAs–critical source areas

the reduction potential of the measures in each river water body, the amount depends on grown crops and the location of CSAs. Thus, the effect of conservation tillage on the P-leaching is stronger in the upper catchment area Jahna than in the lower one due to the higher number of CSAs. The same applies to the other measures on the CSAs: 'conversion of arable land into grassland', 'grassed drainage channels' and 'renunciation of maize and root crop cultivation'. The N-discharge shows greater differences on plots with highest N-leaching due to crop-specific intercropping systems and reduced nitrogen fertilization.

ES Assessment

One way to evaluate and prioritise different measure scenarios provides the *utility analysis* according to Zangemeister (1971). The different target variables can be better compared with each other through their transmission in a common value system. As targets for the WFD implementation, the reduction of N- and P-inputs into the waters and the costs and acceptability of measures plays an important role. Utility functions between 0 (no benefit) and 1 (highest benefit) are defined for these target variables, which in case of the nutrients are determined by environmental quality standards or guidance values (Naumann and Kurzer 2010). The part-worth utilities of the various scenarios were determined for the target variables with these utility functions. The total benefit of the different measure scenarios to be compared was the result of adding the part-worth values of the target variables, whereby a weighting of the target variables was still carried out by the agent (Table 6.15).

By an equal weighting of the target variables conservation tillage on the CSAs represents the measure with the highest total utility in the Jahna catchment area. The reasons for this are the low costs, the relatively low P-concentration by the

Table 6.15 Part and total utility values of target variables of measures scenarios for the river catchment Jahna. (Grunewald and Naumann 2012)

Target variables Measure scenarios	Particulate P		Diffuse N		Costs		Accep- tance		Total utility
Conservation tillage on CSAs (100 %)	0.80	0.2	0.29	0.07	0.90	0.23	0.5	0.13	*0.63*
Conservation tillage on the arable land (100 %)	1.00	0.25	0.57	0.14	0.21	0.05	0.5	0.13	*0.57*
Greened runoff pathways on CSAs	0.40	0.1	0.25	0.06	0.49	0.12	0	0	*0.28*
Water protection stripes (buffer zones)	1.00	0.25	0.27	0.07	0.92	0.23	0	0	*0.55*
Conversion of arable land into grassland on CSAs	1.00	0.25	0.29	0.07	0.72	0.18	0	0	*0.50*
Conversion of arable land into grassland on areas with the highest N-leaching	0.20	0.05	0.45	0.11	0.72	0.18	0	0	*0.34*
Catch crop cultivation 9 %	0.20	0.05	0.57	0.14	0.87	0.22	0.5	0.13	*0.54*
Catch crop cultivation 20 %	0.30	0.08	0.97	0.24	0.60	0.15	0.5	0.13	*0.60*

CSAs critical source areas

modelled PP-input and the mean acceptance of the measure by the farmers. This is followed by the measure 20 % catch crops, whose high total utility is mainly due to the high part of the benefit in N-concentration, and the measure catchment-wide implementation of conservation tillage. If more emphasis is placed on the nutrient input, the extensive conservation tillage farming is the preferred option, followed by catch crop. The high part-worth utility of P-concentration of the conservation tillage and the high part-worth utility of N-concentration of intercropping contribute to this result. If the costs, however, have the highest priority despite the high relative cost water protection strips (buffer zones) gain in importance, as due to the small area the total costs for the watershed Jahna are low. The measures 'greened runoff pathways' and 'conversion of arable land to grassland' on areas with highest N-leaching occupy by all weights only lower ranking positions due to the low modelled contribution to matter input reduction and low acceptance.

The *environmental costs of erosion and nutrients emission* are not precisely known. Likewise, the social benefits of erosion protection and the reduction of nutrient translocation/leaching can only be estimated. In this case it is questioned whether the cost-benefit analysis is sufficiently precised to capture concrete effects of projects, measures and policies. Crop yields on eroded soils are lower than those on protected land as erosion reduces the ES soil fertility and water availability. Erosion affects soil quality and productivity adversely by reducing water infiltration rates, water-holding capacity, nutrients, organic matter, soil biota and soil depth. Moderately eroded topsoils absorb from 10 to 300 mm less water per hectare per year than uneroded soils (correspond to 7–44 % of total rainfall, see Pimentel et al. 1995). A ton of fertile agricultural topsoil typically contains 1 to 6 kg N and 1 to 3 kg P, which can be lost through runoff. These are so-called on-site damages, which land owners and users wants to keep as low as possible. As shown in the previous sections the losses can be significantly reduced by erosion control measures.

The off-site costs of erosion must also be considered. The soil loss not only represents a loss for farmers but also can affect habitats on neighbouring areas adversely or block the public sewage system, which must then be cleaned with financial expense. The hydroecological damages were outlined (sediment and nutrient input, eutrophication, increased water treatment costs, etc.). The real costs of this are not exactly quantifiable and the persons responsible are hard to make liable, even for small, localizable erosion events. Nevertheless, the entity shall presume that both the individual and the society are interested to keep off-site damages (and therefore costs) as low as possible.

Pimentel et al. (1995) estimated the on-site and off-site costs of erosion in the USA to about US $ 100 per hectare per year in the mid-1990s. If one estimates the so-called replacement costs of soil and fertiliser (according to internet research a ton of topsoil costs about 10 € and current fertiliser prices are to be set at about 600 € t^{-1} for N respectively 750 € t^{-1} for P; Source: AMI 2010) and damage costs (cleaning of roads, land, properties after erosion damages, desludging of reservoirs, ponds, canals, etc.), one arrives at a similar monetary magnitude of on-site and off-site damages for the Jahna catchment area (*benefit transfer*).

Accordingly, replacement and damage costs of € 1.4 million per year would be calculated for the nearly 20,000 ha of agricultural land in the Jahna catchment area (with USD/EUR exchange rate of 1.4). A comparison of this order of magnitude with erosion reduction measures revealed a very positive *benefit-cost ratio* in areas with high erosion threats. Pimentel et al. (1995) give this example for the USA with 5 to about 1; thereby reducing soil erosion by water and wind from 17 t ha^{-1} yr^{-1} to 1 t ha^{-1} yr^{-1}. A benefit-cost ratio of about 2 to 1 would result for the society assuming the costs of most effective measures in the watershed Jahna with 760,000 € (100 % conservation tillage on CSAs and catch crops on 20 %, and current funding rates: 85 € ha^{-1} for intercropping, 68 € ha^{-1} for conservation tillage).

The assessment of the benefits is primarily oriented to the objectives of the WFD in the case study. Tangent goals concern soil protection, nature conservation, agricultural productivity and others. An integrated assessment and planning takes the impact of measures on all relevant target dimensions into account. The area under consideration–in this case the catchment area of Jahna–therefore, represents a common field of action for water management, agriculture and nature conservation (Grunewald and Naumann 2012).

Conclusion

A monetary assessment cannot capture all values of an ecosystem. But by applying economically oriented planning methods and usage of a benchmark, such as money exchange values of planning, variants are more visible and more conscious. Changes of characteristics and services of soils and waters can be quantified, modelled and represented in simulated scenarios. Values are assessable using ES approaches. This can lead to new insights in regard of the mediation between land use and conservation interests.

The major criticism of the presented and for the catchment area of the Jahna exemplified ES approach is that the data required for the operationalisation of quantitative models and the monetisation (benefit transfer method in this case) is methodologically uncertain. Only few of the ES have an economic value, which land users can realise on markets. Numerous ES of catchment areas are in economic terms public goods. This means that markets do not adequately reflect the costs and benefits associated with a change in supply. Furthermore, it is unfortunate that social, aesthetic and health values are underrepresented in the ES evaluation and planning.

6.6.3 Economic Valuation of Ecosystem Services–The Case of Wetland Restoration Along the German Elbe River

V. Hartje and M. Grossmann

Introduction

This case study has been developed in the context of the debate on the restoration of riverine wetlands to achieve nature conservation objectives. In Germany, the debate ignited on the question to what extent the relocation of dikes in riverine wetlands as a measure for floodplain restoration has positive effects on flood protection or whether they even would constitute a better flood protection strategy than a programme to construct flood polders. This paper contributes to the debate by comparing a dike relocation programme with a polder construction programme, with a cost-ben-

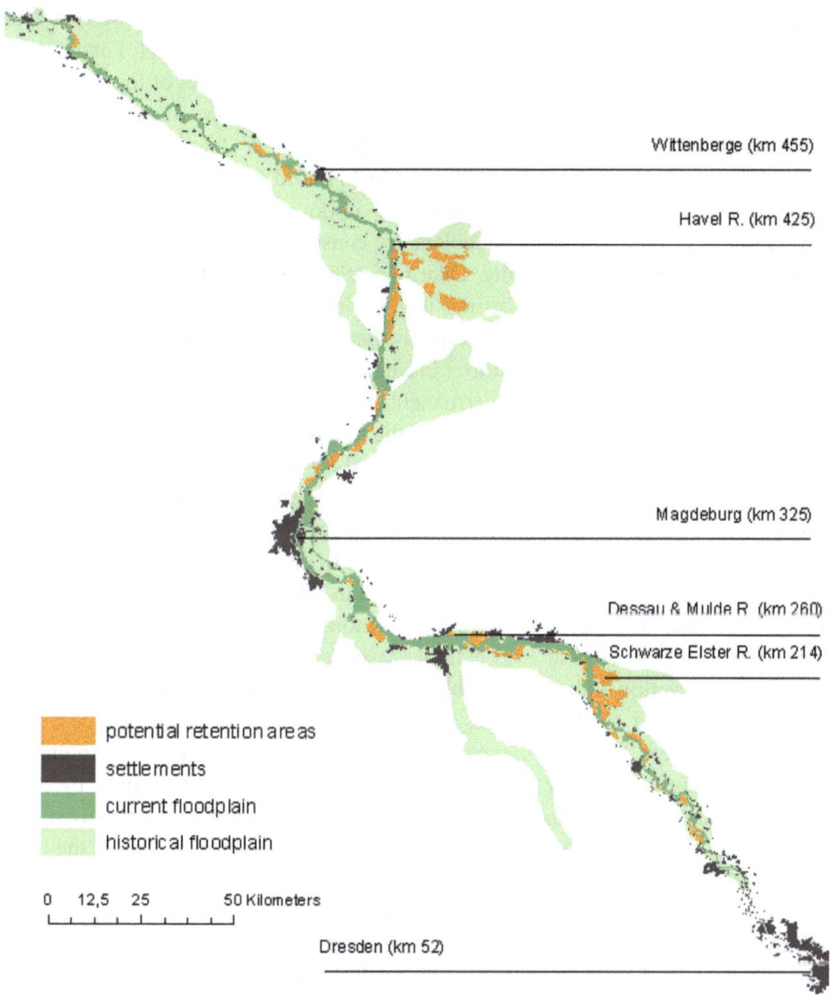

Fig. 6.26 Map of the study area showing the location of the potential retention areas

efit-analysis based on the quantification and monetisation of selected associated changes of ES. The covered ES include the reduction of the risks of flooding, the reduction of nutrient discharges and the cost-benefit-analysis includes the maintenance of wetlands habitats and species. In the following, the alternative programmes of measures are presented before the specific approaches to estimate the changes of the ES and their economic valuation are explained and the results are presented. Finally, the article concludes.

The Relocation of Dikes as a Component of a Wetland Restoration Policy at the Elbe

The focus of this article is an integrated analysis of dike relocation as part of a wetland restoration policy instead of being part of three separate sectoral policies: Preventative flood protection, nature conservation and water quality control. The following chapter summarises the policy debate in Germany and describes the valuation scenarios.

The German part of the Elbe (Fig. 6.26) can be characterised as a lowland river in a broad alluvial valley downstream of the city of Dresden with a high degree of losses of floodplain wetlands. The loss of the wetlands in the floodplain in the upper and middle range of the Elbe varies with the breadth of the river valley. In the narrower valley of the southern part, the losses of the floodplain wetlands have been smaller. When the river enters the broader valley between 50 and 90 % of the historical wetlands have been lost due to diking (Brunotte et al. 2009). Despite these large losses of the riverine wetlands, the Elbe is still one of the larger free-flowing rivers in Central Europe.

The Federal Ministry of the Environment (BMU) and the Federal Office for Nature Conservation (BfN) actively support the concept of integrated management of riverine wetlands (BMU, BfN 2009). With this concept, the mutual advantages of maintaining and restoring riverine wetlands in the area of flood protection, water quality control, nature conservation and climate change mitigation shall be realised.

Already in the 1990s, Neuschulz and Purps (2000, 2003) surveyed potential large area sites at the Elbe with a high nature conservation value. They identified 52 potential areas with a total of 23,249 ha including 11 summer polders. Following the great floods in the Elbe basin in the year 2002, the public debate about wetland restoration via dike relocation intensified considerably. The International Commission for the Protection of the Elbe River developed a Flood Protection Action Plan (IKSE 2004), which included a comprehensive strategy for flood risk management.

Alternative Programmes for Wetland Restoration

Here, seven programmes will be analysed and compared, summarising various surveys and alternative project proposals which were undertaken in the recent past. However, potential relocation sites involving the risks of life or settlement areas with larger real estate assets were not included. From these surveys, only those sites were included where agricultural and forestry land use could be converted into wetlands. The total number, the precise location and the retention volume have been put to an intensive public debate afterwards, and they are subject to changes in the planning phases if they are to be implemented. For this analysis, the proposals for the relocation of dikes and the construction of polders, and the estimates of their retention volume are drawn from a number of sources (Merkel et al. 2002; Ihringer et al. 2003; IKSE 2004; Förster et al. 2005, BfG 2006). If the estimates of the area of the sites differ among the sources, the larger alternative has been included. A detailed description of the sites can be in found in Grossmann et al. (2010).

Here, two options to restore wetlands in the floodplain are considered: Dike relocation and polder. Dike relocations consist of the removal or the cutting of the old dike, and the construction of a new dike further away from the river which may be shorter in a number of cases. Dike relocations involve a permanent change in land use and imply a reintroduction of the functions of the natural wetland. Polders are designed for a controlled flooding of an area enclosed by dikes where the inflow is controlled by weirs. The advantage of polders is that they are built to control the inflow into retention areas which allows them to be used to cut the peak of a flood wave effectively. Furthermore, as the polders are only flooded in extreme cases, continued agricultural land use is possible, but cropping areas have to be converted to grassland uses. Thus, retention polders can be viewed as a partial restoration strategy. The retention polders can be operated in an 'ecological' manner by being opened for regular flood events as well, but in extreme cases they will be closed before the peak of the flood wave arrives. In the case of an 'ecological' operation, the natural habitats of the riverine wetlands and their ecological functions can be restored completely.

In this analysis, the focus is on seven combinations of measures which have been selected to illustrate the scale of the effects of different combinations of sites at varying locations with different areas (◘ Table 6.16). These measures will be compared to a baseline which represents the situation according to the Flood Protection Action Plan of the year 2000. The changes of the recent past are not included.

The programmes of measures have the following characteristics:

- Dike R large: Dike relocation without control of all 60 potential sites at the Elbe between km 117–536 in the database, independent of their origin as a dike relocation. The total restored wetlands cover an area of 34,658 ha with a storage volume of 738 Mio. m^3. The purpose of this scenario is to test the potential effect of a dike relocation programme which is consider-

Table 6.16 Properties of programmes of measures for wetlands restoration

Short Programme title	Description	Operation of polder	Range (Elbe km)	Number of sites	Polder area (ha)	Restored wetland area (ha)
Dike R large	Dike relocation (large area)	–	117–536	60	0	34,658
Dike R small	Dike relocation (small area)	–	120.5–536	33	0	9432
Polder large	Controlled polder (large area)	Flooded at peak of extreme floods	117–427	31	25,576	0
Polder small	Controlled polder (small area)	Flooded at peak of extreme floods	180	5	3248	0
Polder (ecol) small	Controlled polder (small area) with ecological flooding	Flooding with natural water regime	180	5	3248	0
P + Dike R	Multifunctional combination	Flooded at peak of extreme floods	117–536	17	4143	3402
P(ecol) + Dike R	Multifunctional combination	Flooding with natural water regime	117–536	17	4143	3402

ably larger than the 15,000 ha, which has been analysed by Merkel et al. (2002).

- Dike R small: Dike relocations without control of 33 potential sites recommended in the ICPER Flood Protection Action Plan (IKSE 2004) for the section of the Elbe between km 120.5 and 536. The total acreage amounts to 9432 ha with a storage volume of 251 Mio. m^3. The purpose of this scenario is to assess a programme which is considered to be politically realistic to implement.
- Polder large: Controlled operation of 31 potential sites for retention-polder identified by the ICPER (2004) at the Elbe between km 117–427 with a total area of 25,576 ha and a total retention volume of 494 Mio. m^3. The polders in the state of Sachsen-Anhalt have characteristics corresponding to Ihringer et al. (2003), and the included polders at the Havel comply with the specification of Förster et al. (2005). The purpose of this scenario is to assess the hypothetical maximum of damage prevention made possible by increased retention.
- Polder small: Controlled operation of the five largest sites for polder identified by Ihringer et al. (2003) at km 180 with a total area of 3248 ha and a storage volume of 138 Mio. m^3. The intention of this scenario is to assess the contribution of the largest sites compared to a maximum potential programme of polders.
- Polder (ecol) small: This programme includes the sites of Polder small but the flooding imitates the natural flooding regime.
- P + Dike R: This scenario contains a multifunctional programme based on the results of a detailed study of a federal Water Research Institute, the Bundesanstalt fuer Gewaesserkunde (BFG 2006). It includes the construction and controlled operation of six polders at the upper part of the Elbe at km 117–180 with a total acreage of 4143 ha and a storage volume of 92 Mio. m^3. In addition, 11 dike relocations

with an area of 3402 ha are included. The polders are operated only for flood protection criteria to reduce the peak water level.

- P (ecol) + Dike R: This scenario is identical to P + Dike R, only the polders are operated to imitate the natural flooding regime.

Valuation Methods and Data

Cost-Benefit Analysis of Wetland Restoration

The cost-benefit analysis is based on the standard procedure of a comparison of the programme with the no-programme situation, e.g. in this case the comparison of the discounted net benefits of the selected programme of measures with the baseline. The costs of the programmes are the sums of the individual sites and they consist of three components: First, the costs of the programmes contain the costs of investments, their maintenance and operations. The costs of maintaining the existing dikes are identical in the baseline and programme scenarios with two exceptions: The construction and operating costs of the necessary new dike sections are counted as project costs, while the saved rehabilitation and maintenance costs of the existing dikes, which are to be removed or abandoned, are calculated as benefits. An additional cost category are the opportunity costs resulting from the loss of the original land use, the loss of land rents, here from the losses of agricultural and forest land rents.

In this analysis, four categories of benefits are included. First, the saved costs from reduced rehabilitation and maintenance of the dikes to be removed or abandoned are calculated as benefits. In addition, two benefits are based on the change of ES of the restored wetlands; reduction of the flood damages and nutrient retention. Furthermore, the additional value of the biodiversity of the wetlands is calculated as a benefit.

The net benefits are presented as the present value of the sum of the single net benefits of each programme of measures for wetland restoration, compared to the baseline scenario, discounted over the lifespan of the projects. For this analysis, a discount rate of 3% is applied and a lifespan of the measures of 100 years assumed. With the help of a sensitivity analysis, the effects of a lower discount rate of 1% and a shorter lifespan of 30 years of the measures are assessed.

In the following section, the results of the calculation of the individual components are discussed before they are presented as a summary.

Benefits from a Reduction of Flood risk

The benefits of the changed flood risks are measured as avoided average annual flood damages. In the literature, the changes of the expected values of average annual damages are considered as the adequate method to measure the benefits of flood protection within a cost-benefit analysis (Penning-Rowsell et al. 2003; NRC 2000). In a risk-based approach, the flood risk is conceptualised as the product of the physical flood risk (e.g. properties of extreme events and the probability of occurrence) and of the resulting damages. The assessment of the effects of flood protection measures requires a methodology applicable to a large scale watershed. For this paper, a methodology has been used which has been developed specifically for the Elbe which is described in more detail in de Kok and Grossmann (2010). It is a one-dimensional hydraulic model which allows to model the effects of planned (controlled and noncontrolled) and unplanned retention (as a consequence of dike failures) for peak discharge value. This model is complemented by a model of the topping of the dike's crest and by a macro-scale economic model to estimate the resulting damages differentiated according to water level as a consequence of the dike's topping and land-use category. This method permits a comparison of the flood risks for the range of the whole Elbe which has not been possible so far. However, there are limits to the approach, as additional local lowering of the water table for dike relocation cannot be modelled. These models have been applied to calculate the expected average damage changes for each programme. The flood risks have been calculated on the basis of flood events with a return period of 2, 10, 20, 50, 100, 200, 500 and 1000 years at the gauge in Dresden.

The avoided average annual damage is calculated as the difference between the damage risks with the programme and on the basis of the baseline scenario. Comparing the total effects, the large polder programme (Polder large) is the most effective programme as it causes the highest reduction in expected annual flood damages (Table 6.17). If

Table 6.17 Avoided average annual damage per area of different programmes of measures

Programmes of measures	Restored wetland		Avoided average annual damages (€ ha^{-1})
	Total area (ha)	Share of controlled polder (%)	
Dike R large	34,659	0	165
Dike R small	9432	0	68
Polder large	25,576	100	1015
Polder small/P(ecol) small	3248	100	4120
P + Dike R/P(ecol) + Dike R	7545	55	1825

one looks at the specific effects of the larger polder sites upstream which are included in the large polder programme, it becomes clear that the small polder accounts for 50 % of the avoided damages of the larger polder programme. The results show also the economies of scale as the additional storage volume does reduce the avoided damages significantly. Furthermore, it becomes clear that the avoided damages due to the dike relocations are considerably smaller without a control of the retention (Dike R large and Dike R small).

Benefits from Nutrient Retention

For the valuation of the benefits from the retention of nutrient, an indirect method on the basis of avoidance costs is used. The method of avoidance or replacement costs relies on the consideration that the willingness to pay for an improvement of environmental quality is at least equal or higher than the costs that environmental policy is prepared to incur to improve environmental quality. Values estimated on the basis of the replacement costs method do not constitute a direct estimate of the benefits of ES for the society (e.g. the value of cleaner water). But this method can be used as an indirect method of valuing ES if the following conditions are fulfilled (NRC 2005; Turner et al. 2008): (1) The alternative used as an replacement generates a similar ecosystem service as a natural wetland, (2) the alternative used in comparison is the least-cost option and (3) there should be convincing evidence that a social demand for the services exists on the basis of the least-cost alternative.

Here, the results of a model are used which estimated the replacement costs as a shadow price of the retention services of the riverine wetlands on the basis of a cost minimization model to reduce nutrient discharges in a river basin. The study is based on the nutrient quality objectives which have been developed in the context of the management plan of the Elbe to achieve the objectives of the EU Water Framework Directive of achieving a good ecological status of riverine and coastal waters (FGG Elbe 2009). The current objective is formulated to reduce the phosphorus and nitrogen discharges stepwise by a total of 24 %, with each step achieving a third of the total objective in the reporting period ending in 2014, 2021 und 2027. The details of the minimization model and its application to the Elbe are presented in more detail in Grossmann (2012a).

Table 6.18 shows the results of the estimates of the shadow price of the restored riverine wetland. The shadow price reflects the change of the total costs when an additional unit of the "average, annually flooded wetland" becomes available or is missing. The results show the effects of the increasing minimum requirements and the size of the wetland restoration programme on the shadow price: First, the shadow price increases with increasing stringency of the water quality objective and secondly, the shadow price falls with increasing restored wetland area. The shadow price for a marginal increase of the area from the current level with a reduction rate of 5 % with a value of 1716 €/ha increases with the reduction rate up to 35 % up to a value of 52,914 €/ha.

Table 6.18 Shadow price (limit value) of nutrient retention of additional area of restored riverine wetlands of the Elbe (in € per ha for the average annually flooded area)[a]

Additional area[b] (ha)	Reduction objectives for total load of nitrogen and phosphorus			
	5 %	15 %	25 %	35 %
1	1716 € ha^{-1}	12,218 € ha^{-1}	23,416 € ha^{-1}	52,914 € ha^{-1}
1500	1531 € ha^{-1}	11,849 € ha^{-1}	19,809 € ha^{-1}	40,407 € ha^{-1}

[a] with a retention rate of 0,8 kg P ha^{-1} d^{-1} and 1,5 kg N $ha^{-1}d^{-1}$ for the flooded wetland area
[b] wetland area on average flooded annually

The shadow price for an additional marginal increase after the restoration of 1718 ha increases then from 1531 €/ha with a reduction objective of 5 % up to 40,407 €/ha for a reduction objective of 35 %.

Benefits from an Improvement of the Habitat and Biodiversity Function

To value the restoration of the wetland habitats and their biodiversity, the results of WTP studies are used to cover the share of benefits resulting from nonconsumptive uses. Nonconsumptive uses arise when the ecosystem is used without the withdrawal of biomass. They are based on the enjoyment quality of the ecosystem and on recreational activities (such as the enjoyment of a view of a landscape) and on the 'nonuse'-values of maintaining the natural heritage for future generations, independent of individual considerations for own use, e.g. for recreational purposes (Turner et al. 2008). The economic values of nonuse of biodiversity and habitats can be of considerable magnitude, but they are very difficult to measure. Revealed preference methods, as the travel cost methods and hedonic pricing approaches, have not been applied to value changes of biodiversity, as they require a basis on terms of use which is usually related to recreation. But these methods are not able to measure nonuse values of changes in the landscape. Stated preference methods as the WTP or choice experiments are the only techniques appropriate to value changes in biodiversity which contain elements of nonuse (▶ Sect. 4.2).

For the valuation of this benefit component, the results of two studies are used. First, the results of a primary study are used which estimated the willingness to pay for the restoration of the riverine wetlands along the Elbe river. Detailed information has been presented in two publications (Meyerhoff 2003, 2006). These studies estimate the annual WTP of the German population for a programme of riverine wetlands restoration, consisting of the maintenance of 40,000 ha of existing floodplain wetlands and the reconstruction of additional 15,000 ha wetlands via dike relocation. These studies found an average annual willingness to pay per household of 5.30 €. The figure includes protest answers as true zero answers, it is adjusted for outliers and it takes account of the embedding effect.

Assuming that wetlands are a normal good, economic theory proposes that the values of the wetlands depend on the amount of the proposed measures. Therefore, the above mean WTP, which constitutes a point value, is complemented with an estimate of the demand elasticity, based on a meta-analysis of valuation studies of wetlands. This allows to develop a valuation function depending on the size of the programme. To save space, the details of the meta-analytical estimates can be found in Grossmann (2012b). The results show that the estimates of the WTP depend on the size of the proposed measures of wetlands protection, measured in terms of the area of the wetlands, and that the average willingness to pay increases with the size of the area, but with a declining rate. The meta-analysis is specified in a log-linear form, with both the dependent variable (WTP) and the independent variable, the size of the included wetlands, formulated logarithmically. In this case, the coefficients of the variables can be interpreted as the demand elasticity, here estimated to be 0.3. Combined with the point estimate of the WTP from the primary study at the Elbe, the value of the demand elasticity

Table 6.19 Estimated willingness-to-pay for the maintenance and restoration of riverine wetlands habitats and their biodiversity at the Elbe river

	Unit	Restored wetland area (ha)					
		5000	15,000	20,000	35,000	45,000	55,000
WTP per household	€/household[a]	3.1	3.8	4.3	4.7	5.0	5.3
Aggregated WTP per area	€/ha[b]	5142	2142	1461	1134	936	810

[a] on the basis of an average WTP per household of 5.30 € (2005) for an wetland protection programmes of 55,000 ha and a price elasticity of 0.3.
[b] on the basis of a population of 18.5 Mio. in the Elbe river basin and 2.2 persons per household.

used to derive the WTP as a function of the area of the restored wetlands (Table 6.19).

Cost Estimates

The largest share of the financial costs of dike relocations typically consists of the construction costs and the costs of acquiring the land from the current landowners. Here, the economic costs are used as a basis for calculation: the costs of the construction, maintenance and operation of the flood protection infrastructure and the opportunity costs of the resulting land-use changes. The costs of the construction and the maintenance of the flood protection infrastructure are the capital and labour costs for the construction of the new dikes, the cutting of the old dikes and the dike improvement and the weirs of the polder as well as the construction works for the adjusted landscape. Then, the maintenance and operation cost of the weirs, the dikes and landscape changes are added.

The opportunity costs of the change of agricultural land use are differentiated in two categories: First, the permanent change from agricultural land use to nature conservation in the restored wetland and, second, the permanent change from cropping to grassland uses in the polder areas. In the case of the conversion to nature conservation, the opportunity costs of agricultural and forestry land uses are equivalent to the accumulated value of the lost production, less the variable production costs. This corresponds roughly to the purchase price of land. Thus, the purchase prices for forested areas, cropland and grazing land have been used to estimate the opportunity costs of the permanent land-use changes. For the conversion of agricultural land into a polder which will only be flooded in extreme events, an agricultural land use can continue. Here, it is assumed that cropland will be converted into grassland so that the opportunity costs can be calculated on the basis of the lowered contribution margin. Additional opportunity costs arise from the irregular flooding of the polder areas. The loss depends of the expected frequency during the growth period which is assumed to be every 20 years.

The total costs have been calculated for each of the 60 sites on the basis of a dataset which contains information about the size, the share of grassland, of cropland, of forested area, the length of the new dike, the number of the new weirs, the length of the old dikes which can be cut or removed and their status with respect to rehabilitation. The data about the dike infrastructure are based on a survey undertaken by the ICPER (IKSE 2001) which takes into account the status of the infrastructure as of 2000.

Results of the Cost-Benefit Analysis

The most important results of the cost-benefit analysis are presented in Table 6.20. The results are shown as Net Present Value (NPV) and as Benefit Cost Ratio (BNR) from two valuation perspectives: First, only from a flood protection perspective and, second, with an integrated floodplain management perspective. The programmes with the highest NPV have the highest economic net benefit and, thus, the highest contribution to an improvement of potential social welfare. When a restriction of the available funds exists which prevents the whole programme from being implemented, then the benefit

Table 6.20 Net Present Value and Benefit Cost Ratio of the wetlands restoration programmes

Programme	Area	Flood protection only NPV	Integration Ecosystem services, biodiversity NPV	Flood protection only BCR	Integration Ecosystem services, biodiversity, BCR
	(ha)	(Mio. €)[a]	(Mio. €)[a]	(–)	(–)
Dike R large	34,659	–128	2520	0.8	5.8
Dike R small	9432	–69	1465	0.7	7.6
Polder large	25,577	354	354	1.8	1.8
Polder small	3248	331	331	5.0	5.0
P(ecol) small	3248	352	1396	6.6	23.1
P + Dike R	7545	300	1375	2.8	9.0
P(ecol) + Dike R	7545	326	1481	3.2	11.2

[a] discount rate 3 %, lifespan of 100 years

costs ratio is a helpful indicator in ranking the projects for which funds are available.

From a flood protection perspective, the BCR for the small polder programme in the upstream area of the German part of the river is the highest (polder small and P(ecol). The BCR for the programme with the imitation of the natural flooding regime is the highest because the additional benefits add up. The combination of the polder in the upstream area with a dike relocation (P + Dike/ P(ecol) + Dike R) reduces the BCR since the dike relocation programmes (Dike R large; Dike R small) have a BCR smaller than one from a flood protection perspective. The programme with the additional number of polder along the river (Polder large) has a lower BCR as well as the additional reduction of flood damages is relatively small.

The picture changes when the perspective is broadened to an integrated floodplain management perspective, and the economic values of the biodiversity and the ecosystem service nutrient retention are included. First, the BCR of those programmes are higher which generate additional benefits with the restoration of the floodplain wetlands. Second, the ranking changes since they generate biodiversity and nutrient retention benefits additional to the high flood protection net benefits due to their imitation of the natural flooding regime. The programmes with dike relocations (Dike R large; Dike R small; Polder + Dike R) enter the higher ranks because of larger BCR. The programmes which do not offer additional ES (Polder large, Polder small) obtain only lower ranks.

The total net present value gives an indication of the potential absolute increase of social welfare which can be achieved with the various programmes. With the perspective of an integrated floodplain management, the increase is the largest with the large dike relocation programme (Dike R large) with 35,000 ha. The other programmes with dike relocation or other programmes with components to restore wetlands (Dike R small; Polder (ecol) small; Polder + Dike R; P(ecol) + Dike R) with smaller areas, varying between 3200 and 9400 ha, generate high net present values as well.

Finally, a sensitivity analysis of the results has been performed to analyse the influence of assumptions of key variables: (a) discount rate (b) lifespan (c) costs and benefits of the ES. The first interesting result is that all programmes have a positive NPV when the lifespan of the projects are reduced by 30 years, roughly a third of the expected economic life of the dikes. A lower discount rate increases the NPV further. The effects of changes of the assumed costs and of the benefits of ES are disproportional, and they do not cause a change of the NPV to a negative value with one exception: The size of the effect depends on the project types in the programme and the share of the benefit they generate. For the programmes which restore wetlands habitats, halving the benefits from biodiversity maintenance reduces the total results by 24–35 %, a halving of the benefits from nutrient retention reduces the total net benefits by 4–19 % and a halving of the flood protection benefits lowers the results by 1–16 %. An increase of the costs by 50 % reduces the NPV only between 3 and 6 %. Finally, a joint increase of the costs by 50 % and a halving of all ES lead to a reduction of the NPV between 55 and 62 %. The sensitivity analysis shows that the core results of this study concerning the positive economic effects of a floodplain wetland restoration are stable, even under a broad range of alternative assumptions concerning costs and benefits.

Policy Implications and Conclusions

It is possible to include the economic value of biodiversity and ES in a cost-benefit analysis, comparing dike relocations with a polder-based flood protection policy. This potential can be a contribution to include the economic perspective in the decision-making process with efficiency as an objective. The main result of this empirical study is that the large-scale restoration of riverine wetlands leads to an increase of efficiency. The integration of additional benefit categories, as the economic value of biodiversity and ES, allows a better evaluation of the multifunctional quality of projects and to identify priorities for decision-making for an integrated floodplain management approach. The results presented here only cover two ES of wetlands in general and riverine wetlands in particular. The literature lists more ES which should be included for a complete valuation of riverine wetlands (Turner et al. 2008). Especially the benefits from the sequestration of greenhouse gases are expected to be of substantial importance (Jenkins et al. 2010).

Conclusion

It should be clear that the standard cost-benefit analysis commonly applied for flood protection purposes needs to be extended to include the benefits from ES, providing a basis for an integrated floodplain management. This study is an example for the applicability of the existing approaches and methods to a German river basin illustrating the economic advantages of restoring riverine wetlands on a larger scale. But it remains to note that for a broader application in public decision-making an improvement of the methods and the data basis is required as well as the quantification and modelling of the biophysical dimensions of ES and their economic valuation.

6.6.4 Peatland Use in Mecklenburg-Western Pomerania, Germany: Monetarization of the Ecosystem Service Climate Protection

A. Schäfer

Natural water-saturated peatlands bind the carbon contained in dead plants as peat. Peatlands are an important component of the global carbon cycle; they store twice as much carbon as do forests (Joosten and Couwenberg 2008). Drainage and agricultural use rob them of their original ecosystemic function, since the aeration of the soil initiates a process of mineralisation which, together with the peatland subsidence caused by the compaction of the peat, leads to peat depletion. As a result, the nutrients–N and P–bound within the peat, and the humins in adjacent bodies of water, as well as climate-relevant trace gases, are emitted into the atmosphere. The result is that an accumulating ecosystem becomes an ecosystem which releases nutrients into the environment (Koppisch 2001; Augustin 2001; Grunewald et al. 2011; Zak et al. 2011).

Modern agricultural use, which depends on continual drainage, causes an additional degradation of peatland sites, with the result of the level of the peatland surface drops ever further below that of the natural runoff capacity, so that its water-management regulation requirement–drainage in the spring and water input in the summer–becomes ever more costly (Kuntze 1983; Succow 1988). In addition to the other existing site related difficulties, the agricultural production conditions have changed greatly during the past two decades. As a result of breeding and technological progress, and

Table 6.21 Land use and GHG emissions of peatlands in Mecklenburg-Western Pomerania. (Source: own calculations after MSK Mecklenburg-Vorpommern 2009, p. 29)

Land use	ha	TCDE $ha^{-1}a^{-1}$	TCDE
Close to nature management	38,445	2.2	83,142
Unused	51,760	14.4	746,214
Forest	44,178	17.8	787,572
Agricultural usage	171,307	26.0	4,449,789
– Humid grassland, salt meadows	20,790	16.5	343,035
– Grassland (extensive)	17,516	15.0	262,740
– Grassland (intensive)	96,439	24.0	2,314,536
– Field	36,562	43.2	1,579,478
Total area	305,690	20.1	6,158,303

6

the structural transformation in dairy farming, the use of grassland for raw feed production has declined in importance in recent years. Overall, we have for years seen a decline in the stock of cattle and at the same time, increased demands on the quality of the basic fodder in Mecklenburg-Western Pomerania–and not only there. The estimated requirement for grassland, based on the stock of cattle, is considerably lower than the potential of grassland areas available in Mecklenburg-Western Pomerania. Purely arithmetically, an excess of grassland of 66,800 ha existed in 2009 (Benecke 2009).

Due to high requirements for grassland, dairy operations with very high outputs of milk have caused fodder production to shift increasingly to farmland. Extensive animal husbandry with suckler cows, which has been significant in Mecklenburg-Western Pomerania in terms of the area used, is, as an "economically fragile form of land use," highly dependent on subsidies (Müller and Heilmann 2011). Opportunities are seen in the use of grassland plants as energy crops (MLUV MV 2011). However, these opportunities should be viewed very critically on drained fen sites, since drainage-based agriculture overall releases considerably more greenhouse gases than the direct burning of peat. In terms of heating value, the emission factors of peat are 106 g CO_2/MJ. Taking site-related GHG (greenhouse gas) emissions of peat mineralization into account, maize cultivation on drained peatlands for the production of biogas causes emissions of 880 g CO_2/MJ (Couwenberg 2007).

The peatland areas in Mecklenburg-Western Pomerania cover 305,690 ha (MSK Mecklenburg-Vorpommern 2009). A small part of that surface is in near-natural condition and/or not used. Together with the drained forest peatlands, peatland surfaces account for approximately one quarter of GHG emissions in the state, the bulk of which is caused by intensively used grassland and farmland for the cultivation of grass silage and maize (corn) for dairy and biogas production (Table 6.21). Hence, there is a clear trade-off between two welfare-relevant ES: climate protection on the one hand and the productive function for the generation of agricultural products (meat and milk) and biomass for energy crops (biogas) on the other.

Economic Valuation of the ES

An economic valuation of peatland-use-associated ES always involves a balancing of considerations between various alternatives, under conditions of scarcity (► Sect. 4.2). ES are welfare relevant, since they provide utility and reduce scarcity. Prices are an important indicator for scarcity, ensuring thrifty management of goods in short supply. A comprehensive accounting for the monetary effects can ensure that the economic costs and benefits of peatland use and possible measures for climate protection (rewetting and land-use change) are revealed. Thus, the monetarisation of ES is an important foundation for policy decision-making and the formulation of economic-policy control instruments. That is especially important if there are competing use possibilities between various ES.

Many environmental problems arise due to the fact that the goods and services provided by nature have no price–they are cost-free. As a result, the demand exceeds the available supply, resulting in an overuse of scarce resources. According to the principle of the primacy of sustainability, the long-term effects of actions must be taken into account in the case of climate-relevant scarcity problems. According to the concept of sustainable development, the natural absorption capacity of ecosystems should not be exceeded, and the functionality of ecosystems not be eliminated due to economic activity (Geisendorf et al. 1998; Ott and Döring 2004). The concept of sustainable development demands that scarce natural resources are used in an environmentally compatible and economically sensible manner, and that prices express 'ecological truth' (von Weizsäcker 1989).

The current agricultural use of peatlands involves costs, since there are competing or mutually exclusive use possibilities between the two welfare-relevant ES, production or supply ES versus climate protection ES (opportunity costs). Moreover, the use of peatland, dependent as it is on drainage, causes negative external effects in the form of climate damage, and also detracts from other important ecosystem functions (e.g. nutrient input, or the water balance of the landscape).

According to the polluter-pays principle, the costs of environmental use which occur as a result of economic activity are to be assigned to the party causing that activity. While in other economic sectors with suitable market-economic instruments, such as the eco-tax or certificate trading, an internalisation of external costs already exists, the agricultural use of peatland is in fact promoted by subsidies–especially direct payments and allowances under the German Renewable Energies Act. Government support for agricultural peatland-use is not due to any failure of the market (Fritsch et al. 2007), but is clearly a failure of policy, since, due to policy decisions, the polluter-pays principle, which is fundamental to environmental policy, has been turned upside down.

More than 42 million tons of CO_2 equivalents (MTCDE) are emitted in Germany every year due to nonsustainable peatland use. These emissions considerably exceed the reduction obligations incumbent upon Germany's energy and industrial enterprises (15 MTCDE), and the private household and transportation sectors (22 MTCDE) (UBA 2009). This generates external costs, since it means that considerable funds need to be spent in other parts of the economy to achieve climate goals (abatement costs).

Quantification of GHG Emissions and of Reduction Potentials

The external costs of peatland use and the ES for climate protection induced by land-use change can be monetarised with the instruments of cost-benefit analysis (Schäfer and Degenhardt 1999; Schäfer 2009). The monetarisation of the costs of peatland use and the associated environmental damage thus provides important information for the internalization of external effects. The precondition for the estimation of climate-change costs is that the GHG emissions of traditional agricultural peatland use and the potential for GHG reductions as a result of rewetting is estimated with sufficient precision.

The quantification of GHG flows caused by peatland use depends on various site parameters, e.g. water level, temperature, or vegetation growth, which vary considerably on a seasonal basis, as well as from year to year (Roulet et al. 2007; Maljanen et al. 2010). For the establishment of GHG accounting, long-term measurements are needed in order to cover the daily, seasonal and annual variability. Direct measurement procedures for full-scale ascertainment of GHG flows are very costly, and cannot be applied on a full-scale basis. However, they can be used on selected areas in order to develop, calibrate and verify simple and pragmatically applicable models (indicators or proxies), with which GHG flows can then be quantified in practice in much larger areas.

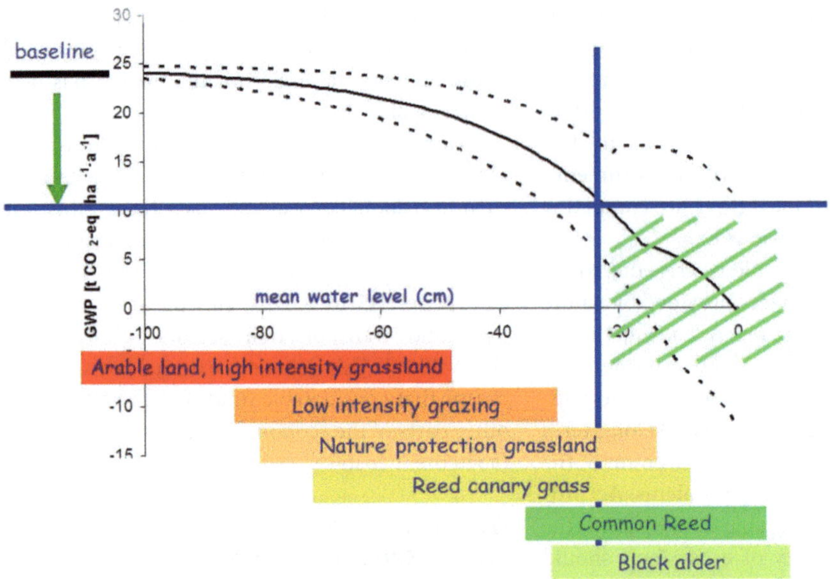

Fig. 6.27 GHG emissions, mean water level and land use. (Modified after Couwenberg et al. 2008)

Meta analyses of the large number of data from all over the world have shown that the mean annual water level is the best single quantum for explaining annual GHG flows from peatlands. Evaluation of the data indicated that the measurement of N_2O flows is more difficult, due to the greater variability over time. Measurement results provided in the literature are to some extent contradictory, and scatter very erratically. Since N_2O emissions always drop after rewetting, nonconsideration of this factor yields a conservative estimate of savings potentials (Couwenberg et al. 2008, 2010).

On the basis of the vegetation-form concept of Koska et al. (2001), Couwenberg et al. (2011) established an overview of all possible vegetation types in central European peatlands. These vegetation types clearly represent the mean water levels of peatland sites. In the context of meta-studies, vegetation types were merged with a large number of greenhouse-gas emissions measurements. The resulting matrix permitted extrapolation and interpolation of measured flow rates along various axes of site conditions. The resulting greenhouse-gas emissions site types (GEST) were based primarily on water levels and presence/absence of plants with coarse aerenchyma; however, nutrient levels, pH values and land use were also observed. With the GEST approach, GHG flows of drained and rewetted peatlands could be quantified, trends and regularities between emissions and site parameters recorded, sites with similar emissions behavior categorised, and hence climate-relevant ES estimated.

Figure 6.27 shows the connection between drainage depth, agricultural and silvicultural use possibilities and GHG emissions. It shows clearly that the mean water level is closely correlated to the use of peatlands (Fig. 6.27).

In case of water levels of more than 20 cm below surface, GHG emissions are determined exclusively by CO_2 emissions. If the water level is higher, CH_4 emissions also occur, and the downward curve of CO_2 emissions is deflected, although it continues to drop overall. The annual emissions of deeply drained peatlands (water stage 2+, mean water levels 50 to 140 cm below surface) amount to some 22–24 TCDE/ha, not counting N_2O emissions. For an extensive pasturing and conservation-appropriate grassland use, mean water levels of 5–35 cm below surface (water stages 3+ through 4+) are required.

The GHG emissions here are approximately 10 to 15 TCDE/ha/yr (Couwenberg 2011).

Site-adapted wet farming procedures are possible at considerably higher mean water levels (water stages 4 + through 5 +; mean water level 5–15 cm below surface). The rewetted peatland sites can be utilised for growing biomass–red, canarygrass, sedge and bulrush stands (Wichtmann and Schäfer 2005; Wichmann and Wichtmann 2011)–both for energy-crop and material use, or for the cultivation of high grade alderwood (Schäfer and Joosten 2005). In this way, fossil fuels can be substituted, and additional climate-relevant GHG emissions reductions realised.

Monetarisation of ES and its External Effects

With respect to the monetarisation of the two welfare-relevant ES which we are focusing on here, the production function and the climate protection function, the question arises from an economic point of view: What monetary value is generated by traditional farming of peatlands, and what climate-damaging effects might that cause? An economic valuation of the external costs accompanying agricultural peatland use can be carried out by means of the value-added method. Under this method, the value added by the production process is compared to the costs engendered by GHG damage. The value-added method does not claim to ascertain the entire economic utility–the total economic value (Pearce 1993; ► Sect. 4.2.2). Rewetting and land-use change are connected with further welfare-relevant use benefits, e.g. biodiversity, stabilisation of the water balance of the landscape, or support for the microclimate. A monetarisation of this additional utility includes not only use-dependent value, but also nonuse-dependent values which can be ascertained by using the suitable methods (e.g. WTP analyses). The avoided damage costs as a result of rewetting and land-use change can thus be interpreted as the lower limit of the utility of the measures.

Damage costs are the current value of climate change damage which the unit (TCDE) of the greenhouse gas emitted today will cause in the future. These costs are marginal costs, i.e. the costs engendered by the emission of additional TCDE, and should not be confused with the overall cost of climate change, or with the average costs of GHG emissions. Damage costs are calculated with the aid of integrated evaluation models, with which climate systems and their interactions with the socio-economic system are modelled in scenarios, and the damage costs ascertained, dependent on the various stabilization goals, GHG emissions and GHG paths.

The amount of the cost depends on the point in time of emissions, the development of the overall GHG emissions, and a number of assumptions, such as time horizons, discount rates, or regional damage distributions. In view of the long-term effects of climate change, the ascertainment of GHG damage costs by means of various methods involves various methodological difficulties (Kuick et al. 2008) and normative assumptions which are difficult to justify (Schelling 1995; Lind and Schuler 1998; Hampicke 2011). Due to the different assumptions, the results also vary within a relatively broad spectrum between € 14 and € 300/TCDE (Clarkson and Deyes 2002; Pearce 2003; Downing et al. 2005). Various authors point to the fact that existing studies underestimate the cost of climate change because they take into account singular and extreme events with serious implications, and the cost of adaptation to climate change are only being minimally taken into account (Tol 2005; Watkiss et al. 2005; Stern 2007).

In spite of the existing insecurities, practical economic policy requires an orientation quantum with which to evaluate the effects of climate change. The methodological convention for the economic valuation of the environmental damage presented by the German Federal Environment Agency requires that the external costs of public investments be incorporated into the decision-making process (UBA 2007). Based on an evaluation of the extensive literature on the cost of climate change, the methodological convention recommends the use of marginal damage costs of € 70/TCDE as the best estimate value. The consistent use of the methodological convention will result in this estimated value also being taken into account for the calculation of external costs for nonsustainable peatland use. Intensive farming and pasturing in the area under investigation cause the emission of 20 MTCDE/yr (Schäfer 2009).

The nationwide farm accountancy data network is an important database to the evaluation of the economic success of agricultural operations (BMELV 2011a). Evaluation of added value based on market prices has the advantage that it can be carried out on the basis of very reliable databases. The economic results of representatively evaluated agricultural enterprises are gathered in the German states by the Federal Ministry of Food, Agriculture and Consumer Protection according to uniform, annually updated methods in the context of its accounting process for testing and requirements. In addition to other economic success indicators, operational revenue is a very well-suited indicator for the value added by an enterprise, since it also contains medium and long-term effective operational expenses from the fixed costs. The added value is the amount available for the payment of all income factors used in the enterprise. In the national accountancy, the operational revenue of the added value of an enterprise corresponds to the gross domestic product.

The enterprises operating on peatland sites in Mecklenburg-Western Pomerania are largely specialised forage growing farms, while the dairy farms can be classified as either primarily milk-cow dairy farms, or as pasturing cattle farms with no particular major emphasis. The classification of an enterprise is determined according to the relative shares of various types of production in the company's overall standard gross margin. Intensively operating dairy farms require deeply drained surfaces (water stage 2+, 2−) with a functioning runoff capacity. Pasturing farms run sites with simply regulatable water conditions (water stage 3+, 3−). Extensive utilisation of hydrologically difficult peatland sites is practiced by farms in order to maintain the minimum care condition that is a precondition for eligibility for premiums (Müller and Heilmann 2011).

According to the evaluation of the farm account data network in Mecklenburg-Western Pomerania (LFA o. J.), intensively operating dairy farms had an average operating income of € 657/ha of land surface in operational use in the period from 2005 through 2010. For an overall economic analysis, allowances and subsidies must be subtracted from that. These farms received an average of € 370/ha in allowances and subsidies during the operating years 2005 through 2010. Without these payments, these farms thus generated an added value of € 287/ha.

For the pasturing farms with no particular major emphasis, average operational income was € 324/ha, considerably less than the intensively operating dairy farms. These operations received € 388/ha in allowances and subsidies, somewhat more than the dairy farms. Without these government payments, the operational value added is negative. However, it should be noted that these payments are to some extent compensation for services rendered in the context of the agriculture/environmental programmes for the implementation of conservation goals, such as the Natura 2000 Network or the Habitats Directive. This involves income from state payments made directly from the public purse which is both product and cost/operationally referenced.

A juxtaposition of the value-added and the external damage costs related to climate change –€ 595/ha/yr. versus € 1680/ha/yr.–shows that the value added by drainage-based agricultural use is considerably higher than the value added by meat and milk production (◘ Table 6.22).

Avoided Damage Costs by Means of Rewetting and Environmentally Compatible Use

The ES situation could be improved by a change in land use. This could reduce the cost of damage due to climate change. It should be noted that the avoidance of the cost of damage is to be interpreted as the lower limit for the utility of the measures; however, they should not be confused with abatement costs. The latter are costs to the national economy oriented towards stipulated minimum goals or reduction obligations. They reflect the opportunity costs of possible alternative uses due to rewetting and land-use change.

After rewetting of the drained peatland sites, various measures for these sites may be considered:

- Rewetting of farmland, grassland or unused fallow land (wilderness)

Table 6.22 GHG emissions of conventional agricultural use on drained peatlands

Utilization category[a]	Water level	GHG emissions[b] (TCDE $ha^{-1}a^{-1}$)	Damage cost[c] (€ $ha^{-1}a^{-1}$)
Dairy cows	2+, 2−	24.0	1680
Yearling bulls, dry and suckler cows	3+, 3−	15.0	1050
Continuous grazing with low stocking rates and meadows	4+ to 3+/3−	8.5	595

[a] Utilization category according to Müller and Heilmann 2011
[b] GHG emissions after Couwenberg et al. 2008
[c] after UBA 2007

- Extensive grassland-use after rewetting of farmland and heavily drained grassland
- Rewetting of farmland, grassland or fallows with environmental use of biomass (reed, canarygrass and sedge beds
- New forest formation by afforestation and/or succession after rewetting

Depending on the initial situation and the intensity of the rewetting achieved, the implementation of the measures yields a broad spectrum of possibilities for GHG reduction potentials (Schäfer 2009). Between 2000 and 2008, some 30,000 ha of peatland were rewetted in Mecklenburg-Western Pomerania. In these areas, an average of 10 TCDE/ha are saved annually (MSK Mecklenburg-Vorpommern 2009).

The cost of the implementation of these measures involves primarily planning and construction costs. The costs for the 33 rewetting measures carried out in Mecklenburg-Western Pomerania prior to 2003 came to an average of € 1070/ha (Schäfer and Joosten 2005). In the Müritz National Park, various measures for the restoration of a near-natural water balance have been carried out in recent years. The area-referenced costs for the implementation of the measures come to € 832/ha (Rowinski and Kobel 2011). The annual costs can be ascertained as a perpetuity. At an interest rate of 2% (4%), these come to € 41.60 (€ 83.20)/ha. Assuming that these measures could save 10 TCDE annually, the abatement costs of € 4.16 (€ 8.32)/ha would be considerably lower than those of other climate-protection measures (Enkvist et al. 2007).

If the area is not to be used further, there may be further costs for the purchase of other land. When ascertaining economic abatement costs at the national level, costs for land purchase can only be taken into account if the price of the land reflects the consumption value of the production factor soil that is consumed. However, since land prices receive a high level of subsidies–often climate-damaging subsidies–this ascertainment of the national economic abatement costs must include the subtraction of subsidies (e.g. single-area payments) and other transfer payments, since they are provided as transfer payments with no quid pro quo.

The abatement costs of site-appropriate use alternatives, on the other hand, are considerably lower, because there are no costs for land purchase. In the case of environmentally compatible high-quality alderwood production, abatement costs, very conservatively estimated, are between € 0 and € 4/TCDE (Schäfer and Joosten 2005). A comparison of abatement costs with other climate-protection measures in bioenergy production (Isermeyer et al. 2008) shows that wetland production procedures are certainly an interesting alternative, even from an economic point of view.

Conclusion

The economic valuation of the ES connected with agricultural peatland use can constitute an important foundation for policy-goal definition and the formulation of economic policy control instruments. The methodological basis for the evaluation is based on an expanded national economic cost-benefit analysis in which external factors are also taken into account.

An important precondition for the evaluation is the physical quantification of GHG emissions, which can be represented with the aid of the GEST approach. The welfare effects of the external effects and of the ES can then be shown in monetary quanta with the aid of value-added amounts and climate-damage costs. In that way, the demands for sustainable land use can be taken into account and the additional use of possible alternatives revealed.

A monetarisation of climate damage according to the methodological convention of the German Environment Agency shows that the marginal damage costs are many times greater than the value-added with traditional agricultural peatland use. The abatement costs of sustainable use alternatives, on the other hand, are relatively low, and associated with additional welfare-relevant use provisions.

6.7 Systematisation of the Case Studies

K. Grunewald and O. Bastian

In the previous chapters eleven case studies with various ES-application aspects have been shown. Together with the examples in ▶ Chap. 3, ▶ Chap. 4 and ▶ Chap. 5 they are definitely representative for the present situation in Germany, but they do not make a claim for completeness.

According to the dominant land-use character in Central Europe the focus was based on agro-, forest- and urban-ecosystems (◘ Table 6.23). Protected areas have been discussed by means of the Natura–2000 network (▶ Sect. 6.6.1). They cover a significant share of the German territory, too, and are of great importance to the provision and conservation of ES and biodiversity as well as to the sustainable land utilization.

In the case studies, the concept and classification system, explained in previous chapters (e.g. EPPS-framework, ▶ Sect. 3.1.2), has been used mainly. Space and time aspects were explicitly applied each whereby the reference units are associated to the local/regional level. Particularly in the assessment of cultural landscape/landscape management (▶ Sect. 6.5) the landscape approach was consciously used. Only in one case (HNV-Grassland in Germany, ▶ Sect. 6.2.4) there was no primary focus on a specific region. Since typical ecosystems and problems have been assessed, fundamental statements and knowledge are applicable context-specific.

According to the specific issues and objectives respective methods, procedures and techniques (▶ Chap. 4 and 5) were selected and used to analyse and assess the ES. However, the case studies had mostly 'pilot or test character' for the ES concept. Therefore it was important to present the data, indicators and linkage rules to be comprehensible. Several case studies (▶ Sect. 6.2.2, ▶ Sect. 6.5.2) have not had the aim evaluating the ES in the first place. However, containing essential facets of the ES concept, they have been part of the discussions.

In 10 out of 11 cases regulation-ES have been the centre of attention like illustrated in ◘ Table 6.23. Also, (socio)cultural ES have been processed in more than half of the examples even though these are economically difficult to capture. Only few individual ES have been evaluated. Consequent analysis of the so-called 'total values' and of all 'trade-offs' seems unrealistic with the selected studies.

Qualitative and quantitative ES assessments have represented the fundamentals of the selected studies of ES investigations which most likely had been imbedded in research projects. Economic/monetary observations of costs and/or benefits of services and values have been taken into account in almost all case studies (◘ Table 6.23). However, the assessments of urban ES (▶ Sect. 6.4) and (socio) cultural ES of the meadow orchards (▶ Sect. 6.5.1) indicate that economic accounting is not necessary in every case.

To achieve practical efficiency and acceptance of the stakeholders it is appropriate to get the public, particularly the users, owners and decision makers involved into the assessment process right from the beginning. In this respect alternative land-use scenarios with their ES or disservices (▶ Sect. 2.1) should be addressed in their effects and space-time structure.

The presented case studies verify that the ES concept can be applied for very different questions in the areas of land use, landscape management and nature-/environment protection. We do not share the opinion that the approach is more suitable for

Table 6.23 Systematisation of the presented ES case studies

Section (author)	Region (size)	Ecosystem type			Considered ES			Economic valuation (monetary)		Stakeholder groups involved in the assessments
		u	a	f	P	R	C	Costs	Benefits	
6.2.2 (Bastian)	Jahna river catchment in Saxony (244 km²)		x		x	x	x	x		Farmer, authorities
6.2.3 (Bastian)	(former) rural community Promnitztal in Saxony (21 km²)		x			x		x		Farmer
6.2.4 (Reutter/Matzdorf)	HNV-Grassland (ca. 10'000 km²) in Germany		x			x		x	x	–
6.3 (Elsasser/Englert)	Model region (17,599 km²) in the North-Eastern German Lowland			x	x	x	x		x	Regional communities/local people
6.4 (Haase)	City of Leipzig in Saxony (ca. 300 km²)	x				x	(x)			–
6.5.1 (Ohnesorge et al.)	Biosphere reserve Swabian Alb (853 km²)		x				x			–
6.5.2 (Grunewald/Syrbe /Bastian)	State of Saxony (18,400 km²)	(x)	x	x		x	x	x		Environmental Authority, landscape maintenance associations (*Landschaftspflegeverbände*)
6.6.1 (Bastian)	Upper range of Ore Mountains (Germany/Czech Republic, ca. 1500 km²)		x	x	x	x	x	x	x	landscape maintenance associations (*Landschaftspflegeverbände*), tourists
6.6.2 (Grunewald)	Jahna river catchment in Saxony (244 km²)		x			x		x	x	Farmer, Environmental Authority
6.6.3 (Hartje)	Floodplains in the Elbe river basin (350 km²)		x		x	x	(x)	x	x	–
6.6.4 (Schäfer)	Peatlands in the State Mecklenburg-Western Pomerania (3057 km²)		x		x	x		x	x	–

Ecosystem type (dominating land use type): *u* urban; *a* agricultural, *f* forestry
Considered ES classes (► Sect. 3.2): *P* provisioning, *R* regulation-, *C* (socio)cultural ES

near-natural ecosystems (Matzdorf et al. 2008) or agro-ecosystems not matching this concept (Haber 2011). Also, ecosystems modified or characterised by humans contain natural components (organism species, soils, water, etc). Ultimately–despite all changes–it trades about agro- or urban ecosystems but still ecosystems and why should not they deliver services?

Food and forage plants grow on arable land and meadows, water seeps and cooling and fresh air is generated. The well-being of urban population is not least depending on ecosystem structures and processes (e.g. climate regulation and air-quality improvement by trees). Urban ecosystems may also show high biodiversity. But the application of the ES concept for sustainable development of urban regions is still in its initial stage (► Sect. 6.4).

Nonetheless, the services rendered in these cases are not exclusively reduced to the ecosystem, but humans have a significant part, e.g. through the cultivation of fields or the preservation of green areas. Therefore it is required (although challenging) to distinguish between the quantification and evaluation of services originated by ecosystems or humans. This has not yet succeeded convincingly.

It has already been mentioned that some dimensions of the nature could and should not be measured in monetary terms (► Sect. 4.2). Many ES are not marketable so that it would lead into a latent undervaluation–even if only the economic perspectives are valid (Mertz et al. 2007; Bayon and Jenkins 2010; de Groot et al. 2010). A monetary evaluation is not considered in case of religious, spiritual and amenity values to be expedient (Spangenberg and Settele 2010). Ecological as well as social reasons are objected to define nature conservation solely via market mechanisms (Ring et al. 2010). Alternatives were shown with (socio)cultural services (case study meadow orchards, ► Sect. 6.5.1).

However, the imagination of exact monetary values may be tempting as they already feature partly significant differing results (according Schweppe-Kraft 2009 versus Seják et al. 2010 to the valuation of mountain meadows, ► Sect. 6.6.1) to fundamental as well as methodical weaknesses–depending upon calculation base. Furthermore, the recovering of destroyed habitats is not always possible, particularly if the conditions relating to location have changed irreversibly or individual species have become extinct.

However, an economic valuation of nature can provide important additional information for nature conservation to decision-makers. However, it is not the appropriate method to define priorities of protection or goals. But economic instruments can be applied as effective incentive instruments for nature and landscape conservation to maintain ES (Spangenberg and Settele 2010). In this way, they will help to design and establish financial efforts for long-term conservation due to expressed expectations towards perceived needs for investigations in conservation either through installation and treatment of protection areas or through traditional economic instruments like taxes, licences or the development of markets and conventions of payments (incentives) for ES (► Sect. 6.5.2).

In nature conservation the interest in social norms and institutions to make decisions is increasing, too. Societies and citizens value certain goods appropriately because of cultural or religious reasons or preferences; either it is a rare species or a special cultural landscape and they will not constantly check the cost-value-calculations. In order to achieve balanced and sustainable solutions processes of discourse are necessary. In addition, there is need for integrated assessments of ecological, social and economic systems as well as the negotiation between stakeholders in consideration of a broader range of objectives as only the economic efficiency (Gómez-Baggethun and Kelemen 2008; de Groot et al. 2010; Ring et al. 2010; Spangenberg and Settele 2010; Oikonomou et al. 2011). Suitable political instruments for biodiversity, nature conservation and ES revert to diverse valuation methods, like ecological value analysis or ranking definition, detection of preferences and intentions as well as citizen participation (Ring and Schröter-Schlaack 2011). An offensive application of the ES concept in the field of nature conservation (e.g. as part of landscape planning) could help these to be better respected in commonplace weighing-up processes, by systematically demonstrating and if possible quantifying the different services of conservation areas. Such evaluation is promising especially for targeted funding of nature conservation, e.g. through agro-environmental schemes.

Our understanding is that analysis and evaluation of ES will play an important role in planning, organisation and use of landscapes to the changing demographical, climatical, energy, political and technological basic conditions. The land use is a key factor in this context. ES are able to help capture, demonstrate, communicate and balance opportunities in either synergies or goal conflicts, for example between enhanced cultivation of energy crops and climate protection/biodiversity aims. This requires complex analysis and evaluation of ES on the basis of easily manageable standards for which equivalent instruments (e.g. ES models) must be developed further.

References

AMI (2010) Marktinfo Düngemittel. Bericht Nr. 4 der Agrarmarkt Informations-Gesellschaft mbH (AMI), Bonn

Augustin J (2001) Emission, Aufnahme und Klimarelevanz von Spurengasen. In: Succow M, Joosten H (eds) Landschaftsökologische Moorkunde. 2. edn. Schweizerbart, Stuttgart, pp 28–37

AuW (2007) Richtlinie des Sächsischen Staatsministeriums für Umwelt und Landwirtschaft zur Förderung von flächenbezogenen Agrarumweltmaßnahmen und der ökologischen Waldmehrung im Freistaat Sachsen. SächsABl. Jg. 2007, Blatt-Nr. 49

Bastian O, Schrack M (eds) (1997) Die Moritzburger Kleinkuppenlandschaft–einmalig in Mitteleuropa! Veröff. Mus. Westlausitz Kamenz, Tagungsband, 118 p

Bastian O, Corti C, Lebboroni M (2007) Determining Environmental Minimum Requirements for functions provided by agro-ecosystems. Agronon Sustain Dev 27:1–13

Bastian O, Haase D, Grunewald K (2012) Ecosystem properties, potentials and services–the EPPS conceptual framework and an urban application example. Ecol Indic 21:7–12. doi:10.1016/j.ecolind.2011.03.014

Bastian O, Lütz M, Unger C, Köppen I, Röder M, Syrbe R-U (2003) Rahmenmethodik zur Entwicklung lokaler Agrar-Umweltprogramme in Europa–1. Das Indikatorkonzept. Landnutzung und Landentwicklung 44:229–237

Bastian O, Lütz M, Unger C, Röder M, Syrbe R-U (2005) Rahmenmethodik zur Entwicklung lokaler Agrar-Umweltprogramme in Europa–2. Ableitung von Agrar-Umweltzielen und -maßnahmen. Wasser-Wirtschaft 95:27–34

Bastian O, Neruda M, Filipová L, Machová I, Leibenath M (2010) Natura 2000 sites as an asset for rural development: the German-Czech Ore Mountains Green Network Project. J Landscape Ecol 3:41–58

Bayon R, Jenkins M (2010) The business of biodiversity. Nature 466:184–185

BBodSchG (1998) Bundes-Bodenschutzgesetz vom 17. März 1998 (BGBl. I S. 502)

BBodSchV (1999) Bundes-Bodenschutz- und Altlastenverordnung. Bundesgesetzblatt, J. 1999, Teil I, Nr. 36, Bonn, 16.7.1999

Benecke R (2009) Grünland im Umbruch. Vortrag an der Internationalen Akademie für Naturschutz auf der Insel Vilm am 28.04.2009. ► www.bfn.de/fileadmin/MDB/documents/ina/vortraege/2009-Gruenland-Benecke.pdf

BfG–Bundesanstalt für Gewässerkunde (2006) Modellgestützter Nachweis der Auswirkungen von geplanten Rückhaltemaßnahmen in Sachsen und Sachsen-Anhalt auf Hochwasser der Elbe. BfG-Bericht 1542, Koblenz

BfN–Bundesamt für Naturschutz (2004) Daten zur Natur. Bundesamt für Naturschutz, Bonn

BfN–Bundesamt für Naturschutz (2008) Endbericht zum F + E-Vorhaben "Entwicklung des High Nature Value Farmland-Indikators" (FKZ 3507 80 800) des Bundesamtes für Naturschutz. Bonn-Bad Godesberg, 107 p

BfN–Bundesamt für Naturschutz (2009) Erfassungsanleitung für die HNV-farmland-Probeflächen. Stand 15.4.2009

BfN–Bundesamt für Naturschutz (2010) Anteile von HNV-Grünland innerhalb der Naturräume. Schriftliche Information

BfN–Bundesamt für Naturschutz (2012a) BfN Forschungsprojekte: "Klima-Benefits" (FKZ: 3508 81 2100) sowie "Ländlicher Raum und naturschutzbezogene Anpassungsstrategien an den Klimawandel" (FKZ: 3508 88 0700)

BfN–Bundesamt für Naturschutz (2012b) Natura 2000. Bundesamt für Naturschutz, Bonn. ► www.bfn.de/0316_natura2000.html. Accessed 25 April 2012

BfN–Bundesamt für Naturschutz (2012c) Daten zur Natur. Bundesamt für Naturschutz, Bonn

Bieling C (2004) Non-industrial private-forest owners: possibilities for increasing adoption of close-to-nature forest management. Eur J Forest Res 123:293–303

Bieling C, Plieninger T (2013) Recording manifestations of cultural ecosystem services in the landscape. Landscape Research 38(5):649–667

BMELV (2008) BMELV-Bericht zum Klimaschutz im Bereich Land- und Forstwirtschaft. ► www.bmelv.de/SharedDocs/Standardartikel/Landwirtschaft/Klima-und-Umwelt/Klimaschutz/BerichtKlimaschutz.html

BMELV (2011a) Die wirtschaftliche Lage der landwirtschaftlichen Betriebe. Buchführungsergebnisse der Testbetriebe. ► http://berichte.bmelv-statistik.de/BFB-0111101-2011.pdf

BMELV (2011b) Rahmenplan der Gemeinschaftsaufgabe "Verbesserung der Agrarstruktur und des Küstenschutzes" für den Zeitraum 2011 bis 2014. Bonn, 111 p

BMU–Bundesministerium für Umwelt, Naturschutz und Reaktorsicherheit (2007) Nationale Strategie zur Biologischen Vielfalt. Berlin, 178 p

BMU–Bundesministerium für Umwelt, Naturschutz und Reaktorsicherheit (2010a) Indikatorenbericht 2010 zur Nationalen Strategie zur biologischen Vielfalt. Berlin

BMU–Bundesministerium für Umwelt, Naturschutz und Reaktorsicherheit (2010b) Die Wasserrahmenrichtlinie–Auf dem Weg zu guten Gewässern. Berlin
BMU–Bundesministerium für Umwelt, Naturschutz und Reaktorsicherheit (2011) Bundesprogramm Biologische Vielfalt. Berlin
BMU/BfN (2009) Außenzustandsbericht. Flussauen in Deutschland. Bundesministerium für Umwelt, Naturschutz und Reaktorsicherheit (BMU) und Bundesamt für Naturschutz (BfN), Berlin, Bonn
Böhnert W, Franz U, Walter S, Kamprad S, Henze A (2005) FFH-Managementplan für das FFH-Gebiet SCI 5248–303, Landesmeldenummer 039E Geisingberg und Geisingwiesen, Weißeritzkreis. Abschlussbericht. Sächsisches Landesamt für Umwelt und Geologie, Dresden
Bolund P, Hunhammar S (1999) Ecosystem services in urban areas. Ecol Econ 29:293–301
Bowler DE, Buyung-Ali L, Knight TM, Pullin AS (2010) Urban greening to cool towns and cities: a systematic review of the empirical evidence. Landscape Urban Plan 97:147–155
Breitschuh G, Eckert H, Kuhaupt H, Gernand U, Sauerbeck D, Roth S (2000) Erarbeitung von Beurteilungskriterien und Messparametern für nutzungsbezogene Bodenqualitätsziele. Anpassung und Anwendung von Kriterien zur Bewertung nutzungsbedingter Bodengefährdungen. Umweltbundesamt Berlin (ed), UBA-Texte 50–00
Breuste J (1999) Stadtnatur–warum und für wen? In: Breuste J (ed) 3. Leipziger Symposium Stadtökologie "Stadtnatur–quo vadis?" Natur zwischen Kosten und Nutzen. UFZ-Bericht 10, Leipzig
Breuste J, Feldmann H, Uhlmann O (eds) (1998) Urban ecology. Springer, Berlin
Breuste J, Niemelä J, Anuchat P, Triet T (2007) Urban Ecosystem Management. Handbook and manual. Helsinki University Press, Helsinki
Briemle G, Oppermann R (2003) Von der Idee zum Programm: Die Förderung artenreichen Grünlandes in MEKA II. In: Oppermann R, Gujer HU (eds) Artenreiches Grünland bewerten und fördern–MEKA und ÖQV in der Praxis. Eugen Ulmer, Stuttgart, pp 65–70
Bronner G, Oppermann R, Rösler S (1997) Umweltleistungen als Grundlage der landwirtschaftlichen Förderung–Vorschläge zur Fortentwicklung des MEKA-Programms in Baden-Württenberg. Nat Schutz Landsch Plan 29:357–365
Brunotte E, Dister E, Günther-Diringer D, Koenzen U, Mehl D (2009) Flussauen in Deutschland–Erfassung und Bewertung des Auenzustandes. Bundesamt für Naturschutz, Naturschutz und Biologische Vielfalt 87, Bonn
Burgess J (1988) People, Parks and Urban Green. A Study of Popular Meanings and Values for Open Spaces in the City. Urban Studies 25:455–473
Carpenter SR, Turner M (2000) Opening the black boxes: Ecosystem science and economic valuation. Ecosystems 3:1–3
CBD–Convention on Biological Biodiversity (2010) Global Biodiversity Outlook 3. CBD Secretariat, Montreal
Chan KMH, Satterfield T, Goldstein J (2012) Rethinking ecosystem services to better address and navigate cultural values. Ecol Econ 74:8–18
Chiesura A (2004) The role of urban parks for the sustainable city. Landscape Urban Plan 68:129–138
Churkina G (2008) Modeling the carbon cycle of urban systems. Ecol Model 216:107–113
Clarkson R, Deyes K (2002) Estimating the Social Cost of Carbon Emissions. Government Economic Service Working Paper 140. HM Treasury, London
Comber A, Brunsdon C, Green E (2008) Using a GIS-based network analysis to determine urban green space accessibility for different ethnic and religious groups. Landscape Urban Plan 86:103–114
Cooney R, Dickson B (2005) Precautionary principle, precautionary practice: lessons and insights. In: Cooney R, Dickson B (eds) Biodiversity and the precautionary principle: risk and uncertainty in conservation and sustainable use. Earthscan, London, pp 287–298
Costanza R, d'Arge R, de Groot RS, Farber S, Grasso M, Hannon B, Limburg K, Naeem S, O'Neill RV, Paruelo J, Raskin RG, Sutton P, van den Belt M (1997) The value of the world's ecosystem services and natural capital. Nature 387:253–260
Couwenberg J (2007) Biomass energy crops on peatlands: on emissions and perversions. IMCG-Newsletter 2/2007:12–14
Couwenberg J (2011) Vegetation as a proxy for greenhouse gas fluxes–the GEST approach. In: Tanneberger F, Wichtmann W (eds) Carbon credits from rewetting. Schweizerbart, Stuttgart, pp 37–42
Couwenberg J, Augustin J, Michaelis D, Wichtmann W, Joosten H (2008) Entwicklung von Grundsätzen für eine Bewertung von Niedermooren hinsichtlich ihrer Klimarelevanz. DUENE e. V., Greifswald
Couwenberg J, Dommain R, Joosten H (2010) Greenhouse gas fluxes from tropical peatlands in South-East Asia. Glob Change Biol 16:1715–1732
Couwenberg J, Thiele A, Tanneberger F, Augustin J, Bärisch S, Dubovik D, Liashchynskaya N, Michaelis D, Minke M, Skuratovich A, Joosten H (2011) Assessing greenhouse gas emissions from peatlands using vegetation as a proxy. Hydrobiologia 674:67–89
Czybulka D, Hampicke U, Litterski B, Schäfer A, Wagner A (2009) Integration von Kompensationsmaßnahmen in die landwirtschaftliche Produktion. Nat Schutz Landsch Plan 41:245–255
Daily GC (ed) (1997) Nature's Services: Societal Dependence on Natural Ecosystems. Island Press, Washington DC
Daly HE (1992) Allocation, distribution, and scale: towards an economics that is efficient, just, and sustainable. Ecol Econ 6:185–193
DEHSt–Deutsche Emissionshandelsstelle (2012) Versteigerung von Emissionsberechtigungen in Deutschland. Periodischer Bericht, Januar 2012, 8 p

de Groot RS, Alkemade R, Braat L, Hein L, Willemen L (2010) Challenges in integrating the concept of ecosystem services and values in landscape planning, management and decision making. Ecol Complexity 7: 260–272

de Kok JL, Grossmann M (2010) Large-scale assessment of flood risk and the effects of mitigation measures along the River Elbe. Nat Hazards 52: 143–166

Diercks R, Heitefuss R (1990) Integrierter Landbau: Systeme umweltbewußter Pflanzenproduktion. BLV Verlagsgesellschaft, München

Dieter M, Elsasser P (2002) Carbon stocks and carbon stock changes in the tree biomass of Germany's forests. Forstwiss Centralbl 121:195–210

Döring J (2005) Hinweise zur Landschaftspflege. Materialien zu Naturschutz und Landschaftspflege. Sächsisches Landesamt für Umwelt und Geologie (ed) Abteilung Natur, Landschaft, Boden

Downing TE, Anthoff D, Butterfield B, Ceronsky M, Grubb M, Guo J, Hepburn C, Hope C, Hunt A, Li A, Markandya A, Moss S, Nyong A, Tol RSJ, Watkiss P (2005) Social cost of carbon: a closer look of uncertainty. DEFRA, London

Dröschmeister R (1998) Aufbau von bundesweiten Monitoringprogrammen für Naturschutz–welche Basis bietet die Langzeitforschung? Schriftenreihe Landschaftspflege Naturschutz 58:319–337

Druckenbrod C, van Elsen T, Hampicke U (2011) Produktionsintegrierte Kompensation: Umsetzungsbeispiele und Akzeptanz. Nat Landsch 43:111–116

DVL–Deutscher Verband für Landschaftspflege e. V. (2007) Natura 2000–Lebensraum für Mensch und Natur–Leitfaden zur Umsetzung. DVL-Schriftenreihe "Landschaft als Lebensraum", Heft 11

Eckert H, Breitschuh G, Sauerbeck D (2000) Criteria and standards for sustainable agriculture. J Plant Nutr Soil Sci 163:337–351

EEA–European Environment Agency (2004) High nature value farmland characteristics, trends and policy challenges. EEA Report No 1, 26 p

Ehrlich PR, Ehrlich AH (1981) Extinction: the causes and consequences of the disappearance of species. Random House, New York

Eiter S (2010) Landscape as an area perceived through activity: implications for diversity management and conservation. Landscape Res 35:339–359

Elmqvist T, Fragkias M, Güneralp B, et al (2013) (eds) Global urbanisation, biodiversity and ecosystem services: challenges and opportunities. Springer, New York

Elsasser P, Englert H, Hamilton J (2010a) Landscape benefits of a forest conversion programme in North East Germany: results of a choice experiment. Ann Forest Res 53:37–50

Elsasser P, Englert H, Hamilton J, Müller HA (2010b) Nachhaltige Entwicklung von Waldlandschaften im Nordostdeutschen Tiefland: Ökonomische und sozioökonomische Bewertungen von simulierten Szenarien der Landschaftsdynamik. Hamburg: von-Thünen-Institut. Arbeitsbericht vTI-OEF 2010/1, 96 p

Endlicher W, Jendritzky G, Ficher J, Redlich JP (2008) Heat waves, urban climate and human health. In: Marzluff J et al (eds) Urban ecology–an international perspective on the interaction between humans and nature. Springer, New York

Enkvist PA, Nauclér T, Rosander J (2007) A cost curve for greenhouse gas reduction. The McKinsey Quarterly 1:35–45

EU Kommission (2008) Natura 2000. Newsletter "Natur" der europäischen Kommission GD Umwelt 24:4

EU Kommission (2011) Our life insurance, our natural capital: an EU biodiversity strategy to 2020–COM (2011) 244 final–Brüssel, 20 p

Farber S, Costanza R, Childers DL, Erickson J, Gross K, Grove M, Hopkinson CS, Kahn J, Pincetl S, Troy A, Warren P, Wilson M (2006) Linking ecology and economics for ecosystem management. Bioscience 56(2):117–129

FGG Elbe (2009) Hintergrundpapier zur Ableitung der überregionalen Bewirtschaftungsziele für die Oberflächengewässer im deutschen Teil der Flussgebietseinheit Elbe für den Belastungsschwerpunkt Nährstoffe. Flussgebietsgemeinschaft Elbe, Magdeburg

Fisher B, Turner K, Morling P (2009) Defining and classifying ecosystem services for decision making. Ecol Econ 68:643–653

Förster S, Kneis D, Gocht M, Bronstert A (2005) Flood risk reduction by the use of retention areas at the Elbe River. Int J River Basin Management 3:21–29

Fritsch M, Wein T, Ewers HJ (2007) Marktversagen und Wirtschaftspolitik. Mikroökonomische Grundlagen staatlichen Handelns. 7. edn. Vahlen, München

Fritz P (ed) (2006) Ökologischer Waldumbau in Deutschland. Fragen, Antworten, Perspektiven. Oekom, München

Gantioler S, Rayment M, Bassi S, Kettunen M, McConville A, Landgrebe R, Gerdes H, ten Brink P (2010) Costs and socio-economic benefits associated with the Natura 2000 network. Final report to the European Commission, DG Environment. Institute for European Environmental Policy/GHK/Ecologic, Brussels

Geisendorf S, Gronemann S, Hampicke U (1998) Die Bedeutung des Naturvermögens und der Biodiversität für eine nachhaltige Wirtschaftsweise. Möglichkeiten und Grenzen ihrer Erfassbarkeit und Wertmessung. Erich Schmidt Verlag, Berlin

Geisler E (1995) Grenzen und Perspektiven der Landschaftsplanung. Nat Schutz Landsch Plan 27:89–92

Getzner M, Jost S, Jungmeier M (2002) Naturschutz und Regionalwirtschaft: Regionalwirtschaftliche Auswirkungen von Natura-2000-Gebieten in Österreich. Peter Lang, Frankfurt a. M.

Gill S, Handley JF, Ennos AR, Pauleit S (2007) Adapting cities for climate change: the role of the green infrastructure. Built Environ 33:115–33

Glugla G, Fürtig G (1997) Dokumentation zur Anwendung des Rechenprogramms ABIMO. Bundesanstalt für Gewässerkunde, Mimeograph, Berlin

Gómez-Baggethun E, de Groot R, Lomas PL, Montes C (2010) The history of ecosystem services in economic theory and practice: from early notions to markets and payment schemes. Ecol Econ 69:1209–1218

Gómez-Baggethun E, Kelemen E (2008) Linking institutional change and the flows of ecosystem services. Case studies from Spain and Hungary. Proc. Book of the THEMES Summer School, Vysoké Tatry, Slovakia, pp 118–145

Gottschall M (2011) Energiegewinnung und Landschaftspflege–Ein Entwicklungskonzept für das Föhrenried. Diplomarbeit (unpublished), TU Dresden

Grossmann M (2012a) Economic value of the nutrient retention function of restored floodplain wetlands in the Elbe River basin. Ecol Econ 83:108–117

Grossmann M (2012b) Economic Valuation of Wetland Ecosystem Services. Case studies from the Elbe River Basin. Dissertation, Berlin (opus4.kobv.de/opus4-tuberlin/files/3311/grossmann_malte.pdf. Accessed 12.12.2014)

Grossmann M, Hartje V, Meyerhoff J (2010) Ökonomische Bewertung naturverträglicher Hochwasservorsorge an der Elbe. Bundesamt für Naturschutz, Naturschutz und Biologische Vielfalt 89, Bonn

Gruehn D, Budinger A, Baumgarten H (2012) Bedeutung des Stadtgrüns für den Wert von Grundstücken und Immobilien. Stadt und Grün 61:9–13

Gruehn D, Kenneweg H (2002) Wirksamkeit der örtlichen Landschaftsplanung im Kontext zur Agrarfachplanung. BfN-Skripten 59, 156 p

Grunewald K (1997) Großräumige Bodenkontaminationen–Wirkungsgefüge, Erkundungsmethoden und Lösungsansätze. Springer, Berlin

Grunewald K, Mannsfeld K (1999) Landschaftsökologische Ansätze für Konfliktlösungsstrategien im interdisziplinären Dialog–Erfordernisse, Möglichkeiten, Grenzen. In: Böhm H-P, Dietz J, Gebauer H (eds) Nachhaltigkeit als Leitbild für die Wirtschaft? TU Dresden, ZIT, pp 149–157

Grunewald K, Naumann S (2012) Bewertung von Ökosystemdienstleistungen im Hinblick auf die Erreichung von Umweltzielen der Wasserrahmenrichtlinie am Beispiel des Flusseinzugsgebietes der Jahna in Sachsen. Nat Landsch 1:17–23

Grunewald K, Scheithauer J, Sudbrack R, Heiser A, Freier K, Andreae H (2011) Untersuchungen zum Wasser- und Stoffhaushalt in Einzugsgebieten mit degradierten Hochmooren im oberen Erzgebirge, Talsperre Carlsfeld. TELMA 41:171–190

Grunewald K, Syrbe R-U (2013) Bilanzierung der Landschaftspflege in Sachsen. Schriftenreihe des Landesamtes für Umwelt, Landwirtschaft und Geologie, Heft 17, Dresden. ► https://publikationen.sachsen.de/bdb

Grunewald K, Syrbe R-U, Bastian O (2014) Landscape management accounting as a tool for indicating the need of action for ecosystem maintenance and restoration–exemplified for Saxony. Ecol Indicators 37(Part A):241–251

Grunewald K, Syrbe R-U, Wachler C (2012) Analyse der ästhetischen und monetären Wertschätzung der Landschaft am Erzgebirgskamm durch den Tourismus. GEOÖKO 33:34–65

Guo Z, Zhang L, Li Y (2010) Increased dependence of humans on ecosystem services and biodiversity. PLoS One 5:1–7

Güthler W, Oppermann R (2005) Agrarumweltprogramme und Vertragsnaturschutz weiter entwickeln. Naturschutz und Biologische Vielfalt 13, Bonn, 226 p

Güthler W, Orlich I (2009) Naturschutzförderung in Deutschland im Rahmen der EU-Agrarpolitik. Analyse der Mittelausstattung in den Bundesländern. Nat Schutz Landsch Plan 41:133–138

Haase D (2003) Holocene floodplains and their distribution in urban areas–functionality indicators for their retention potentials. Landscape Urban Plan 66:5–18

Haase D (2008) Urban ecology of shrinking cities: an unrecognised opportunity? Nature and Culture 3:1–8

Haase D (2009) Effects of urbanisation on the water balance–a long-term trajectory. Environ Impact Assess Rev 29:211–219

Haase D (2011) Urbane Ökosysteme IV-1.1.4. Handbuch der Umweltwissenschaften. Wiley-VCH, Weinheim

Haase D, Larondelle N, Andersson E et al (2014) A quantitative review of urban ecosystem services assessment: concepts, models and implementation. AMBIO 43(4):413–433

Haase D, Nuissl H (2007) Does urban sprawl drive changes in the water balance and policy? The case of Leipzig (Germany) 1870–2003. Landscape Urban Plan 80:1–13

Haase D, Nuissl H (2010) The urban-to-rural gradient of land use change and impervious cover: a long-term trajectory for the city of Leipzig. Land Use Sci 5:123–142

Haase D, Seppelt R, Haase A (2007) Land use impacts of demographic change–lessons from eastern German urban regions. In: Petrosillo I, Müller F, Jones KB, Zurlini G, Krauze K, Victorov S, Li BL, Kepner WG (eds) Use of landscape sciences for the assessment of environmental security. Springer, Dordrecht, pp 329–344

Haber W (2011) Umweltpolitikberatung–eine persönliche Bilanz. Studienarchiv Umweltgeschichte 16:15–25. ► www.iugr.net

Habitat Directive (1992) COUNCIL DIRECTIVE 92/43/EECof 21 May 1992 on the conservation of natural habitats and of wild fauna and flora (OJ L 206, 22.7.1992, p. 7)

Halbfaß S, Grunewald K (2003) Räumliche Variabilität der Phosphorgehalte im Oberboden landwirtschaftlich genutzter Flächen in kleinen Einzugsgebieten. Plant Nutr Soil Sci 166:197–203

Halbfaß S, Grunewald K (2008) Ermittlung räumlich verteilter SDR-Faktoren zur Modellierung von Sedimenteinträgen in Fließgewässer im mittleren Maßstab. Wasser-Wirtschaft 3:31–35

Hampicke U (2006) Anreiz. Ökonomie der Honorierung ökologischer Leistungen. BfN-Skripten 179, Bonn

Hampicke U (2011) Climate change economics and discounted utilitarianism. Ecol Econ 72:45–52

Hampicke U, Horlitz T, Kiemstedt H, Tampe K, Timp D, Walters M (1991) Kosten und Wertschätzung des Arten- und Biotopschutzes. Berichte des Umweltbundesamtes Nr. 3/91, Erich Schmidt Verlag, Berlin

Haslett JR, Berry PM, Bela G, Jongman RHG, Pataki G, Samways MJ, Zobel M (2010) Changing conservation strategies in Europe: a framework for integrating ecosystem services and dynamics. Biodivers Conserv 19:2963–2977

Hauer S, Ansorge A, Zöphel U (2009) Atlas der Säugetiere Sachsens. Sächsisches Landesamt für Umwelt, Landwirtschaft und Geologie, Dresden

Häusler A, Scherer-Lorenzen M (2002) Nachhaltige Forstwirtschaft in Deutschland im Spiegel des ganzheitlichen Ansatzes der Biodiversitätskonvention. BfN-Skripten 62, Bonn

Haygarth PM, Hepworth L, Jarvis SC (1998) Forms of phosphorus transfer in hydrological pathways from soil under grazed grassland. Eur J Soil Sci 49:65–72.

Heathwaite AL, Quinn PF, Hewett, CJM (2005) Modelling and managing critical source areas of diffuse pollution from agricultural land using flow connectivity simulation. J Hydrol 304:446–461

Hensher DA, Rose JM, Greene WH (2005) Applied choice analysis: a primer. Cambridge University Press, Cambridge

Herminghaus H (2012) CO_2-Vermeidungskosten für Wasserkraft, Windenergie, Biomasse, Photovoltaik. ► www.co2-emissionen-vergleichen.de/Vermeidungskosten/Vergleich-CO2-Vermeidungskosten.html. Accessed 14 March 2012

Herzog F (2000) The importance of perennial trees for the balance of northern European agricultural landscapes. Unasylva 200:42–48

Hettwer C, Malt S, Schulz D, Warnke-Grüttner R, Zöphel U (2009) Berichtspflichten zur europäischen Fauna-Flora-Habitat-Richtlinie in Sachsen. Naturschutzarbeit in Sachsen 51:36–59

Hutyra LR, Yoon B, Alberti M (2011) Terrestrial carbon stocks across a gradient of urbanization: a study of the Seattle, WA region. Glob Change Biol 17:783–797

IEEP (ten Brink P, Monkhouse C, Richartz S) (2002) Promoting the socio-economic benefits of Natura 2000. Background Report, Institute for European Environmental Policy (IEEP), Europ. Conference, Brussels, November 28–29, 2002

Ihringer J, Büchele B, Mikovec R (2003) Untersuchung von Hochwasserretentionsmaßnahmen entlang der Elbe im Bereich der Landkreise Wittenberg und Anhalt-Zerbst. Studie im Auftrag des Landesbetriebes Hochwasserschutz und Wasserwirtschaft Sachsen-Anhalt. Institut für Wasserwirtschaft und Kulturtechnik, Universität Karlsruhe, Karlsruhe

IKSE–Internationale Kommission zum Schutz der Elbe (2001) Bestandsaufnahme des vorhandenen Hochwasserschutzniveaus im Einzugsgebiet der Elbe. Internationale Kommission zum Schutz der Elbe, Magdeburg

IKSE–Internationale Kommission zum Schutz der Elbe (2004) Aktionsplan Hochwasserschutz. Internationale Kommission zum Schutz der Elbe, Magdeburg

Isermeyer F, Otte A, Christen O, Dabbert S, Frohberg K, Grabski-Kieron U, Hartung J, Heißenhuber A, Hess J, Kirschke D, Schmitz PM, Spiller A, Sundrum A, Thoroe C (2008) Nutzung von Biomasse zur Energiegewinnung. Landwirtschaftsverlag, Münster-Hiltrup

Jedicke E (ed) (1996) Praktische Landschaftspflege–Grundlagen und Maßnahmen (2. edn.). Eugen Ulmer, Stuttgart

Jenkins A, Murray B, Kramer R, Faulkner S (2010) Valuing ecosystem services from wetlands restoration in the Mississippi Alluvial Valley. Ecol Econ 69:1051–1061

Jessel B (2011) Ökosystemdienstleistungen, In: BBN (ed) Frischer Wind und weite Horizonte. Jb Natursch Landschaftspfl 58:72–87

Jessel B, Tschimpke O, Walser M (2009) Produktivkraft Natur. Hoffmann & Campe, Hamburg

Jin M, Dickinson RE, Zhang DL (2005) The footprint of urban areas on global climate as characterized by MODIS. J Clim 18:1551–1565

Job H, Metzler D (2005) Regionalökonomische Effekte von Großschutzgebieten. Nat Landsch 80:465–471

Jones M (2003) The concept of cultural landscape: discourse and narratives. In: Palang H, Fry G (eds) Landscape interfaces: cultural heritage in changing landscapes. Kluwer Academic, Dordrecht, pp 21–51

Joosten H, Couwenberg J (2008) Peatlands and carbon. In: Parish F, Sirin A, Charman D, Joosten H, Minaeva T, Silvius M (eds) Assessment on peatlands, biodiversity and climate change. Global Environment Centre, Kuala Lumpur and Wetlands International, Wageningen, pp 99–117

Kabisch N, Haase D (2011) Gerecht verteilt? Grünflächen in Berlin. Zeitschrift für amtliche Statistik Berlin Brandenburg 6:2–7

Kersebaum KC, Matzdorf B, Kiesel J, Piorr A, Steidl J (2006) Model-based evaluation of agri-environmental measures in the Federal State of Brandenburg (Germany) concerning N pollution of groundwater and surface water. J Plant Nutr Soil Sci 169:352–359

Keienburg T, Most A, Prüter J (2006) Entwicklung und Erprobung von Methoden für die ergebnisorientierte Honorierung ökologischer Leistungen im Grünland Nordwestdeutschlands. NNA-Berichte 19:3–19

Kettunen M, Bassi S, Gantioler S, ten Brink P (2009) Assessing socio-economic benefits of Natura 2000–a toolkit for practitioners (September Edition). Output of the European Commission project Financing Natura 2000: cost estimate and benefits of Natura 2000. Institute for European Environmental Policy (IEEP), Brüssel

Kluge HG, Ley F, Wittberg V (2010) Gutachten zur Abschätzung der Verwaltungskosten. ► www.unserboden.at/files/gutachten_zur_abschaetzung_der_verwaltungskosten_final.pdf. Accessed 28 Feb 2012

Knickel K, Berhold J, Schramek J, Käppel K (2001) Naturschutz und Landwirtschaft: Kriterienkatalog zur "Guten fachlichen Praxis". Angew Landschaftsökologie 41, 152 p

Knoke T, Ammer C, Stimm B, Mosandl R (2008) Admixing broadleaved to coniferous tree species: a review on yield, ecological stability and economics. Eur J Forest Res 127:89–101

Koch A, Deussen M, Hüttinger A, Mathaj M, Goldberg R (2011) Naturschutzförderung im Freistaat Sachsen–Umsetzungsstand und erste Ergebnisse zur Wirksamkeit. In: Naturschutzarbeit in Sachsen 2011:20–37. ► https://publikationen.sachsen.de/bdb/artikel/15192

Kollmann F (1982) Technologie des Holzes und der Holzwerkstoffe (2. edn.). Springer, Berlin

Köppen I (2002) Die ökologische Wirksamkeit des Agrarumwelt-Programms gemäß EU-Verordnung 1257/1999: Anwendung und Umsetzung nach dem Gießkannenprinzip oder regionalspezifisch angepasst? Untersuchung am Beispiel des Freistaates Sachsen. Diplomarbeit TU Berlin, Institut f. Landschafts- u. Umweltplanung, Sächsische Akademie der Wissenschaften zu Leipzig, Ast. Dresden

Koppisch D (2001) Torfbildung. In: Succow M, Joosten H (eds) Landschaftsökologische Moorkunde (2. edn.). Schweizerbart, Stuttgart, pp 8–17

Koska I, Succow M, Clausnitzer U, Timmermann T, Roth S (2001) Vegetationskundliche Kennzeichnung von Mooren (topische Betrachtung). In: Succow M, Joosten H (eds) Landschaftsökologische Moorkunde (2. edn.). Schweizerbart, Stuttgart, pp 112–184

Kottmeier C, Biegert C, Corsmeier U (2007) Effects of urban land use on surface temperature in Berlin: Case Study. J Urban Plan Dev, pp 128–137

Kramer M, Brendel J, Gebel M, Grunewald K, Haubold F, Kaulfuß W (1999) Ableitung von Bodenfunktionenkarten für Planungszwecke aus dem Fachinformationssystem Boden. Dresdener Geographische Beiträge, Heft 8, Im Selbstverlag der TU Dresden, Institut für Geographie, 109 p

Kühbauch W (1995) Grenzen für die Verwertung von Stickstoff durch die Grasnarbe. Vorträge der 47. Hochschultagung der Landwirtschaftlichen Fakultät der Universität Bonn vom 21. Februar 1995 in Bonn. Landwirtschaftsverlag GmbH, Münster-Hiltrup

Kuick O, Buchner B, Catenacci M, Goria A, Karakaya E, Tol RSJ (2008) Methodological aspects of recent climate change damage cost studies. Integr Assess J 8:19–40

Kuntze H (1983) Probleme bei der landwirtschaftlichen Moornutzung. TELMA 13:137–152

Lamlom SH, Savidge RA (2003) A reassessment of carbon content in wood: variation within and between 41 North American species. Biomass Bioenerg 25:381–388

LFA Mecklenburg-Vorpommern, laufende Testbetriebsergebnisse des Landes Mecklenburg-Vorpommern. ► www.landwirtschaft-mv.de/cms2/LFA_prod/LFA/content/de/Fachinformationen/Betriebswirtschaft/index.jsp?&artikel=970

LfUG–Landesamt für Umwelt und Geologie, Abteilung Natur- und Landschaftsschutz (1999) Landschaftspflegekonzeption für den Freistaat Sachsen. Dresden (unpublished)

Liebe U, Preisendörfer P, Meyerhoff J (2006) Nutzen aus Biodiversitätsveränderungen. In: Meyerhoff J, Hartje V, Zerbe S (eds) Biologische Vielfalt und deren Bewertung am Beispiel des ökologischen Waldumbaus in den Regionen Solling und Lüneburger Heide. Forschungszentrum Waldökosysteme (Selbstverlag), Göttingen, pp 101–155

Lind B, Stein S, Kärcher A, Klein M (2008) Where have all the flowers gone? Grünland im Umbruch. Hintergrundpapier und Empfehlungen des BfN, Bonn, 16 p

Lind RC, Schuler RE (1998) Equity and discounting in climate-change decisions. In: Nordhaus WD (eds) Economics and policy issues in climate change. Resources for the future, Washington DC, pp 59–96

Lorance Rall ED, Haase D (2011) Creative intervention in a dynamic city: a sustainability assessment of an interim use strategy for brownfields in Leipzig, Germany. Landscape Urban Plan 100:189–201

LUBW–Landesanstalt für Umwelt, Messungen und Naturschutz Baden-Württemberg (2009) Naturraumsteckbrief Nr. 101– Mittleres Albvorland. Landesanstalt für Umwelt, Messungen und Naturschutz Baden-Württemberg, Stuttgart

Lucke R, Silbereisen R, Herzberger E (1992) Obstbäume in der Landschaft. Ulmer, Stuttgart

Lütke-Daldrup E (2001) Die perforierte Stadt. Eine Versuchsanordnung. Stadtbauwelt 150:40–45

Lütz M, Bastian O (2000) Vom Landschaftsplan zum Bewirtschaftungsentwurf. Z Kulturtech Landentw 41:259–266

Lütz M, Bastian O (2002) Implementation of landscape planning and nature conservation in the agricultural landscape–a case study from Saxony. Agric Ecosyst Environ 92:159–170

Lütz M, Bastian O, Weber C (2006) Rahmenmethodik zur Entwicklung lokaler Agrarumweltprogramme in Europa–Akzeptanz und Monitoring von Agrarumweltmaßnahmen. Wasser-Wirtschaft 10:34–40

Macke S, Schweppe-Kraft B (2011) Ökonomisches Denken für den Naturschutz–ein Plädoyer für gemeinsame Argumente. Nat Landsch 86:146–147

Mährlein A (1993) Kalkulationsdaten für die Grünlandbewirtschaftung unter Naturschutzauflagen. KTLB Arbeitspapier 179, Darmstadt

Maljanen M, Óskarsson H, Sigurdsson BD, Guðmundsson J, Huttunen JT, Martikainen PJ (2010) Greenhouse gas balances of managed peatlands in the Nordic countries–present knowledge and gaps. Biogeoscience 7:2711–2738

Mannsfeld K (1972) Das Naherholungsgebiet Moritzburg. Sächs Heimatblätter 2:49–56

Marschall I (1998) Wer bewegt die Kulturlandschaft? Vol 1, 2, Bauernwissenschaft, ABL Bauernblatt Verlags-GmbH, Rheda-Wiedenbrück

Masten SE (2000) Transaction-Cost Economics and the Organization of Agricultural Transactions. Ind Organ 9:173–195

Matzdorf B, Kaiser T, Rohner MS (2008) Developing biodiversity indicator to design efficient agri-environmental schemes for extensively used grassland. Ecol Indic 8:256–269

MEA–Millennium Ecosystem Assessment (2003) Ecosystems and human well-being. A framework for assessment. Island Press, Washington DC, 212 p

MEA–Millennium Ecosystem Assessment (2005) Ecosystems and human well-being. Synthesis. Island Press, Washington DC

Merkel U, Helms M, Büchele B, Ihringer J, Nestmann F (2002) Wirksamkeit von Deichrückverlegungsmaßnahmen auf die Abflussverhältnisse entlang der Elbe. In: Nestmann F, Büchele B (eds) Morphodynamik der Elbe. Institut für Wasserwirtschaft und Kulturtechnik, Universität Karlsruhe, Karlsruhe

Mertz O, Ravnborg HM, Lövei GL, Nielsen I, Konijnendijk CC (2007) Ecosystem services and biodiversity in developing countries. Biodivers Conserv 16:2729–2737

Meyerhoff J (2003) Verfahren zur Korrektur des Embedding-Effektes bei der Kontingenten Bewertung. Agrarwirtschaft 52:370–378

Meyerhoff J (2006) Stated willingness to pay as hypothetical behaviour: can attitudes tell us more? J Environ Plan Manage 49:209–226

Meyerhoff J, Hartje V, Zerbe S (eds) (2006) Biologische Vielfalt und deren Bewertung am Beispiel des ökologischen Waldumbaus in den Regionen Solling und Lüneburger Heide. Forschungszentrum Waldökosysteme (Selbstverlag), Göttingen, 240 p

Meyerhoff J, Angeli D, Hartje V (2010) Social benefits of implementing a national strategy on biological diversity in Germany. Paper presented to the 12th International BIOECON Conference "From the Wealth of Nations to the Wealth of Nature: Rethinking Economic Growth" Venedig, 27.–28. September. ► www.bioecon.ucl.ac.uk/12th_2010/Angeli.pdf

Meyerhoff J, Angeli D, Hartje V (2012) Valuing the benefits of implementing a national strategy on biological diversity–the case of Germany. Environ Sci Pol 23:109–119

MLR–Ministerium für Ernährung und Ländlichen Raum Baden-Württemberg (2008) Maßnahmen- und Entwicklungsplan Ländlicher Raum Baden-Württemberg 2007–2013. Ministerium für Ernährung und Ländlichen Raum Baden-Württemberg, Stuttgart

MLR–Ministerium für Ernährung und Ländlichen Raum Baden-Württemberg (2009) Streuobstwiesen in Baden-Württemberg. Daten, Handlungsfelder, Maßnahmen, Förderung. Ministerium für Ernährung und Ländlichen Raum Baden-Württemberg, Stuttgart

MLR–Ministerium für Ernährung und Ländlichen Raum Baden-Württemberg (2010) Halbzeitbewertung "Maßnahmen- und Entwicklungsplan Ländlicher Raum Baden-Württemberg 2007–2013 (MEPL II)" nach der VO (EG) 1698/2005. Ministerium für Ernährung und Ländlichen Raum Baden-Württemberg, Stuttgart

MLUV MV–Ministerium für Landwirtschaft, Umwelt und Verbraucherschutz Mecklenburg-Vorpommern (2011) Agrarbericht des Landes Mecklenburg-Vorpommern (Berichtsjahr 2009–2010). Schwerin

Moreira F, Queiroz AI, Aronson J (2006) Restoration principles applied to cultural landscapes. J Nat Conserv 14:217–224

MSK–Ministerium für Landwirtschaft, Umwelt und Verbraucherschutz Mecklenburg-Vorpommern (2009) Konzept zum Schutz und zur Nutzung der Moore. Fortschreibung des Konzeptes zur Bestandssicherung und zur Entwicklung der Moore. Ministerium für Landwirtschaft, Umwelt und Verbraucherschutz Mecklenburg-Vorpommern, Schwerin

Müller J (2005) Landschaftselemente aus Menschenhand. Biotope und Strukturen als Ergebnis extensiver Nutzung. Elsevier, München

Müller J, Heilmann H (2011) Stand und Entwicklung der agrarischen Nutzung von Niedermoorgrünland in Mecklenburg-Vorpommern. TELMA, Beiheft 4:235–248

Naumann S, Kurzer HJ (2010) Etablierung eines Entscheidungshilfesystems zur Erstellung von Managementplänen auf Einzugsgebietsebene. Schriftenreihe des LfULG (Heft 7):144 pp

NE (2007) Richtlinie des Sächsischen Staatsministeriums für Umwelt und Landwirtschaft für die Förderung von Maßnahmen zur Sicherung der natürlichen biologischen Vielfalt und des natürlichen ländlichen Erbes im Freistaat Sachsen. SächsABl. 2008(5):218

Neef E (1962) Der Reichtum der Dresdener Landschaft. Geogr Berichte 24:256–269

Neidlein HC, Walser M (2005) Natur ist Mehr-Wert. Zum ökonomischen Nutzen des Naturschutzes. Bundesamt für Naturschutz, BfN-Skripten 154

Neuschulz F, Purps J (2000) Rückverlegung von Hochwasserschutzdeichen zur Wiederherstellung von Überflutungsflächen. In: Friese B, Witter G, Miehlich G, Rode M (eds) Stoffhaushalt von Auenökosystemen. Springer, Berlin

Neuschulz F, Purps J (2003) Auenregeneration durch Deichrückverlegung–ein Naturschutzprojekt an der Elbe bei Lenzen mit Pilotfunktion für einen vorbeugenden Hochwasserschutz. Naturschutz und Landschaftspflege in Brandenburg 12:85–91

Niedersächsische Landesregierung (1991) Langfristige ökologische Waldentwicklung in den Landesforsten. Programm der Landesregierung Niedersachsen (2. edn.). Niedersächsisches Ministerium für Ernährung, Landwirtschaft und Forsten, Hannover, 49 p

Nitsch H, Osterburg B, Roggendorf W, Laggner B (2012) Cross compliance and the protection of grassland–illustrative analyses of land use transitions between permanent grassland and arable land in German regions. Land Use Pol 29:440–448

NKGCF–Nationales Komitee für Global Change Forschung (2011) Ökosystemare Dienstleistungen und Biodiversität 2020: ökologische, soziale und ökonomische Entwicklungen und Gestaltungsoptionen im globalen Wandel. ► http://www.nkgcf.org/files/downloads/ESS_Biodiv2020_Final_inclAnhang.pdf. Accessed 16 July 2012

Nowak DJ, Crane DE (2002) Carbon storage and sequestration by urban trees in the USA. Environ Pollut 116:381–389

NRC–National Research Council (2000) Risk and uncertainty in flood damage reduction studies. National Academy Press, Washington DC

NRC–National Research Council (2005) Valuing ecosystem services. Toward better environmental decision-making. National Academy Press, Washington DC

Oikonomou V, Dimitrakopoulos PG, Troumbis AY (2011) Incorporating ecosystem function concept in environmental planning and decision making by means of multi-criteria evaluation: the case-study of Kalloni, Lesbos, Greece. Environ Manage 47:77–92

Osterburg B, Rühling I, Runge T, Schmidt TG, Seidel K, Antony F, Gödecke B, Witt-Altfelder P (2007) Kosteneffiziente Maßnahmenkombinationen nach Wasserrahmenrichtlinie zur Nitratreduktion in der Landwirtschaft. In: Osterburg B, Runge T (eds) Maßnahmen zur Reduzierung von Stickstoffeinträgen in die Gewässer–eine wasserschutzorientierte Landwirtschaft zur Umsetzung der Wasserrahmenrichtlinie. Landbauforschung Völkenrode, Sonderheft 307

Ott K (2010) Umweltethik zur Einführung. Junius, Hamburg

Ott K, Döring R (2004) Theorie und Praxis starker Nachhaltigkeit. Metropolis, Marburg

PAN–Planungsbüro für angewandten Naturschutz, IFAB–Institut für Agrarökologie und Biodiversität, INL–Institut für Landschaftsökologie und Naturschutz (2011) Umsetzung des High Nature Value Farmland-Indikators in Deutschland–Ergebnisse eines Forschungsvorhabens (UFOPLAN FKZ 3508 89 0400) im Auftrag des Bundesamtes für Naturschutz. Bearbeitung durch: PAN, IFAB, INL. München, Mannheim, Singen, 54 p

Pearce DW (1993) Economic values and the natural world. MIT Press, London

Pearce DW (2003) The social cost of carbon and its policy implications. Oxford Rev Econ Policy 19:1–32

Penning-Rowsell EC, Johnson C, Tunstall S, Tapsell S, Morris J, Chatterton J, Coker A, Green C (2003) The Benefits of flood and coastal defence: techniques and data. Flood Hazard Research Center, Middlesex University, United Kingdom

Peters W (2009) Biomassepotenziale aus der Landschaftspflege in Sachsen. FuE-Bericht im Auftrag des Sächsischen Landesamtes für Umwelt, Landwirtschaft und Geologie (unveröffentlicht), Bosch & Partner, Berlin

Pimentel D, Harvey C, Resosudarmo P et al (1995) Environmental and economic costs of soil erosion and conservation benefits. Science 267:1117–1123

PLENUM (2008) Regionalentwicklungskonzept PLENUM Schwäbische Alb 2008–2013. PLENUM, Reutlingen

Plieninger T, Schleyer C (2010) Konzept der Ökosystemleistungen im Kontext der Europäischen Landwirtschaft. Agrarische Rundschau 6:12–15

Plieninger T, Höchtl F, Spek T (2006) Traditional land-use and nature conservation in European rural landscapes. Environ Sci Policy 9:317–321

Popp D, Hage G (2003) Großschutzgebiete als Träger einer naturverträglichen, nachhaltigen Regionalentwicklung. Nat Landsch 78:311–316

Qui Z (2009) Assessing critical source areas in watersheds for conservation buffer planning and riparian restoration. Environ Manage 44:968–980. doi:10.1007/s00267-009-9380-y

Ridder B (2008) Questioning the ecosystem services argument for biodiversity conservation. Biodivers Conserv 17:781–790

Ring I, Schröter-Schlaack C (eds) (2011) Instrument mixes for biodiversity policies. POLICYMIX Report, Issue No. 2/2011, Helmholtz Centre for Environmental Research, UFZ, Leipzig. ► http://policymix.nina.no. 01 Aug 2011

Ring I, Hansjürgens B, Elmqvist T, Wittmer H, Sukhdev P (2010) Challenges in framing the economics of ecosystems and biodiversity: the TEEB initiative. Curr Opin Environ Sustain 2:15–26

Ringler A, Rossmann D, Steidl I (1997) Hecken und Feldgehölze. BayStMLU, ANL (eds.)

Roulet NT, Lafleur PM, Richard PJH, Moore TR, Humphreys ER, Bubier J (2007) Contemporary carbon balance and late Holocene carbon accumulation in a northern peatland. Glob Change Biol 13:397–411

Rounsevell MDA, Dawson TP, Harrison PA (2010) A conceptual framework to assess the effects of environmental change on ecosystem services. Biodivers Conserv 19:2823–2842

Rowinski V, Kobel J (2011) Erfassung, Bewertung und Wiedervernässung von Mooren im Müritz-Nationalpark. TELMA, Beiheft 4:49–72

Santos LD, Martins I (2007) Monitoring urban quality of life–the Porto experience. Soc Indic Res 80:411–425

Schäfer A (2009) Moore und Euros–die vergessenen Millionen. Arch Forstwes Landschaftsökol 43:156–160

Schäfer A, Degenhardt S (1999) Sanierte Niedermoore und Klimaschutz–Ökonomische Aspekte. Arch Nat Schutz Landsch Forsch 38:335–354

Schäfer A, Joosten H (eds) (2005) Erlenaufforstung auf wieder vernässten Niedermooren. DUENE e. V., Greifswald

Schaich H, Bieling C, Plieninger T (2010) Linking ecosystem services with cultural landscape research. GAIA 19:269–277

Schelling TC (1995) Intergenerational discounting. Energy Policy 23:395–401

Schetke S, Haase D, Kötter T (2012) Innovative urban land development–a new methodological design for implementing ecological targets into strategic planning of the City of Essen, Germany. Environ Impact Assess Rev 32:195–210

Schleyer C, Plieninger T (2011) Obstacles and options for the design and implementation of payment schemes for ecosystem services provided through farm trees in Saxony, Germany. Environ Conserv 38:454–463

Schrack M (eds) (2008) Der Natur verpflichtet. Projekte, Ergebnisse und Erfahrungen der ehrenamtlichen Naturschutzarbeit in Großdittmannsdorf. Veröff. Mus. Westlaus. Kamenz, Sonderheft, 180 p

Schwartz MW, Bringham CA, Hoeksema JD, Lyons KG, Mills MH, van Mantgem PJ (2000) Linking biodiversity to ecosystem function: implications for conservation ecology. Oecologia 122:297–305

Schwarz N (2010) Urban form revisited–selecting indicators for characterising European cities. Landscape Urban Plan 96:29–47

Schwarz N, Bauer A, Haase D (2011) Assessing climate impacts of local and regional planning policies—quantification of impacts for Leipzig (Germany). Environ Impact Assess Rev 31:97–111

Schweppe-Kraft B (2009) Natural capital in Germany–state and valuation; with special reference to Biodiversity. In Döring R (eds) Sustainability, natural capital and nature conservation. Beiträge zur Theorie und Praxis starker Nachhaltigkeit (Vol 3). Metropolis, Marburg, pp 193–216

Seják J, Dejmal I et al (2010) Method of monetary valuation of territorial ecological functions. Mscr., J. E. Purkyne University, Ústí nad Labem, Tschechische Republik

Seto KC, Fragkias M, Güneralp B, Reilly MK (2011) A meta-analysis of global uban land expansion. PLoS One 6: e23777. doi:10.1371/journal.pone.0023777

Simoncini R (1998) The IUCN/European Sustainable Specialist Group/Agricultural Working Group Pan-European Project. Manuscript (unpublished), Florence

SMUL–Sächsisches Staatsministerium für Umwelt und Landwirtschaft (2009a) Programm zur Biologischen Vielfalt im Freistaat Sachsen des Sächsischen Staatsministeriums für Umwelt und Landwirtschaft. Dresden

SMUL–Sächsisches Staatsministerium für Umwelt und Landwirtschaft (2009b) Waldzustandsbericht 2009. Sächsisches Staatsministerium für Umwelt und Landwirtschaft (ed) Sachsenforst, Dresden

SMUL–Sächsisches Staatsministerium für Umwelt und Landwirtschaft (2009c) Agrarbericht in Zahlen 2009, Mai 2010. ► www.smul.sachsen.de/umwelt/.../BioDiv_Prog_ Mrz09_fin.pdf. Accessed 21 Oct 2009

SMUL–Sächsisches Staatsministerium für Umwelt und Landwirtschaft (2010) Maßnahmenplan des Sächsischen Staatsministeriums für Umwelt und Landwirtschaft zur Biologischen Vielfalt im Freistaat Sachsen. Dresden,18 p

Spangenberg JH, Settele J (2010) Precisely incorrect? Monetising the value of ecosystem services. Ecol Complex 7:327–337

Staub C, Ott W et al (2011) Indikatoren für Ökosystemleistungen: Systematik, Methodik und Umsetzungsempfehlungen für eine wohlfahrtsbezogene Umweltberichterstattung. Bundesamt für Umwelt, Umwelt-Wissen Nr. 1102, Bern

Steffens R, Saemann D, Größler K (1998) Die Vogelwelt Sachsens. Fischer, Jena, Stuttgart

Stegmann H, Zeitz J (2001) Bodenbildende Prozesse entwässerter Moore. In: Succow M, Joosten H (edn) Landschaftsökologische Moorkunde (2. edn). Schweizerbart, Stuttgart, pp. 47–57

Stern N (2007) The Economics of Climate Change. Cambridge University Press, Cambridge

StMUG–Bayerisches Staatsministerium für Umwelt und Gesundheit (2011) Mehr Geld für Vertragsnaturschutz und Landschaftspflege. Nat Landsch 86:172

Strohbach MW, Haase D (2012) Estimating the carbon stock of a city: a study from Leipzig, Germany. Landsc Urban Plan 104:95–104

Strohbach MW, Arnold E, Haase D (2012) The carbon mitigation potential of urban restructuring–a life cycle analysis of green space development. Landsc Urban Plan 104:220–229

Succow M (1988) Landschaftsökologische Moorkunde. Gustav Fischer, Jena

Syrbe R-U, Grunewald K (2013) Restrukturierungsbedarf für regionaltypische Landschaftselemente und Biotopstrukturen am Beispiel Sachsens. Natur und Landschaft 88(3):103–111

TEEB–The Economics of Ecosystems and Biodiversity (2009) TEEB for National and International Policy Makers. Rewarding benefits through payments and markets. ► www.teebweb.org/LinkClick.aspx?fileticket=vYOqLx-i7aOg%3d&tabid=1019&language=en-US. Accessed 28 June 2010

TEEB–The Economics of Ecosystems and Biodiversity (2010) Mainstreaming the economics of nature: a synthesis of the approach, conclusions and recommendations of TEEB. ► www.teebweb.org/TEEBSynthesisReport/tabid/29410/Default.aspx

Termorshuizen JW, Opdam P (2009) Landscape services as a bridge between landscape ecology and sustainable development. Landsc Ecol 24:1037–1052

The Council of the European Union (2005) Council Regulation (EC) No 1698/2005 of 20 September 2005 on support for rural development by the European Agricultural Fund for Rural Development (EAFRD). Official Journal of the European Union, L 277/1. 21.10.2005

The Council of the European Union (2009) Council Regulation (EC) No 74/2009 of 19 January 2009 amending Regulation (EC) No 1698/2005 on support for rural development by the European Agricultural Fund for Rural Development (EAFRD). Official Journal of the European Union, L 30/100. 31.1.2009

Tol RSJ (2005) The marginal damage costs of carbon dioxide emissions: an assessment of uncertainties. Energy Policy 33:2064–2074

Tratalos J, Fuller RA, Warren PH, Davies RG, Gaston KJ (2007) Urban form, biodiversity potential and ecosystem services. Landsc Urban Plan 83:308–17

Turner K, Georgiou S, Fisher B (2008) Valuing ecosystem services. the case of multifunctional wetlands. Earthscan, London

UBA–Umweltbundesamt (2007) Ökonomische Bewertung von Umweltschäden. Methodenkonvention zur Schätzung externer Umweltkosten. Umweltbundesamt, Berlin

UBA–Umweltbundesamt (2009) National Inventory Report for the German Greenhouse Gas Inventory under the United Nations Framework Convention on Climate Change 1990–2007. Umweltbundesamt, Dessau

UBA–Umweltbundesamt (2010) Nationaler Inventarbericht Deutschland–2010. Berichterstattung unter der Klimarahmenkonvention der Vereinten Nationen, 668 p. ► www.waldundklima.net/klima/klima_docs/uba_nir_dtl_2010.pdf. Accessed 16 june 2012

UBA–Umweltbundesamt (2011) Daten zur Umwelt. Umwelt und Landwirtschaft, Bonn, 98 p

Valsangiacomo A (1998) Die Natur der Ökologie: Anspruch und Grenzen ökologischer Wissenschaften. vdf, Zürich

Vos W, Meekes H (1999) Trends in European cultural landscape development: perspectives for a sustainable future. Landsc Urban Plan 46:3–14

Watkiss P, Anthoff D, Downing T, Hepburn C, Hope C, Hunt A, Tol R (2005) The social cost of carbon (SSC). review–methodological approaches for using SCC estimates in policy assessment. Final Report. AETA Technology Environment, Harwell

Weizsäcker EU von (1989) Erdpolitik–Ökologische Realpolitik an der Schwelle zum Jahrhundert der Umwelt. Wissenschaftliche Buchgesellschaft, Darmstadt

Weller F (2006) XI–2.11. Streuobstwiesen. In: Konold W, Böcker R, Hampicke U (eds) Handbuch Naturschutz und Landschaftspflege. Wiley-VCH, Weinheim, pp 1–42

Wichmann S, Wichtmann W (2011) Paludikultur: Standortgerechte Bewirtschaftung wiedervernässter Moore. TELMA, Beiheft 4:215–234

Wichtmann W, Schäfer A (2005) Energiegewinnung von ertragsschwachen Ackerstandorten und Niedermooren. Nat Landsch 80:421–425

WRRL (2000) Richtlinie 2000/60/EG des Europäischen Parlaments und des Rates vom 23. Oktober 2000 zur Schaffung eines Ordnungsrahmens für Maßnahmen der Gemeinschaft im Bereich der Wasserpolitik. Amtsblatt der Europäischen Gemeinschaft vom 22.12.2000 L 327/1

WuF (2007) Richtlinie des Sächsischen Staatsministeriums für Umwelt und Landwirtschaft zur Förderung der naturnahen Waldbewirtschaftung, forstwirtschaftlicher Zusammenschlüsse und des Naturschutzes im Wald im Freistaat Sachsen. SächsABl. Jg. 2007, Bl.-Nr. 44, p 1449

Yli-Pelkonen V, Nielema J (2005) Linking ecological and social systems in cities: urban planning in Finland as a case. Biodivers Conserv 14:1947–1967

Zahrnt V (2010) Eine Gemeinsame Agrarpolitik für europäische Gemeingüter: Deklaration einer Gruppe führender Agrarökonomen. ▶ www.reformthecap.eu/Declaration-on-CAP-reform. Accessed 02 Nov 2010

Zak D, Augustin J, Trepel M, Gelbrecht J (2011) Strategien und Konfliktvermeidung bei der Restaurierung von Niedermooren unter Gewässer-, Klima- und Naturschutzaspekten, dargestellt am Beispiel des nordostdeutschen Tieflandes. TELMA, Beiheft 4:133–150

Zangemeister C (1971) Nutzwertanalyse in der Systemtechnik. Eine Methodik zur multidimensionalen Bewertung und Auswahl von Projektalternativen (4. edn). Dissertation, TU Berlin, 1970, München, 370 p

Zeddies J, Kazenwadel G, Löthe K (1997) Auswirkungen einer Steuer auf mineralische Stickstoffdünger. Agrarwirtschaft 46:214–224

ZMP (2009) ZMP-Marktberichterstattung der Jahre 2005 und 2006. ▶ www.zmp.de. Accessed 24 April 2009

Recommendations and Outlook

K. Grunewald, O. Bastian (eds.), *Ecosystem Services – Concept, Methods and Case Studies*
DOI 10.1007/978-3-662-44143-5_7, © Springer-Verlag Berlin Heidelberg 2015

7.1 Work Steps for the Analysis and Evaluation of ES

O. Bastian and K. Grunewald

"You would not find out whether something is successful or not by thinking about it; only by trying it out."

Methodological recommendations for action are extremely helpful for the practical application of the ES concept, for example in the form of a guideline which identifies the most important work steps in order, and provides and explains suitable approaches to solutions for each one.

However, it would hardly be realistic to stipulate a schematic, generally applicable process plan, like a cooking recipe, for the ascertainment and/or evaluation of ES. The respective problems and the substantive ecological, economic, sociocultural and spatial configurations and contexts are too different. For that reason, the guideline presented here (◘ Table 7.1) is based on the EPPS framework methodology (▶ Sect. 3.1), and should first of all be seen as an orientation which indicates important work steps and points out significant aspects and/or requirements, without being able to go into any detail. It should also be noted that completeness need not always be achieved, but that rather the program for the investigation should be tailored very specifically to the particular task at hand, among other things in order to keep the effort required for the work within limits.

This guideline is based both on the experiences of the authors and the evaluation of the following sources from the literature: Bastian and Schreiber (1999); OECD (2008); Haines-Young and Potschin (2009); Kettunen et al. (2009); Grunewald and Bastian (2010); TEEB (2010); UNEP-WCMC (2011); Bastian et al. (2012); Burkhard et al. (2012); Seppelt et al. (2012).

▪ 1) Definition of the Task

First of all, *the purpose of the investigation* needs to be defined, and the concrete task definition clearly formulated. The question as to why the ES need to be evaluated in the first place should be answered, and also which advantages this would have in comparison with traditional approaches in the concrete case. Without a deeper purpose, it will in most cases hardly be possible to address the more high-effort tasks, particularly the quantitative evaluation of ES. The latter may for example be useful in protected areas, especially if these have a high socio-economic or a significant development potential, if a pollution or danger thereof due to inappropriate land-use practices exists, or if alternatives are sought or protective goals are to be formulated (especially with regard to land-use changes, e.g. forest restructuring; ▶ Sect. 6.3). The calling of ES to attention is useful if stakeholders are to be won over to the protective effort, and if financial resources need to be secured. A clear setting of goals is also useful in order to find suitable indicators and to avoid mistaken interpretations with regard to ES as much as possible.

▪ 2) Characteristics of the Area to be Investigated

Ideally, we should start by obtaining an *overview* of the area under investigation: its size, for example in order to establish the suitable scale for the ascertainment; properties and natural capacity of important ecosystems, its dominant land uses and the overall socio-economic situation; known uses based on the ecosystem and the beneficiaries of those uses; existing conflicts, problems or pollution situations; expired or expected changes or trends, existing planning procedures, or pending decisions.

The fundamental factor is knowledge of the *data situation*: Have ES already been recorded in the area, or are ES-relevant data available? Which data sources are available (e.g. maps, databases remote sensing, GIS, landscape plans, local knowledge, expert knowledge or ascertainment of resources)? Existing information gaps should also be identified.

▪ 3) Clear Definition of the Terminology

In order to develop a clear and comprehensible investigative approach, to correctly interpret and communicate the results, and to avoid having the experts from different disciplines and scientific schools, and the practitioners, talk past each other, a definition and explanation of the *terminology* to be used is indispensable (▶ Sect. 2.1). Otherwise, misunderstandings may arise, for to this day there is no system of terminology for the ES area which is clear in detail and generally accepted. That includes the term ES itself, and also the term 'function,' among others. Moreover, to date there is no

Table 7.1 Work steps for the analysis and evaluation of ES (explanation in the text)

Point	Work step
1	Definition of the task
2	Characteristics of the area to be investigated
3	Clear definition of the terminology
4	A balanced selection of the ES
5	Selection of suitable evaluation procedures and indicators
6	Selection and processing of ecological/biophysical assessment approaches
7	Realisation of monetary evaluation–if possible and necessary
8	Differentiated view of ES and cost/benefit
9	Consideration of dangers, risks, limit values, trade-offs
10	Consideration for space/time aspects
11	Identification of stakeholders and institutions
12	Analysis of drivers and scenarios
13	Conveying knowledge, communication about ES
14	Recommendations for need for action and ES management
15	Monitoring ES

The numbers do not indicate a mandatory sequence; however, Points 1 to 5 can be attributed to the initial phase, Points 6 to 12 to the main processing phase, and Points 13 to 15 to the final phase.

consensus regarding the most useful *classification* of ES. It can therefore be noted as clearly as possible why a certain classification system has been decided upon. The corresponding proposals have been presented in ▶ Sect. 3.2.

▪ 4) A Balanced Selection of the ES

A *representative selection* should be made from the wide variety of possible ES. What are the most important ES upon which society depends and/or needs? Which ES are endangered, and which are subject to changes, or will be in the foreseeable future? If only for reasons of the work effort involved, it would appear to be hardly possible to process a very large number of ES, let alone the entire spectrum (see case studies, ▶ Chap. 6). However, the representative selection of ES should be considered important–and not only provisioning ES such as food and fibres should be represented, but also regulatory and sociocultural ES, including those relevant for the preservation of biological diversity.

Functioning properly ecosystems often yield a whole *bundle of different ES*. Their shares vary from one ecosystem to the other, from place to place and from time to time. It is important to keep track of the totality of ES and their linkages, in order, e.g. to avoid establishing financial incentives for the benefit of a single ES at the expense of others, which are then damaged. This often occurs in the interaction between provisioning ES and regulatory ES, e.g. energy crop cultivation vs. biodiversity. The consideration of a broad spectrum of ES is also necessary in order to arrive at statements on sustainability.

▪ 5) Selection of Suitable Evaluation Procedures and Indicators

When ascertaining or evaluating ES, the question frequently arises: What are the most appropriate procedures for the concrete situation? As explained in ▶ Sect. 4.2, using the example of the travel cost method, the inappropriate application of a procedure may lead to nonsensical results.

Minimum demands must be placed on evaluation procedures:

1. The basic precondition is that the procedure be logically structured, clear and of significant explanatory power.
2. The selected procedure should be equivalent with the spatial segment under examination, the criteria of evaluation and the precision of the results.
3. The latest state of information and the current value criteria should be taken into account.
4. The input values and ecological contexts should be scientifically secured to the maximum extent possible (validity).
5. It should be possible to gather the basic data within a reasonable period of time.
6. The ascertainment and processing of the data should be transparent.

7. The procedure should be understandable and flexible.
8. The evaluation results should be clearly and comprehensibly presented.

ES evaluations should not ignore risks and uncertainties or *knowledge gaps* with regard to the effect of people on ecosystems and ES, and their significance for human well-being.

On the one hand, it is necessary to simplify, e.g. in order to communicate with political leaders or the broad public; on the other, misguided simple solutions to complex problems should be avoided, in order not to cause more harm than good, thus for example causing mistakes in the decision-making process. The most useful solution appears to be a medium level of complexity, or differentiation according to various levels of complexity (as in the model InVEST, ▶ Sect. 4.4).

The determination of suitable *indicators* (▶ Sect. 4.1) has proven to be helpful. These should have a high level of explanatory potential for the problem to be solved, and also be politically relevant in order to be able to correctly interpret and apply the results of the investigations.

▪ 6) Selection and Processing of Ecological/Biophysical Assessment Approaches

Every ES evaluation starts with a compilation of existing knowledge, the ascertainment/measurement needed to obtain the necessary basic data, and a qualitative evaluation which is generally, but not always followed by a quantification. Not all ES are easy to quantify. Also suitable are qualitative measures (see the example of the orchard meadows, ▶ Sect. 6.5.1). Where no direct measures are available, and/or possible, and where there are also no exact data, it may be necessary to work with estimated values. These steps do not however constituted evaluation in the strict sense.

The principle of the *distinction between measurement and evaluation*, or between the factual level (the structures and current processes existing within ecosystems), and the value level is often overlooked. Both descriptive and normative work steps are necessary in ES investigations, and must be embedded into a broader socio-ecological context.

The pyramid of ES valuation methods designed by ten Brink (2008) depicts the step-by-step narrowing down of the quantum of investigation from the qualitative overview to the monetary evaluation; the latter is a very high-effort procedure, and hence not always used (◘ Fig. 7.1).

The *ecological ascertainment* of ecosystem structures and processes, e.g. based on data, maps, fieldwork, experiments, measurements and modelling, is necessary in order to gain an understanding of how ES are generated, and in order to provide a scientifically based framework within which the actual valuation will take place.

▪ 7) Realisation of Monetary Evaluation: If Possible and Necessary

The *pros and cons* of economic (monetary) valuation methods have already been addressed (▶ Sect. 4.2). They require not only economic expertise, but are also usually very high-effort, and can be usefully applied only in certain situations. Economic valuations can however contribute to a greater consideration for ES in economic calculations, balancing of interests and planning processes, for instance in cost-benefit analyses, and also in the internalisation of environmental impacts.

Certain services attributed to ecosystems are not provided by them alone, but require human input. For example, the crops grown on farmland are the product of ES, dependent upon site conditions and potentials; they provide utility for people and contain value. They are planted, cared for, fertilised, irrigated, etc., in order to secure increased yields. If appropriate, reference should be made to such *environmental services* (▶ Sect. 2.1, 4.2, 6.2.4).

In general, valuation methods–and not only economic ones–should be seen as part of a *broad spectrum of diagnostic instruments* and of institutional and policy mechanisms (including legal stipulations, participatory methods, and governance; ▶ Sect. 4.3, 5.1, 5.4) which facilitate an understanding of complex socio-ecological systems and provide decision-makers with the necessary background knowledge. ES evaluations require a strong *interdisciplinary perspective* which integrates not only ecological and economic factors, but also a wide variety of natural, planning and social-scientific disciplines.

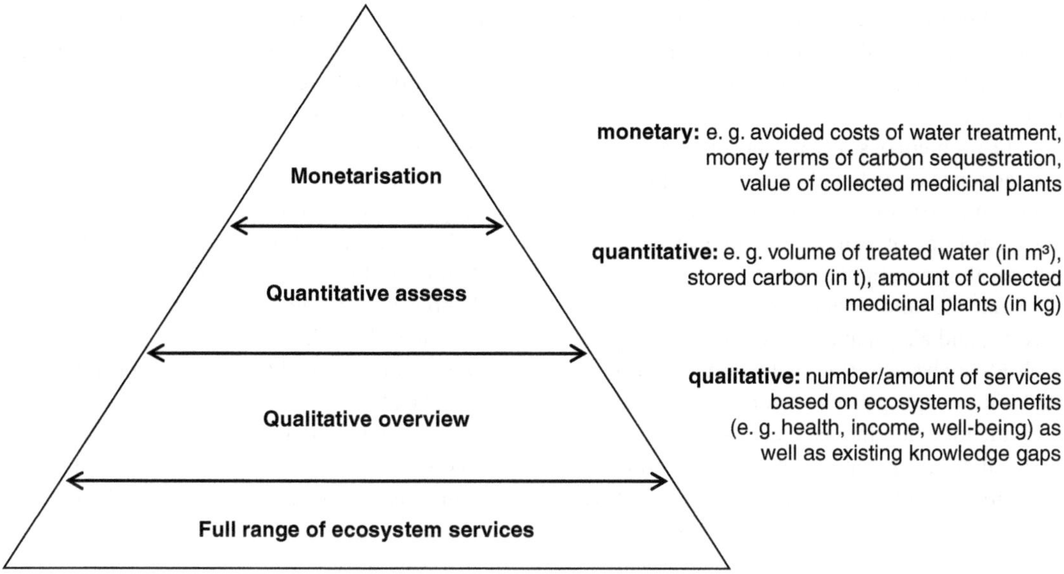

Fig. 7.1 Basic approaches of ecosystem services evaluation. Adapted from ten Brink 2008

▪ 8) A Differentiated View of ES and cost/ Benefit

ES are the link between ecosystems/landscapes and the utility and/or values which they provide to people. It is important to understand whether a service is simply being 'supplied' by the ecosystem, or whether this service is actually being 'demanded' by people. In the former case, the service is a *potential* of the ecosystem and/or of the landscape (► Sect. 3.1). Whether the utility corresponds to the potential, whether overuse exists which constitutes a burden on the ecosystem, or whether there is leeway for further-reaching utilisation is helpful for the evaluation and planning process.

There are also *values* which cannot be assigned to a certain ES, but which should nonetheless not be neglected, e.g. the existence of rare animal and plant species, regardless of their role with respect to ES.

The effect of ES on *human well-being* should be demonstrated: Is that well-being being affected by the increase or decrease in ES, and what policy abilities to ES provide to increase that well-being? How, for example, would an improvement of the percolation and water-retention capacity of soil in the watershed of a river increase the safety of the inhabitants by reducing the risk of flooding, or preventing it altogether?

▪ 9) Consideration of Dangers, Risks, Limit Values and Trade-Offs

Important questions in this context include:

- Which limit and threshold values are known, and need to be addressed?
- Which causal contexts are there between certain ES?
- Are any of the ascertained uses endangered, declining, or subject to serious risks? Knowledge of such matters could help establish direct or long-term measures in order to secure the maintenance of ES.
- What are possible trade-offs between various uses which have to be considered? A focus on increasing the level of certain ES and of their associated utility can have negative effects on other ES. The ascertainment of existing and potential trade-offs can help make a determination as to which uses or utilities could be supported and which should not, in accordance, too, with the principle of sustainability.

▪ 10) Consideration for Space/Time Aspects

For ES, there is a high level of *relevance of scales* with respect to space, time and complexity. That is true of data ascertainment, models, evaluations, utilities and values, institutions and economic and political processes alike. The dependency on scales should always be taken into account, including interactions between scales and their subordinated hierarchies.

The problem of scales is reflected for example in the reference units for ES and the compatibility of scales and measures. Also to be taken into account are transitions of scales, including upscaling and downscaling and tipping points.

Since ES and economic values are context-specific, and also spatially and temporally specific, each ES analysis and/or evaluation should be carried out in appropriate temporal and spatial scales relevant for scientific ascertainment, and also for political decision-making and for policy measures.

Important *spatial aspects* include (▶ Sect. 3.3):
- Area requirement: A minimum area for the supply of ES, with a specific quality (structure, abiotic characteristics, biodiversity)
- Spatial composition: Land cover diversity, patch richness
- Spatial configuration: Zonation of protected areas (core areas, buffers), land-use gradients
- Functional connections: Supply–transfer–demand relationships, likelihood of transmission and transfer (e.g. habitat networks, river-floodplain relationships).

Far too few ES studies to date are taking the so-called off-site effects into consideration: the provision of ES in certain area can be affected by decisions made in other areas or at another level;regional, national or global.

Time aspects especially refer to:
- Time requirements: Minimal process time, regeneration times of ecosystems and ES
- Temporal sequences: Changes in ecosystems and ES (trends)
- Time lags: Precautionary measures, risks, option values, intergenerational time lags (the present generation benefits, the next pays for the environmental damage).

▪ 11) Identification of Stakeholders and Institutions

The ascertainment/evaluation of ES typically does not stop with the analysis of natural-scientific facts or at the provision of services (potentials), but rather also raises the issue of the often complex structures and interactions of ES-relevant stakeholders and institutions (▶ Sect. 4.3, 5.1). If the utility for the user provided by ES is known, suitable options can be ascertained for maintaining these ES by involving these beneficiaries.

Important questions include:
- Who is dependent on which ES? Who profits? Who pays for or suffers from disadvantages? Who is responsible for maintaining ES? Who has caused the deterioration?
- Which institutional factors–laws, ordinances, standards and rules, incentive systems, property relationships, traditions and customs, decision-making structures, etc.–affect the condition of the ecosystems and of the ES?
- Which actors are relevant at the various levels of decision-making?
- Who can or should contribute financially to maintaining ES? How can stakeholders be incorporated into the ES concept, e.g. through the identification and evaluation of relevant ES, or the development of management options and their practical implementation?

▪ 12) Analysis of Drivers and Scenarios

Changes in ecosystems and/or ES are triggered by direct and indirect *drivers*, which should be identified. Important drivers include globalisation, demographic change, climate change and also policy decisions, incentive mechanisms, legislation, authorities and institutions.

ES evaluations should be placed into the context of contrasting *scenarios* (▶ Sect. 4.3, 6.2.4, 6.3), with the aid of which questions such as the following can be answered:
- What could the future of the ecosystems or the ES concerned look like (development of so-called storylines)
- Would the economic development and well-being of various stakeholders in the area be influenced by an increase or a decrease of ES?
- How will various policy options affect development drivers? How will changes in the drivers affect ES?
- How might the state of knowledge and the appreciation of the values of ES change in the future?

13) Conveying Knowledge and Communications Regarding ES

One key limiting factor in the preservation of natural capital is the widespread ignorance about how ecosystems function and contribute to human well-being. These deficits could be overcome by targeted and continual educational activity in general, and by appropriate information regarding particular projects, such as renaturation, protected-area certification, etc., and particularly by cooperation on the basis of trust between various stakeholders. In ► Sect. 4.5, difficulties and also opportunities, such as knowledge transfer, campaigns, discourses etc. are discussed in this regard.

14) Recommendations for Need for Action and ES Management

One of the most important goals in ES evaluation involves optimising and improving the ES in an area. *Instruments* possible for this purpose include:

- Adoption of suitable legislation and other regulations
- Payments for ecosystem services (PES): Demonstrated socio-economic utility can persuade beneficiaries to agree to new models for assuming the cost of protection (► Sect. 5.2)
- Elimination of harmful policies and incentive mechanisms which favour the degradation of ES
- Compensation mechanisms for losses due to impacts
- Dialogues with stakeholders
- Establishment of protected areas
- Development of capacities (jobs, funding), e.g. conservation and landscape care
- Support for research to improve the ascertainment and evaluation methodology.

The ES management should meet certain criteria and comply with certain *principles*:

- Effectivity and efficiency, political and economic viability
- Risks concerning ES, application of the precautionary principle
- Possible balancing of interests between different ES, or else prioritisation (but not only taking supply ES into account); possibly reweighting of priorities from supply towards regulatory services; sustainable level of 'utilisation'; avoidance of conflict between protective and use aspects
- Trade-offs: Balancing the improvement of one ES against the possible deterioration of another ES
- Equal rights among stakeholders, different dependencies of stakeholders on ES, and use in property rights with regard to ES
- Incorporation of relevant stakeholders at the local, regional, national and global levels in the formulation of management goals for the implementation of measures

15) Monitoring ES

An effective and efficient monitoring process is necessary in order to observe changes in ecosystems and ES, contract implementation and success (or failure) of measures, and, if necessary, to draw the necessary conclusions, e.g. adaptive management, or modification of control instruments.

For the modern meaning of changes with regard to ES, biodiversity, economic and social goals and requirements, suitable key indicators are necessary. The frequency of monitoring should correspond to the respective factual situation and the issues at one hand, and it should be sufficiently flexible at the other hand, so as to be able to permit modifications, e.g. in the measurement process.

Conclusion

Finally, let us emphasise: This guideline is designed to convey an overview of important aspects of ES ascertainment and evaluation, without any claim of completeness. It should also be noted that there are no evaluation methods suitable for all possible cases. Ultimately, the specific task at hand in the respective context of the investigations will determine which specifications of the procedure are necessary and sensible.

7.2 Future Challenges Regarding ES

K. Grunewald and O. Bastian

"For human beings, 'existence' in the sense of merely 'subsisting' on earth is not enough. People have their desires, and often, their greed knows no limits. Humankind must experience its existence. In pain, in ecstasy, in failure, in triumph (after Zeh 2009)."

We have formulated the following key issue of the ES concept (▶ Chap. 1): How can the services of nature be ascertained? What are the use claims of people with regard to the services which nature can provide, and how can these claims be ascertained and integrated into rational action?

Unlike animals, humans can at least to some extent raise themselves above the constraints of nature. However, ever more growth does not necessarily bring more well-being, for the 'good life' is not definable solely by means of material parameters. Even if traditional thought patterns, manners of behaviour and economic models demonstrate greater capacity for resilience, the end of the non-sustainable path to development is foreseeable; however that will require time, and the actors to carry out change. In this context, the line of thought associated with the concept of ES can be seen both as a political and as an economic concept for action.

The goal of the ES approach is to obtain a higher appreciation for the value of our natural foundations for life. The concept 'ecosystem services' is relatively new, a 'fashionable term' (▶ Chap. 1); however, the intents and approaches behind it are not, to a large extent (▶ Chap. 2). Since the concept is not a revolutionary new one, it can hardly be expected to solve fundamental conflicts in the relationship between humankind and nature. The ES concept has no patent remedy to offer policy-makers, nor can it relieve them of the burden of decision-making.

Ever more voices (e.g. Fatheuer 2012; Löschmann 2012; Schröter et al. 2014) are sounding the warning against providing further support for the commercialisation of nature under the disguise of a 'green economy' to be increasingly subjugated to the crisis-prone, non-sustainable and non-future-capable global turbo capitalist financial market system. Does not the ES approach correspond with precisely those mechanisms and instruments which have shaken the financial system and essentially caused the debt crisis, which is so in need of reform? The question is how do we avoid having a new business sector opened up which would reduce nature to the role of a provider of services under the signboard of ES? Should we not be developing and supporting extra-economic approaches in decision-making and alternative patterns of thinking, such as communication, the rules of sustainable lifestyles, and free spaces for creativity, independent of the interests of added value and of profit? How are cultural traditions, political ideals, obligations towards future generations, ideas about the moral self-value of natural beings, a spiritual view of nature, existential attitudes, such as fascination amazement or respect, and, yes, the multiplicity of environmental-ethical arguments and the totality of the wealth of humankind's relationship with nature (Ott 2012) to obtain the high degree of attention beyond thinking in mere economic categories that they deserve? How can we avoid the term ES from becoming a meaningless shell, due to careless overuse, as has already happened, for instance, to the term 'sustainability'?

Overall, terms, concepts, indicators and skills in the context of the human-environment-technology debate are multilayered and multifaceted, and not sufficiently coordinated with one another. Holistic integrative views in general, and those on ES in particular, are necessary in order to observe interactions to direct one's view towards the essential, and to avoid irrationality. However, how can an integrative perspective in an extremely complex world be realised despite the specialisation of the sciences?

The idea that we can live 'in harmony with nature' (Succow 2011) is, according to Haber (2011) an illusion, and the efforts to 'ecologise society' will probably not be very successful. Why are the term and the concept of ES nonetheless attaining a high level of political and scientific attention, both nationally and internationally? Is it the attractiveness of the large sums of money with which the values of nature and biodiversity are suddenly being measured? Is it the possibility of making the costs and benefits of alternative courses of action clearer by way of the ascertainment of ES? More economics in conservation (Schweppe-Kraft 2010)–is that the right way?

The fact is that the ES concept can demonstrate and valuate the dependency of human beings on nature and the landscape more clearly than has hitherto been the case. Thus, we at least have a vision of 'making the world a better place', for wrong decisions are a problem, since knowledge of ES is insufficient (Armsworth et al. 2007; TEEB 2010). What is new in this context is that compared with previous evaluations (e.g. Bastian and Schreiber 1999), the demand side (utility, users, beneficiaries), i.e. the desires and sensitivities of people, have been moved more strongly into focus. This requires a degree of social-scientific competence, because ES can contribute to a more sustainable societal development (Jetzkowitz 2011).

In order to enhance the effectiveness of the ES concept, however, suitable basic conditions will be needed, which means a functional economic, financial and monetary system not exclusively oriented towards growth. Although people develop their environment primarily according to economic and to some extent sociocultural aspects (Adam 1996), this necessarily leads to changes in ecological characteristics and the integrity of ecosystems and the landscape. It is therefore necessary to perceive and to administer them in an economic context, since these impacts generate numerous socio-economic effects. This also affects all categories of protected areas. But not everything can be evaluated according to aspects of economic viability. For example, one should not, in this context, overlook spontaneous development of ecosystems and landscape.

Knowledge about action or inaction must be understood and considered very precisely. In general, action in nature is successful if efficient goals are achieved (Ott 2010). Shrinkage and nonuse cannot be economically exploited, but they can be assessed and evaluated. However, the ES concept is not primarily about nonuse, but rather about pricing ancillary and follow-up effects. For that reason future developments will increasingly have to ensure that:

- The costs and benefits of measures in ecosystems and the landscape are internalised as much as possible.
- The incentives for economic use are formulated in such a way that they serve the goals of biodiversity and conservation in equal measure.
- Market distortions which have a negative effect on biodiversity are reduced.

The economic evaluation methods (▶ Sect. 4.2) are particularly based on alternatives for decisions around the provision and use of ES, and are translated into changes in human well-being by means of monetary values. This occurs, among other things, via surveys of citizens on their willingness to pay for public goods. However, do the results really reflect their true preferences? The willingness to pay for a little-known species may be very low, and yet its value may not be low at all. For that reason, the use of economic evaluation methods for assessing the value of assets remote from the market, such as aspects of biodiversity, or the beauty of landscapes, can only play a supportive role, but cannot be the sole standard for options for action. Here, caution, control and a sense of measure are appropriate (▶ Sect. 4.2 and 5.2). It is no question that cost-benefit analyses improve the comparability of alternative options. In some cases, however, where monetary values can be assigned only with difficulty or not at all, it is better to ascertain ES in some other manner, such has been shown in the case of the sociocultural ES orchard meadows (▶ Sect. 6.5.1). But monetary and non-monetary measures, such as participatory approaches and the negotiation of priorities, are important for the decision-making process.

The various different perspectives shown and the possibility of participation and formulation are essentially what make the ES concept attractive. Ecologists and economists quantify, map, model and evaluate ES (▶ Chap. 4); political scientists and planners address the issues of institutions, ES control instruments, management and governance; and financial experts handle PES (▶ Chap. 5). And participation, coordination, ethics, communications and implementation are the factors that we definitely cannot do without.

'The consistent engagement of the environmental movement has sensitised our awareness for the destruction of nature and for overexploitation. We've happily accepted the restrictions on our personal freedom in order to improve conditions. And with success: Today, our rivers are cleaner, our air is clearer and our forests are healthier than ever', writes Ebert (2011), not someone whom anyone would suspect of being an ecologist or an environmental scientist. This positive environmental conclusion for

Germany–it could be supplemented by such other success stories as toxic site cleanup–should not blind us to the fact that we have not succeeded in stopping fragmentation of the landscape, loss of biodiversity and land consumption for housing and infrastructure. The omission of climate-destructive gases is continuing, the German 'energy-policy turnaround' still needs to be completed, and the growing recreational and tourism sector is certainly not always 'compatible with nature'. The same is true of agriculture. Here, the ambivalence between the above-described efficiency, biological diversity and food security and/or consumer protection are particularly evident (▶ Sect. 6.2; Trepl 2012).

Preservation of nature and the landscape starts at the middle level, in the consciousness, in the heads of people, where we 'learn to feel our homeland as part of the world as a whole, as something that gives us a place in the world' (Václav Havel, quoted in Weinzierl 1999). We need to understand that many ecosystems and landscapes are part of our cultural heritage. Even the 'Green Charter of Mainau Island' (*Grüne Charta von der Mainau*) of 1961 saw 'the dignity of humankind threatened wherever its natural environment is impaired', and demanded 'reorientation of thinking of the entire population through an enhanced education of the public regarding the importance of the landscape, both urban and rural, and the dangers threatening it' (DRL 1997).

The goal of dealing responsibly with natural resources in the interest of humankind itself is the basis of the model of sustainable development adopted by the community of nations at the 1992 Conference on Environment and Development in Rio de Janeiro. In 1994, the German Parliament officially made sustainability a goal of the state under the constitution, and all federal governments and many interest groups have since then expressly confirmed their support for it. However, about 20 years later, where do we stand (Rio + 20)?

The model of sustainable development is certainly true and important for the twenty-first century. However, it has been accepted by the broad mass of the population only very hesitantly. Who knows the sustainability indicators in detail? Who keeps track of the changes in them? The political sphere as a generator of impulses and as the instance to which sustainability, biodiversity or energy concepts are submitted, is itself heterogeneous, split into parties, departments, lobbies, etc. Nonetheless, we are convinced that the ES concept can reinforce the understanding of broad sections of the population for protection of nature and the environment.

The implementation of the model of 'sustainable development' should be oriented towards ES and multifunctional landscapes, in other words, one of the things it must do is analyse which ES are compatible and/or combinable with one another, where services can be integrated, where conflicts exist, and where they can be resolved. That particularly requires farsighted land-use policy, control of land use and the creation of societal awareness for the development of ecosystems and landscapes. The conveyance of knowledge is an important social component of sustainability in this regard.

■ What New Impulses can be Ascertained?

In May 2011, the German Federal Agency for Nature Conservation found in the Competence Center 'Natural Capital Germany'. The international TEEB study (TEEB 2010) is to be implemented at the national level. This involves particularly the further development of the ascertainment and evaluation of ecosystem services and biodiversity for the economy and society. The explanation of concepts, terms, methods and regional case studies in this book can provide a basis for that.

German scientists are also strongly involved in the international scientific community, e.g. in the context of the Ecosystem Services Partnership (▶ www.fsd.nl/esp). The fact that Bonn was selected as the site of the office of the Secretariat of the Intergovernmental Science-Policy Platform on Biodiversity and Ecosystem Services (IPBES) in 2012 can be seen as a milestone. The main task of the IPBES is to provide policy decision-makers with reliable, independent and credible information on the state and the development of biodiversity, in order to support their decision-making processes. However that alone is not enough to achieve biodiversity goals. Implementation and measure-oriented approaches based on the ES concept are needed for that.

Finally, the establishment of a national research centre for the area of biodiversity at the associated

Universities of Halle, Jena and Leipzig by the German Research Foundation (DFG) and the current scientific activities initiated for implementation by the Federal Agency for Nature Conservation and the Federal Ministry for Education Research (BMBF) provide clear evidence that the ES concept will in the future assume a prominent role for the sustainable activation of the potentials of landscapes in Germany. This will involve building upon the high level of the ecological and environmental/economic and spatial-planning sciences, and also taking into account the complex, change-resistant systems of business, finance, administration and governance.

▪ What Must be Taken into Account in the Future?

The concept of ES has become established in the cognitive orientation of both the scientific community and the practitioners, and the awareness in the population for the problem that our lives and our well-being depend on nature is being strengthened. The approach has been positively received by the public; that fact should be put to use by the 'sustainability trinity', and not watered down by semantical or turf battles. It may be irrelevant which exact label we place on ES; what is not trivial is the distinction between functions and services (▶ Chap. 2). In case of doubt, one should always explain what exactly is meant. Dealing with metamorphous, multi-meaning terms such as ES–or ecosystem and landscape, natural capital, biodiversity, sustainability, and resilience or governance–is difficult, controversial in the scientific community and often confusing for policymakers and practitioners.

Conceptually, the framework of assessment/evaluation of ES has been staked out (▶ Sect. 3.1). Ecosystems provide services for people which need to be translated into the categories of use and welfare. The claim and evaluation of services are manifested especially in the manner and intensity of land use, and feeds back upon the structures and processes of ecosystems, which in turn affects their potential capacity to render services. Portraying this complex interaction in terms of its causes, effects and consequences, and controlling it 'correctly' is the actual challenge.

The assessment of ES is carried out for certain services, which are generally classed in the categories provisioning, regulatory and sociocultural ES (▶ Sect. 3.2). They are subjected to selection and weighing in accordance with the task at hand, and the possibilities. The foundations for investigation are provided by geobiophysically based ecological and–if appropriate–economic/monetary classifications. In this regard, ES-specific standards (procedures, techniques, models) have not yet been sufficiently developed and tested (▶ Sect. 4.1, 4.2, 4.3, 4.4). That involves particularly the comprehensive taking into account of interactions between various ES (trade-offs), and the delimitation of ecosystemic and human/technological services.

Within the framework of the Ecosystem Services Partnership and other networks, working groups are currently being formed in order to prepare and collect regional, and especially type-referenced ES values (e.g. the SERVES-database, ▶ www.esvaluation.org). This also includes databases with ES values and valuations currently being established. In TEEB (2010), some 1300 monetary values were calculated for 11 biomes worldwide. For each biome, e.g. a wetland or a meadow, 22 ES were included in the data collection–in theory; in the case studies, however, as few as a single ES might then be evaluated (▶ www.teebweb.org). This indicates a high degree of regional heterogeneity–for example, the ES values assigned to lakes and rivers vary between $ 13,448 and $ 1779 per hectare and year–and are hardly transferable to the local conditions in central Europe. What, for example, does such a monetary value for the ES, e.g. for a lake ecosystem, actually tell us? Do not such vague monetary statements feed the fundamental critique of the economisation of nature? Should not the search for alternative value measurement systems be accelerated?

Numerous ES case studies on methodological development and procedural application are currently being realised. Since many facts relevant for conservation can be attributed to the properties of nature and land use, the case studies in ▶ Chap. 6 have been selected accordingly. The framework conditions for land use/land-use decision-making are a key to the formulation of the future and for securing ES. For that, a corresponding further development of the ecological planning approach, the environment and welfare accounting, the financial and subsidy practice in the context of value discussions, and comparison with alternatives are needed.

Simply reforming the highly complex environmental, planning and tax law in Germany with regard to new framework conditions and requirements, and not making it even clumsier, more complicated and harder to handle byte integrating the ES concept into it, is a task which is difficult to solve, yet worthwhile. Ideas for how to do that, have been provided at various points (► Chap. 5).

Options for action have to be negotiated in public discourse. The analysis and evaluation of ES with scenarios decision-making alternatives can be very helpful in this process of balancing of interests. The task of scientists is not to make society's decisions for it. What they do have a right to do is to make recommendations and to participate in action in the context of their citizens' rights.

Even though ES valuations are not patent recipes, they can help overcome the lack of economic perceptibility of nature, which has often led to wrong political and economic decisions, and ultimately to the destruction of nature, of ecosystems and of biodiversity, and will continue to do so in the future.

References

Adam T (1996) Mensch und Natur: das Primat des Ökonomischen. Entstehen, Bedrohung und Schutz von Kulturlandschaften aus dem Geiste materieller Interessen. Nat Landsch 71:155–159

Armsworth PR, Chan KMA, Daily GC, Ehrlich PR, Kremen C, Ricketts TH, Sanjayan MA (2007) Ecosystem-Service science and the way forward for conservation. Conserv Biol 21:1383–1384

Bastian O, Schreiber KF (Eds) (1999) Analyse und ökologische Bewertung der Landschaft. Spektrum Akad. Verlag, Heidelberg

Bastian O, Grunewald K, Syrbe RU (2012) Space and time aspects of ecosystem services, using the example of the EU water framework directive. Int J Biodivers Sci Ecosyst Serv Manage 8:5–16

Burkhard B, de Groot R, Costanza R, Seppelt R, Jørgensen SE, Potschin M (2012) Solutions for sustaining natural capital and ecosystem services. Ecol Indic 21:1–6

DRL–Deutscher Rat für Landespflege (1997) Betrachtungen zur "Grünen Charta von der Mainau" im Jahre 1997. Schriftenreihe d. Deutschen Rates f. Landespflege, Heft 68

Ebert V (2011) Machen Sie sich frei. Rowohlt Taschenbuch, Reinbek

Fatheuer T (2012) In Rendite vereint–Ökonomie und Natur: Die Geschichte einer schwierigen Beziehung. Böll. Thema 1 (Magazin der Heinrich-Böll-Stiftung, Berlin), pp 5–7

Grunewald K, Bastian O (2010) Ökosystemdienstleistungen analysieren–begrifflicher und konzeptioneller Rahmen aus landschaftsökologischer Sicht. GEOÖKO 31:50–82

Haber W (2011) Umweltpolitikberatung–eine persönliche Bilanz. Studienarchiv Umweltgeschichte 16:15–25 (► www.iugr.net)

Haines-Young RH, Potschin MB (2009) Methodologies for defining and assessing ecosystem services. Final Report, JNCC, Project Code C08-0170-0062, 69 pp

Jetzkowitz J (2011) Ökosystemdienstleistungen in soziologischer Perspektive. In: Groß M (Eds) Handbuch Umweltsoziologie. VS Verlag für Sozialwissenschaften, Wiesbaden, pp 303–324

Kettunen M, Bassi S, Gantioler S, ten Brink P (2009) Assessing socio-economic benefits of natura 2000–a toolkit for practitioners (September 2009 Edition). Output of the European Commission project Financing Natura 2000: Cost estimate and benefits of Natura 2000. Institute for European Environmental Policy (IEEP), Brüssel, 191 pp

Löschmann H (2012) Die Geister, die ich rief–Das riskante Spiel auf dem Geld- und Finanzmarkt. Böll. Thema 1 (Magazin der Heinrich-Böll-Stiftung, Berlin), pp 9–11

OECD (2008) Strategic environmental assessment and ecosystem services. DAC Network on Environment and Development Co-operation (ENVIRONET), Paris, 26 pp

Ott K (2010) Umweltethik zur Einführung. Junius, Hamburg

Ott K (2012) Ein wenig mehr Bescheidenheit, bitte!–Über die Verführung, Kosten-Nutzen-Kalküle zur höchsten menschlichen Weisheit zu erklären. Böll. Thema 1 (Magazin der Heinrich-Böll-Stiftung, Berlin), pp 33–35

Schröter M, van der Zanden EH, van Oudenhoven APE, Remme RP, Serna-Chavez HM, de Groot RS, Opdam P (2014) Ecosystem services as a contested concept: a synthesis of critique and counter-arguments. Conservation Letters. doi:10.1111/conl.12091

Schweppe-Kraft B (2010) Ökosystemdienstleistungen: ein Ansatz zur ökonomischen Bewertung von Natur. Local land & soil news 34/35 II/10, The Bulletin of the European Land and Soil Alliance (ELSA) e. V., pp 11–14

Seppelt R, Fath B, Burkhard B, Fisher JL, Grêt-Regamey A, Lautenbach S, Pert P, Hotes S, Spangenberg J, Verburg PH, Van Oudenhoven APE (2012) Form follows function? Proposing a blueprint for ecosystem service assessments based on reviews and case studies. Ecol Indic 21:145–154

Succow M (2011) "… warum mir der Naturschutz so am Herzen liegt!" Nat. Landsch. 1:19–23

TEEB (2010) The economics of ecosystems and biodiversity: ecological and economic foundations. In: Kumar P (Eds) Earthscan, London
ten Brink P (2008) Ecological losses to economic losses. Presentation at the workshop on the economics of the global loss of biological diversity, 5–6 March, Brussels
Trepl L (2012) Biodiversitätsbasierte Ökosystemdienstleistungen. ▶ www.scilogs.de/chrono/blog/landschaft-oekologie/biodiversitat-und-aussterben/2012–02–20/biodiversit-tsbasierte-kosystemdienstleitungen. Accessed 22 Feb 2012
UNEP-WCMC (2011) Developing ecosystem services indicators: experiences and lessons learned from sub-global assessments and other initiatives. CBD Technical Series 58
Weinzierl H (1999) Land und Leute. Die Zukunft der Landnutzung in Bayern. Nationalpark 4:39–43
Zeh J (2009) Corpus Delicti: Ein Prozess. Schöffling und Co., Frankfurt a. M.

Index

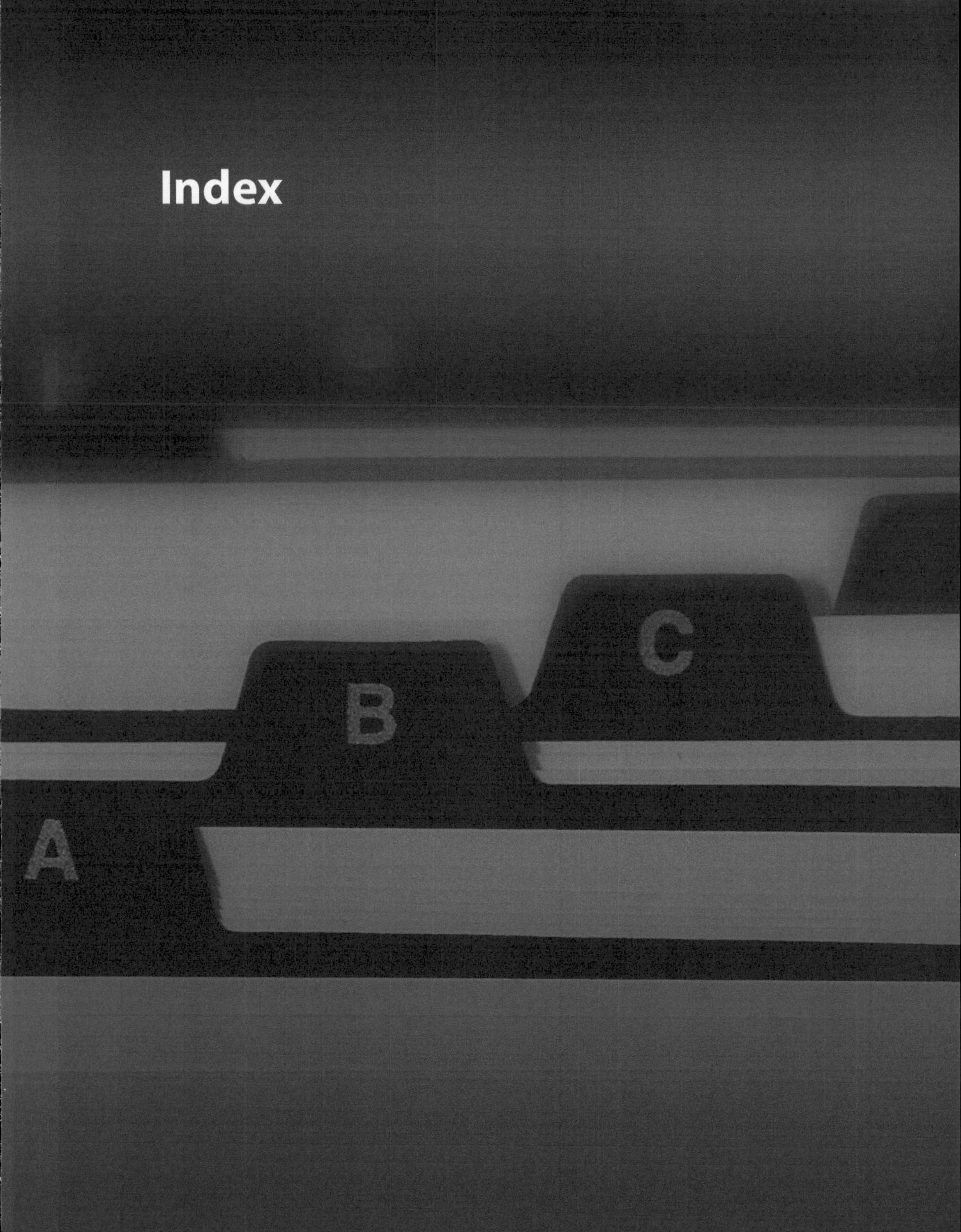

K. Grunewald, O. Bastian (eds.), *Ecosystem Services – Concept, Methods and Case Studies*,
DOI 10.1007/978-3-662-44143-5, © Springer-Verlag Berlin Heidelberg 2015

A

B

C

D

E

F

G

H

I

MIX
Papier aus verantwortungsvollen Quellen
Paper from responsible sources
FSC® C105338

If you have any concerns about our products,
you can contact us on
ProductSafety@springernature.com

In case Publisher is established outside the EU,
the EU authorized representative is:
Springer Nature Customer Service Center GmbH
Europaplatz 3, 69115 Heidelberg, Germany

Printed by Libri Plureos GmbH
in Hamburg, Germany